TRUFFLES

Diversity, Ecology and Biotechnology

<u>*Series*</u>

Progress in Mycological Research

- Fungi from Different Environments (2009)
- Systematics and Evolution of fungi (2011)
- Fungi from Different Substrates (2015)
- Fungi: Applications and Management Strategies (2016)
- Advances in Macrofungi: Diversity, Ecology and Biotechnology (2019)
- Advances in Macrofungi: Industrial Avenues and Prospects (2021)
- Advances in Macrofungi: Pharmaceuticals and Cosmeceuticals (2021)
- Fungal Biotechnology: Prospects and Avenues (2022)
- Bioprospects of Macrofungi: Recent Developments (2023)
- Ecology of Macrofungi: An Overview (2024)
- Advances in *Cordyceps* Research Prospects and Avenues (2025)
- The Role of Entomopathogenic fungi in Agriculture (2025)

Series: Progress in Mycological Research

TRUFFLES

Diversity, Ecology and Biotechnology

Editors

Kandikere R. Sridhar
Department of Biosciences
Mangalore University, Mangalore, India

Sunil K. Deshmukh
Scientific Advisor
Greenvention Biotech Pvt. Ltd., Pune, India

CRC Press
Taylor & Francis Group
Boca Raton London New York

CRC Press is an imprint of the
Taylor & Francis Group, an **informa** business
A SCIENCE PUBLISHERS BOOK

First edition published 2025
by CRC Press
2385 NW Executive Center Drive, Suite 320, Boca Raton FL 33431

and by CRC Press
4 Park Square, Milton Park, Abingdon, Oxon, OX14 4RN

CRC Press is an imprint of Taylor & Francis Group, LLC

Library of Congress Cataloging-in-Publication Data (applied for)

ISBN: 978-1-032-81992-1 (hbk)
ISBN: 978-1-032-82735-3 (pbk)
ISBN: 978-1-003-50592-1 (ebk)

DOI: 10.1201/9781003505921

Typeset in Times New Roman
by Prime Publishing Services

Preface

Mycology, being an important field of biology, has a major impact on many human affairs due to the fungal benefits to health, medicine, agriculture, the environment, and industry. Fungi are known to govern the global economy owing to their significance in nutrition, health, biogeochemical cycles, and economically valued products (e.g., antibiotics, beverages, biopesticides, drugs, enzymes, fermentation, natural compounds, nutraceuticals, organic acids, pigments, polysaccharides, and proteins) (Hyde et al., 2019, 2024; Meyer et al., 2020; Niego et al., 2023).

Among macrofungi, mycorrhizal fungi are of paramount interest owing to their mutualistic association with the roots of herbaceous plants and tree species. Truffles have been classified under the category of ectomycorrhizal or ectendomycorrhizal ascomycetous fungi. The genomics of desert truffles revealed new insights into the lifestyle of plant-fungal interaction (Marqués-Gálves et al., 2020). They are hypogenous, have a wide geographical distribution, and each species is known to emit pleasant aromas and flavours. So far, 180–220 species of truffles are known worldwide, and up to 20 species are commercialized (Bonito et al., 2010, 2013). Their diversity is high in Western North America, Mediterranean Europe, and Australia (Trappe and Claridge, 2015). Black truffles are well known for their medicinal values, wehich combat many lifestyle diseases. Truffles are among the highly prized macrofungi owing to their nutritional and medicinal values. However, their cultivation is time-consuming and demands thorough technical knowledge.

As gourmet mushrooms, truffles have been cherished worldwide due to their exceptional aroma. Food industries ventured to transfer the aromatic compounds of truffles to other food products (e.g., cheese, oil, and butter) to uplift their quality (Tejedor-Calvo et al., 2023). Each truffle is capable of producing a blend of different volatile aromatic compounds (e.g., alcohols, aldehydes, dimethyl sulphide, dimethyl disulphide, and ketones). The truffle industry is flourishing due to the nutraceutical as well as cosmeceutical value of several truffles.

The book *Truffles: Diversity, Ecology and Biotechnology* projects many facets of investigation into truffles. 18 chapters written by 30 authors from 15 countries probe into different aspects of truffle research. This book highlights fundamental and applied aspects of truffles in five subdivisions: 1) Overview; 2) Diversity and Ecology; 3) Bioactive Potential; 4) Nutraceutical and Economic Perspectives; 5) Industrial Applications. The first subsection introduces truffles in different geographic locations, including deserts, bibliometric analysis and the truffle industry. The second subsection encompasses truffle diversity, ecology, morphology, vegetation support for truffles, and cultivation schemes. In the third subsection, the bioactivity, health, aroma and medicinal significance of truffles have been dealt with. The fourth subsection includes the ethnic, nutraceutical and socioeconomic prospects of truffles in Morocco. The fifth subsection delves into industrial applications with small-scale cultivation, industrial products, and the marketing of truffle products.

The editors are optimistic that the contributed chapters on various aspects of basic and applied fields of truffles will be beneficial for readers with a wide interest in mycology and allied fields (applied biology, biotechnology, botany, ecology, environmental biology, field biology, forestry, and microbiology).

The editors are indebted to the contributors and reviewers for the submission of chapters and evaluation at the stipulated time, respectively. We are thankful to Dr. Mahadevakumar, S. Scientist, Botanical Survey of India, Port Blair, for technical support. We are obliged to the CRC Press, Boca Raton, Florida, USA, for its productive collaboration to fulfil the official formalities to be followed to publish this book.

Mangalore, India
Pune, India
August 26, 2024

Kandikere R. Sridhar
Sunil K. Deshmukh

References

Bonito, G.M., Gryganskyi, A.P., Trappe, J.M. and Vigalys, R.A. (2010). Global meta-analysis of *Tuber* ITS rDNA sequence: Species diversity, host associations and long-distance dispersal. Mol. Ecol., 19: 4994–5008. https://doi.org/10.1111/j.1365-294X.2010.04855.x

Bonito, G., Smith, M.E., Nowak, M., Healy, R.A. and Guevara, G. (2013). Historical Biogeography and Diversification of Truffles in the *Tuberaceae* and Their Newly Identified Southern Hemisphere Sister Lineage. PLoS One, 8: e52765. https://doi.org/10.1371/journal.pone.0052765

Hyde, K.D., Xu, J., Rapior, S., Jeewon, R., Lumoung, S. et al. (2019). The amazing potential of fungi: 50 ways we can exploit fungi industrially. Fungal Divers., 97: 1–136. https://doi.org/10.1007/s13225-019-00430-9

Hyde, K.D., Baldrin, P., Chen, Y., Chethana, K.W.T., De Hong, S. et al. (2024). Current trends, limitations and future research in the fungi? Fungal Divers., 125: 1–71. https://doi.org/10.1007/s13225-023-00532-5

Marqués-Gálves, J.E., Miyauchi, S., Paolocci, F., Navarro-Ródenas, A., Arenas, F. et al. (2020). Desert truffle genomes reveal their reproductive modes and new insights into plant-fungal interaction and ectendomycorrhizal lifestyle. New Phytol., 229: 2917–2932. https://doi.org/10.1111/nph.17044

Meyer, V., Basenko, E.Y., Benz, J.P., Braus, G.H., Caddick, M.X. et al. (2020). Growing a circular economy with fungal biotechnology: a white paper. Fungal Biol. Biotechnol., 7: 1–23. https://doi.org/10.1186/s40694-020-00095-z

Neigo, A.G.T., Lambert, C., Mortimer, P., Thongklang, N., Rapior, S. et al. (2023). The contribution of fungi to the global economy. Fungal Divers., 121: 95–137. https://doi.org/10.1007/s13225-023-00520-9

Tejedor-Calvo, E., Gracía-Barreda, S., Felices-Mayordomo, M., Blando, D., Sánchez, S. and Morco, P. (2023). Truffle flavored commercial products veracity and sensory analysis from truffle and non-truffle consumers. Food Control, 245: 109424. https://doi.org/10.1016/j.foodcont.2022.109424

Trappe, J.M. and Claridge, A.W. (2015). The hidden life of truffles. Scientific American, 24(2). 10.1038/scientificamericanfood0615-22

Contents

Bioactive Potential

Nutraceutical and Economic Perspectives

Industrial Applications

List of Contributors

Abo-Atya, Doaa M.
Department of Chemistry, Faculty of Science, Menoufia University, Shebin El-Kom, Egypt.

Akyüz, Mehmet
Department of Biology, Faculty of Arts and Science, Bitlis Eren University, Bitlis, Türkiye.

Allen, Kirsten
New Use Agriculture and Natural Plant Products Program (NUANPP), Rutgers University, Department of Plant Biology-Foran Hall, New Jersey, USA.

Aouali, S.
Forest Pathology and Mycology Laboratory, Forest Protection Division, Forest Research National Institute, Algeria.

Dedhia, Nirali
Department of Food Engineering and Technology, Institute of Chemical Technology, Mumbai, India.

El-Seedi, Hesham R.
Department of Pharmaceutical Biosciences, Uppsala University, Biomedical Centre, Uppsala, Sweden.
International Research Center for Food Nutrition and Safety, Jiangsu University, Zhenjiang, China.
Chemistry Department, Faculty of Science, Islamic University of Madinah, Madinah, Saudi Arabia.

El Enshasy, Hesham Ali
Innovation Centre of Agritechnology in Advanced Bioprocessing (ICA), Universiti Teknologi Malayisa (UTM), Pagoh Campus, Johor, Malaysia.
Faculty of Chemical Engineering, Universiti Teknologi Malaysia, UTM Skudai, Johor, Malaysia.
City of Scientific Research and Technology Applications, New Burg Al Arab, Alexandria, Egypt.

Hakkou, S.
Faculty of Letters and Human Sciences, Rabat, Morocco.

Hanapi, Siti Zulaiha
Department of Deputy Vice Chancellor (Research & Innovation), Universiti Teknologi Malaysia.

Hussain, Faiq H.S.
Medical Analysis Department, Faculty of Applied Science, Tishk International University, Erbil, Iraq.

Jamali, Samad
Department of Plant Protection, College of Agriculture, Razi University, Kermanshah, Iran.

Kaddouri, Hicham
Cadi Ayyad University, Laboratory of Microbial Biotechnologies, Agrosciences and Environment (BioMAgE), Morocco.
University of Sultan Moulay Slimane Morocco, Faculty of Sciences and Techniques, Beni Mellal, Morocco.

Karim, Rozaliana Abdul
Innovation Centre of Agritechnology in Advanced Bioprocessing (ICA), Universiti Teknologi Malayisa (UTM), Pagoh Campus, Johor, Malaysia.

Khalifa, Shaden
Department of Molecular Biosciences, The Wenner-Gren Institute (MBW), Stockholm University, Stockholm, Sweden.

Kırbağ, Sevda
Department of Biology, Faculty of Science, Fırat University, Elazığ, Türkiye.

Ljubojević, Srdjan
The Independent University of Banja Luka, Faculty for Ecology and Institute for Scientific Research, Bosnia and Herzegovina.

Machouri, N.
Faculty of Letters and Human Sciences, Rabat, Morocco.

Marathe, Sandesh J.
Division of Hematology and Oncology, Department of Medicine, College of Medicine, University of Tennessee Health Science Center (UTHSC), Memphis, USA.
UTHSC Center for Cancer Research College of Medicine (UTSHC), Memphis, USA.

Mohamed, Intan Baizura
Innovation Centre of Agritechnology in Advanced Bioprocessing (ICA), Universiti Teknologi Malayisa (UTM), Pagoh Campus, Johor, Malaysia.

Nurjayadi, Muktiningsih
Chemistry Department, Faculty of Mathematics and Natural Sciences, Universitas Negeri Jakarta, Indonesia.

Ouahmane, Lahcen
Cadi Ayyad University, Laboratory of Microbial Biotechnologies, Agrosciences and Environment (BioMAgE), Morocco.

Peng, Tony
GK Bio International Sdn. Bhd., Petaling Jaya, Selangor, Malaysia

Popović-Djordjević, Jelena
University of Belgrade-Faculty of Agriculture, Department of Food Technology and Biochemistry, Belgrade, Serbia.

Radovac, Tihomir
Society for the Environment and Sustainability on Earth, Banja Luka, Bosnia and Herzegovina.

Sabir, M.
National School of Forest Engineers, Sale, Morocco.

Sukmawati, Dalia
Biology Department, Faculty of Mathematics and Natural Sciences, Universitas Negeri Jakarta, Indonesia.

Tounsi, Abdessamad
University of Sultan Moulay Slimane Morocco, Faculty of Sciences and Techniques, Beni Mellal, Morocco.

Vasilišin, Ladislav
Faculty of Technology, University of Banja Luka, Bosnia and Herzegovina.

Vučić, Goran
Faculty of Technology, University of Banja Luka, Bosnia and Herzegovina

Zhang, Tiehua
College of Food Science and Engineering, Jilin University, Changchun, China.

Overview

CHAPTER 1

Global Diversity and Ecology of Truffles

Hakkou, S.,[1,*] *Sabir, M.*[2] and *Machouri, N.*[1]

1. Introduction

Truffles are hypogeous fungi that live entirely underground and belong to the Ascomycetes (Morte et al., 2008; Trappe et al., 2008). There are two types: true truffles and desert truffles known as "*terfes*". True truffles are those belonging to the *Tuber* genus, associated with trees, and highly prized for their intense, excludable aroma. The best-known truffles are: *Tuber aestivum, T. borchii, T. brumale, T. indicum, T. macrosporum, T. magnatum, T. melanosporum, T. mesentericum,* and *T. uncinatum*. Desert truffles are those that grow in the typically arid and semi-arid climates of the Mediterranean region, some of which are endemic. They mainly refer to species of the genera *Terfezia*, *Tirmania*, and *Tuber*, but also to lesser-known species such as the genera *Loculotuber*, *Mattirolomyces*, and *Picoa* (Alsheikh, 1994; Morte et al., 2008; Rodriguez, 2008; Trappe et al., 2008).

The emergence and growth of truffles rely on the interplay of three key factors: soil, climate and the host plant (Grente and Chevalier, 1980; Dexheimer et al., 1985). Species distribution and variability are mainly influenced by regional differences in these factors (De Laplane, 2010). There are distinct differences between the conditions required for desert truffles and those for true truffles. Truffles are found in a large number of countries around the world, with a diversity of species and a remarkable difference in their numbers. This differentiation may be due to the availability of key factors for their production: soil, climate, and host plant species.

---

[1] Faculty of Letters and Human Sciences, Rabat, Morocco.
[2] National School of Forest Engineers, Sale, Morocco.
* Corresponding author: soukaina.hakkou@gmail.com

The distribution of truffles involves a large number of geographic regions, with a diversity of species and a remarkable difference in their numbers. Some countries harbor only a single species, while others, such as China, boast over sixty. The United States as a whole has 50 species; we have treated them as independent units, thus, each state is presented with its own species. Some countries are not listed as truffle-growing, despite being adjacent to those that are, and sharing the same climatic and edaphic conditions. This can be explained by the fact that some countries are more advanced than others in research in this field. It is also possible that truffles exist there, but that they do not arouse the same interest, or that they aren't so abundant as to be the subject of research. It is worth mentioning that the discovery of new species of truffles continues across the world. The aim of this chapter is to present the diversity and distribution of truffles around the world and how environmental factors influence truffle production.

2. Taxonomy of Truffles

The morphology of truffles, in terms of shape, size, and diameter, varies greatly from one species to another, but this does not prevent the majority from having common characteristics, such as the characteristics of the peridium (appearance, and color), spores (number, shape, and ornamentation), and other complementary characters (shape or size of fruiting bodies, gleba coloration, vein arrangement, odor, and habitat). It is these characteristics that enable truffle taxonomy (Trappe, 1971, 1979). The current taxonomy of truffles has gone through several stages and changes (Table 1).

2.1 The Genus Tuber

The first nomenclature of the *Tuber* genus dates back to 1729 by Micheli, grouping together all European truffle species and those of greatest interest in truffle growing for their high economic value and geographical distribution (Urban et al., 2004; Mello et al., 2005; Le Tacon et al., 2006; Laessoe and Hansen, 2007; Murat et al., 2008). The *Tuberaceae* family, which includes the genus *Tuber,* was created in 1822 by Dumort (Urban et al., 2004; Laessoe and Hansen, 2007). In the early classification by Korf in 1973, the Tuberaceae family belonged to the order Tuberales, a classification later abandoned by Trappe in 1979, and the family was subsequently transferred to the order Pezizales (Riousset et al., 2001; Laessoe and Hansen, 2007).

2.2 The Genera Terfezia and Tirmania

The taxonomic history of *Terfezia* began in 1829 with Moris, who attributed the first described species of *Terfezia* to the genus *Tuber* (Alsheikh, 1994). While the Tulsane brothers did not consolidate all African truffles into the genus *Terfezia leonis* until 1851 (Fortas, 1990) Chatin's subsequent work in 1892 led to the creation of the genus *Tirmania.* Chatin described several species and varieties of *Terfezia* based on extensive studies of these fungi (Fortas, 1990; Loizides et al., 2012). In 1906, Maire documented a novel species discovered in Algeria: *Tirmania pinoy.* Subsequently,

Table 1. Classification schemes of truffles (orders, classes and genera) according to three authors.

Korf (1973)	**Trappe (1979)**	**Laessoe and Hansen (2007)**
Order: Tuberal	**Order: Pezizales**	**Order: Pezizales**
Elaphomycetaceae *Elaphomyces*	**Pezizaceae** *Amylascus* *Mycoclelandia* (as *Clelandia*) *Hydnotryopsis* *Peziza* spp. *Tirmania*	**Pezizaceae** *Amylascus* *Cazia* *Eremiomyces* *Hydnobolites* *Hydnotryopsis* *Kalaharituber* *Mattirolomyces* *Mycoclelandia* *Pachyphloeus* *Peziza* spp. *Ruhlandiella* *Sphaerozone* *Terfezia* *Termania*
Terfeziaceae *Carbomyces* *Delastria* *Mukagomyces* *Paradoxa* *Picoa* *Terfezia* *Tirmania*	**Terfeziaceae** *Choiromyces* *Delastria* *Hydnobolites* *Pachyphloeus* *Terfezia*	**Helvellaceae** *Balsamia* *Barssia* *Helvella astieri*
Tuberaceae *Barssia* *Balsamia* *Caulocarpa* *Choiromyces* *Elderia* *Fischerula* *Lespiaultinia* *Labyrinthomyces* *Hydnobolites* *Hydnoplicata* *Hydnotrya* *Pachyphloeus* *Phymatomyces* *Piersonia* *Protogenea* *Pseudobalsamia* *Stephensia* *Tuber*	**Helvellaceae** *Hydnotrya* *Dingleya* *Fischerula* **Balsamiaceae** *Balsamia* *Barssia* *Picoa*	**Tuberaceae** *Choiromyces* *Dingleya* *Labyrinthomyces* *Paradoxa* *Reddelomyces* *Tuber*
Geneaceae	*Tuberaceae* *Paradoxa* *Tuber*	**Morchellaceae/Discinaceae** *Gymnohydnotrya* *Hydnotrya* *Fischerula* *Leucangium*

Table 1 contd. ...

Table 1 contd. ...

Korf (1973)	Trappe (1979)	Laessoe and Hansen (2007)
	Pyronemataceae *Geopora cooperi* *Hydnocystis* *Labyrinthomyces* *Paurocotylis* *Petchiomyces* *Sphaerozone* *Stephensia* *Geneaceae* *Genea*	**Pyronemataceae** *Genabea* *Genea* *Geopora cooperi*

in 1971, this species was reclassified into three distinct species - *Terfezia arenaria, T. terfezioides*, and *Tirmania nivea*-according to research by Splivallo and Bonfante (Splivallo and Bonfante, 2006).

In 1897, Fischer first proposed the *Terfeziaceae* family status and included it in the Tuberales, a classification that continued with Korf in 1973. In 1979, Trappe modified the family concept and later transferred it to the order Pezizales, so *Terfezia* and *Tirmania* belong to the families *Terfeziaceae* and *Pezizaceae*, respectively, in the order Pezizales. The latest change in the classification of desert truffles was made in 2007 by Laessoe and Hansen, who included the genera *Terfezia* and *Tirmania* in the family *Pezizaceae*, while the genus *Picoa* was placed in the family *Pyronemataceae*, a classification based on recent molecular phylogenetic data combined with morphological characters (Laessoe and Hansen, 2007).

3. What Factors Govern Truffle Production?

The emergence and development of truffles depend on the interplay of three primary factors: soil, vegetation, and climate. Rainfall intensity and seasonal distribution significantly affect their yield and abundance. The soil's physicochemical characteristics also play a crucial role in its growth (Fortas, 2009; Al Qarawi and Mridha, 2012; Boufeldja, 2017). The distribution and variability of truffle species and genera are primarily influenced by regional differences in these factors (De Laplane, 2010).

3.1 Desert Truffles

Desert truffles have been reported from several geographic locations throughout the world (Table 2). The highest number of desert truffles was found in Morocco in Africa (18 spp.), Turkey in Asia (17 spp.), Central Australia (11 spp.), Spain in Europe (40 spp.), and California in North America (9 spp.). Their occurrence has been governed by climatic conditions, soil edaphic factors and the specificities of host plant species. Identification of desert truffles was highest in Europe (152), followed by Asia (109) and Africa (84) (Table 3).

Table 2. Distribution patterns of desert truffles and true truffles around the world (*, number of desert truffles/true truffles).

Continent: Country/ Province	Desert truffles	True truffles (*Tuber*)	Reference
Africa			
Algeria (11/1)*	*Picoa carthusiana, P. juneperi, P. lefebvrei, Terfezia arenaria, T. boudieri, T. claveryi, T. crassiverrucosa, T. eliocrocae, T. leptoderma, Tirmania nivea* and *T. pinoyi*	*Tuber moretii*	Fortas, 1990; Alsheikh, 1994; Dib-Bellahouel, 2012; Zitouni-haouar, 2016; Zitouni-haouar et al., 2018; Kirk, 2020
Botswana (3/0)	*Eremiomyces echinulatus, Kalaharituberpfeilii* and *Mattirolomyces austroafricanus*	-	Alsheikh, 1994; Trappe et al., 2008
Canary Islands (5/0)	*Eremiomyces innocentii, Terfezia boudieri, T. claveryi, T. leptoderma* and *T. olbiensis*	-	Alsheikh, 1994
Egypt (7/0)	*Picoajuniperi, P. lefebvrei, Terfezia arenaria, T. boudieri, T. claveryi, Tirmania nivea* and *T. pinoyi*	-	Alsheikh and Trappe, 1983a; Omer et al., 1994; Feeney, 2002
Kenya (1/0)	*Delastria rosea*	-	Kirk, 2020
Libya (6/0)	*Picoa lefebvrei, Terfezia boudieri, T. claveryi, T. metaxasi, Tirmania nivea* and *T. pinoyi*	-	Ahmed et al., 1981; Alsheikh and Trappe, 1983a,b; Alsheikh, 1994; Feeney, 2002; Hansen et al., 2006
Madagascar (2/0)	*Terfezia pfeilii* and *T. decaryi*	-	Roth-Bejerano et al., 2004; Ferdman et al., 2005; Trappe et al., 2008, 2010

Table 2. contd. ...

Table 2. contd. ...

Continent: Country/ Province	**Desert truffles**	**True truffles (*Tuber*)**	**Reference**
Mauritania (3/0)	*Carbomyces emergens, Terfezia boudieri* and *T. eremita*	-	Alsheikh, 1994
Morocco (18/13)	*Delastria rosea,Delastria javieri, Picoa juniperi, Eremiomyces innocentii, Reddellomyces donkii, Terfezia arenaria, T. boudieri, T. cistophila, T. claveryi, T. leptoderma, T. magnusii, T. olbiensis, T. pini, Tirmania nivea, T. pinoyi, Tuber oligospermum, T. asa* and *T. gennadii*	*Tuber aestivum, T. borchii* var. *sphaerospermum, T. brumale, T. buendiae, T. dryophilum, T. excavatum, T. lacunosum, T. melanosporum, T. malenconii, T. melosporum, T. nitidum, T. puberulum* and *T. rufum*	Trappe, 1971; Malençon, 1973; Alsheikh, 1994; Rodriguez, 2008; Khabar, 2016; Abourouh, 2011, 2020; Healy et al., 2016; Kirk, 2020; Hakkouet al., 2022
Mozambique (1/0)	*Eremiomyces echinulatus*	-	Kirk, 2020
Namibia (5/0)	*Eremiomyces echinulatus, Kalaharituberpfeilii, Mattirolomyces austroafricanus, Terfezia claveryi* and *T. pfeilii*	-	Roth-Bejerano et al., 2004; Ferdman et al., 2005; Trappe et al., 2008, 2010
Sierra Leone (4/0)	*Terfezia boudieri, T. claveryi, Tirmanianivea* and *T. pinoyi*	-	Kirk, 2020
South Africa (6/1)	*Eremiomyces echinulatus, Kalaharituberpfeilii, Mattirolomyces austroafricanus, Terfezia arenaria, T. decaryi* and *T. pfeilii*	*Tuber rapaeodorum*	Alsheikh, 1994; Trappe et al., 2008; Kirk, 2020

Tunisia (11/0)	*Picoa carthusiana, P. juniperi, P. lefebvrei, Reddellomyces parvulosporus, Terfezia arenaria, T. boudieri, T. claveryi, T. metaxasi, T. schweinfurthii, Tirmania nivea* and *T. pinoyi,*	-	Alsheikh and Trappe, 1983a; Alsheikh,1994; Slama and Neffati, 2004; Slama et al., 2006, 2010; Sbissi, 2013
Zimbabwe (1/0)	*Kalaharituber pfeilii*	-	Ferdman et al., 2005
Asia			
Armenia (1/2)	*Terfezia claveryi*	*Tuber rapaeodorum* and *T. scruposum*	Alsheikh, 1994; Lancellotti et al.,2016; Kirk, 2020
Azerbaijan (1/1)	*Terfezia arenaria*	*Tuber aestivum*	Bagi and Fekete, 2010; Boufeldja, 2017
Bahrain (4/0)	*Picoa lefebvrei, T. boudieri, T. claveryi* and *Tirmania nivea,*	-	Alsheikh, 1994; Moubasher, 1995 ; Mandeel and Al-Laith, 2007
Bhutan (0/1)	-	*Tuber anniae*	Kirk, 2020
China (11/76)	*Choiromyces cerebriformis, C. helanshanensis, C. meandriformis,C. setchellii, Delastria rosea,Hydnobolites cerebriformis,Pachyphloeus virescens, Terfezia arenaria, T. gigantea, T.terfezioides* and *Tuber oligospermum*	*Tuber aestivum, T. alboumbilicum, T. anniae, T. badium, T. beyerlei, T. bomiense, T. borchii, T.californicum, T. calosporum, T. caoi, T. crassitunicatum, T. depressum, T. dryophilum, T. excavatum, T. foetidum, T. formosanum, T. glabrum, T. griseolivaceum T. himalayense, T.hubeiense, T. huidongense, T. huizeanum, T. indicum, T. jinshajiangense, T. nitidum, T. latisporum, T. liaotongense, T. liyuanum, T. lijiangense, T. lishanenese, T. liui, T. magnatum, T. maculatum, T. melanosporum, T. microsphaerosporum, T. microspiculatum, T. microverrucosum, T. neoexcavatum, T. panzhihuanense, T. parvomurphium, T. piceatum, T. polymorphosporum, T. polyspermum, T. pseudobrumale, T. pseudoexcavatum,*	Fortas, 1990; Alsheikh, 1994; Callot, 1999; Du et al., 2014; Bonito and Smith, 2016; Le Tacon et al., 2006; Chen and Liu, 2007; Jeandroz et al., 2008; Wangand Liu, 2009; Al-Thani, 2010; Fan et al., 2011, 2012, 2014; Wang, 2013; Weden et al., 2013; Ammarellou et al., 2014; Su et al., 2014; Healy et al., 2016; Liu and li, 2017; Van et al., 2018; Kirk, 2020;Wan et al., 2021; Kinoshita et al., 2022

Table 2. contd. ...

Table 2. contd. ...

Continent: Country/ Province	**Desert truffles**	**True truffles (*Tuber*)**	**Reference**
		T. pseudohimalayense, T.pseudomagnatum, T. pseudosphaerosporum, T.puberulum, T. qujingense, T. rapaeodorum, T. rufum, T. shidianense, T. shii, T. sinensis, T. sinoalbidum, T. sinoaestivum, T. sinoexcavatum, T. sinomonosporum, T.sinoniveum, T. sinopuberulum, T.sinosphaerosporum, T. songlu, T.subglobosum, T. taiyuanense, T. texense, T. turmericum, T. umbilicatum, T. verrucosivolvum, T. vesicoperidium, T.wanglangense, T. wenchuanense, T. wumengense, T. xanthomonosporu, T. xuanhuaense, T. yigongense and *T. zhongdianensis*	
India (3/7)	*Eremiomyces echinulatus, Mattirolomyces spinosus* and *Terfezia* sp.,	*Tuber beyerlei, T. bomiense, T. caoi, T. huidongense, T. indicum, T. pseudoexcavatum* and *T. whetstonense*	Le Tacon et al., 2006, Chen et al., 2016; Kirk 2020
Indonesia (2/1)	*Choiromyces* sp. and *Terfezia indica,*	*Tuber foetidum*	Alsheikh, 1994 ; Kirk, 2020
Iran (10/13)	*Kalaharituber pfeilii, Picoa juniperi, P. lefebvrei Terfezia arenaria, T. boudieri, T. claveryi, T. metaxasi, Tirmania nivea, T. pinoyi* and *Tuber oligospermum*	*Tuber aestivum, T. borchii, T. brumale, T. dryophilum, T. excavatum, T. fulgens, T.iranicum, T. lucidum, T. macrosporum, T. melosporum, T. puberulum, T. rapaeodorum* and *T. rufum*	Alsheikh, 1994; Jamali and Banihashemi, 2012; Ammarellou et al., 2014; Mostowfizadeh-Ghalamfarsa et al., 2014; Lancellotti et al., 2016; Pulgia et al., 2021; Sepahvand et al., 2021
Irak (9/0)	*Picoa lefebvrei, Terfezia arenaria, T. boudieri, T. claveryi, T. crassiverrucosa T. hafizi, T. metaxasi, Tirmania nivea* and *T. pinoyi,*	-	Alsheikh and Trappe, 1983b; Ewaze and Al-Naama, 1989; Alsheikh, 1994; Zitouni-Haouar, 2016

Japan (3/19)	*Delastria*sp., *Terfezia gigantea* and *Tirmania* sp.	*Tuber aestivum, T. anniae, T. borchii, T. foetidum, T. formosanum, T. himalayense, T.hiromichii, T. indicum, T. japonicum, T. liyuanum, T. longispinosum T. macrosporum, T. maculatum, T. magnatum, T. melanosporum, T. pseudoexcavatum, T. subglobosum, T. taiyuanense* and *T. torulosum*	Alsheikh, 1994; Hashimoto et al., 2012; Du et al., 2014; Healy et al., 2016; Le tacon, 2017; Kirk, 2020; Kinoshita et al., 2022
Jorcan (1/0)	*Terfezia claveryi*	-	Alsheikh, 1994
Kuwait (6/0)	*Picoa lefebvrei, Terfezia arenaria, T. boudieri, T.claveryi, Tirmania nivea* and *T. pinoyi*	-	Alsheikh and Trappe, 1983b; Moustafa, 1985 ; Alsheikh, 1994
Kyrgyzstan (0/1)	-	*Tuber melosporum*	Kirk, 2020
Lebanon (3/0)	*Terfezia boudieri, T. claveryi* and *Tirmania* sp.	-	Alsheikh, 1994; Mandeel et Al-Laith, 2007
Nepal (0/1)	-	*Tuber* sp.	Taxonomic distribution of occurrences in nepal (gbif. org)
Oman (1/1)	*Terfezia arenaria*	*Tuber* sp.	Alsheikh, 1994; Boufeldja, 2017
Pakistan (2/2)	*Terfezia lahorensis* and *T. terfezioides*	*Tuber bomiense* and *T. melosporum*	Alsheikh, 1994; Kirk, 2020
Palestine (9/2)	*Picoa lefebvrei, Reddellomyces westraliensis, Terfezia arenaria, T. boudieri, T. claveryi, T. olbiensis, T. metaxasi, Tirmania nivea* and *Tuber oligospermum*	*Tuber aestivum* and *T. melanosporum*	Alsheikh and Trappe, 1983b; Roth-Bejerano et al., 1990; Alsheikh, 1994; Pinkas et al., 2000; Dib-Bellahouel, 2012
Qatar (3/1)	*Picoalefebvrei, Terfezia claveryi* and *Tirmania nivea,*	*Tuber melosporum*	Moubasher, 1995; Al-Thani, 2010; Kirk, 2020

Table 2. contd. ...

Table 2. contd. ...

Continent: Country/ Province	Desert truffles	True truffles (*Tuber*)	Reference
Russia (3/18)	*Balasamia* sp., *Choiromyces helanshanensis* and *C. meandriformis*	*Tuber aestivum, T. anniae, T. californicum, T. dryophilum, T. exiguum, T. foetidum, T. furfuraceum, T. himalayense, T. huidongense, T. indicum, T. intermedium, T. liaotongense, T. maculatum, T. melosporum, T. michailowskianum, T. nitidum, T. puberulum* and *T. rufum*	Ceruti et al., 2003; Bidaud and Vanvooren, 2008; Lancellotti et al., 2016; Kirk, 2020
Saudi Arabia (6/0)	*Picoa lefebvrei, Terfezia arenaria, T. boudieri,T.claveryi, Tirmania nivea* and *T. pinoyi*	-	Alsheikh, 1994; Hussain and Ruqaie, 1999
South Korea (1/3)	*Delastria* sp.	*Tuber aestivum, T. himalayense* and *T.indicum*	Hall et al., 2008
Sri Lanka (1/0)	*Choiromices* sp.	-	Taxonomic distribution of occurrences in Sri Lanka(gbif.org)
Syria (5/0)	*Terfezia arenaria, T. boudieri, T.claveryi, Tirmania nivea* and *T. pinoyi*	-	Awameh and Alsheikh, 1980; Alsheikh,1994
Taiwan (0/3)	-	*Tuber furfuraceum, T. taiyuanense* and *T. zhongdianense*	Healy et al., 2016; Kirk, 2020
Thailand (1/7)	*Delastria* sp.	*Tuber candidum, T. lannaense, T.pseudomagnatum, T. rufum, T. thailandicum, T. turmericum* and *T. whetstonense*	Yan et al., 2018; Kirk, 2020; Kinoshita et al., 2022
Turkey (17/14)	*Balsamia gunerii, Balsamia hellenica, Balsamia vulgaris, Choiromyces meandriformis, Picoa juniperi, P. lefebvrei, Reddellomyces westraliensis,R. parvulosporus, Terfezia albida, T. arenaria, T. boudieri,*	*Tuber aestivum, T. borchii, T. brumale, T. excavatum, T. fulgens, T. furfuraceum, T. macrosporum, T.melanosporum, T. melosporum, T. mesentericum, T. nitidum, T. puberulum, T. rapaeodorum* and *T. rufum*	Alsheikh, 1994; Ceruti et al., 2003; Gucin et al., 2010; Akyuz et al., 2012; Castellano and Turkoglu, 2012; Kagan-Zur and Akyuz, 2014; Türkoglu and Castellano, 2014; Türkoglu et al.,2015; Olivier et al., 2018; Alli and Doğan, 2019; Akata et al., 2022

	R. cistophila, T. claveryi, T. leptoderma, T. olbiensis, Tirmania pinoyi and *Tuber oligospermum*		
Turkmenistan (2/0)	*Terfezia arenaria* and *T. claveryi*	-	Alsheikh, 1994
United Arab Emirates (4/0)	*Terfezia arenaria* and *T. boudieri, T.claveryi, Tirmania nivea*	-	Barseghyan and Wasser, 2010
Yemen	*Terfezia boudieri*	-	Boufeldja, 2017
Australia			
Central Australian desert (11/12)	*Delastria rosea, Elderia arenivaga,Horakiella watarrkana, Kalaharituber pfeilii, Mattirolomyces mulpu, Mycoclelandia arenacea, M. bulundari, Reddellomyces magnisporus, R. parvulosporus R. westraliensis* and*Ulurua nonparaphysata*	*Tuber aestivum, T. borchii, T. brumale, T. californicum, T. dingleya, T. hiromichii, T. indicum, T. maculatum, T. melanosporum, T. protium, T. rapaeodorum* and *T. reddellomyces*	Ferdman et al., 2005; Trappe et al.2008, 2010; Reyna and Garcia-Barreda, 2014; Bonito and Smith, 2016; Kirk, 2020
Europe			
Andorra (0/1)	-	*Tuber rapaeodorum*	Kirk, 2020
Austria (3/15)	*Choiromyces meandriformis, Picoacarthusiana* and *Tuber oligospermum*	*Tuber aestivum, T. borchii, T. brumale, T. dryophilum, T. excavatum, T. foetidum, T. fulgens, T. macrosporum, T. maculatum, T.mesentericum, T. nitidum, T. puberulum, T.rapaeodorum, T. rufum* and *T. scruposum*	Bonito and Smith, 2016; Lancellottiet al., 2016; Kirk, 2020
Belgium (1/9)	*Choiromyces meandriformis*	*Tuber aestivum, T. borchii, T. dryophilum, T. excavatum, T. maculatum, T. nitidum, T. puberulum, T. rapaeodorum* and *T. rufum*	Kirk, 2020

Table 2. contd. ...

Table 2. contd. ...

Continent: Country/ Province	**Desert truffles**	**True truffles (*Tuber*)**	**Reference**
Belarus (0/1)	-	Tuber anniae	Kirk, 2020
Bulgaria (0/9)	-	*Tuber aestivum, T. borchii, T. brumale, T. macrosporum, T. magnatum, T.melanosporum, T. mesentericum, T. puberulum* and *T. pulchrosporum*	Bidaud and Van Vooren, 2008; Letacon, 2017; Olivier et al. 2018 ; Polemis et al., 2019
Croatia (3/6)	*Delastria* sp., *Terfezia arenaria* and *Tuber oligospermum,*	*Tuber aestivum, T. brumale, T. macrosporum, T. magnatum, T.melanosporum* and *T. puberulum*	Mello et al., 2005; Trappe et al., 2010; Lancellotti et al., 2016; Boufeldja,2017; Olivier et al., 2018
Cyprus (3/1)	*Terfezia aphroditis, T. arenaria* and *T.boudieri*	*Tuber* sp.	Chatin, 1897; Alsheikh, 1994; Kirk, 2020
Czech Republic (2/16)	*Choiromycesmeandriformis* and *Tuber oligospermum,*	*Tuber aestivum, T. anniae, T. borchii, T. brumale, T. dryophilum, T. excavatum, T. foetidum, T. huidongense, T. indicum, T. maculatum, T. melanosporum, T. mesentericum, T. puberulum, T. Pustulatum, T. rapaeodorum* and *T. rufum*	Gryndler et al., 2013; Lancellotti et al., 2016; Le Tacon, 2017; Kirk, 2020
Danemark (3/16)	*Balsamia platyspora, Choiromycesmeandriformis* and *Terfezia* sp.	*Tuber aestivum, T. anniae, T. borchii, T. brumale, T. dryophilum, T. excavatum, T. foetidum, T. fulgens, T. levissimum, T. maculatum, T. macrosporum, T. mesentericum, T. puberulum, T. rapaeodorum, T. rufum* and *T. scruposum*	Bidaud and Van Vooren, 2008, Kirk, 2020; Søgeresultat - Danmarks Svampeatlas (databasen. org)
Estonia (3/11)	*Balsamia platyspora, Reddellomyces* sp. and *Terfezia* sp.	*Tuber aestivum, T. anniae, T. borchii, T. dryophilum, T. foetidum, T. maculatum, T. puberulum, T. rapaeodorum, T. rufum, T. scruposum* and *T. rufum*	Du et al., 2014; Kirk, 2020
Finland (5/16)	*Choiromyces meandriformis, Terfezia alsheikhii, T. claveryi, T. lusitanica* and *Tuber gennadii*	*Tuber anniae, T. borchii, T. californicum,* T. *dryophilum, T. foetidum, T. himalayense, T. indicum, T. maculatum, T. melosporum, T. oregonense, T. pseudobrumale, T. puberulum, T. rapaeodorum, T. rufum, T. scruposum* and *T. vesicoperidium*	Trappe et al., 2010; Lancellotti et al.,2016; Le Tacon, 2017; Kirk, 2020; Browse and Search | Finnish BiodiversityInformation Facility (laji.fi)

France (13/27)	*Balsamia vulgaris, Choiromyces meandriformis, Delastria rosea, Picoa carthusiana, P. juniperi, P. lefebvrei, Terfezia arenaria, T. boudieri, T. claveryi, T. leptoderma, T. olbiensis, T. oligosperma* and *T. terfezioides*	*Tuber aestivum, T. albidum, T. asa- foetida, T. bellonae, T. borchii var. rioussettorum, T. brumale, T. dryophilum, T. excavatum, T. ferrugineum, T. fulgens, T. hiemalbum, T.* maculatum, *T. magentipuncatum, T. malenconii, T. macrosporum, T. magnatum, T.malacodermum, T. melanosporum, T. mesentericum, T. mutabile, T. nitidum, T. panniferum, T. puberulum, T. pustulatum, T. rapaeodorum, T. rufum* and *T. sphaerospermum*	Chevalier et al., 1984, 2004; Fortas and Chevalier, 1988; Alsheikh, 1994; Callot, 1999; SMNF, 2006; Bidaud and Van Vooren, 2008; Garcia-Monteroet al., 2008; Tabouret, 2011; Chevalier, 2014; Benucci et al., 2016;Leonardi et al., 2019; Olivier et al., 2018
Georgia (1/5)	*Terfezia boudieri*	*Tuber anniae, T. brumale, T. fulgens, T.puberulum* and *T. rufum*	Alsheikh, 1994; Colgan and Trappe, 1997; Kirk, 2020
Germany (3/24)	*Choiromyces meandriformis, Terfezia oligosperma* and *Tuber gennadii*	*Tuber aestivum, T. belonei, T. bernardinii, T. borchii, T. brumale, T. canaliculatum, T. dryophilum, T. excavatum, T.exiguum, T. ferrugineum, T. foetidum, T. fulgens, T. gibbosum, T. macrosporum, T. maculatum, T. melanosporum, T. mesentericum, T. murinum, T. nitidum, T. puberulum, T. rapaeodorum, T. reganum, T. rufum* and *T. scruposum*	Bidaud and Van Vooren, 2008; Stobbe et al., 2012; Benucci et al., 2016; Dämmrich, 2016; Lancellotti et al., 2016; Ratti et al., 2016; Olivier et al., 2018; Kirk, 2020;
Greece (8/15)	*Terfezia arenaria, T. cistophila, T. gennadii, T. grisea, T. olbiensis, T. oligosperma, T. pseudoleptoderma* and *T.terfezioides*	*Tuber aestivum, T. bituminatum, T. beyerlei, T. borchii, T. brumale, T. excavatum, T. melanosporum, T. melosporum, T. mesentericum, T. panniferum, T. puberulum, T.pulchrosporum, T. rufum, T. rufum* var. *apiculatum* and *T. rufum* var. *nitidum*	Alsheikh, 1994; Chevalier et al., 2004; Diamandis and Perlerou, 2008; Bordallo et al., 2015; Le Tacon, 2017; Polemis et al., 2019
Hungary (5/19)	*Choiromycesmeandriformis, Terfezia leptoderma, T. magnusii, T.terfezioides* and *Tuber oligospermum*	*Tuber aestivum, T. borchii, T. brumale, T. cryptobrumale, T. dryophilum, T. excavatum, T. ferrugineum, T. foetidum, T. fulgens, T. maculatum, T. macrosporum, T.magentipuncatum, T. melanosporum, T. mesentericum, T. nitidum, T. puberulum, T.rapaeodorum, T. regianum* and *T. rufum*	Alsheikh, 1994; Le Tacon et al., 2006; Bidaud and Van Vooren, 2008; Rodriguez, 2008; Healy et al., 2016; Benucci et al., 2016; Olivier et al., 2018 ; Polemis et al., 2019; Kirk, 2020

Table 2. contd. ...

Table 2. contd. ...

Continent: Country/ Province	**Desert truffles**	**True truffles (*Tuber*)**	**Reference**
Irland (0/6)	-	*Tuber aestivum, T. borchii, T.dryophilum, T. puberulum. T. rapaeodorum* and *T. rufum*	Ceruti et al., 2003; Olivier et al., 2018;Kirk, 2020
Island (1/0)	*Terfezia arenaria* and *T. boudieri*	-	Kirk, 2020
Italy (19/36)	*Balsamia vulgaris, Delastria rosea, Choiromyces meandriformis, C. gangliformis, Picoa juneperi, Reddellomyces donkii, Terfezia arenaria, T. boudieri, T. cistophila, T. claveryi, T. gennadii, T. leptoderma, T. magnusii, T. olbiensis, T. oligosperma, T. pini, T.pseudoleptoderma, T. terfezioides* and *Tirmania nivea*	*Tuber aestivum, T. albidum, T. anniae, T. bellonae, T. bernardinii, T. bituminatum, T. borchii, T. brumale, T. buendiae, T. dryophilum, T. excavatum, T. ferrugineum, T. foetidum, T. fulgens, T. gibosum, T. hiemalbum, T. lacunosum, T. lucidum, T. macrosporum, T.magentipuncatum, T. magnatum, T. malenconii, T. melanosporum, T. melosporum, T. mesentericum, T. monosporum, T. mutabile, T. nitidum, T. panniferum, T. puberulum, T. rapaeodorum, T. regianum, T. rufum, T.scruposum, T. sphaerospermum* and *T. stramineum*	Janex-Favre et al., 1988; Alsheikh,1994; Bratek et al., 1996; Callot,1999; Chevalier et al., 2004; Roux, 2006; Bidaud and Van Vooren, 2008; Garcia-Montero et al., 2008; Benucci et al., 2016; Healy et al., 2016; Lancellotti et al., 2016; Olivier et al., 2018 ; Polemis et al., 2019
Lativa (1/5)	*Terfezia* sp.	*Tuber anniae, T. foetidum, T. maculatum, T. puberulum* and *T. rufum*	Bidaud and Van Vooren, 2008; Kirk,2020
Lituania (2/2)	*Delastria* sp. and *Terfezia* sp.	*Tuber anniae* and *T. maculatum*	Kirk, 2020
Luxembourg (1/4)	*Choiromyces meandriformis*	*Tuber excavatum, T. melanosporum, T.puberulum* and *T. rufum*	Kirk, 2020; Dämmrich, 2022
Montenegro (0/2)	-	*Tuber puberulum* and *T. rufum*	Kirk, 2020
Netherlands (0/8)	-	*Tuber aestivum, T. borchii, T. excavatum, T. maculatum, T. melosporum, T. mesentericum, T. puberulum* and *T. rufum*	Lancellotti et al., 2016

Northern Macedonia (2/17)	*Balsamia vulgaris* and *Choiromycesmeandriformis*	*Tuber aestivum, T. belonei, T. borchii, T. brumale, T. brumale* var. *moschatum, T. dryophilum, T. excavatum, T. foetidum, T. fulgens, T. macrosporum, T. magnatum, T. melanosporum, T. mesentericum, T. puberulum, T. rufum f. ferrugineum, T. rufum f.lucidum* and *T. rufum f. nitidum*	Karadelev et al.,2019
Norway (1/9)	*Balsamia platyspora*	*Tuber anniae, T. aestivum, T. borchii, T. dryophilum, T. foetidum, T. maculatum, T. puberulum, T. rapaeodorum* and *T. rufum*	Le Tacon et al., 2006; Du et al., 2014; Molinier et al., 2016; Kirk, 2020
Poland (4/17)	*Balsamia platyspora, B. vulgaris,Choiromyces meandriformis,Eremiomyces* sp.	*Tuber aestivum, T. albidum, T. borchii, T. brumale, T. dryophilum, T. excavatum, T. foetidum, T. fulgens, T. macrosporum, T. maculatum, T. magnatum, T. melanosporum, T. mesentericum, T. nitidum, T. puberulum, T. rapeodorum* and *T. rufum*	Percudani et al., 1999; Bidaud and VanVooren, 2008; Hilszczanska et al., 2014; Lancellotti et al., 2016; Le Tacon, 2017; Olivier et al., 2018; Kirk, 2020
Portugal (12/8)	*Choiromyces meandriformis, Delastria rosea, Terfezia arenaria, T. alsheikhii, T. berberiodora, T. fanfanii, T. hafizi, T. leptoderma, T. lusitanica, T. magnusii, T.olbiensis* and *Tuber oligospermum,*	*Tuber aestivum, T. borchii, T. excavatum, T. lacunosum, T. maculatum, T. melanosporum, T. puberulum* and *T. rufum*	Alsheikh, 1994; Ceruti et al., 2003; Bordallo et al., 2013; Molinier et al., 2016; Kirk, 2020
Romania (2/10)	*Choiromycesmeandriformis* and *Terfezia arenaria*	*Tuber bernardinii, T. borchii, T. brumale, T. excavatum, T. macrosporum, T.magentipuncatum, T. melanosporum, T. puberulum, T. rapaeodorum* and *T. regianum*	Alsheikh, 1994; Benucci et al., 2016; Le Tacon, 2017; Merényi et al., 2017
Serbia (1/15)	*Terfezia terfezioides*	*Tuber aestivum, T. excavatum, T. borchii, T. brumale, T. ferrugineum, T. foetidum, T. fulgens, T. maculatum, T. macrosporum, T. magnatum, T. melanosporum, T. nitidum, T. puberulum, T. petrophilum* and *T. rufum*	Kovacs et al., 2001; Kovacs, 2002; Healy et al., 2016; Le Tacon, 2017; Merényi et al., 2017; Olivier et al., 2018

Table 2. contd. ...

Table 2. contd. ...

Continent: Country/ Province	Desert truffles	True truffles (*Tuber*)	Reference
Slovakia (2/7)	*Terfezia olbiensis* and *T. terfezioides*	*Tuber aestivum, T. brumale, T. macrosporum, T. magentipuncatum, T.puberulum, T. regianum* and *T. rufum*	Alsheikh, 1994; Glejdura and Kunca, 2012; Benucci et al., 2016; Robin et al., 2016; Polemis et al., 2019
Slovenia (2/16)	*Balsamia vulgaris* and *Choiromycesmeandriformis*	*Tuber aestivum, T. borchii, T. brumale, T. dryophilum, T. excavatum, T. ferrugineum, T. fulgens, T. hiemalbum, T. gulosorum, T. macrosporum, T. magnatum T. melanosporum, T. mesentericum, T. nitidum, T. puberulum* and *T. rufum*	Ceruti et al., 2003; Hall et al., 2008; Benucci et al., 2016; Merényi et al., 2017; Olivier et al., 2018; Kirk, 2020
Spain (40/38)	*Balsamia vulgaris, Choiromyces gangliformis, C. meandriformis, Delastria evae, D. faustinia, D. javieri, D. liebanensis, D. rosea, D. supernova, Picoa juniperi, P. lefebvrei, Reddellomyces donkii, R. westraliensis, Terfezia albida, T. Alsheikhii, T. arenaria, T. berberiodora, T. boudieri, T. claveryi, T. canariensis, T. crassiverrucosa, T. eliocrocae, T.extremadurensis, T. fanfani, T. gennadii, T. grisea, T. hispanica, T. honrubiae, T. leptoderma, T. lusitanica, T. magnusii, T.morenoi, T. olbiensis, T. oligosperma, T. pallida, T. pini, T. pseudoleptoderma, T. terfezioides, T. Trappei* and *Tuber gennadii*	*Tuber aestivum, T. albidum, T. alcaracense, T. anniae, T. excavatum, T. bituminatum, T. borchii, T. brumale, T. buendiae, T. cistophilum, T. dryophilum, T.ferrugineum, T. foetidum, T. fulgens, T. himalayense, T. indicum, T. lacunosum, T. levissimum, T. lucentum, T. lusitanicum, T. maculatum, T. macrosporum, T. magnatum, T. malacodermum, T. malenconii, T. maresa, T. melanosporum, T. melosporum, T. mesentericum, T. multimaculatum, T. nitidum, T. panniferum, T. puberulum, T. pustulatum, T. rapaeodorum, T. regianum, T. rufum* and *T. sphaerospermum*	Castro and Freire, 1982; Alvarez et al.,1992; Alsheikh, 1994; Callot, 1999; Moreno et al., 2000; Calonge et al., 2003; Bidaud and Van Vooren, 2008; Garcia-Montero et al., 2008; Rodriguez, 2008; Kovács et al., 2009; Bordallo et al., 2013; Bordallo and Rodriguez, 2014; Chevalier, 2014; Du et al., 2014; Garcia-montero et al.,2014; Menta et al., 2014; Zambonelli et al., 2014; Bordallo et al., 2015; Benucci et al., 2016; Lancellotti et al., 2016; Leonardi et al., 2019; Kirk, 2020

Sweden (2/11)	*Balsamia platyspora* and *Choiromycesmeandriformis*	*Tuber aestivum, T. anniae, T. excavatum, T. foetidum, T. maculatum, T. melanosporum, T. mesentericum, T. puberulum, T. rapaeodorum, T. rufum* and *T. tequilanum*	Pegler et al., 1993; Le Tacon et al., 2006; Reyna and Garcia-barreda, 2014; Kirk, 2020
Switzerland (4/16)	*Balsamia vulgaris, Choiromyces meandriformis, Picoa carthusiana* and *Terfezia magnusii*	*Tuber aestivum, T. anniae, T. borchii, T. brumale, T. dryophilum, T. excavatum, T. foetidum, T. fulgens, T. maculatum, T. macrosporum, T. magnatum, T. melanosporum, T. mesentericum, T. nitidum, T. puberulum* and *T.rufum*	Bidaud and Van Vooren, 2008; Benucci et al., 2016; Kirk, 2020
Ukraine (0/5)	-	*Tuber aestivum, T. excavatum, T. macrosporum, T. rapaeodorum* and *T. rufum*	Benucci et al., 2016, Electronic Distribution Maps of Ukrainian Fungi (cybertruffle.org.uk)
United Kingdom (3/15)	*Balsamia platyspora, Choiromyces meandriformis* and *Terfezia arenaria*	*Tuber aestivum, T. bituminatum, T. borchii, T. brumale, T. dryophilum, T. excavatum, T. foetidum, T. macrosporum, T. maculatum, T. melanosporum, T. mesentericum, T. nitidum, T.puberulum, T. rapaeodorum, T. rufum* and *T. scleroneuron*	Mohamed-benkada, 1999; Lancellotti et al., 2016; Robin et al., 2016; Merényi et al., 2017; Olivier et al., 2018; Kirk, 2020
North America			
Alaska (0/1)	-	*Tuber anniae*	Alsheikh, 1994; Callot, 1999; Trappe et al., 2001; Healy, 2003; Le Tacon et al., 2006; Bidaud and van Vooren, 2008; Guevara et al., 2008; Trappe et al., 2010; Zitouni-haouar, 2010; Kovacs et al., 2011; Moreno et al., 2011; Guevara et al., 2013; Trappe et al., 2014; Bonito and Smith, 2016; Healy et al., 2016; Lancellotti et al., 2016 ; Riccioni et al., 2016; Merényiet al., 2017; Guevara-Guerrero, 2018; Van et al., 2018; Watkinson, 2018; Leonardi et al., 2019; Kirk, 2020; Taxonomic Distribution of Occurrences in Costa Rica (gbif.org); Taxonomic Distribution of Occurrences in Cuba (gbif.org)

Table 2. contd. ...

Table 2. contd. ...

Continent: Country/ Province	**Desert truffles**	**True truffles (*Tuber*)**	**Reference**
Appalachia (1/0)	*Terfezia gigantea*	-	
Arizona (1/2)	*Terfezia spinosa*	*Tuber beyerlei* and *T. tequilanum*	
California (9/21)	*Balsamia alba, B. magnata, B. nigrans, Carbomyces emergens, C. gilbertsonii, C. alveolatus, C. compactus, C. setchellii* and *Delastria rosea*	*Tuber argentum, T. bellisporum, T. beyerlei, T. californicum, T. candidum, T. castellanoi T. citrinum, T. gardneri, T. gibbosum, T. Harknessii, T. irradians, T. lauryi, T.levissimum, T. linsdalei, T. monticola, T. oregonense, T. pacificum, T. quercicola, T. rufum, T. separans* and *T. whetstonense*	
Canada (6/22)	Balsamia platyspora, B. vulgaris,Choiromyces meandriformis, *Choiromyces* sp., *Delastria* sp. and *Picoa carthusiana*	*Tuber anniae, T. arnoldianum, T. besseyi, T. beyerlei, T. bisporum, T. borchii, T. brennemanii, T. canaliculatum, T. dryophilum, T. foetidum, T. gibbosum, T. lyonia, T. maculatum, T. magnatum, T. malacodermum, T. mesenterium T. mexiusanum, T. monticola, T. pacificum, T. puberulum, T. rapaeodorum* and *T. walkeri*	
Connecticut (0/1)	-	*Tuber canaliculatum*	
Colorado (1/0)	*Balsamia magnata*	-	
Costa Rica (0/1)	-	*Tuber* sp.	
Cuba (1/1)	*Delastria* sp.	*Tuber* sp.	
Florida (0/4)	-	*Tuber beyerlei, T. brennemanii, T. floridanum* and *T.lyonii*	
Georgia (0/2)	-	*Tuber lyonii* and *T. texense*	
Greenland (0/1)	-	*Tuber brennemanii*	

Guatemala (0/1)	-	Tuber tequilanum	
Idaho (3/3)	*Balsamia vulgaris, Choiromyces alveolatus* and *Picoa carthusiana*	*Tuber anniae, T. monticola* and *T. nitidum*	
Indiana (0/3)	-	*Tuber lyonii, T. mexiusanum* and *T. walkeri*	
Iowa (3/2)	*Mattirolomyces tiffanyae, Terfeziaspinosa* and *T. oligosperma*	*Tuber lyonii* and *T. mexiusanum*	
Kansas (0/1)	-	*Tuber rufum*	
Louisiana (2/0)	*Terfezia arenaria* and *T. spinosa*	-	
Maryland (0/1)	-	*Tuber spinoreticulatum*	
Mexico (6/22)	*Terfezia longii, T. mexicana. T. spinosa, Carbomyces emergens, C. gilbertsonii* and*Delastria* sp.	*Tuber anniae, T. aztecorum, T. beyerlei, T. bonitoi, T. brunneum, T. castilloi, T. gadneri, T. gibosum, T. guevarae, T. guzmanii, T. incognitum T. irradians, T. lyonia, T.maculatum, T. malacodermum, T. miquihuanense, T. mexiusanum, T. pseudoseparans T. regimontanum, T. separans, T. tequilanum* and *T. theleascum,*	
Michigan (1/10)	*Tuber oligospermum*	*Tuber arnoldianum, T. brennemanii, T. canaliculatum, T. candidum, T. mexiusanum T. murinum, T. sphaerospermum, T. separans, T. texense* and *T. walkeri*	
Minnesota (0/4)	-	*Tuber gardneri, T. mexiusanum, T. rapaeodorum* and *T. walkeri*	
Missouri (0/2)	-	*Tuber lyonii* and *T. whetstonense*	
Nebraska (0/2)	-	*Tuber besseyi* and *T. rufum*	
Nevada (0/1)	-	*Tuber bellisporum*	

Table 2. contd. ...

Table 2. contd. ...

Continent: Country/ Province	**Desert truffles**	**True truffles (*Tuber*)**	**Reference**
New Mexico (4/4)	*Carbomyces emergens, C. gilbertsonii, C. longii* and *Terfezia longii*	*Tuber bellisporum, T. brennemanii, T. lyonii* and *T. tequilanum*	
North Carolina (1/4)	*Terfezia gigantea*	*Tuber borchii, T. canaliculatum, T. magnatum, T.mexiusanum* and *T. separans*	
Ohio (0/6)	-	*Tuber canaliculatum, T. lyonii, T. longisporum, T. puberulum, T. separans* and *T. texense*	
Oregon (5/21)	*Balsamia alba, B. magnata, B. nigrans, B. platyspora* and *Picoa carthusiana*	*Tuber anniae, T. bellisporum, T. beyerlei, T. candidum, T. castellanoi, T. gibbosum, T. irradians, T. lauryi, T. levissimum, T. luomai, T. lyonii, T. oregonense, T. nitidum, T. pacificum, T. phlebodermum, T. quercicola, T. separans, T. shearii, T. sphaerospermum, T. wheeleri* and *T. whetstonense*	
Panama (0/1)	-	*Tuber puberulum*	
Pennsylvania (1/2)	*Terfezia gigantea*	*Tuber canaliculatum* and *T. dryophilum*	
Texas (0/3)	-	*Tuber floridanum, T. lyonii* and *T. texense*	
Utah (1/4)	*Choiromyces alveolatus*	*Tuber candidum, T. dryophilum, T. levissimum* and *T.nitidum*	
Virginie (0/3)	-	*Tuber canaliculatum, T. shearii* and *T.sphaerospermum*	
Washington (3/8)	*Balsamia nigrans, Choiromyces cookei* and*Picoa carthusiana*	*Tuber anniae, T. californicum, T. gadneri, T.gibosum, T. oregonense, T. pacificum, T. phlebodermum* and *T. sphaerospermum*	
Wisconsin (0/1)	-	*Tuber gardneri*	
Oceania			

Fiji (1/0)	*Delastria* sp.	-	Taxonomic Distribution of Occurrences in Fiji (gbif.org)
New Zealand (1/12)	*Terfezia* sp.	*Tuber aestivum, T. borchii, T. brumale, T.californicum, T. canaliculatum, T. dryophilum, T. foetidum, T. levissimum, T. maculatum, T. melanosporum, T. rufum* and *T. separans*	Alexander, 2008; Du et al., 2014; Reyna and Garcia-Barreda, 2014; Bonito and Smith, 2016; Healy et al., 2016; Chen et al., 2016; Lancellottiet al., 2016; Olivier et al., 2018; Kirk, 2020
Papua-Guinea (1/1)	*Eremiomyces echinulatus*	*Tuber* sp.	Ferdman et al., 2005
South America			
Argertina (2/7)	*Eremiomyces innocentii* and *Terfezia longii*	*Tuber argentinum, T. borchii, T. californicum, T.maculatum, T. melanosporum, T. melosporum* and *T. separans*	Alsheikh, 1994; Cortes et al., 2011; Kirk, 2020
Bolivia (0/1)	-	*Tuber rufum*	Kirk; 2020
Brazil (1/3)	*Terfezia arenaria*	*Tuber floridanum, T. magnatum* and *T. melosporum*	Kirk, 2020
Chile (0/2)	-	*Tuber magnatum* and *T. melanosporum*	Santelices et al., 2004; Perez et al., 2007
Colombia (0/2)	-	*Tuber borchii* and *T. rufum*	Kirk, 2020
Uruguay (0/1)	-	*Tuber aestivum*	Kirk, 2020

3.1.1 Climatic Conditions

Desert truffles are adapted to the climates of semi-arid, arid, and Saharan zones, which is why they are called desert truffles. They thrive in climates with humid summers and winters, with their development closely linked to autumn and winter rainfall. These regions typically receive annual rainfall between 50 and 380 mm (Trappe, 1971; Malençon, 1973; Awameh, 1981; Feeney, 2002; Kagan-Zur and Roth-Bejerano, 2008). According to Rebière (1981), desert truffles require a minimum of 80 mm of well-distributed rain between October and March to ensure their production and the abundance and density of host plants. Heim (1969) asserts that good truffle production depends on abundant rainfall in November and December.

In Morocco, truffles thrive in a temperate climate with distinct seasons. Their production is dependent on adequate rainfall, with early autumn rains favoring some species and winter and spring rains benefiting others (Khabar, 2016). In the Maâmora forest, optimal truffle production requires significant precipitation before March, with 240 mm being sufficient for a good yield (Khabar, 2002). In the eastern region, El Aji (1999) notes that the years 1990–1991 and 1993–1994 were particularly productive, with rainfall totals of 222.6 mm and 214 mm, respectively.

In Algeria, truffles thrive in arid, semi-arid, and Saharan zones. They require a warm climate with good autumn or winter rainfall, followed by a period of drought. For optimal production, three rainy periods are essential: the stormy rains of late summer and early autumn, and the rains of October, November, and March. Favorable rainfall amounts to 40 to 60 mm during October and November, and 5 to 15 mm in March, followed by a dry period in April (Bouchareb, 1994; Tadja, 1996; Fortas, 2004; Boufeldja, 2017; Benattia and Messaoudi, 2018). Rainfall from September to April facilitates the germination of truffle ascospores and mycelial growth, while spring rainfall aids in their ripening (Derbali, 2021).

In Kuwait, truffle development requires at least 180 mm of rainfall between October and March (Awameh and Alsheikh, 1979). According to Alsheikh (1994), truffles emerge after 3 to 4 months of adequate autumn and winter rainfall, which coincides with the completion of the host plant's life cycle. Abdaal and Al-Bahooh (1987) noted that in Saudi Arabia, a truffle field irrigated during a drought season yielded 110 metric tons over an area of 100 km^2. Even small amounts of rain are essential for the transport, dispersal, and germination of ascospores. However, excessive or poorly distributed rainfall, as well as prolonged periods of extreme cold or heat, can disrupt the truffle's biological cycle and lead to spore rot (Feeney, 2002; Bradai et al., 2013, 2015; Khabar, 2016).

3.1.2 Edaphic Factors

Desert truffles thrive in sandy, gypsiferous, or gravelly-gypsiferous soils that are well-aerated and light, which facilitate better circulation of mineral elements. These soils are relatively rich in limestone but poor in organic matter and phosphorus, with most truffles preferring alkaline or near-neutral pH levels. The pH and humidity are higher just below the surface in the immediate terfez environment compared to the surface, while the organic matter content is lower (Alsheikh and Trappe, 1983;

Fortas, 1990; Fortas and Chevalier, 1992a; Hashem and Al-Obaid, 1996; Bonifacio and Morte, 2014).

In Morocco, truffles require well-structured, aerated sandy soils that allow for the circulation of mineral elements. In the Maâmora region, analyzed soils consist of coarse sands and are slightly acidic, with a water pH between 6.7 and 4.6 and a potassium chloride pH between 5.4 and 4.2. The organic matter levels are extremely low, ranging from 0.57% upstream to 0.19% downstream (Khabar, 2016; Hakkou et al., 2023). In the eastern region of Morocco, particularly in Bouarfa, truffle soils are composed of sand with a low organic matter content (Tahiri, 1997; Hakkou, 2023).

In Algeria, truffles grow in slightly alkaline soils with a pH range of 7.85 to 8.57. These are sandy or sandy-loamy calcareous soils dominated by clay and silt, well-aerated, and light, allowing for better mineral circulation. These soils are relatively low in organic matter, rich in magnesium, well-supplied with potassium and calcium, and low in phosphorus (Fortas, 1990; Fortas and Chevalier, 1992b; Bradai, 2006; Zitouni and Zelmat, 2007; Aibeche, 2008; Benattia and Messaoudi, 2018). The soil conditions can vary between coastal and steppe regions (Derbali, 2021). In Kuwait, truffles grow on gypsum or salty gravelly-gypsum soils (Halwagy and Halwagy, 1974a, b).

3.1.3 Host Plant Species

Truffles live in symbiotic association with plants belonging to the *Cistaceae* family, generally *Helianthemums* (Awameh and Alsheikh., 1980; Alsheikh and Trappe, 1983; El Aji, 1999; Moreno et al., 2000; Diez et al, 2002; Zambonelli et al, 2014). In Morocco, they are collected in the vicinity of herbaceous plants of the genus *Helianthemum* (*H. apenninum, H. apertum, H. croceum, H. glaucum, H. guttatum, H. hirtum, H. ledifolium*, and *H. Lipii*), genus *Cistus* (*Cistus halimifolius, C. ladaniferus, C. monspeliensis, C. salicifolius,* and *C. salvifolius*) and also pines (*Pinus halepensis* and *P. pinaster* var. *atlantica*) (Khabar, 1988, 2016; Tahiri, 1997; Hakkou et al, 2023).

In Algeria, truffles live in symbiosis with *Helianthemum guttatum, H. hirtum* var. *deserti, H. ledifolium, H. lipii, H. salicifolium, H. squamatum*, and *Plantago albicans* (Fortas, 1990; ZitouniHaouar et al., 2014; Zitouni-Haouar, 2016). In Kuwait, they live in symbiosis with *H. ledifolium* and *H. salicifolium* (Awameh and Alsheikh, 1979). Other plants have been reported in other countries, *Artemisia herba-alba*, *Helianthemum almeriense, H. ovatum,* and *Plantago albicans* in France (Janex-Favre et al., 1988; Gutierrez et al, 2003; Kovacs et al, 2003).

Truffle production can be influenced by factors other than rainfall, soil, and the host plant. These include certain human practices. The Egyptian truffle, for example, has been almost wiped out in Muqattam by urban sprawl. Consequently , the product has faded into obscurity, unknown even to the local inhabitants, and remains unrecorded in the existing documents housed within the Egyptian Museum of Agriculture (Feeney, 2002).

According to Alsheikh (1994), the Gulf War of 1990–1999 ruined some of Kuwait's already fragile truffle-growing areas. Military operations took place mainly in the Al-Dibdibah formation, the main truffle-growing area in the desert. They disturbed vast areas and destroyed much of the vegetation. One of the main shrubs destroyed is *Thymelaea hirsuta*, which grows in desert truffle habitats and is associated with *Terfezia claveryi*. The oil falls congealed with the sand to form a brittle crust that affected soil drainage and aeration and reduced biological activity. People refused to eat the truffles because of the oil contamination. The only solution was to import truffles from Saudi Arabia and Libya to compensate for the lack of local trade and the loss of their spring harvest. There have been cross border influxes of Kuwaitis to try their luck in the truffle-growing areas of Saudi Arabia.

3.2 True Truffles

Similar to desert truffles, the occurrence of *Tuber* spp. in different geographic locations (Table 2) is dependent on climatic conditions, soil edaphic factors, and host plant preferences. The highest number of *Tuber* spp. was found in Morocco in Africa (13 spp.), China in Asia (76 spp.), Central Australia (12 spp.), Spain in Europe (38 spp.), and Canada and Mexico (22 spp. each). Identification of true truffles was highest in Europe (439), followed by Asia (173) and North America (165) (Table 3).

3.2.1 Climatic Conditions

These truffles grow in various climates, depending on the species, and they require alternating periods of heat and humidity for their development. Frost is a limiting factor, as it can interrupt ascocarp growth and ripening and degrade or kill the mushroom fruit when it affects the first five to twenty centimeters of soil (Callot, 1999; Bernadach, 2008; Olivier et al., 2018). Excessive rainfall in autumn can lead to truffle rot if the soil is poorly or moderately filtered, while persistent drought in October can also cause significant damage (Olivier et al., 2018).

Tuber melanosporum, known as the Périgord truffle, thrives in Mediterranean climates but can also tolerate sub-continental, oceanic, or sub-oceanic climates in sunny locations. It has limited water requirements in June and July, while thunderstorms in August are highly beneficial. A severe drought in May and June can impact truffle production, which occurs in April and May. *Tuber brumale*, known as the musk truffle, is adapted to the Mediterranean climate but also grows in oceanic and sub-oceanic climates. *Tuber uncinatum*, known as the Burgundy truffle, grows in a wide range of climates, including oceanic, semi-continental, continental, and Mediterranean mountain climates. It benefits from abundant summer rainfall in June, July, and August (60−130 mm/month) and matures before the coldest weather to avoid the risk of frost (Callot, 1999; Bernadach, 2008).

Callot (1999) reports that *Tuber magnatum* prefers warm, temperate climates. The *Tuber aestivum* species found in Iran thrives in humid temperate climates with annual rainfall ranging from 350 to 676 mm and maximum daily rainfall from 65.2 to 90 mm (Puliga et al., 2021). In Morocco, *Tuber* species are found in subhumid and humid climates (Khabar, 2016).

3.2.2 Edaphic Factors

Truffles thrive in calcareous soils containing fine elements such as clay that are well-aerated, draining, and easily warmed with intense biological activity. Clay plays a vital role in soil structure and moisture regulation, while biological activity

Table 3. Number of identifications of desert truffles and true truffles in different continents.

Continent		Desert truffles	True truffles
Africa	Total	84	15
	Mean per province	5.6	1.0
	Range	1–18	0–13
Asia	Total	109	173
	Mean per province	3.6	5.8
	Range	0–17	0–19
Australia	Total	11	12
Europe	Total	152	439
	Mean per province	4.1	11.9
	Range	0–40	0–38
North America	Total	43	165
	Mean per province	1.2	4.6
	Range	0–9	0–22
Oceania	Total	2.0	13
	Mean per province	1.0	6.5
	Range	1	0–12
South America	Total	3	16
	Mean per province	0.5	2.7
	Range	0–2	1–7

provides essential nutrients and aerates the soil. However, permanent humidity is not conducive to truffle development (Bernadach, 2008; Olivier et al., 2018).

According to Callot (1999), *Tuber magnatum* prefers moist, well-drained limestone environments with very low bulk density. *Tuber melanosporum* is typically found in areas with draining limestone soils and low moisture content. *Tuber. uncinatum* thrives in organic clay soils, often without litter. *Tuber indicum* bears fruit in calcareous soils with a pH between 5.5 and 8.5. In Iran, *Tuber aestivum* production areas are characterized by high pH sandy soils with high $CaCO_3$ content and low organic matter (Jamali, 2017). Truffle species harvested in Morocco are typically found in highly calcareous soils with silty-clay textures, rich in potassium, nitrogen, and magnesium (Khabar, 2016).

3.2.3 Host Plant Species

Tubers live entirely in association with the roots of certain trees, mainly oaks (Rodriguez, 2008). There are a number of factors that favor a tree being truffle-bearing, including soil, climate, and biotic factors. Trees traditionally considered truffle-bearing include holm oak, downy oak, kermes oak, pedunculate oak, faggoty oak, cork oak, hazelnut, and lime (*Corylus avellana*, *Quercus coccifera, Q. faginea, Q. ilex, Q. pubescens, Q. robur, Q. suber,* and *Tillia*). Other trees can be truffle-bearing if conditions are favorable, such as hornbeam, Aleppo pine, black pine, Atlas cedar, and chestnut (*Carpinus, Castanea, Pinus halepensis,* and *P. nigra,*) (Olivier et al., 2018). *Tuber magnatum* lives in association with pubescent oak, pedunculate oak, lime, hazelnut, poplar (*Populus*) and willow (*Salix*). *Tuber melanosporum* and *T. aestivum* are associated with oak, hazel, hornbeam, lime, olive (*Olea europaea*), and conifers such as pine and cedar (*Cedrus*). Whereas *Tuber indicum* is harvested under oak (*Quercus incana*) and pine with a shrub layer consisting of *Berberis, Piracantha* sp., *Thymus*, and *Vibernum* (Callot, 1999). *Tuber* species existing in Morocco are harvested under holm oak (Khabar, 2016).

5. Conclusion

Truffles are hypogeous fungi of the order Ascomycetes that live in symbiosis with host plants. They come in two types: true truffles of the *Tuber* genus and desert truffles adapted to arid and semi-arid climates. They are highly diversified and spread across all continents with a predominance in the countries of Asia, America, and Europe. Their discovery depends on the importance given to them in each country. They are also found in the countries of the Mediterranean basin. Their production depends on three essential factors: the soil, which ensures growth; the climate, especially rainfall, which influences yield; and the host plant, which is indispensable for the growth of the fungus, which is unable to synthesize organic substances or use solar energy. These factors vary according to the production region and truffle species. These factors control the distribution of truffles around the world.

References

Abdaal, M. and M. Al-Bahooh. (1987). Internal Report, Division of Food Resources, Department of Hadroproduction, Kuwait Institute for Scientific Research, Kuwait, p. 11.

Abourouh, M. (2011). Truffes du désert du Maroc: diversité et modes d'exploitation. 6ème Rencontre de Micosylva , Mértola (ADPM), Portugal, 15–18 Mars, p. 23.

Ahmed, A.A., Mohamed, M.A. and Hamid, M.A. (1981). Libyan truffles "*Terfezia boudieri* Chatin": Chemical composition and toxicity. J. Food. Sci., 46 (3): 927–929.

Aïbèche, C. (2008). Caractérisation écologiques et mycologiques d'une espèce de terfez du littoral Ouest algérien. Essai de mycorhization contrôlée avec sa plante-hôte naturelle *Helianthemum guttatum*. Mémoire de Magister. Univ d'Oran. Es-Sénia, p. 85.

Akata, I., Şen, I., Sevindik, M., Kabaktepe, S. (2022). Truffle checklist of Turkey II with a new record. Turk. J. Agric. Food Sci. Technol., 10(10): 1913–1920.

Akyuz, M., Kirbag, S. and Kursat, M. (2012). Ecological aspects of the arid and semi-arid truffle in Turkey: Evaluation of soil characteristics, morphology, distribution, and mycorrhizal relationships. Turk. J. Bot., 36: 386–391.

Alexander, B.J.R. (2008). *Tuber brumale*, a new truffle in New Zealand. *In:* Abstract of the 9th International Congress of Plant Pathology, Torino, 24–29 August.

Alli, H. and Doğan, H-H. (2019). A new genus (*Balsamia*) addition for Turkish mycota. The Journal of Fungus 10(1): 23–25.

Al-Qarawi, A.A and Mridha, M.A.U. (2012). Status and need of research on desert truffles in Saudi Arabia. J. Pure Appl. Microbiol., 6(3): 1051–1062.

Alsheikh, A.M. (1994). Taxonomy and mycorrhizal ecology of the desert truffles in the Genus *Terfezia*. Ph.D. thesis, Oregon State University, Oregon, USA, p. 237.

Alsheikh, A.M. and Trappe, J. M. (1983). Desert truffles: The genus *Tirmania*. Trans. Br. Mycol. Soc., 81: 83–90.

Al-Thani, R.F. (2010). Survey of macrofungi (including truffles) in Qatar. KBM J. Biol., 1 (2): 26–29. 10.5147/kbmjb.2010.0009

Alvarez, I.F., Parlade, J., Trappe, J.M. (1992). *Loculotuber gennadii* gen. et comb. nov. and *Tuber multimaculatum* sp. nov. Mycologia, 84(6): 926–929.

Ammarellou, A., Wang, Y., Nematzadeh, G. and Tajick, M. (2014). Non-Mediterranean Asian desert countries. pp. 173–192. *In:* Kagan-Zur, V., Roth-Bejerano, N., Sitrit, Y. and Morte, A. (eds.). Desert Truffles Phylogeny, Physiology, Distribution and Domestication, Springer, Berlin, Germany.

Awameh, M.S. (1981). Response of *Helianthemum salicifolium* and *H. ledifolium* to infection by the desert truffle *Terfezia boudieri*. Mushroom science. pp. 843–583. *In:* Nair, N.G. and Clift, A.D. (eds.). Proceedings of the Eleventh International Scientific Congress on the Cultivation of Edible Fungi, Australia.

Awameh, M.S. and Alsheikh, A. (1979). Laboratory and field study of four kinds of truffle (Kamah), *Terfezia* and *Tirmania* species, for cultivation. Mush. Sci., 10: 507–517.

Awameh, M.S. and Alsheikh, A. (1980). Ascospore germination of black kame (*Terfezia boudieri*). Mycologia, 72(1): 50–54.

Bagi, I. and Fekete, O. (2010). Identification of *Tuber aestivum* habitats in the South Caucasus, Azerbaidjan. *In:* Communication at the 2nd Congress of the *Tuber aestivum/uncinatum*, European Scientific Group, Juva, Finland, August 20–22.

Barseghyan, G.S and Wasser, S.P. (2010). Species diversity of hypogeous ascomycetes in Israel. Mycobiology, 38(3): 159–165.

Benattia, Z. and Messaoudi, S. (2018). Etude de Les activités antimicrobiennes et dosage des composés phénoliques des extraits de terfez (*Terfezia arenaria* Chatin). Mémoire de master. Université Mohamed Boudiaf, M'sila, p. 97.

Benucci, G.M.N., Csorbai, A.G., Falini, L.B., Marozzi, G., Suriano, E. et al. (2016). Taxonomy, biology and ecology of *Tuber macrosporum* Vittad. and *Tuber mesentericum* Vittad. pp. 69–86. *In:* Zambonelli, A., Lotti, M. And Murat, C. (eds.). True Truffle (*Tuber* spp.) in the World. Springer Cham.

Bernadach, P. (2008). Forêts de France, 518: p. 16.

Bidaud, A. and Van Vooren, N. (2008). *Tuber puberulum*, récolté en Isère. Bull. Mycol. Bot. Dauphiné-Savoie, 189: 45–48.

Bonifacio, E., Morte A. (2014). Soil Properties. pp. 57–67. *In:* Kagan-Zur, V., Roth-Bejerano, N., Sitrit, Y. and Morte, A. (eds.). Desert Truffles Phylogeny, Physiology, Distribution and Domestication, Springer, Berlin, Germany.

Bonito, G.M and Smith, M.E. (2016). General systematic position of the truffles: evolutionary theories. pp. 3–18. *In*: Zambonelli, A., Lotti, M. and Murat, C. (eds.). True Truffle (*Tuber* spp.) in the World, Soil Biology, Springer Cham.

Bordallo, J.-J. and Rodríguez, A. (2014). Cryptic and New Species. pp. 39–53. *In:* Kagan-Zur, V., Roth-Bejerano, N., Sitrit, Y. and Morte, A. (eds.). Desert Truffles Phylogeny, Physiology, Distribution and Domestication. Springer, Berlin, Germany.

Bordallo, J.J., Rodríguez, A., Muñoz-Mohedano, J.M., Suz, L.M., Honrubia, M et al. (2013). Five new *Terfezia* species from the Iberian Peninsula. Mycotaxon, 124: 189–208.

Bordallo, J-J., Rodriguez, A., Kaounas, V., Camello, F., Honrubia, M et al. (2015). Two new *Terfezia* species from Southern Europe. Phytotaxa, 230(3): 239–249.

Bouchareb, F. (1994). Etude écologique des terfez, cas de la région d'Ain Sefra (Wilaya de Naâma). Mém. Ing. D'état. Agron., I.N.F.S.A. Mostaganem, p. 81.

Boufeldja, W. (2017). Evaluation du potentiel nutritionnel et antioxydant de quelques variétés de Truffes du Sud-ouest algérien. Effets antimicrobien et antiinflammatoire. Thèse de doctorat. Université Djillali Liabes, p. 112.

Bradai, L. (2006). Contribution à l'étude bioécologique de la truffe blanche de désert (*Terfezia* sp): Cas de la région d'Oued Mya (Ouargla). Dissertation, Université de Ouargla, Algérie, p. 131.

Bradai, L., Bissati, S. and Chenchouni, H. (2013). Étude mycologique et bioécologique de la truffe blanche du désert (*Tirmania nivea* Desf. Trappe 1971) dans la région de Oued M'ya (Ouargla, Sahara Algérien). Revue des BioResources, 3(1): 6–14.

Bradai, L., Neffar, S., Amrani, Kh., Bissati, S. and Chenchouni, H et al. (2015). Ethnomycological survey of traditional usage and indigenous knowledge on desert truffles among the native Sahara Desert people of Algeria. J. Ethnopharmacol., 162: 31–38.

Bratek, Z., Jakucs, E. and Boka, K. (1996). Mycorrhizae between black locust (*Robinia pseudoacacia*) and *Terfezia terfezoïdes*. Mycorrhizae, 6: 271–274.

Callot, G. (1999). La truffe, la terre, la vie. Ed. INRA., Paris, p. 209.

Calonge, F.D., García, A., Miguel, S. and Bastardo, J. (2003). Some interesting fungi found in Spain, with special reference to the province of Valladolid. Mycol. Ital., 2: 45–52.

Castellano, M.A. and Turkoğlu, A. (2012). New records of truffle taxa in *Tuber* and *Terfezia* from Turkey. Turk. J. Bot., 36: 295–298.

Castro, M.L and Freire, L. (1982). *Terfezia berberiodora* y *T. oligosperma* encontradas en Galicia, nuevas para Espagna. Collect Bot Esp., 13(2): 441–444.

Ceruti, A., Fontana, A. and Nosenzo, C. (2003), Le specie europee del genere *Tuber*. Una revisione storica. Museo Regionale di Scienze Naturali, Monografie XXXVII. Regione Piemonte, Torino, p. 467.

Chatin, M.A. (1897). Un Nouveau Terfas (*Terfezia Aphroditis*) de L'Ile de Chypre. Bulletin de la Société Botanique de France, 44(4): 290–292.

Chen, J. and Liu, P.G. (2007). *Tuber latisporum* sp. nov. and related taxa, based on morphology and DNA sequence data. Mycologia, 99(3): 475–481.

Chen, J., Murat, C., Oviatt, P., Wang, Y. and Le Tacon, F et al. (2016). The black truffles *Tuber melanosporum* and *Tuber indicum*. pp. 19–32. *In:* Zambonelli, A., Lotti, M. and Murat, C. (eds.). True Truffle (*Tuber* spp.) in the World, Soil Biology, Springer Cham.

Chevalier, G. (2014). The European desert truffles. pp. 121–141. *In:* Kagan-Zur, V., Roth-Bejerano, N., Sitrit, Y. and Morte, A. (eds.). Desert Truffles Phylogeny, Physiology, Distribution and Domestication, Springer, Berlin, Germany.

Chevalier, G., Riousset, L., Dexheimer J. and Dupre, C. (1984). Synthèse mycorhizienne entre *Terfezia leptoderma* Tul et diverses Cistacées. Agronomie, 4: 210–211.

Chevalier, G., Riousset, L., Riousset, G. and Dupré, C. (2004). Truffes et terfez communes à la France et au Maroc. Premier symposium sur les champignons hypogés du bassin méditerranéen, Rabat (Maroc), p 23.

Colgan, W. and Trappe, J.M. (1997). NATS truffle and truffle-like fungi. 7. *Tuber anniae* sp. nov. (Ascomycota). Mycotaxon, 64: 437–442

Cortes, G., Barroetavena, C. and Rajchenberg, M. (2011). Argentina the new frontiers for truffles. In le VI th international workshop on edible mycorhizal mushrooms (IWEMM6). Rabat, Maroc, 6–10 avril 2011, p. 31.

Dämmrich, F. (2016). Taxon list of fungi and fungal-like organisms from Germany compiled by the DGfM. Staatliche Naturwissenschaftliche Sammlungen Bayerns. https://doi.org/10.15468/gtvmjw

De Laplane, D. (2010). Des truffes sauvages aux truffes cultivées en loudunais. Mémoires de l'Académie des Sciences, Arts et Belles-Lettres de Touraine, 23: 215–241.

Derbali, S. (2021). Les paramètres bioécologiques des truffes du désert. Mémoire de Master. Faculté des sciences exactes et des sciences de la nature et de la vie. Université de Biskra, p. 53.

Dexheimer, J., Gerard, J., Leduc, J.P. and Chevalier, G. (1985). Comparative ultrastructural study of symbiotic mycorrhizal associations between Helianthemum salicifolium-Terfezia claveryi and *Helianthemum salicifolium-Terfezia leptoderma*. Can. J. Bot., 63: 582–591.

Diamandis, S. and Perlerou, C. (2008). Recent records of hypogeous fungi in Greece. Acta Mycol., 43(2): 139–142.

Dib-Bellahouel, S. (2012). Etude du pouvoir antimicrobien et mycorhizien de deux espèces de Terfez: *Tirmania pinoyi* (Maire) Malençon et *Terfezia leptoderma* Tul. Thèse de doctorat. Université d'ORAN, p. 200.

Diez, J., Manjon, J.L. and Martin F. (2002). Molecular phylogeny of the mycorrhizal desert truffles (*Terfezia* and *Tirmania*), host specificity and edaphic tolerance. Mycologia, 94(2): 247–259.

Du, M.H., Huang, S.M. and Wang, J.W. (2014). The volatiles from fermentation product of *Tuber formosanum*. Open J. For., 4: 426–429.

El Aji, A. (1999). Contribution à l'étude des Terfez des parcours de l'oriental marocain

Ewaze, J.O. and Al-Naama, M.M. (1989). Studies on nitrogen metabolism of *Terfezia* spp. and *Tirmania* spp. New Phytol., 112: 419–422.

Fan, L., Cao, J.Z. and Li, Y. (2012). *Tuber microsphaerosporum* and *Paradoxa sinensis* spp. nov. Mycotaxon, 120: 471–475.

Fan, L., Feng, S. and Cao, J.Z. (2014). The phylogenetic position of *Tuber glabrum* sp. nov. and *T. sinomonosporum* nom. nov., two *Paradoxa*-like truffle species from China. Mycol. Prog. 13(2): 241–246.

Fan, L., Hou, C.L. and Cao, J.Z. (2011). *Tuber sinoalbidum* and *T. polyspermum* - new species from China. Mycotaxon, 118: 403–410.

Feeney, J. (2002). Desert truffles galore. Saudi Aramco World, pp. 22–27.

Ferdman, Y., Aviram, S., Roth-Bejerano, N., Trappe, J.M. and Kagan-Zur, V et al. (2005). Phylogenetic studies of *Terfezia pfeilii* and *Choiromyces echinulatus* (Pezizales) support new genera for southern African truffles: Kalaharituber and Eremiomyces. Mycol. Res., 109: 237–245.

Fortas, Z. (1990). Etude de trois Terfez. Caractères culturaux et cytologie du mycélium isolé et associé à l'*Helianthemum guttatum*. Thèse Doctorat d'état, université d'Oran Es-Sénia, pp. 9–68

Fortas, Z. (2004). Ecologie et production naturelle des terfez d'Algérie. Premier symposium sur les champignons hypogés du bassin méditerranéen, Rabat (Maroc), p 24-25.

Fortas, Z. (2009). Diversité des espèces de terfez (truffes des sables) des zones arides algériennes. Séminaire international: Protection et préservation des écosystèmes sahariens. Ouargla, p 51.

Fortas, Z. and Chevalier, G. (1988). Effet des conditions de culture sur la mycorhization, d'*Helianthemum guttatum* par trois espèces du genre *Terfezia* et *Tirmania* (truffes du désert). 2ème congresso internazionale sul. Tartufo spoleto, 197–203.

Fortas, Z. and Chevalier, G. (1992a). Caractéristiques de la germination des ascospores de *Terfezia arenaria* (Moris) Trappe, récoltée en Algérie. Cryptog. Mycol., 13: 21–29.

Fortas, Z. and Chevalier, G. (1992b). Effet des conditions de culture sur la mycorhization de l'Helianthemum guttatum par trois espèces de terfez des genres Terfezia et *Tirmania* d'Algérie. Can. J. Bot., 70(12): 2453–2460.

Garcia-Montero, L.G., Diaz, P., Martín-Fernandez, S. and Casermeiro, M-A. (2008). Soil factors that favor the production of *Tuber melanosporum* carpophores over other truffle species: a multivariate statistical approach. Acta Agric. Scand., 58(4): 322–329.

Glejdura, S. and Kunca, V. (2012). First record of *Mattirolomyces terfezioides* from Slovakia-the northernmost locality in Europe. Catathelasma, 14: 5–9.

Grente, J. and Chevalier G. (1980). La trufficulture, un or noir, pour la Région Méditerranéenne. Forêt Méditerranéennes t., 1(2): 151–162.

Gryndler, M., Trilcová, J., Hrselová, H., Streiblová, E., Gryndlerová, H et al. (2013). *Tuber aestivum* Vittad. mycelium quantified: Aadvantages and limitations of a qPCR approach. Mycorrhiza, 23(5): 341–348.

Gucin, F., Kaya, A., Soylu, M.K. and Uzun, Y. (2010). *Picoa* Vittad., a new truffle genus record for Turkey. Biol. Diver. Conserv., 3: 23–25.

Guevara, G., Bonito, G., Cázares, E., Rodríguez, J., Vilgalys, R et al. (2008). *Tuber regimontanum*, una nueva especie de trufa de México. Rev. Mex. Mic., p. 26.

Guevara, G., Bonito, G., Trappe, J.M., Cázares, E., Williams, G. et al. (2013). New North American truffles (*Tuber* spp.) and their ectomycorrhizal associations. Mycologia, 105(1): 194–209.

Guevara-Guerrero, G., Bonito, G., Smith, ME., Healy, R., Grupe, A.C. et al. (2018). *Tuber aztecorum* sp. nov., a truffle species from Mexico belonging to the Maculatum clade (Tuberaceae, Pezizales). MycoKeys, 30: 61–72.

Gutiérrez, A., Morte, A. and Honrubia, M. (2003). Morphological characterization of the mycorrhiza formed by *Helianthemum almeriense* Pau with *Terfezia claveryi* Chatin and Picoa lefebvrei (Pat.) Maire. Mycorrhiza, 13: 299–307.

Hakkou, S. (2023). Les truffes marocaines: étude socio-géographique et environnementale. Faculté des lettre set des sciences humaines, Université Mohamed V. p. 478.

Hakkou, S., Sabir, M. and Machouri, N. (2022). Le Maroc truffier: la répartition géographique des truffes et truffes du désert et leur productivité. Rev. Mar. Sci. Agron. Vét., 10(3): 395–407.

Hakkou, S., Sabir, M. and Machouri, N. (2023). Contribution to the study of the production factors of Moroccan truffles (soil, climate, and vegetation). J. Mater. Environ. Sci., 14(1): 13–30.

Hall, I.R., Brown, G. and Zambonelli, A. (2008). Taming the truffle: The history, lore and science of the ultimate mushroom. Timber Press, Oregon USA, p. 304.

Halwagy, R. and Halwagy, M. (1974a). Ecological studies on the desert of Kuwait I. The physical environment. J. The Univ. Kuwait (Sci.) 1: 75–86.

Halwagy, R. and Halwagy, M. (1974b). Ecological Studies on the desert of Kuwait: II the vegetation. J. Univ. Kuwait (Sci.) 1: 85–95.

Hansen, K., Lobuglio, K.F. and Pfister, D.H. (2006). Evolutionary relationships of the cup-fungus genus *Peziza* and Pezizaceae inferred from multiple nuclear genes: RPB2, beta-tubulin, and LSU rDNA. Mol. Phylo. Evol., 36:1–23.

Hashem, A.R. and Al-Obaid, A.M. (1996). Mineral composition of soil and wild desert truffles in Saudi Arabia. J. King Saud. Univ., 8(1): 5–10.

Hashimoto, Y., Fukuwawa, S., Kunishi, A., Suga, H., Richard, F. et al. (2012). Mycoheterotrophic germination of *Pyrola asarifolia* dust seeds reveals convergences with germination in orchids. New Phytol., 195(3): 620–630.

Healy R.A. (2003). *Mattirolomyces tiffanyae*, a new truffle from Iowa, with ultrastructural evidence for its classification in the Pezizaceae. Mycologia, 95(4): 765–772.

Healy, R., Bonito, G.M. and Smith, M.E. (2016). A brief overview of the systematics, taxonomy, and ecology of the *Tuber rufum* clade. pp. 125–136. *In*: Zambonelli, A., Lotti, M. and Murat, C. (eds.). True Truffle (*Tuber* spp.) in the World. Springer Cham.

Heim, R. (1969). Champignons d'Europe. E- Généralité - Ascomycètes - Basidiomycètes. Edité par Bougée and Cie: 223–235.

Hilszczanska, D., Rosa-Gruszecka, A. and Szmidla, H. (2014). Characteristic of *Tuber* spp. localities in natural stands with emphasis on plant species composition. Acta Mycol., 49(2): 267–277.

Hussain, G. and Al-Ruqaie, I.M. (1999). Occurrence, chemical composition and nutritional value of truffles: an overview. Pak. J. bio. Sci., 2(2): 510–514.

Jamali, S. (2017). First report of identification and molecular characterization of *Tuber aestivum* in Iran. Agrofor. Sys., 91: 335–343.

Jamali, S. and Banihashemi, Z. (2012). Fungi associated with ascocarps of desert truffles from different parts of Iran. J. Crop. Protect., 1: 41–47.

Janex-Favre, M.C., Parguey-Leduc, A. and Riousset, L. (1988). L'ascocarpe hypogé d'une terfez française (*Terfezia leptoderma* Tul., Tubérales, Discomycètes). Bull. Soc. Mycol. Fr., 104: 145–178.

Jeandroz, S., Murat, C., Wang, Y.J., Bonfante, P. and Le Tacon, F. et al. (2008). Molecular phylogeny and historical biogeography of the genus *Tuber*, the true truffles. J. Biogeogr., 35(5): 815–829.

Kagan-Zur, V. and Akyuz, M. (2014). Asian Mediterranean desert truffles. pp. 159–172. *In:* Kagan-Zur, V., Roth-Bejerano, N., Sitrit, Y. and Morte, A. (eds.). Desert Truffles Phylogeny, Physiology, Distribution and Domestication. Springer, Berlin, Germany.

Kagan-Zur, V. and Roth-Bejerano, N. (2008). Desert truffles. Fungi 1(3): 32–37.

Karadelev, M., Rusevska, K., Kajevska, I. and Kopanja, D.-M. (2019). Checklist of larger ascomycetes in the Republic of Macedonia. Section of Natural, Mathematical and Biotechnical Sciences, MASA 40(2): 239–253.

Khabar L. (2016). Les Terfess et les truffes du Maroc, biodiversité et valorisation. Éd: Univ. Européennes, p. 276.

Khabar, L. (1988). Le genre Terfezia Tull. (Terfass) de la Forêt de la Maâmora (Région de Salé): étude systématique, écologique, morphologique, cytologique et ultra structurale. Université Mohammed V, Rabat. Thèse de troisième cycle, p. 150.

Khabar, L. (2002). Etudes pluridisciplinaires des truffes du Maroc et perspectives pour l'amélioration de production des "Terfess" de la forêt de la Maâmora. Doctorat d'Etat Es- Sciences, Université Mohamed V, Faculté des Sciences, Rabat, Maroc, p. 167.

Kinoshitaa, A., Yamamotob, K., Tainakad, T., Handae, T. and Yamada A et al. (2022). *Tuber torulosum*: A new truffle species decorated with moniliform cystidia from Japan. Mycoscience, 62: 1–7. org/10.47371/mycosci.2021.10.005

Kirk, P. (2020). Species Fungorum Plus. *In:* Bánki, O., Roskov, Y., Döring, M., Ower, G., Hernández Robles, D.R. et al. Catalog of Life Checklist (April 2024). Royal Botanic Gardens, Kew, UK.

Kovács, G.M. (2002). Study of mycorrhizae on the Great Hungarian Plain. Ph.D. dissertation, University of Szeged, Hungary.

Kovács, G.M., Balazs, T.K., Calonge, F.D. and Martin, M.P. (2011). The diversity of *Terfezia* desert truffles: new species and a highly variable species complex with intrasporocarpic nrDNA ITS heterogeneity . Mycologia, 103(4): 841–853.

Kovács, G.M., Martín, M.P. and Calonge, F.D. (2009). First record of Mattirolomyces terfezioides from the Iberian Peninsula: its southern- and westernmost locality. Mycotaxon, 110: 325–330.

Kovács, G.M., Rudnoy, S., Vagvolgyi, C., Lasztity, D., Racz, I. et al. (2001). Intraspecific invariability of the internal transcribed spacer region of the rDNA of the truffle *Terfezia terfezoïdes* in Europe. Folia Microbial., 46 (5): 423–426.

Kovács, G.M., Vagvolgyi, C. and Oberwinkler, F. (2003). In vitro interaction of the truffle *Terfezia terfezoïdes* with *Robinia pseudoacacia* and *Helianthemum ovatum*. Folia Microbiol., 48(3): 369–378.

Laessoe, T. and Hansen, K. (2007). Truffle trouble: what happened to the Tuberales, Mycol. Res., 111: 1075–1099.

Lancellotti, E., Iotti, M., Zambonelli, A. and Franceschini, A. (2016). The Puberulum group Sensu Lato (Whitish Truffles). pp. 105–124. *In*: Zambonelli, A., Lotti, M. and Murat, C. (eds.). True Truffle (*Tuber* spp.) in the World, Soil Biology, Springer Cham.

Le Tacon, F. (2017). La domestication des truffes. Académie de Stanislas, p. 21.

Le Tacon, F., Jeandroz, S., Murat, C. and Wang, Y. (2006). Répartition et histoire des truffes dans l'hémisphère Nord. Automne, 128., 11–15.

Leonardi, M., Paz-Conde, A., Guevara, G., Salvi, D. and Pacioni, G et al. (2019). Two new species of *Tuber* previously reported as *Tuber malacodermum*. Mycologia, 111(4): 676–689.

Liu, C. and Li, Y. (2017). Evaluation of the volatile profile of *Tuber liyuanum* by HSSPME with GC-MS. Nat. Prod. Res., 31: 972–975.

Loizides, M., Hobart, C., Konstandinides, G. and Yiangou, Y. (2012). Desert truffles: The mysterious jewels of antiquity. Field Mycology, 13(1): 17–21.

Malençon, G. (1973). Champignons hypogés du Nord de l'Afrique. I. Ascomycètes. Persoonia, 7: 261–288.

Mandeel, Q.A. and Al-Laith A.A.A. (2007). Ethnomycological aspects of the desert truffle among native Bahraini and non-Bahraini peoples of the Kingdom of Bahrain. J. Ethnopharmacol., 110: 118–129.

Mello, A., Murat, C., Vizzini, A., Gavazza, V. and Bonfante, P et al. (2005). *Tuber magnatum* Pico, a species of limited geographical distribution: its genetic diversity inside and outside a truffle ground. Environ. Microbiol., 7(1): 55–65.

Menta, C., Gonzaga, L., Stefania, G-M., Federica, P., Conti, D. et al. (2014). Does the natural microcosm created by *Tuber aestivum* affect soil microarthropods? A new hypothesis based on Collembola in truffle culture. Appl. Soil Ecol., 84: 31–37.

Merényi, Z., Varga, T., Hubai, A.G., Pitlik, P., Erős, A. et al. (2017). Challenges in the delimitation of morphologically similar species: a case study of *Tuber brumale* agg. (Ascomycota, Pezizales). Mycol. Prog., 16(6): 613–624.

Mohamed-Benkada, M. (1999). Extraction et essai d'isolement des principes antimicrobiens de *Terfezia claveryi* Chat. Thèse de Magister, Univ. D'Oran, Es-Sénia, p. 83.

Molinier, V., Peter, M., Stobbe, U. and Egli, S. (2016). The burgundy truffle (*Tuber aestivum* syn. *uncinatum*): a truffle species with a wide habitat range over Europe. pp. 33–48. *In:* Zambonelli, A., Lotti, M. and Murat, C. (eds.). True Truffle (*Tuber* spp.) in the World, Soil Biology, Springer Cham.

Moreno, G., Alvarado, P. and Manjón J.L. (2011). Phylogenetic affiliation of *Choiromyces magnusii* Tuberaceae Ascomycota) from Spain. Mycol. Prog., 11: 463–471. https://doi.org/10.1007/s11557-011-0762-1

Moreno, G., Diez, J. and Manjon, J.L. (2000). *Picoa lefebvrei* and *Tirmania nivea*, two are hypogeous fungi from Spain. Mycol. Res., 104(3): 378–381.

Morte, A., Honrubia, M. and Gutierrez, A. (2008). Biotechnology and cultivation of desert truffles. *In:* Varma A, ed. Mycorrhiza: state of the art genetics and molecular biology, eco-function, biotechnology, eco-physiology, structure and systematics. Berlin, Heidelberg: Springer Verlag. P. 467–483.

Mostowfizadeh-Ghalamfarsa, R., Banihashemi, Z. and Jamali, S. (2014). Specific primers for identification and detection of three major desert truffle Species in Iran with special reference to *Terfezia claveryi*. Iran Agric. Res., 33(1): 29–42.

Moubasher A.H. (1995). Soil fungi are an active partner of our ecosystem, their biodiversity and activities should be appreciated. Qatar Univ. Sci. J., 15(1): 239–247.

Moustafa A.F. (1985). Taxonomic studies on the fungi of Kuwait III: Ascomycotina (Plectomycetes and Discomycetes). J. Univ. Kuwait. (Sci.), 12: 79–100.

Murat, C., Mello, A., Abbà, S., Vizzini, A. and Bonfante, P et al. (2008). Edible mycorrhizal fungi: Identification, life cycle and morphogenesis. Varma, A. (ed.). Mycorrhiza, Springer-Verlag, Berlin, Heidelberg, 707–732.

Olivier, J.-M., Savignac, J.-C. and Sourzat, P. (2018). Truffe and trufficulture. Périgueux: Editions Fanlac, p. 352.

Omer, E.A., Smith, D.L., Wood, K.V. and El-Menshawi, B.S. (1994). The volatiles of desert truffle: *Tirmania nivea*. Pl. Foods Hum. Nutr., 45(3): 247–249.

Pegler, D.N., Spooner, B.M. and Young, T.W.K. (1993). British Truffles – A revision of British Hypogeous Fungi. Royal Botanic Gardens, Kew, p. 224.

Percudani, R., Trevisi, A., Zambonelli, A. and Ottonello, S. (1999). Molecular phylogeny of truffles (Pezizales: Terfeziaceae, Tuberaceae) derived from nuclear rDNA sequence analysis. Mol. Phyl. Evol., 13(1): 169–180.

Perez, F., Palfner, G. and Brunel, N. (2007). Synthesis and establishment of *Tuber melanosporum* Vitt. Ectomycorrhizae on two *Nothofagus* species in Chile. Mycorrhiza, 17: 627–632.

Pinkas, Y., Maimon, M., Shabi, E., Elisha, S., Shmulewich, Y et al. and Free-man, S. (2000). Inoculation, isolation and identification of *Tuber melanosporum* from old and new oak hosts in Israel. Mycol. Res., 104: 472–477.

Polemis, E., Konstantinidis, G., Fryssouli, V., Slavova, S., Tsampazis, T. et al. (2019). *Tuber pulchrosporum* sp. nov., a black truffle of the Aestivum clade (Tuberaceae, Pezizales) from the Balkan peninsula. MycoKeys, 47: 35–51.

Puliga, F., Illice, M., Iotti, M., Leonardi, P, Baghdadi, A. et al. (2021). True truffle diversity in Iran. Ital. J. Mycol., 50: 52–56.

Ratti, C., Iotti, M., Zambonelli, A. and Terlizzi, F. (2016). Mycoviruses infect true truffles. pp. 333–349. *In*: Zambonelli, A., Lotti, M. and Murat, C. (eds.). True Truffle (*Tuber* spp.) in the World, Soil Biology Systematics and Biochemistry. Springer Cham.

Rebière, J. (1981). La truffe du Périgord. Ed. Pierre Fanlac, Périgueux (France), p. 210.

Reyna, S. and Garcia-Barreda, S. (2014). Black truffle cultivation: a global reality. For. Syst., 23(2): 317–328.

Riccioni, C., Rubini, A., Belfiori, B., Gregori, G. and Paolocci, F et al. (2016). Tuber magnatum: The special one. what makes it so different from the other *Tuber* spp.? pp. 87–104. *In*: Zambonelli, A., Lotti, M. and Murat, C. (eds.). True Truffle (*Tuber* spp.) in the World, Soil Biology, Springer Cham.

Riousset, L., Riousset, G., Chevalier, G. and Bardet, M.C. (2001). Truffes d'Europe et de Chine. Ed. INRA, 181 p.

Robin, Ch., Pousse, N. and Le Tacon, F. (2016). Soil characteristics for *Tuber aestivum* (syn. *T. uncinatum*). pp. 211–231. *In:* Zambonelli, A., Lotti, M. and Murat, C. (eds.). True Truffle (*Tuber* spp.) in the World, Soil Biology, Springer Cham.

Rodriguez, A. (2008). Desert truffles. Sur http://www.trufamania.com/desert-truffles.htm

Roth-Bejerano, N., Livne, D. and Kagan-Zur, V. (1990). *Helianthemum-Terfezia* relations in different growth media. New phytol., 114(2): 235–238.

Roth-Bejerano, N., Mendlinger, S. and Kagan-Zur, V. (2004). Effect of calcium on growth of submerged *Terfezia boudieri* mycelium. Mycoscience, 45: 30–34.

Roux, P. (2006). Mille et un champignons. Editions Roux Sainte Sigolène, France, p. 1223.

Santelices, R., Perez, F., Reyna, Y.S. and Ramirez, R. (2004). Introduction cultive de Tuber melanosporum Vitt. En Chile. Jornadas Técnicas Forestales y Ambientales, 9: 135–140.

Sbissi, I. (2013). Phylogénie des truffes du désert en Tunisie Méridionale. Thèse de Doctorat, Univ. Tunis El-Manar, Tunisie, p. 120.

Sepahvand, D., Shirvany, A., Etemad, V., Javan-Nikkhah, M., DeLuca, Th et al. (2021). The first report of *Choiromyces venosus* and a new report of *Tuber aestivum* from North Hyrcanian Forest of Iran. Cent. Asian J. Pl. Sci. Innov., 1(3): 151–159.

Slama, A. and Neffati, M. (2004). Les truffes de la Tunisie méridionale: Etude écologique et mycologique. Revue des régions arides, 15: 3–52.

Slama, A., Fortas, Z., Boudabous, A. and Neffati, M. (2010). Cultivation of an edible desert truffle (*Terfezia boudieri* Chatin). Afr. J. Microbiol, Res., 4(22): 2350–2356.

Slama, A., Neffati, M., Fortas, Z., Khabar, L. and Boudabous, A et al. (2006). Etude taxinomique de quelques ascomycota hypogés (terfeziaceae) de la Tunisie Méridionale. Bull. Soc. Mycol. Fr., 122(2–3): 187–195.

SMNF (2006). *Tuber maculatum*. Société Mycologique du Nord de la France https://www.smnf.fr/wpcontent/Photos_SMNF/Photos_SMNF_T/Tuber_maculatum.htm.

Splivallo, R. and Bonfante, P. (2006). Truffles volatiles and exudates investigation of their role in truffle/plant interactions. Cebiovem. Ipp-cnr., 1: 10–14.

Stobbe, U., Buntgen, U., Sproll, L., Tegel, W., Egli, S et al. (2012). Spatial distribution and ecological variation of re-discovered German truffle habitats. Fungal Ecol., 5(5): 591–599.

Su, R., Tay, L. and Diener, Ed. (2014). The Development and validation of the comprehensive inventory of thriving (CIT) and the brief inventory of thriving (BIT). Appl. Psychol. Health Well-Being, 6(3): 251–279. 10.1111/aphw.12027

Tabouret, P. (2011). Exclusif! Description d'un site français producteur de truffe blanche d'Italie. Le Trufficulteur 78: 18.

Tadja, A. (1996). Etude écologique de deux espèces de *Terfezia* du Sud-Ouest Algérien. Essai de leur mycorhization sur trois espèces céréalières. Magister thesis, National Institute of Agronomy, El-Harrach, Algeria, p. 57.

Tahiri, A. (1997). Relation mycorhizienne entre *Helianthemum ledifolium* (L.) Mill, et des espèces de Terfez du Maroc. Mémoire de troisième cycle en agronomie, Ecole nationale d'agriculture, Meknès, p. 80.

Trappe, J.M. (1979). The orders, families, and genera of hypogeous Ascomycotina (truffles and their relatives) "*Clelandia arenacea, Dingleya verrucosa, Choiromyces aboriginum, Peziza stuntzii*", new taxa, Fungi. Mycotaxon, (9): 297–340.

Trappe, J.M. and Weber, N.S. (2001). North American desert truffles: the genus *Carbomyces* (Ascomycota, Carbomycetaceae). Harvard Pap. Bot., 6(1): 209–214.

Trappe, J.M., Claridge, A.D., Arora, D. and Smit, W.A. (2008). Desert truffles of the African Kalahari: Ecology, ethnomycology, and taxonomy. Econ. Bot., 62: 521–529.

Trappe, J.M., Gábor, M., Kovács, A. and Claridge, W. (2010). Comparative taxonomy of desert truffles of the Australian outback and the African Kalahari. Mycol. Prog., 9: 131–143.

Trappe, J.M., Gábor, M., Kovács, A. and Claridge, W. (2010). Validation of the new *Mattirolomyces austroafricanus*. Mycol. Progr., 9: 145. 10.1007/s11557-009-0631-3

Trappe, J.M., Kovacs, G.M. and Weber, N.S. (2014). Ecology and Distribution of Desert Truffles in Western North America. pp. 106–120. *In*: Kagan-Zur, V., Roth-Bejerano, N., Sitrit, Y. and Morte, A. (eds.). Desert Truffles Phylogeny, Physiology, Distribution and Domestication. Springer, Berlin, Germany.

Trappe. J.M. (1971). A synopsis of the Carbomyceteaceae and Terfeziaceae (Tuberales). Trans. Br. Mycol. Soc., 57: 85–92.

Turkoğlu, A. and Castellano, M.A. (2014). New records of some Ascomycete truffle fungi from Turkey. Turk. J. Bot., 38: 406–416.

Turkoğlu, A., Castellano, M.A., Trappe, J.M. and Güngör M.Y. (2015). Turkish truffles I: 18 new records for Turkey. Turk. J. Bot., 39: 359–376.

Urban, A., Neuner-Plattner, I., Krisai-Greilhuber, I. and Haselwandter, K. (2004). Molecular studies on terricolous microfungi reveal novel anamorphs of two *Tuber* species. Mycol. Res., 108: 749–758.

Wan, Sh., Liu, J., Huang, L., Qin, X., Liu, W et al. (2021). *Tuber qujingense* and *T. songlu*, two new species from Yunnan, China. Phytotaxa, 527(4): 248–256. 10.11646/phytotaxa.527.4.2

Wang, X. (2013). Truffle cultivation in China. pp. 227–240. *In:* Zambonelli, A. and Bonito, G. (eds.). Edible Ectomycorrhizal Mushrooms: Current Knowledge and Future Prospects. Springer Berlin, Heidelberg, Germany.

Wang, Y. and Liu, P. (2009). Achievements and challenges of research in truffles in China. Acta Bot. Yunnanica, 31(Suppl. 16): 1–9

Watkinson, M.L. (2018). Two new truffle species discovered in Florida pecan orchards. Florida Musium Research news. https://www.floridamuseum.ufl.edu/science/two-new-truffle-species-discovered-in-florida-pecan-orchards/

Weden, C., Chevalier, G. and Backlund, A. (2013). The first production results of *Tuber melanosporum* in Sweden. Abstracts of the 1st international congress of trufficulture, *Tuber* 2013, Teruel, March 5–8.

Yan, X., Cao, J. and Fan, L. (2018). Four new *Tuber* species added to the from China based on morphological and molecular evidence. Mycologia, 110(4): 771–1779. 10.1080/00275514.2018.1490120.

Zambonelli, A., Donnini, D., Rana, G.L., Fascetti, S., Benucci, G.M.N. et al. (2014). Hypogeous fungi in Mediterranean maquis, arid and semi-arid forests. Plant Biosys., 148(2): 392–400.

Zitouni, F.H. (2010). Etude des associations mycorhiziennes entre quatre espèces de terfez et diverses plantes Cistacées et ligneuse en conditions contrôlées. Mém. Magister, Univ. Oran, p. 262.

Zitouni, F.H. (2016). Etude de la diversité des truffes du désert et de leurs associations mycorhiziennes. Thèse de doctorat. Université d'Oran 1 Ahmed Ben Bella, p. 95.

Zitouni, F.H. and Zelmat, F.Z. (2007). Caractérisation écologique et pédologique de trois sites à terfez de la steppe centrale d'Algérie. Essai de mycorhization de *Tirmania pinoyi* avec *Pinus halepensis* en conditions génotoxiques. Mém. Ing. Biotech, Univ. Oran, p. 62.

Zitouni-Haouar, F., Carlavilla, J.R., Moreno, G., Manjon, J.L. and Fortas, Z et al. (2018). Genetic diversity of the genus *Terfezia* (Pezizaceae, Pezizales): new species and new record from North Africa. Phytotaxa, 334(2): 183–194.

Zitouni-Haouar, F.E., Fortas, Z. and Chevalier, G. (2014). Morphological characterization of mycorrhizae formed between three *Terfezia* species (desert truffles) and several Cistaceae and Aleppo pine. Mycorrhiza, 24(5): 397–403.

Chapter 2

Unveiling Global Trends and Implications

A Bibliometric Analysis of Research on Desert Truffles

Samad Jamali

1. Introduction

Desert truffles occupy an important place in the history of human consumption in their regions of origin. These fungi are valued for both their nutritional and gustatory properties, offering a unique profile. The protein content of desert truffles is remarkably high, averaging 20% of their dry weight. This is considerably more than what is found in most vegetables and other mushrooms. One of the most fascinating aspects of desert truffles is their antioxidant properties. They have a strong ability to inhibit lipid oxidation, a process that is detrimental to cellular health. The study by Murcia et al. (2002) has shown that *Terfezia claveryi* and *Picoa lefebvrei* exhibit a higher percentage of oxidation inhibition than conventional food antioxidants like tocopherol and propyl gallate. This remarkable property is retained even after industrial freezing and canning processes. Raw desert truffles exhibit strong antioxidant activity and effectively scavenge various oxygen species. This finding is promising for the potential substitution of synthetic antioxidants with natural truffle extracts in food preservation. Apart from their fascinating chemical composition and antioxidant activity, desert truffles have also shown promising antibiotic properties (Janakat et al., 2005). Remarkably, no recognized toxic compounds have been found in them (Ahmed et al., 1981). In certain regions, *T. claveryi* is used in traditional

Department of Plant Protection, College of Agriculture, Razi University, Kermanshah, Iran.
Email: jamali454@yahoo.com

medicine to treat eye conditions such as trichiasis. *Terfezia* and *Tirmania* species are also used to treat skin and eye diseases. The study by Chellal and Lukasova (1995) has shown that fractions of the methanolic extract of *Terfezia* exhibit activity against certain gram-positive bacteria, such as *Bacillus subtilis* and *Staphylococcus aureus*. Desert truffles have a certain mystique attached to them, partly because of their price. Despite their captivating characteristics, our understanding of the basic biology of desert truffles remains limited. Although research efforts continue, it is estimated that more than 300 publications on desert truffles have been published to date. However, many important biological aspects still require further research.

Bibliometric studies can be a valuable tool to advance research on desert truffles. Bibliometric studies on desert truffles can help us understand research trends, identify knowledge gaps, and promote international collaboration in this field. By analyzing the volume and citation patterns of scientific publications, bibliometrics can provide insights into the most studied aspects of desert truffles, the most productive research institutions and countries, and the most influential authors. This information can guide future research efforts, identify areas requiring further research, and promote collaboration among researchers from different countries.

2. Methods

The author has chosen to use Scopus (https://www.scopus.com/), the world's leading abstract and citation database covering peer-reviewed literature (books, conference proceedings, and journals), as the primary data source for the period 1891–2023 (Elsevier, 2017, 2019). To identify relevant publications, the author conducted a search in the combined fields of title, abstract, and keywords using the following terms: "Desert truffle", "Terfezia", "Tirmania", "Carbomyces", "Elderia", "Eremiomyces", "Kalaharituber", "Mattirolomyces", "Mycoclelandia", "Picoa", "Stouffera» or "Ulurua". Search results were downloaded and included information on various topics, such as the type and name of the document (research article, review, book chapter, and book), title, abstract, keywords, authors, year of publication, and references. We accept publications in any language. All documents were downloaded in CSV format, which enables citation analysis using VOSviewer (version 1.6.18) (Van Eck & Waltman, 2010) to create bibliometric networks that allow visualization of literature hotspots and trends. We used the default parameters of VOSviewer. Each identified term was assigned a label and a circle, with the font size of the label and the diameter of the circle proportional to its frequency. In the networks created with VOSviewer, terms displayed in the same color tend to cluster together, with the shorter distance between two terms indicating a stronger relationship (Van Eck and Waltman, 2010).

2.1 Evolution of Research on Desert Truffles

In the period from 1891 to 2023, 328 documents with the keywords "desert truffle", "Terfezia", "Tirmania", "Carbomyces", "Elderia", "Eremiomyces", "Kalaharituber", "Mattirolomyces", "Mycoclelandia", "Picoa", "Stouffera" or "Ulurua" were found in Scopus. Between 1891 and 2000 (109 years), the number of publications per year was less than 5. The number of publications presented an increasing trend from 2012 onwards (Fig. 1), and the highest number of documents was published in 2022 (n=34). From 1981 onwards, the average number of publications per year was 2.4 and ranged from 0 to 34 (in 2022). Among 328 articles, more than half (53%) were published in the nine-year period 2015–2023 (Fig. 1).

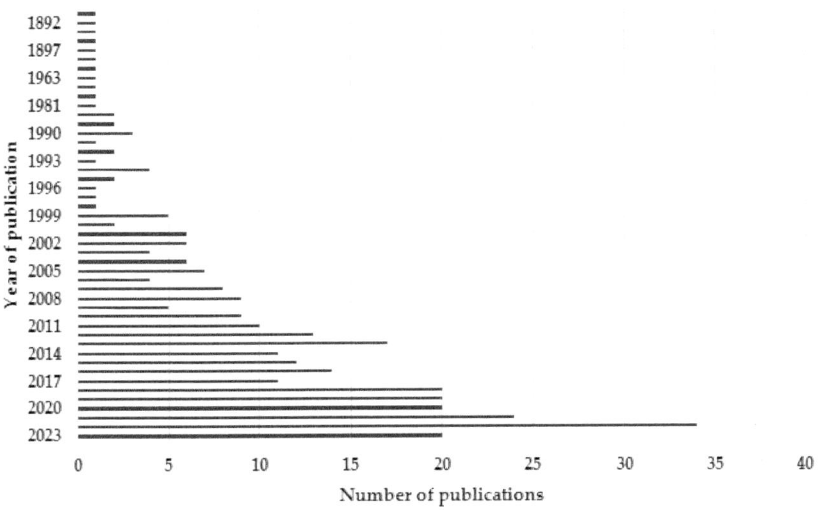

Fig. 1. Annual number of publications in Scopus related to the keywords: *Desert truffle, Carbomyces, Elderia, Eremiomyces, Kalaharituber, Mattirolomyces, Mycoclelandia, Picoa, Stouffera, Terfezia, Tirmania,* and *Ulurua* during the period 1891–2023. Scopus has identified 328 publications.

2.2 Document Types and Language of Publication

Most of the publications found were research articles (298; 92%), followed by reviews (10; 3%), book chapters (8; 2%), conference papers (6; 2%), and notes (2; 1%) (Fig. 2). The first article entitled "Contribution à bhistoire botanique de la truffe: Kamé de damas (*Terfezia claveryi*)" (Chatin, 1891) was published in the Bulletin de la Société Botanique de France (38: 332–335). This article deals with the botanical history of the truffle species *T. claveryi*. In this article, Chatin describes the truffles from Syria and Algeria. He comes to the conclusion that T. boudieri var. arabica is a synonym of *T. claveryi*. Chatin also concludes that *T. claveryi* is a common truffle species in both Syria and Algeria. In total, 315 out of the 328 articles were in English, 6 in French, 2 in Chinese, 2 in Persian, 2 in Turkish, and 1 in Spanish.

2.3 Journal Analysis

Figure 3 displays the various topics identified by Scopus in the 328 publications containing the terms desert truffle, *Carbomyces, Elderia, Eremiomyces, Kalaharituber, Mattirolomyces, Mycoclelandia, Picoa, Stouffera, Terfezia, Tirmania,* and *Ulurua,*

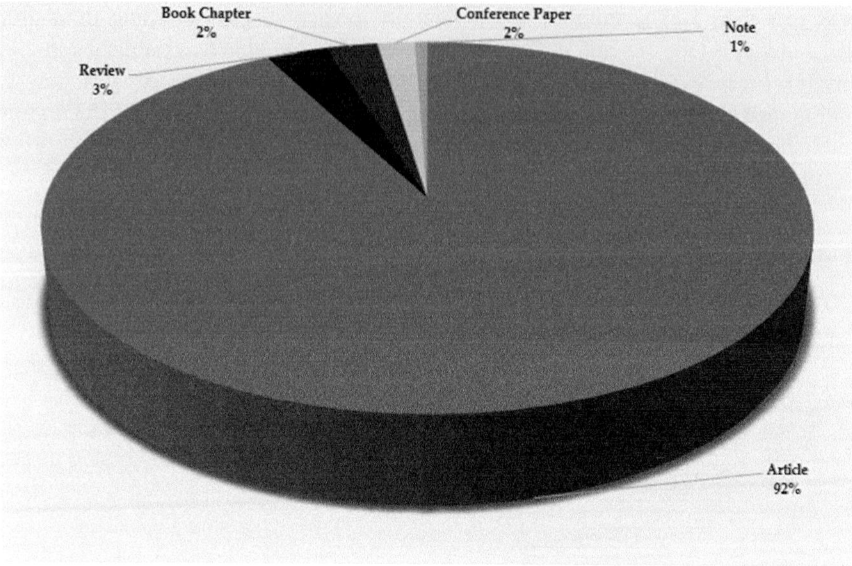

Fig. 2. The languages of the publication about desert truffles obtained from the Scopus.

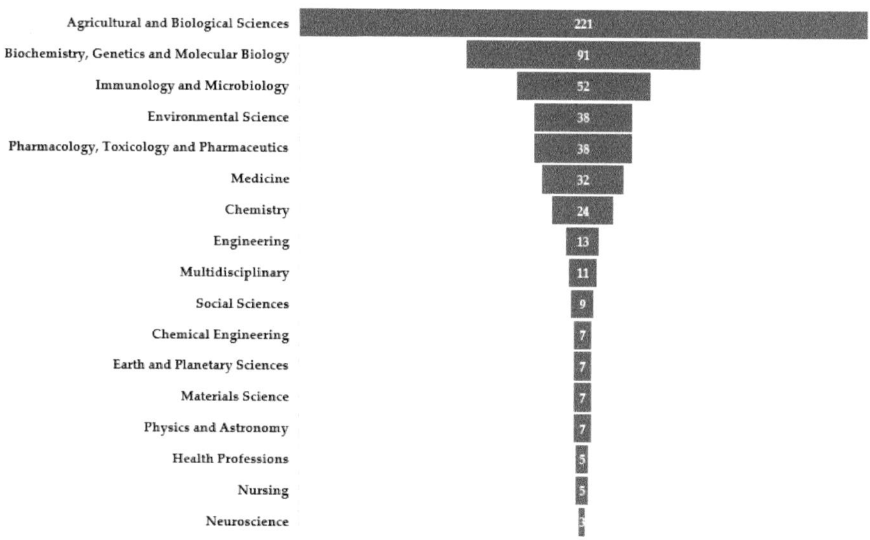

Fig. 3. Chart depicting the subject areas for the 328 articles retrieved by Scopus for the period 1891–2023, using the keyword: *Desert truffle*, *Carbomyces*, *Elderia*, *Eremiomyces*, *Kalaharituber*, *Mattirolomyces*, *Mycoclelandia*, *Picoa*, *Stouffera*, *Terfezia*, *Tirmania* and *Ulurua*.

with the three predominant subject areas being "Agriculture and Biological Sciences" (39%), "Biochemistry, Genetics, and Molecular Biology" (16%), and "Immunology and Microbiology" (9%). The highest number of articles published in a single journal appeared in Mycorrhiza (n=12), followed by the International Journal of Medicinal Mushrooms (9) and Mycologia (8). The most cited article in Mycorrhiza was by

Morte et al. (2000), and their study reported on the effects of drought stress on the growth and water relations of the mycorrhizal association with *Helianthemum almeriense-Terfezia claveryi*. The most studied edible desert truffle species are *Picoa lefebvrei*, *P. juniperi*, *Terfezia boudieri*, *T. claveryi*, *Tirmania pinoyi*, and *T. nivea*. Recently, researchers have been concentrating on the cultivation of desert truffles on synthetic media, quantifying T. claveryi mycorrhizae in truffle production, studying the bioecology of desert truffles, exploring antimicrobial and antioxidant activities, and researching taxonomy and mycorrhizal ecology. Enhancing our understanding of the biology of these fungi and the molecular mechanisms that regulate their production and interaction at the host-fungus level is crucial for truffle conservation in natural habitats and for truffle plantation owners.

2.4 Citation Analysis

It goes without saying that highly cited articles have a noteworthy impact on the topic globally. The most frequently cited article (135 times) was published in 2002 under the title "Antioxidant activity of edible fungi (truffles and mushrooms): Losses during industrial processing" in the Journal of Food Protection. Three of the most cited articles were research articles that dealt with the effects of drought stress on the growth and water relations of the mycorrhizal association (*Helianthemum almeriense-Terfezia claveryi*), as well as the antioxidant components and antioxidant/antiradical activities of desert truffles.

2.5 Country, Author and Institution Analysis

The chosen group of studies included 160 individual authors from 74 countries across six continents. The five countries with the highest number of authors were Spain (69 articles), the United States (33 articles), Saudi Arabia (32 articles), Algeria (29 articles), Italy (27 articles), Turkey (26 articles), and Iraq (25 articles), which accounted for 80% of the total (Fig. 4A). Identifying the most frequent authors helps to identify potential mentors and reviewers. Five scientists appear in the list of the most prolific authors: Morte, A. (47 articles), Navarro-Ródenas, A. (25 articles), Kagan-Zur, V. (17 articles), Pérez-Gilabert, M. (15 articles), and Roth-Bejerano, N. (14 articles). The leading organizations were the Universidad de Murcia (Spain), Consiglio Nazionale delle Ricerche (Spain), and Istituto per la Protezione Sostenibile delle Piante, Consiglio Nazionale delle Ricerche (Spain), with 44, 7, and 6 publications, respectively.

2.6 Bibliometric Analysis of Co-Authorship

Authors from 77 countries were selected based on their ability to present at least five documents and at least one citation each. Thirty-three met the threshold, and these formed five clusters. Spain had the highest number of documents (n=68), with 1838 citations and a total link strength of 105. Spain's main collaborating partners were Italy and Morocco. The United States was in second place with 33 documents and 1251 citations. The link strength for the United States was 102. The United

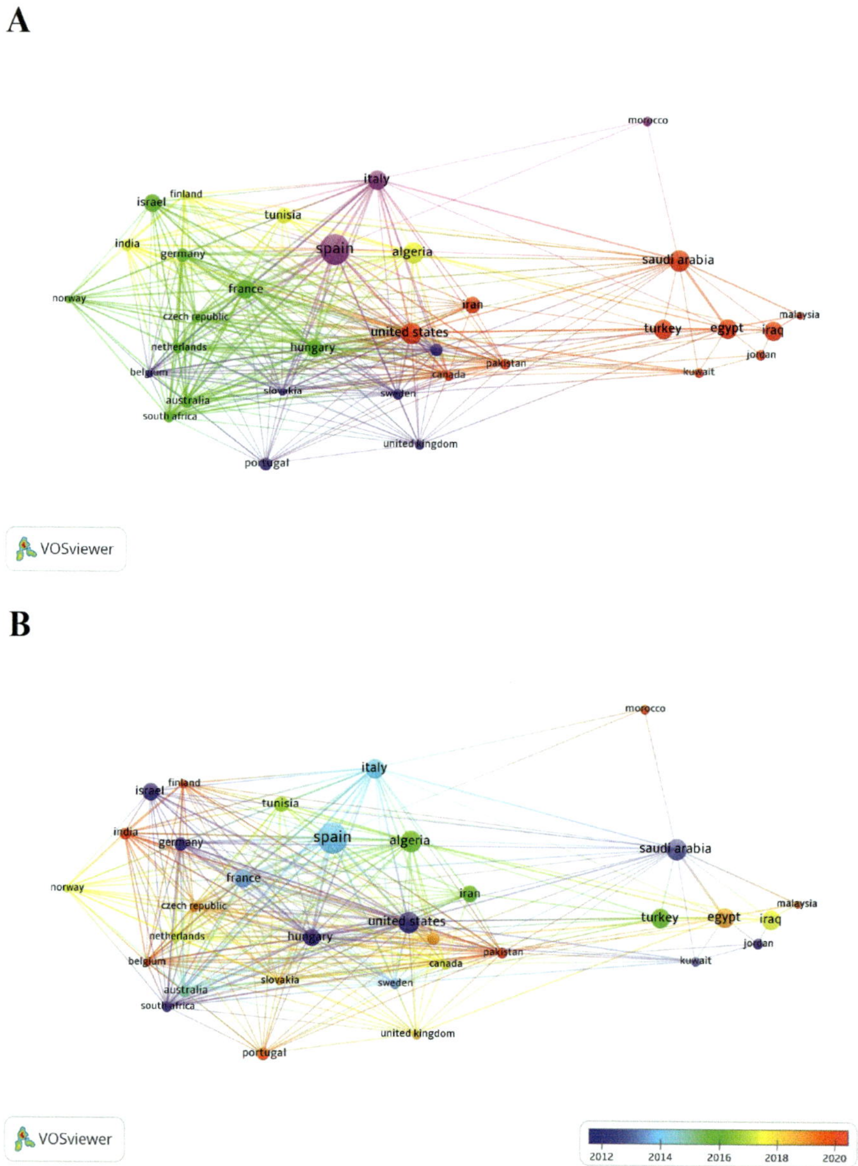

Fig. 4. The network map of co-authorship based on the affiliation of authors belonging to different countries (A). The timeline of the countries exploring desert truffles (B).

Kingdom›s most important cooperation partners included Saudi Arabia, Turkey, Pakistan, Kuwait, Jordan, Iraq, and Iran. Saudi Arabia was in third place with 32 documents, 406 citations, and 38 link strengths. Out of the 814 organizations involved in the research, nine collaborated, with a minimum threshold of three documents and three citations.

2.7 Timeline of Countries Exploring Desert Truffles

Figure 4B illustrates a timeline of the countries exploring desert truffles. The countries are all depicted in different colors and connected by lines. The timeline starts in 2012 and ends in 2020. In 2012, the countries researching desert truffles included the United States, Saudi Arabia, Hungary, Israel, Germany, South Africa, Jordan, and Kuwait. By 2014, the number of countries researching desert truffles had increased to 13. In 2014, new countries engaged in desert truffle research included Spain, Italy, France, Australia, and Sweden. By 2016, the number of countries researching desert truffles had risen to 17. In 2016, the countries that conducted research on desert truffles were Turkey, Algeria, Iran, and Tunisia. By 2018, the number of countries researching desert truffles had risen to 21. In 2018, new countries that explored desert truffles included Iraq, Norway, the Netherlands, and Canada. By 2020, the number of countries researching desert truffles had risen to 33. In 2020, new countries conducting research on desert truffles included Egypt, the Czech Republic, Slovakia, India, Pakistan, Portugal, Finland, and Belgium. Figure 4B illustrates that the number of countries researching desert truffles has steadily increased over time. This is likely due to a number of factors, including increased awareness of the value of desert truffles, the adoption of new legislation to protect desert truffles, and the development of new technologies to locate and harvest desert truffles. The figure also shows that desert truffles are found worldwide. Countries on every continent except Antarctica are researching desert truffles. This underscores the potential of desert truffles to enhance living conditions and promote sustainable development in arid regions worldwide.

2.8 Article Title Text Mining

Figure 5A was generated using VOSviewer and illustrates the connections between the research topics discussed in the 328 publications identified by Scopus on the terms: desert truffle, *Carbomyces*, *Elderia*, *Eremiomyces*, *Kalaharituber*, *Mattirolomyces*, *Mycoclelandia*, *Picoa*, *Stouffera*, *Terfezia*, *Tirmania*, and *Ulurua* from 1891 to 2023. 212 research topics were identified, each appearing in at least five publications, and together they formed four clusters or themes (Fig. 5A). The five largest nodes represented, in decreasing order, the keywords Ascomycota, *Terfezia*, non-humans, *T. claveryi*, and *T. boudieri* (Fig. 5A). The first cluster (red) contained 88 terms mainly related to uncontrolled studies, non-humans, *T. boudieri*, *Tirmania nivea* and unclassified drugs. The second cluster (green) contained 58 terms mainly related to *Terfezia*, phylogeny, genetics, and taxonomy. The third cluster (blue) contained 46 terms mainly dealing with *T. claveryi*, mycorrhiza, Cistaceae, and *Helianthemum*. The fourth cluster (yellow) contained 17 terms related to desert truffles, polymerase chain reaction (PCR), nucleotide sequence, internal transcribed spacer, and genetic variability.

2.9 Evolving Landscape of Desert Truffle Research

Figure 5B illustrates a timeline of research on desert truffles. The map is color-coded, with each color representing a different research topic. The size of each node on the

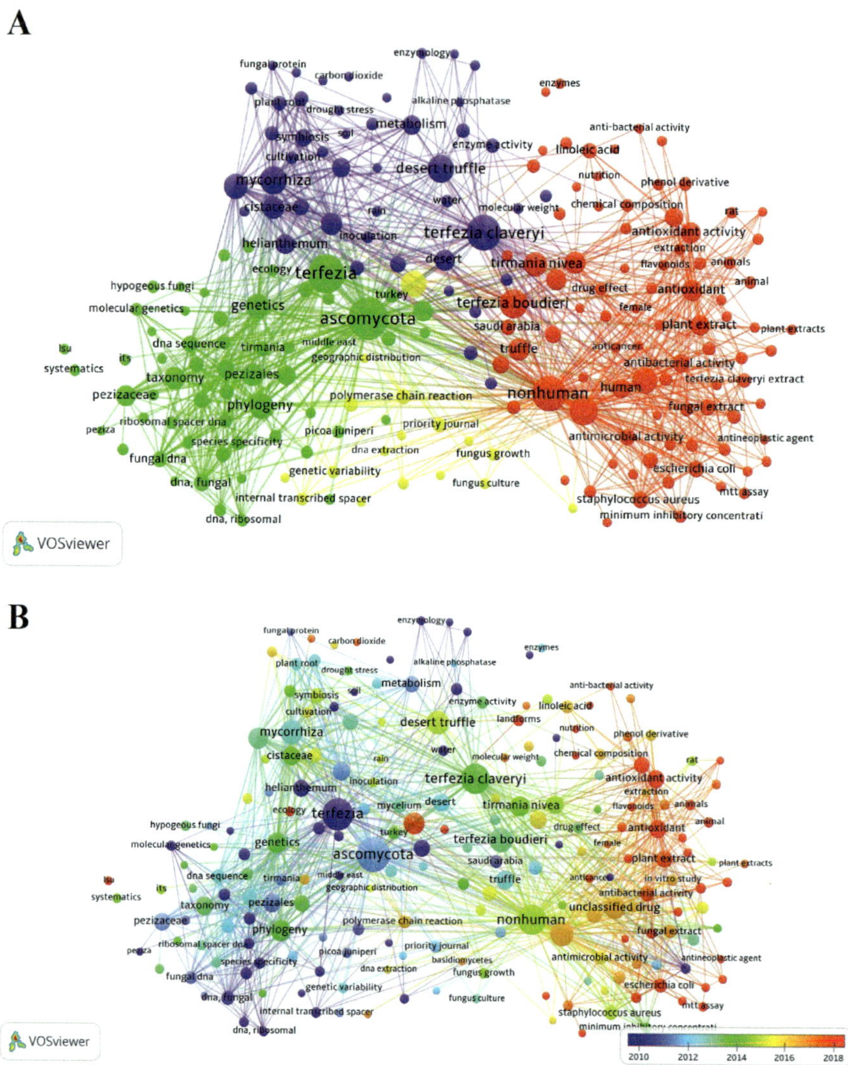

Fig. 5. Network visualization is based on the co-occurrence of keywords. Each node represents a keyword. The size of the node represents the co-occurrence between keywords. Each of the five clusters represents a particular theme. Nodes in a cluster together have a common theme, and their relationships with other nodes are represented by links (A). Overlay visualization of the research-topic map of studies related to desert truffles between 1891 and 2023 (Note: the minimum number of occurrences of a keyword is 5) (B).

map indicates the number of papers published on that topic, and the lines between the nodes represent the connections between these topics. The timeline at the bottom of the map shows annual publication trends. At the beginning (2010–2012), research focused on the basic classification and evolutionary relationships (taxonomy and phylogeny) of desert truffles. This was followed by a growing interest in its genetic nature (molecular genetics) from 2012 to 2014. Between 2014 and 2016, the focus shifted to understanding the ecological role of desert truffles. Since 2016, research on the potential medicinal applications of desert truffles has increased significantly. The timeline highlights the dynamic growth of research on desert truffles. While there is still much to discover about these fascinating fungi, their potential applications in medicine (e.g., anti-cancer) and agriculture (cultivation) are promising. Additionally, they could play a role in environmental protection.

3. Research on Truffles in Iran

3.1 Early Studies of Desert Truffles in Iran

Many researchers have studied various aspects of desert truffles in the Middle East, but very few studies have been conducted in Iran. The first underground ascomycete species reported from Iran (collected from Shiraz) was *Terfezia hafizi* before 1895. In 1895, another species of underground ascomycetes named *T. hanotauxii* was described by Chatin from Iran (collected from Tehran) (Chatin, 1895). *Terfezia claveryi* was reported from Iran in 1973 by Malençon (Malençon, 1973). In 2003, two species, *T. leonis* and *Tirmania pinoyi*, were reported by Daneshpajuh in the southern and northwestern regions of Iran (Daneshpajuh, 1991). In 2007, the protein profile analysis of *T. claveryi* was conducted in Iran. In this experiment, 13 protein bands were observed on an agarose gel, ranging in size from 15 to 205 kDa (Ammarellou, 2007).

3.2 Finding Desert Truffles in Iran: The Hunter's Techniques

In Iran, skilled truffle hunters use several methods to identify desert truffles. They look for characteristic plants that often grow alongside desert truffles, as well as specific soil formations and cracks. As the desert truffles mature underground, they expand and push upwards, which leads to the formation of cracks and small mounds on the surface. Trained truffle hunters are able to recognize these subtle disruptions with their experienced eyes. Each desert truffle is connected to the surface by a basal connection. This connection is a strong and compact structure made up of fungal threads (hyphae), soil particles, and plant roots. It plays a crucial role in linking the truffle to its surrounding environment. The lower part of the connection, which can range from 2 to 10 centimeters in length, attaches to fungal threads that extend from nearby plant roots (Jamali and Banihashemi, 2012a).

3.3 Terfezia claveryi: King of the Iranian Desert Truffles

Examinations carried out in various rangelands in Iran have revealed that *T. claveryi* is the most commonly found desert truffle. This species thrives in numerous regions across Iran, including Fars, Lorestan, West and East Azarbaijan, Kermanshah, Kerman, Kordestan, Hormozgan, Yasuj, Zanjan, and Sistan and Baluchestan. An impressive 60% of the desert truffles collected in Iran belong to the *T. claveryi* species (Jamali and Banihashemi, 2012a).

3.4 Precipitation Patterns and Desert Truffle Abundance

Field studies have also shown a critical link between rainfall patterns and the abundance of desert truffles in Iran. The success of a desert truffle hunt depends on two key factors related to rainfall: the total amount of precipitation and when it occurs during the preceding season (November to March). Several years during the survey emerged as particularly fruitful harvest seasons: 2005, 2006, 2007, 2009, and 2010. All of these years had one common feature: they received a cumulative rainfall exceeding 220 millimeters. On the other hand, the year 2008 yielded no truffles due to a complete absence of rainfall between November 2007 and March 2008. This finding highlights the reliance of desert truffles on sufficient winter and early spring rains.

3.5 The Desert Truffle Habitat

In Iran, desert truffles are typically found in locations with sandy or loamy sand soil. The pH of the soil is alkaline, and it has low organic matter and high calcium carbonate (CaCO3) content. Statistical tests have demonstrated that there is minimal variation in these properties among different truffle-producing sites. Based on these characteristics, the habitat of the truffles can be classified as basic and non-saline. The moisture content, protein, fiber, ash, fat, and carbohydrate levels of Iranian desert truffles were tested.

3.6 Nutritional Content of Iranian Desert Truffles

The results for *Picoa juniperi*, *P. lefebvrei*, *Terfezia claveryi*, *Tirmania pinoyi*, and *T. nivea* were as follows: moisture content: 62%, 66.25%, 70.25%, 38.57%, and 49.3%, respectively; protein: 13.88%, 10.096%, 10.062%, 21.98%, and 15.3%, respectively. Fiber content ranged from 3.09% to 6.36%, ash from 1.64% to 4.47%, fat from 1.48% to 6.45%, and carbohydrates from 12.278% to 25.38%. *Terfezia claveryi* also contained 306.3 mg/100 g of phosphorus, 202.5 mg/100 g of calcium, and 2.13 mg/100 g of iron. *Picoa juniperi* had higher levels of protein and carbohydrates compared to the other truffles. These variations in nutrient levels among the truffles may be attributed to differences in their genes or storage conditions.

3.7 Recent Discoveries: New Species and Mycorrhizal Relationships

In 2008, based on morphological characteristics, the species *T. boudieri* was reported alongside the plant species *Kobresia bellardii* and a species of *Picoa* from Iran (Ammarellou and Trappe, 2007; Ammarellou and Saremi, 2008). In 2010, two species, *Picoa lefebvrei* and *Tirmania nivea*, were reported from Iran (Jamali and Banihashemi, 2010). In 2011, the phylogenetic status of *P. lefebvrei* in Iran was investigated (Ammarellou et al., 2011). In 2012, five species of desert truffles, namely *T. claveryi*, *T. pinoyi*, *T. nivea*, *P. lefebvrei*, and *P. juniperi*, were documented in different regions of Iran based on their morphological and molecular characteristics. In this study, the relationship between different species of *Helianthum* and these fungi was reported for the first time in Iran (Jamali and Banihashemi, 2012a). In a study, the fungal ascocarps of desert truffles in Iran were investigated (Jamali and Banihashemi, 2012b). In this study, 32 species belonging to 21 genera were identified, with *Penicillium* and *Aspergillus* species being the most common associates of desert truffle ascocarps in Iran.

In 2013, *T. claveryi* was detected in the roots of *Helianthum* plants using species-specific primers, and two species, *H. lippii* and *H. salicifolium*, were confirmed as hosts of *T. claveryi* (Jamali and Banihashemi, 2013a). The detection of two species, *P. lefebvrei* and *P. juniperi*, was also confirmed in the roots of *H. ledifolium* var. *ledifolium* using specific primers (Jamali and Banihashemi, 2013b). In addition to *Helianthemum* species, the mycorrhizal association of *P. lefebvrei* with *Carex stenophylla* from the Cyperaceae family was first confirmed using species-specific primers (Jamali, 2015).

Studies have shown a symbiotic relationship between *T. claveryi* and *T. pinoyi* with perennial plants from the Cistaceae family, including *Helianthemum* species (*H. almeriense* and *H. lippii*) and *Cistus* species (*C. ladanifer* and *C. laurifolius*) (Sheibani and Jamali, 2023a, b). The results of this study showed that both *T. claveryi* and *T. pinoyi* form a mycorrhizal symbiosis (external-internal relationship) with all four plant species, with each species exhibiting varying degrees of pod development. In this study, the plant species *H. lippii* exhibited a high mycorrhization rate (87%) and relative mycorrhizal dependency (RMD) of 90% with *T. claveryi* and 48.48% and 57.85% with *T. pinoyi*, respectively, suggesting it as a particularly suitable host plant for these desert truffle species. The growth of inoculated plants, as measured by parameters like plant height and weight, showed statistically significant improvements compared to control plants. The mycorrhizal association between *T. claveryi* and inoculated plant species was confirmed using the Nested-PCR method, which detected fungal DNA within plant roots (Sheibani and Jamali, 2023a, b).

3.8 Beyond Desert Truffles: Exploring Other Tuber Species in Iran

In addition to desert truffles, several tuber species, including *Tuber aestivum* and *T. aestivum* var. *uncinatum*, have recently been reported from Iran (Ammarellou and Alvarado, 2017; Jamali, 2017). Puliga et al. (2020) reported the importation of large quantities of *Tuber aestivum* ascomata from Iran to Italy. During the inspection of these commercial batches, they discovered and described a new species, *Tuber iranicum*. Although mycorrhizal sequences of *Tuber brumale* have been previously identified in Iran (Bahram et al., 2012), the presence of fruiting bodies (ascomata) has not yet been documented. In a study by Puliga et al. (2021), six truffle species, namely *Tuber borchii*, *T. brumale*, *T. excavatum*, *T. fulgens*, *T. macrosporum*, and *T. rufum* f. *lucidum,* were identified in a shipment imported from Iran to Italy. These species have not been previously documented by Iranian researchers. Further investigations are warranted to confirm the presence of these species in Iran.

4. Conclusion

A bibliometric analysis of desert truffle research reveals an intriguing overview of the development, geographical scope, and thematic focus of the field over time. This analysis presents a detailed picture of scientific productivity, showing a significant increase in publication volume since 2010. This increase reflects the growing interest in the cultivation of desert truffles, exploring their medicinal potential, and aligning with the broader trend towards functional foods and phytomedicine. The appeal of the antioxidant and antimicrobial properties of desert truffles further enhances this interest. The study also emphasizes the geographical concentration of research, with Spain emerging as a key contributor. This concentration may be due to ecological factors that support desert truffle growth in these regions, or it could be attributed to local culinary traditions and economic opportunities that drive scientific interest. The intercountry collaborations identified in the bibliometric analysis indicate a developing network of expertise that could lead to future advancements in the field. The significant research output in journals focusing on mycorrhiza and medicinal mushrooms indicates the need for an interdisciplinary approach to desert truffle research, involving mycology, agriculture, pharmacology, and even biotechnology. The importance of the topic is highlighted by the most frequently cited articles, which emphasize the role of desert truffles in reducing oxidative stress, a crucial issue in the advancement of functional foods and nutraceuticals. Despite the growing body of literature, there are still gaps in knowledge. For example, the basic biology of desert truffles, while receiving some attention, requires further exploration to fully harness their ecological and economic potential. Future research should address these gaps in our understanding of desert truffle biology. Experimental studies focusing on optimal growth conditions to enhance the symbiotic relationship between desert truffles and plants could improve cultivation practices. Furthermore, exploring the nutritional and medical benefits of desert truffles could be a promising area, especially given the growing interest in natural and sustainable alternatives to synthetic compounds in various industries. The study also sheds light on research dynamics in Iran, which is emerging as a significant contributor to the field, underscoring the importance of desert truffles in the Middle Eastern context. An examination of the regional propagation of these fungi along with their cultural and economic implications could provide a comprehensive understanding of desert truffles in both global and local contexts. Additionally, studies exploring the potential use of desert truffles as a bioindicator of climate change›s impact on arid environments or as a tool for restoring degraded landscapes can delve into the ecological aspects of these intriguing organisms. In summary, the current discussion on desert truffles reveals a diverse research landscape. Leveraging bibliometric strategies has revealed historical trends, geographic research distribution, and thematic clusters that can guide future studies. By generating data-driven insights, this analysis provides a strategic roadmap aimed at enhancing collaborative efforts, diversifying research topics, and driving impactful advancements in desert truffle research. In addition to the research conducted in Iran, commercial cultivation of desert truffles has not yet been achieved. Future efforts should focus on this aspect.

References

Ahmed, A.A., Mohamed, M.A. and Hami, M.A. (1981). Libyan Truffles "*Terfezia boudieri* Chatin:" chemical composition and toxicity. J. Food Sci., 46(3): 927–929.

Ammarellou, A. (2007). Protein profile analysis of desert truffle (*Terfezia claveryi* Chatin). J. Food Agric. Environ., 5(2): 62–64.

Ammarellou, A. and Alvarado, P. (2017). First report of *Tuber aestivum* var. *uncinatum* from Iran based on morphological and molecular characteristics. Rostaniha, 18(2): 227–228.

Ammarellou, A. and Saremi, H. (2008). Mycorrhiza between *Kobresia bellardii (*All.) Degel and *Terfezia boudieri* Chatin. Tur. J. Bot., 32(1): 17–23.

Ammarellou, A. and Trappe, J.M. (2007). A first Ascomycete genus (Picoa sp.) record for the fungi flora of Iran. Pakistan Journal of Biological Sciences 10 (10): 1772.

Ammarellou, A., Smith, M.E., Tajick, M.A. and Trappe, J.M. (2011). The phylogenetic placement of *Picoa,* with a first report on *Picoa lefebvrei (*Pat.) Maire (= *Phaeangium lefebvrei)* from Iran. Int. J. Environ. Res., 5(2): 509–514.

Bahram, M., Põlme, S., Kõljalg, U., Zarre, S. and Tedersoo, L et al. (2012). Regional and local patterns of ectomycorrhizal fungal diversity and community structure along an altitudinal gradient in the Hyrcanian forests of northern Iran. New Phytologist, 193, 465–473.

Chatin, M.A. (1895). Truffe (Domalan) De Smyrne. Bulletin de la Société Botanique de France, 42(1): 30–33.

Chellal, A.I.C.H.A. and Lukasova, E. (1995). Evidence for antibiotics in the two Algerian truffles *Terfezia* and *Tirmania.* Pharmazie, 50: 228–229.

Daneshpajuh, B. (1991). Two new records of Ascomycetes (Truffles) from Iran. Proceedings of the tenth Plant Protection Congress of Iran 1–5 September, Kerman (Iran Islamic Republic), p. 123.

Elsevier (2017). Scopus: Content coverage guide. The Largest Database of Peer-Reviewed Literature.

Elsevier (2019). https://www.elsevier.com/solutions/scopus

Jamali, S. (2015). Specific PCR-primers for detection of *Picoa lefebvrei* desert truffle in *Carex stenophylla* roots. Rostaniha, 16: 206–207.

Jamali, S. (2017). First report of identification and molecular characterization of *Tuber aestivum* in Iran. Agrofor. Sys., 91: 335–343.

Jamali, S. and Banihashemi, Z. (2010). Two new species of hypogeous ascomycetes, *Tirmania nivea* and *Picoa lefebvrei* from Iran. 19th Iranian Plant Protection Congress.

Jamali, S. and Banihashemi, Z. (2012a). Hosts and distribution of desert truffles in Iran, based on morphological and molecular criteria. J. Agric. Sci. Technol., 14: 1379–1396.

Jamali, S. and Banihashemi, Z. (2012b). Fungi associated with ascocarps of desert truffles from different parts of Iran. J. Crop Prot., 1(1): 41–47.

Jamali, S. and Banihashemi, Z. (2013a). Nested-PCR for detecting *Terfezia claveryi* in roots of *Helianthemum* species in field and greenhouse conditions. J. Agric. Sci. Technol., 15: 377–387.

Jamali, S. and Banihashemi, Z. (2013b). Species-specific ITS primers for the identification of *Picoa juniperi* and *Picoa lefebvrei* and using nested-PCR for detection of *P. juniperi* in planta. Mol. Biol. Rep., 40: 5701–5712.

Janakat, S.M., Al-Fakhiri, S.M. and Sallal, A.K. (2005). Evaluation of antibacterial activity of aqueous and methanolic extracts of the truffle *Terfezia claveryi* against *Pseudomonas aeruginosa.* Saudi Medical Journal, 26(6): 952–955.

Malençon, G. (1973). Champignons hypoges du Nord de l'Afrique. I. Ascomycètes. Persoonia, 7(2): 261–288.

Morte, A., Lovisolo, C. and Schubert, A. (2000). Effect of drought stress on growth and water relations of the mycorrhizal association Helianthemum almeriense-Terfezia claveryi. Mycorrhiza, 10: 115–119.

Murcia, M.A., Martínez-Tomé, M., Jiménez, A.M., Vera, A.M., Honrubia, M. et al. (2002). Antioxidant activity of edible fungi (truffles and mushrooms): losses during industrial processing. J. Food Prot., 65(10): 1614–1622.

Puliga, F., Illice, M., Iotti, M., Baldo, D. and Zambonelli, A. et al. (2020). *Tuber iranicum,* sp. nov., a truffle species belonging to the Excavatum clade. Mycologia, 112(5): 932–940.

Puliga, F., Mirko, I., Mirco, I., Leonardi, P., Baghdadi, A. et al. (2021). True truffle diversity in Iran. Ital. J. Mycol., 50: 52–62.

Sheibani, M. and Jamali, S. (2023a). Detection of *Tirmania pinoyi* in roots of inoculated Cistaceae plant species by Nested polymerase chain reaction. Range. Ecol. Manag., 88: 100–109.

Sheibani, M. and Jamali, S. (2023b). Mycorrhizal association of *Terfezia claveryi* with several Cistaceae plant species. Rostaniha, 24(1): 21–38.

Van Eck, N. and Waltman, L. (2010). Software survey. VOSviewer, a computer program for bibliometric mapping. Journal of Scientometrics, 84(2): 523–538.

Chapter 3

Desert Truffles: An Appraisal

Aouali, S.

1. Introduction

Among mycologists, the terms "truffles", "true truffles", "false truffles" and "desert truffles" have been often discussed. In a simplified approach, the term "truffles" is used in its broadest sense to designate sequestrate fungi that produce their spores within a subterranean, or at ground level, differentiated fruiting structure, with a simultaneous loss of active spore dispersal. Species forming sporocarps in or on the litter, such as *Glaziella* and *Paurocotylis,* will still be grouped with truffles as long as they have lost active spore dispersal (Læssøe and Hansen, 2007). It is generally accepted that within this group of fungi, two categories are commonly considered: the true truffles, which are truffles belonging to the phylum Ascomycota, and the false truffles, which belong to the phylum Basidiomycota (Bonito and Smith, 2016). However, some authors restricted the term "true truffles" to only species belonging to the genus *Tuber* (Bonito et al., 2013). Desert truffles, also referred to as "white truffles" or "sand truffles", are a group of truffles growing in arid and semi-arid habitats (Maublanc 1946, 1952; Honrubia et al., 1992; Moreno et al., 2014).

The above aspect is based on historical nomenclature. Moreno et al. (2014) stated that member species of this group of fungi constitute independent branches of many unrelated families belonging to Ascomycota, Basidiomycota, Glomeromycota, and Mucoromycota. In their detailed review of the hypogeous desert fungi, the above authors described the phylogenetics, listed the member families and genera, and provided practical keys for each taxon. According to Moreno et al. (2014), multiple edible hypogeous fungi growing in arid areas throughout the world can be considered "desert truffles". The authors, define 'hypogeous' fungi as sequestrate species that

Forest Pathology and Mycology Laboratory, Forest Protection Division, Forest Research National Institute, Algeria.
Email: souhila.aouali@gmail.com

are globose in shape and either totally or partially immersed under the soil surface. These species are dispersed by animals feeding on them, such as mammals, birds, and arthropods (Fogel and Peck, 1975; Maser et al., 1978; Alsheikh and Trappe, 1983). When achieving maturity, they can become semi-hypogeous when partially or totally uncovered.

Kovacs and Trappe (2014), by reviewing the definitions of both truffles and deserts, stated that defining precisely what a desert truffle is, is a difficult task, mainly because deserts are primarily defined by their aridity (Ward, 2008). Furthermore, truffles are extremely variable, and so is their definition. Consequently, they delimited desert truffles to a restricted number of hypogeous ascomycetes generally considered desert truffles, namely the genera *Carbomyces, Elderia, Eremiomyces, Kalaharituber, Mattirolomyces, Mycoclelandia, Picoa, Stouffera, Terfezia, Tirmania,* and *Ulurua.* In this chapter, I focused on the broadly known edible desert truffles belonging to the Ascomycota.

2. Biology, Systematics and Evolution

2.1 Biology

Some fungi became intimately associated with plant roots, improving their nutrient acquisition abilities. This symbiotic association, known as mycorrhizae, has evolved to form significant structural changes to the plant roots and to provide different ecological services to various plant groups (Dighton, 2009). Desert truffles are known to establish mycorrhizal symbiosis of ectendomycorhizal type (EEM) with plant species belonging to the family Cistaceae, particularly *Helianthemum* spp., known as rock roses (Table 1). According to Diez et al. (2002), this is mainly due to the fact that *Helianthemum* species have a wide edaphic tolerance, which explains the occurrence of several desert truffle species hosted by *Helianthemum* spp. in both acidic and basic soils. For example, *Terfezia claveryi* establishes a mycorrhizal partnership with *Helianthemum almeriense* (Morté et al., 2010) and *H. lippii* (Baradai et al., 2014). It has been largely demonstrated that mycorrhizas play an important role in plant host protection against drought stress (Omirou et al., 2013; Wu et al., 2013; Darro et al., 2022; Bhardwaj et al., 2023). This role intervenes primarily by altering water rate movements, impacting significantly tissue hydration and plant physiology (Augé 2001; Marjanović et al., 2005; Morte et al., 2000; Ruiz-Lozano et al., 2008). The fungus mycorrhizas, during the drought seasons, supports the survival of its host plant by modifying its physiology and its nutrition by increasing the acquisition of the three essential elements: phosphorus, nitrogen, and potassium (Morte et al., 2000; Morte et al., 2010). Furthermore, this complex partnership leads to a fine-tuned regulation of both plant and fungal aquaporin expression and the hydrogen peroxide content in the roots (Aroca et al., 2007; Porcel et al., 2006; Ruiz-Lozano and Aroca, 2010; Navarro-Rodenas et al., 2013; Marques-Galvez et al., 2019, 2020).

Further studies have shown that this intimate interaction is governed by the combined action of the levels of essential elements, such as organic phosphorus and water (Navarro-Ródenas et al., 2012), and the fungal auxin involved in the ectendomycorhiza formation (Turgeman et al., 2015). While the development of the intercellular mycorrhizal structure is promoted by high levels of the above-cited elements, low levels favor the intracellular structure type (Gutierrez et al., 2003; Zaretsky et al., 2005; Navarro-Rodenas et al., 2012, 2013). Thus, it is suggested that drought adaptation of the host plant, through morphological and physiological changes, could be the result of the combination of a high rate of intracellular colonization and the fine-tuned expression of the aquaporins (Navarro-Rodenas et al., 2013). A comprehensive review of the basic and applied research on desert truffle cultivation was compiled by Morte et al. (2017). In this review, where the association "*Terfezia claveryi-Helianthemum almeriense*" was used as a model, essential bioecological aspects were treated in detail, including the diversity of host plants, the enzymes involved in the primary and secondary metabolisms, and the soil pH requirements. However, the molecular mechanisms taking part in this intimate symbiosis were still to be explored. Indeed, a considerable amount of knowledge has been gained on aspects regarding the molecular mechanisms governing the ectendomycorrhizal interaction and lifestyles, along with the reproductive modes of desert truffles, only in 2021, when Marquez-Galvez et al. published their results on the two most appreciated and economically important desert truffles, *Terfezia claveryi* and *Tirmania nivea*. Through their study, they have shed light on the pertinent singularities of desert truffles in comparison to other mycorrhizal fungi.

As for the characterization of the reproductive modes of the desert truffle species *Terfezia boudieri*, *T. claveryi,* and *Tirmania nivea*, genome sequencing was privileged and revealed their genomic signature as heterothallic species in the presence of a single MAT gene. According to the above authors, fundamental molecular information is revealed through the characterization of the structure and organization of the MAT locus and its surrounding genomic regions. This information has important applications in the optimization of the cultivation strategies for these fungi, in modelling the genomic rearrangements within and around this locus, and in tracing the evolution of the breeding systems within Pezizomycetes. Marquez-Galvez et al. (2021) also explored the genus and species-specific micorhiza-induced small secreted proteins (MiSSPs). They concluded that the presence of a common set of MiSSPs suggests the existence of conserved mechanisms among the desert truffles that allow them to interact with their hosts. They also indicate that the specific *T. claveryi* MiSSPs, that are induced when in symbiosis with *H. almeriense* point to a possible species-specific mechanism.

Regarding the transcriptomic features induced upon symbiosis in *Helianthemum almeriense* plants subjected or not to water stress, the same study revealed the presence of downregulated and upregulated genes among a set of 50 strongly regulated genes. The alternative oxidase (AOX) gene was downregulated. Earlier studies have shown that AOX function in metabolic and signaling homeostasis is particularly important during stress. These include abiotic stresses such as low temperature, drought, and nutrient deficiency, as well as biotic stresses such as bacterial infection (Vanlerberghe, 2013). It is currently known that the AOX pathway is an alternative to the cytochrome oxidase (COX) pathway for electron transport in the respiration chain, and is uncoupled from ATP production. AOX function is particularly important during stress when it is involved in the optimization of respiratory metabolism, which facilitates the continuous turnover of tricarboxylic acid and acts to prevent the excess generation of reactive oxygen species (ROS) (Li et al., 2013; Vanlerberghe, 2013). The decrease of AOX mRNA levels in *T. claveryi* mycorrhizal roots also concurred with previous studies that show the suppression of AOX activity induced by arbuscular mycorrhizal colonization, a condition in which the COX pathway seems to be the predominant one (Liu et al., 2015; Del-Saz et al., 2018; Marquez-Galvez et al., 2021). On the other hand, upregulated genes were mainly related to carbohydrate and lipid metabolisms, as well as to thaumatin-like genes. It is suggested that the upregulation of the genes responsible for the above metabolisms has a direct implication on the metabolic changes produced during the symbiosis and the exchange of nutrients between the plant and the fungus (Wang et al., 2017), whereas the upregulation of the thaumatin-like genes, which are known to play a role in pathogen defense (Liu et al., 2010), might limit the colonization by *T. claveryi* as a direct defense mechanism, or be, like in arbuscular mycorrhizal plants, an enhanced defense mechanism against pathogens triggered by *T. claveryi* colonization (Pozo et al., 2009).

Under dry conditions, *H. almeriense* roots are colonized by singularly differentiated symbiotic structures of *T. claveyri*, totally different from those produced by ectomycorrhizal fungi species. This mycorrhizal colonization is enhanced by drought stress, and as mentioned earlier, environmental conditions have a significant impact on the degree of colonization, and the ability to form either inter- or intracellular structures (Navarro-Rodenas et al., 2013; Marques-Galvez et al., 2019; Marques-Galvez et al., 2021). Recent studies performed by Marques-Galvez et al. (2021), have shown that the higher degree of mycorrhizal colonization in dry conditions, is correlated to *T. claveryi* RNA reads. However, and interestingly, the authors noticed a change in the mycorrhizal structures formed under drought conditions. *T. claveryi* does not generally penetrate into host cells under well-watered conditions; however, their results showed that approximately 40% of the observed mycorrhizal roots involved intracellular hyphal structures. Further investigations shed more light on this observation and revealed that pectin-targeting enzymes are thought to play a key role in the formation of the intercellular Hartig net (Veneault-Fourrey et al., 2014). Fungal genes coding for pectin-targeting enzymes (CE8, GH28, and PL4_1) were downregulated under the drought condition, suggesting that the activity of these enzymes might be dispensable when intracellular colonization is favoured. Whereas in ericoid mycorrhiza, intracellular penetration is mainly attributed to PCWDEs because of their increased expression in mycorrhizal symbiosis (Martino et al., 2018), the generalized downregulation of PCWDEs shown in this work suggests that this may not be the case for EEM desert truffles (Marques-Galvez et al., 2020).

2.2 Systematics and Evolution

Traditionally, describing desert truffles is mainly based on morphological and microscopic characters (peridium morphology, gleba colour, organoleptic characters, and spore ornamentation). This limited set of characters and their homoplasy, are often insufficient to identify them at the species level (Diez et al., 2002). According to Trappe (1979), the parallel evolution of independent lineages of epigeous and hypogeous Pezizales resulted in homoplastic features of their fruiting bodies. The author then transferred many species that were artificially classified in the Tuberales order to the Pezizales order and modified some families in order to include related hypogeous species. Pezizaceae Fries sensu Korf were also modified by the same author in order to contain a related group of hypogeous species with passive spore discharge, such as *Tirmania*. In addition, based on a comparative morphological study, Trappe in 1979, integrated *Terfezia* into the family Terfeziaceae and *Picoa* into the family Balsamiaceae.

The careful examination of the nomenclature and taxonomy of most desert truffles shows that there are a large number of old species names, and it was only with the advent of molecular biological tools that misidentifications were discovered and corrected. Furthermore, the high volume of molecular phylogenetic data led to a constant revision of the hypogeous Ascomycetes (Gargas and Taylor, 1995; Landvik et al., 1997; Harrington et al., 1999). In 1997, O'Donnell et al. were able to provide support for the existence of independent evolution of lines of epigeous or hypogeous fruit bodies in the Pezizales, which has subsequently greatly changed the classification of desert truffles. Later in 1999, Percudani et al. highlighted the close relationship between *Terfezia* and the family Pezizaceae through the analysis of the 18S rDNA sequences.

Based on the literature, it appears that no studies have been carried out on the history of the evolution of desert truffles, and molecular phylogenetic studies are required, in order to clearly demonstrate that morphological species indeed represent phylogenetic species, as suggested by morphological characters. In this matter, Dièz et al. (2002) have proposed an evolutionary history, based on the results of an interesting phylogeny study they undertook on the two genera *Terfezia* and *Tirmania*, coupled with a set of data on the morphology and ecology of these two species. In the study, the researchers used the internal transcribed spacer (ITS) of the nuclear rDNA, supplemented by restriction fragment length polymorphism (RFLP) analysis by sequence analysis. Their results indeed confirmed that *Terfezia* and *Tirmania* are two separate phylogenetic species. At the intra- and interspecies level, distance values were low, which suggests, according to the authors, that these species diverged only recently.

Considerable knowledge has been gained lately in the taxonomy of some desert truffle species, such as *Terfezia* and *Tirmania* species. A morphological and phylogenetic analysis of the large ribosomal subunit (28S rDNA) and internal transcribed spacer (ITS rDNA) conducted by Zitouni-Haouar et al. (2018), on *Terfezia* spp. has allowed the description of a new species, *Terfezia crassiverrucosa*. This species had a well characterized phylogenetic position amongst the five other studied species (*Terfezia arenaria, T. boudieri, T. claveryi, T. eliocrocae*, and *T. olbiensis*), and showed distinctive morphological characters. According to the authors of the study, it is important to undergo extensive sampling from new areas (North Africa or the Middle East) in order to build a determination key that will allow the delineation of the spiny-spored *Terfezia* species. The authors add that comparing *Terfezia* and its plant host, *Helianthemum* phylogenetics, will surely contribute to the understanding of the processes of speciation. Furthermore, genome sequencing of both desert truffles and their plant hosts might reveal the presence of special genes that will facilitate their identification. Additionally, and according to Marquès-Galvez et al. (2021), the genes' functional analyses of both desert truffles and their plant hosts, will help to better understand the interactions that occur between desert truffles and their host plants. In 2019, Moreno et al. described *Terfezia honrubiae* as a new species from Spain. The species was growing in acidic soil, in association with *Tuberaria guttata*.

The phylogenetic relationships among 41 *Tirmania* samples collected from Algeria and Spain were revealed by Zitouni-Haouar et al. (2023), by using the internal transcribed spacer (ITS) and large subunit (LSU) regions of the nuclear-encoded ribosomal DNA (rDNA). The molecular morphological, and bioclimatic data were compared. The results showed the presence of four separated morphological species, supported by four distinct lineages, and the description of a new species, *Tirmania sahariensis*. The authors suggest that the speciation process of *Tirmania* may have been shaped by the prevailing bioclimatic constraints in the Mediterranean and Middle East.

3. Distribution and Ecology

Desert truffles are diversified and are distributed across almost all continents. However, according to Trappe (1979), many of their ecological aspects are similar. Most desert truffle species are native to Mediterranean countries, and establish mycorrhizal symbioses, especially with *Helianthemum* species, but also with other Cistaceae species (Dexheimer et al., 1985; Fortas and Chevalier, 1992). This association, "desert truffles-plants", is essential to prevent soil erosion and desertification through the preservation of Mediterranean shrublands and xerophytic grasslands (Honrubia et al., 1992). Desert truffles were grouped by Diez et al. (2002) according to the soil type and climate in which they grow in their area of origin. The authors found that *Helianthemum* host species have the ability to grow in various soils, and this explains the association of several truffle species with them in both acidic and basic soils. Climate, soil, and vegetation, are factors that significantly influence the distribution and abundance of desert truffles (Henkrar et al., 2023). Climatic parameters are among the most important abiotic factors that effectively control the occurrence of desert truffles (Trappe et al., 2008a, b). Desert truffles fruit only in years with adequate and properly distributed rainfall (Trappe et al., 2008a). Heavy rains (>200 mm annually) in mild to warm weather, followed by heavy evening or early morning dewfalls, are optimal conditions for the growth of desert truffles (Henkar et al., 2023; Owaid, 2017; Baradai et al., 2014). However, studies have shown that the annual precipitation allowing the development of desert truffles may range from 50 mm to 380 mm, depending on the region (Al-Qarawi and Mridha, 2012). The number of desert truffles increases depending on the amount of rainfall (Owaid, 2017). It has also been stated that both thunder and lightning are considered essential requirements for truffle formation (Mandeel and Al-Laith, 2007). As for the substrate, soil properties are very similar to those reported in some truffle autoecological studies, whether in North Africa (Fortas, 1990; Khabar et al., 2001; Slama et al., 2006; Baradai et al., 2014) or in the Middle East (Hashem and Al-Obaid, 1996; Al-Ruqaic, 2002; Mandeel and Al-Laith, 2007). Ascomycetous desert truffles include nine genera with a total number of 26 species, however, not all of them are edible. Some edible species and their distribution are summarized in Table 1.

Table 1. Common edible desert truffles and their distribution.

Family	Genus	Species	Country	Reference
Pezizaceae	*Carbomyces*	*Carbomyces emergens*	USA	Trappe and Weber (2001)
		Carbomyces longii	USA	Trappe and Weber (2001)
		Carbomyces gilbertsonii	USA	Trappe and Weber (2001)
	Delastria	*Delastria rosea*	Algeria, Morocco, Turkey, and Mexico	Malençon (1973); Akata et al. (2023); Gomez-Reyes et al. (2017)
		Delastria liebanensis	Spain	Paz et al. (2018)
	Elderia	*Elderia arenivaga*	Australia	Trappe et al. (2010a)
	Eremiomyces	*Eremiomyces echinulatus*	South Africa, Botswana, and Namibia	Ferdman et al. (2005)
	Mattirolomyces	*Mattirolomyces spinosus*	USA	Kovács et al. (2011)
		Mattirolomyces mexicanus	Mexico	Kovács et al. (2011)
		Mattirolomyces mulpu	Australia	Trappe et al. (2010b)
		Mattirolomyces austroafricanus	South Africa, Botswana, and Namibia	Trappe et al. (2010b)
		Mattirolomyces terfezioides	France, Italy, Hungary, India, South Korea, and China	Lawrynowicz et al. (1997); Wang et al. (2017); Lu et al. (2022)
	Terfezia	Terfezia arenaria	France, Spain, Portugal, Italy, former Yugoslavia, Greece, Cyprus, Romania, Algeria, Morocco, Lybia, Tunisia, Turkey, Egypt, and Azaibijan	Alsheikh (1994); Fortas and Chevalier (1992a); Chatin (1896a); Malençon (1973); Khabar et al. (2001, 2005); Patouillard (1894); Oder (1988); El-Kholy (1989); Seyidova and Huseinm (2012)

Table 1. contd. ...

Table 1. contd. ...

Family	Genus	Species	Country	Reference
Pezizaceae		*Terfezia boudieri*	France, Italy, Spain, Cyprus, Algeria, Tunisia, Mrocco, Lybia, Syria, Turkey, Iran, and Saudi Arabia	Moreno et al. (2002); Chatin (1891); Khabar et al. (2005); El-Kholy and Assim (1991); Awameh and Alsheikh (1980); Gucin (1990); Ammarellou et al. (2007); Bokhary (1987)
		Terfezia claveryi	France, Spain, Italy, Algeria, Mrocco, Tunisia, Egypt, Lybia, Lebanon, Syria, Turkey, Iran, Saudi Arabia, Kuwait, and Iraq	Alsheikh (1994); Chatin (1891); Khabar et al. (2005); Malençon (1973); El-Kholy and Assim (1991); Farhat (2006); Awameh and Alsheikh (1980); Bekci et al. (2011); Laessøe and Hansen (2007); Sawaya et al. (1985); Bokhary et al. (1987); Awameh and AlSheikh (1979); Dahham et al. (2016)
		Terfezia crassiverrucosa	Algeria	Zitouni-Haouar et al. (2018)
		Terfezia eliocrocae	Algeria	Zitouni-Haouar et al. (2018)
		Terfezia leptoderma	Portugal, Spain, France, Italy, Hungary, Morocco, and Turkey	Alsheikh (1994) ; Montecchi and Sarasini (2000); Khabar et al. (2001); Castellano and Turkog˘A (2012);
		Terfezia olbiensis	Portugal, Spain, France, Italy and Czechoslovakia	Alsheikh (1994)
		Terfezia transcaucasica	Azerbaijan	Azerbaijan National Report (2010).
	Tirmania	*Tirmania nivea*	Spain, Italy, Syria, and Saudi Arabia	Diez et al. (2002); Alsheikh (1994); Alsheikh and Trappe (1983); Hashem and Al-Obaid (1996); Bokhary (1987)

Table 1. contd. ...

Table 1. contd. ...

Family	Genus	Species	Country	Reference
Pezizaceae		*Tirmania pinoyi*	Morocco, Algeria, Tunisia, Lybia, Egypt, Syria, Turkey, and Saudi Arabia	Chatin (1891); Khabar et al. (2001); Khabar et al. (2005); El-Kholy and Assim (1991); Alsheikh and Trappe (1983); Ersel and Solak (2004); Janakat et al. (2005); Bokhary (1987)
	Kalaharituber	*Kalaharituber pfeilii*	South Africa, Botswana, and Namibia	Ferdman et al. (2005)
Pyronemataceae	*Picoa*	Picoa juniperi	France, Italy, Spain, Algeria, Morocco, Tunisia, Turkey, and Saudi Arabia	Montecchi and Sarasini (2000); Honrubia et al. (1992); Zitouni-Haouar et al. (2018); Khabar et al. (2001); Khabar et al. (2005); Turkoglu and Yaguz (2012); Bawadekji et al. (2012)
		Picoa carthusiana	Tunisia, Egypt	Khabar et al. (2005); Pacioni and El-kholy (1994).
		Picoa lefebvrei	Algeria, Turkey, and Saudi Arabia	Zitouni-Haouar et al. (2018); Gucin et al. (2012); Bawadekji et al. (2012)

In Kuwait, species of *Terfezia* and *Tirmania* require a minimum rainfall of about 180 mm, well distributed from October through March (Awamah and Alsheikh, 1979). Morte et al. (2008) reported that good yields occur when rainfall ranges from 70 to 120 mm in North African countries and from 100 to 350 mm in countries in Southern Europe. According to the same authors, rainfall distribution is as critical as the amount; indeed, rain is necessary no later than the beginning of December in North African and Middle Eastern countries and no later than the beginning of October in the countries in South Europe. In Saudi Arabia, desert truffles from the eastern provinces fruit in November through December, depending on the amount of precipitation, and in Kuwait, they appear from November to January (Iddison, 2016; Owaid, 2016). In Iran, they harvested from January to March (Ammarellou et al., 2007).

In Algeria, for example, favorable ecosystems where desert truffles grow are usually small beds of Wadis and Dayas (desert depressions), since these geomorphological zones accumulate rainwater, which promotes the growth of both truffles and their host plant (Baradai et al., 2013; Fortas et al., 2021). Baradai et al. (2013) have conducted a study focusing on the bioecology of *Tirmania nivea* in the Oued M'ya (Algeria). The results revealed that the studied species grows in heterogeneous soils of sandy texture, moderately calcareous, slightly alkaline, and low organic matter and phosphorus contents. The development of desert truffles is closely linked with high rainfall occurring during the fall and/or winter. However, in Algeria, Baradai et al. (2014), found that the total and specific productivities of *Terfezia arenaria* and *T. claveryi*, along with *Tirmania nivea*, were intimately related to autumnal precipitations that occur during October through December. Desert truffles were found to grow even under hyper-arid climatic conditions in the Algerian desert.

Henkrar et al. (2023), in their update on the ecology and the distribution of wild truffles in Morocco, described the ecological characteristics of Moroccan desert truffles and their host plants in the Mamora forest and oriental regions of Morocco. Their study showed that desert truffle distribution is typically linked to the presence of *Helianthemum sp.*, and their fructification depends mainly on the precipitation frequency. In terms of annual rainfall and soil type, the two main desert truffle species of the Mamora forest, *Terfezia arenaria* and *Tuber oligospermum,* require on average 435 mm and a slightly acidic soil. Whereas *Terfezia boudieri*, *T. claveryi*, and *Tirmania* spp. that grow in the Oriental and Highlands require, on average, 123 to 267 mm of rainfall and soil with a high $CaCO_3$ content. In Turkey, *T. boudieri* was found to grow in sand-clay and limey soils with low organic matter (1.85%), a pH of 7.12, and a nitrogen content of 0.12% (Akyuz et al., 2012).

In the Kalahari Desert, pans of compact calcareous sand, characterized by unusually uniform zoned vegetation, are common substrates for desert truffles (Trappe et al., 2008a). Data published by Taylor et al. (1995) indicate that the soil pH values of desert truffle habitats in the Kalahari Desert are low, ranging from 5.5 to 6.5. Their total $CaCO_3$ content is also low, ranging from 0.3% to 3.1%. This difference with North African and Middle East habitats is probably due to multiple dissimilarities in regional landscape types such as geomorphology, hydrology, and the type of habitat, climate patterns (temperature, precipitation, and seasonality), soil traits (soil type and evolution), and the type of host plants (Díez et al., 2002; Baradai et al., 2014).

4. Damage to Desert Truffles by Biotic and Abiotic Factors

Nearly all organisms on earth are subjected to attacks by parasites, or are damaged by abiotic factors such as frost, high temperatures, high moisture, and heavy rains. Based on literature, very few interests have been put on the sanitary aspects of desert truffles, compared to the true truffles (Genus *Tuber*), for example. Multiple reasons could be at the root of this situation. From an economic point of view, true truffles are more commercialized have a higher economic value than desert truffles. On the other hand, and from a biological point of view, this group of fungi is extremely perishable, which makes them short-lived products that need immediate consumption. Moreover, desert truffles contain many bioactive molecules with antioxidant, anti-viral, and anti-microbial activities (Casarica et al., 2016) that might prevent parasites from developing in their tissues. Another aspect to consider is that desert truffles, in contrast to true truffles, are not as widely cultivated but rather collected in the wild, where they naturally grow.

However, according to Ahmed-Emhemed (2021), desert truffles in Lybia are often subject to rot before and after harvesting, which constitutes a serious problem for their commercialization. This author investigated the causal agents of these rots in *Tirmania* species of Lybia. He also stated that the rotting process is complex and involves fungi and bacteria. Amongst fungi, the author isolated three species from rotten *Tirmania* spp. fruitbodies: *Alternaria chlamydospore, Fusarium solani*, and *Rhizopus stolonifera*. These fungi, according to the results of the same study, also revealed the presence of the Gram-positive bacteria *Bacillus pumilus*, which has been reported to cause potato tuber decay, and the Gram-negative bacteria *Citrobacter youngae*., known as an enteric pathogen (Chen et al., 2013). These bacteria are commonly found in the roots and rhizospheres of some plants (Dos Santos et al., 2015). They produce toxic secondary metabolites that cause neonatal sepsis and meningitis (Chen et al., 2013). The author of the study explained the presence and dominance of *Citrobacter youngae* in infected *Tirmania* samples, by the fact that the sampling was performed in a pastoral area where sheep and goats were grazing.

Our literature search brought up the lack of research on desert truffle pathogens and insect pests. However, due to the fact that desert truffles grow naturally in the wild, they constitute a natural wild food for many mammals such as meerkats, baboons, bat-eared foxes, and hyenas (Trappe et al., 2014), birds (Alsheikh and Trappe, 1983), rabbits (Bordallo and Rodriguez, 2014), and insects (Murcia et al., 2014), and therefore cannot be considered “pests” or causing “damage”. The main sanitary issues encountered with desert truffles are rots, which occur after harvest and during storage and transportation and affect the quality of the fruitbody, by degrading its structure, smell, and taste. As a consequence, more efforts should be made to improve storage and transportation conditions to preserve the inner quality of the fruitbodies.

5. Conclusion

Desert truffles are a valuable food source for both humans and wild animals, especially in the desert where food is scarce. These fungi also play an important role in preserving desert soils from erosion and, thus, preserving the natural vegetation and habitats. They grow naturally in the wild and are often associated with *Helianthemum* species. Although they are diverse and geographically spread over different deserts in the world, they still have in common many biological and ecological features. Indeed, their growth and development are tributaries of well-distributed rainfall and in sufficient quantity, which makes their harvest uncertain. This irregularity of rainfall and harvest uncertainty has prompted research 25 years ago with focused efforts on their cultivation.

As a consequence of global climate change and the human exploitation of both desert truffles and the areas where they grow (for grazing), the durability of these fungi in the wild is uncertain in the arid and semi-arid regions of the world, which might negatively affect their economic growth and an increase in their price. Hence, their cultivation outside these regions may be the solution. However, due to their very short shelf life, more efforts have to be made in order to find better preservation methods to ensure their marketability. Indeed, desert truffles are prone to various rots after harvest, during storage, and during transportation. These rots are damaging the texture, taste and smell, and thus compromise desert truffles marketability. Unfortunately, research on the sanitary aspects of desert truffles is very scarce, if not absent. A better knowledge of these rot agents might help develop better conservation methods and prevent economic losses.

References

Ahmed-Emhemed, M. (2021). Identification of some pathogenic organisms involved in rotting of the libyan tirmania truffles. Libyan J. Basic Sci., 14(1): 109–116. https://doi.org/10.36811/ljbs.2021.110072

Akata, I., Şen, S., Sahin, E., Çöl, B. and Keskin, E. (2023). *Delastria*, a new genus of hypogeous fungi record for the Turkish mycobiota. Trakya Univ. J. Nat. Sci., 24(2): 85–90. https://doi.org/10.23902/trkjnat.1331537

Akyüz, M., Kirbağ, S. and Kurşat, M. (2012). Ecological aspects of the arid and semi-arid truffle in Turkey: evaluation of soil characteristics, morphology, distribution, and mycorrhizal relationships. Turkish J. Bot., 36(4): 386–391. https://doi.org/10.3906/bot-1103-4

Alqarawi, A.A. and Mridha, M.A. (2012). Status and need of research on desert truffles in Saudi Arabia. J. Pure Appl. Microbiol., 6(3): 1051–1062.

Al-Ruqaie, I. (2009). Effect of treatment process and preservation method on shelf life of truffles: II. Non-conventional methods (radiation). Int. J. Biol. Chem., 3(3): 126–131. https://doi.org/10.3923/ijbc.2009.126.131

Alsheikh, A.M. (1994). Taxonomy and mycorrhizal ecology of the desert truffles in the genus *Terfezia*. Ph.D. thesis, Oregon State University, Oregon, USA, p. 239.

Alsheikh, A.M. and Trappe, J.M. (1983). Taxonomy of *Phaeangium lefebvrei*, a desert truffle eaten by birds. Can. J. Bot., 61(7): 1919–1925. https://doi.org/10.1139/b83-204

Ammarellou, A., Saremi, H. and Gucin, F. (2007). Evaluation of morphology, cytology and mycorrhizal relationships of desert truffles (*Terfezia boudieri*) in Iran. Pak. J. Biol. Sci., 10: 1486–1490.

Aroca, R., Porcel, R., and Ruiz-Lozano, J.M. (2007). How does arbuscular mycorrhizal symbiosis regulate root hydraulic properties and plasma membrane aquaporin in Phaseolus vulgaris under drought, cold or salinity stresses? New Phytol., 173: 808–816. http://doi.org/10.1111/j.1469-8137.2006.01961.x

Augé, R.M., Stodola, A.J.W., Tims, J.E. and Saxton, A.M. (2001). Moisture retention properties of a mycorrhizal soil. Plant Soil, 230: 87–97. http://doi.org/10.1023/A:1004891210871

Awamah, M.S. and A. Alsheikh, (1979). Laboratory and field study of four kinds of truffle (Kamah), *Terfezia* and *Tirmania* species, for cultivation. Mushroom Sci., 10: 507–517.

Azerbaijan National Report (2010). Country study on biodiversity of Azerbaijan Republic. Fourth National Report to Convention of Biological Diversity. Baku, Azerbaijan, P. 77. 10.13140/RG.2.1.3763.6246

Baradai, L., Bissati, S. and Chenchounil, H. (2013). Étude Mycologique et Bio-Ecologique de la Truffe Blanche du Desert (*Tirmania nivea* Desf. Trappe 1971) dans la Région de Oued M'ya (Ouargla, Sahara Algérien). Revue Des Bioressources, 3(1) : 6–14. https://doi.org/10.12816/0008856

Baradai, L., Bissati, S., and Chenchouni, H. (2014). Desert truffles of the North Algerian Sahara: Diversity and bioecology. Emirates J. Food Agric., 26(5): 425–435. https://doi.org/10.9755/ejfa.v26i5.16520

Bawadekji, A., Gargano, M.L., Saitta, A. and Venturella, G.A. (2012). New record of the desert truffle *Picoa lefebvrei* in Saudi Arabia. Mycotaxon, 122(1): 243–247.

Bekçl, H., Altınsoy, B., Sarıkaya, S., Onbasılı, D. and Çelik, G.Y. (2011). Antimicrobial activity of some macrofungi collected from Kastamonu province. Kastamonu University Journal of Forestry Faculty, 11: 187–190.

Bhardwaj, A.K., Chandra, K.K. and Kumar, R. (2023). Water stress changes on AMF colonization, stomatal conductance and photosynthesis of *Dalbergia sissoo* seedlings grown in entisol soil under nursery condition. For. Sci. Technol., 19(1): 47–58. https://doi.org/10.1080/21580103.2023.2167873

Bokhary, H., Suleiman, A., Basalah, M. and Parvez, S. (1987). Chemical composition of desert truffles from Saudi Arabia. Can. Inst. Food Sci. Technol. J., 20(5): 336–341. https://doi.org/10.1016/s0315-5463(87)71328-5

Bonito, G., Smith, M.E., Nowak, M., Healy, R.A., Guevara, G. et al. (2013). Historical biogeography and diversification of truffles in the Tuberaceae and their newly identified southern hemisphere sister lineage. PLoS ONE, 8(1) : e52765. https://doi.org/10.1371/journal.pone.0052765

Bonito, G., Trappe, J.M., Rawlinson, P. and Vilgalys, R. (2010). Improved resolution of major clades within *Tuber* and taxonomy of species within the *Tuber gibbosum* complex. Mycologia, 102(5): 1042–1057. https://doi.org/10.3852/09-213

Bonito, G.M. and Smith, M.E. (2016). General systematic position of the truffles: Evolutionary theories. pp. 3–18. In: Zambonerlli, A., Lotti, M. and Murat, C. (eds.). Tru Truffle (*Tuber* spp.) in The World. Springer International Publishing, Switzerland.

Bordallo, J.J. and Rodríguez, A. (2014). Cryptic and new species. Soil Biol. Desert Truffles, 39–53. https://doi.org/10.1007/978-3-642-40096-4_3

Casarica et al., A purified extract from brown truffles of the species *Terfezia claveryi* chatin and its antimicrobial activity. Farmacia, 64, 298 (2016).

Casarica, A., Moscovici, M.G., Daas, M.S., Nicu, I., Panteli, M. and Rasit, I. (2016). A purified extract from brown truffles of the species *Terfezia claveryi* Chatin and its antimicrobial activity. Farmacia, 64(2): 298–301.

Castellano, M.A. and Türkoğlu, A. (2012). New records of truffle taxa in *Tuber* and *Terfezia* from Turkey. Tur. J. Bot., 36(3): 295–298. https://doi.org/10.3906/bot-1106-10

Chatin A (1896a) Truffes (terfas) de Mesrata, en Tripolitaine. Bull. Soc. Bot. Fr., 43: 139–143.

Chatin, A. (1891). Contribution à l'histoire botanique de la truffe. Deuxième note: Terfas ou truffes d'Afrique (et d'Arabie), genres *Terfezia* et *Tirmania*. Compte Rendu de l' Académie des Sciences, Paris, série 3. 112: 136–141.

Chen, K., Chen, T. and Sue, Y. (2013). *Citrobacter youngae* and *Pantoea agglomerans* peritonitis in a peritoneal dialysis patient. Perit. Dial. Int., 33(3): 336–337. https://doi.org/10.3747/pdi.2012.00151

Dahham, S.S., Al-Rawi, S.S., Ibrahim, A.H., Abdul Majid, A.S. and Abdul Majid, A.M.S. (2018). Antioxidant, anticancer, apoptosis properties and chemical composition of black truffle *Terfezia claveryi*. Saudi J. Biol. Sci., 25(8): 1524–1534. https://doi.org/10.1016/j.sjbs.2016.01.031

Darro, H., Swamy, S.L., Kumar, R. and Bhardwaj, A.K. (2022). Comparison of physico-chemical properties of soils under different forest types in dry tropical forest ecosystem in Achanakmar-Amarkantak Biosphere Reserve, India. Ecol. Environ. Conser., 28(8), S163–S169. https://doi.org/10.53550/eec.2022.v28i08s.026

Del-Saz, N.F., Romero-Munar, A., Cawthray, G.R., Palma, F., Aroca, R. et al. (2018). Phosphorus concentration coordinates a respiratory bypass, synthesis and exudation of citrate, and the expression of high-affinity phosphorus transporters in *Solanum lycopersicum*. Pl. Cell Environ., 41(4) : 865–875. https://doi.org/10.1111/pce.13155

Dexheimer, J., Gerard, J., Leduc, J.P. and Chevalier, G. (1985). Étude ultrastructurale comparée des associations symbiotiques mycorhiziennes *Helianthemum salicifolium - Terfezia claveryi* et *Helianthemum salicifolium - Terfezia leptoderma*. Can. J. Bot., 63(3): 582–591. https://doi.org/10.1139/b85-073

Diez, J., Manjon, J.L. and Martin, F. (2002). molecular phylogeny of the mycorrhizal desert truffles (*Terfezia* and *Tirmania*), host specificity and edaphic tolerance. Mycologia, 94(2): 247–259. https://doi.org/10.2307/3761801

Dighton, J. (2009). Mycorrhizae. Encyclopedia of Microbiology, Elsevier Inc., Amsterdam, 153–162. https://doi.org/10.1016/b978-012373944-5.003278

Dos Santos, G.S., Solidônio, E.G., Costa, M.C., Melo, R.O., de Souza, I.F. et al. (2015). Study of the Enterobacteriaceae Group CESP (*Citrobacter, Enterobacter, Serratia, Providencia, Morganella* and *Hafnia*): A review. pp. 794–805. In: Méndez-Vilas, A. (ed.). The Battle Against Microbial Pathogens: Basic Science, Technological Advances and Educational Programs, Federal University of Pernambuco, Recife, Brazil.

El-Kholy, H. (1989). Genetical and physiological studies on truffles. Ph.D. Thesis, Faculty of Agriculture., Cairo University, Egypt.

El-Kholy, H.K. and Assim, M.M.A. (1991). Truffles in Egypt: Survey and identification. Micol. Vegetatione Mediterr., 7(1): 46.

El-Kholy, H.K., Assim, M.M.A. (1991). Truffles in Egypt: Survey and identification. Micol Vegetatione Mediterranea 7(1): 46

Ersel, F.Y. and Solak, M.H. (2004). Contributions to the Macrofungi of I˙zmir Province. Tur. J. Bot., 28: 487–490.

Farhat, C.C. (2006). Analysis of indigenous nutritional knowledge, cultural importance and nutritional content of wild edible plants. Master of Science Dissertation, American University of Beirut, Beirut, Lebanon, p. 112.

Ferdman, Y., Aviram, S., Roth-Bejerano, N., Trappe, J.M. and Kagan-Zur, V. (2005). Phylogenetic Studies of *Terfezia pfeilii* and *Choiromyces echinulatus* (Pezizales) support new genera for southern African truffles: Kalaharituber and Eremiomyces. Mycol. Res., 109(2): 237–245. https://doi.org/10.1017/s0953756204001789

Fogel, R. and Peck, S.B. (1975). Ecological studies of hypogeous fungi. I. Coleoptera associated with Sporocarps. Mycologia, 67(4) : 741–747. https://doi.org/10.1080/00275514.1975.12019804

Fortas, Z. (1990). Étude de trois espèces de terfez: caractères culturaux et cytologie du mycélium isolé et associé à *Helianthemum guttatum*. Thèse de Doctorat. Université d'Oran Es-Senia, Algérie.

Fortas, Z. and Chevalier, G. (1992b). Effet des conditions de culture sur la mycorhization de *l'Helianthemum guttatum* par trois espèces de terfez des genres *Terfezia* et *Tirmania* d'Algérie. Can. J. Bot., 70(12): 2453–2460. https://doi.org/10.1139/b92-303

Fortas, Z., and Chevalier, G. (1992a). Caracteristiques de la germination des Ascospores de *Terfezia arenaria* (Moris) Trappe, récoltée en Algerie. Crypt. Mycol., 13: 21–29.

Gargas, A. and Taylor, J.W. (1995). Phylogeny of discomycetes and early radiations of the apothecial Ascomycotina inferred from SSU rDNA sequence data. Exp. Mycol., 19(1): 7–15. https://doi.org/10.1006/emyc.1995.1002

Gómez-Reyes, V.M., Gómez-Peralta, M. and Guevara Guerrero, G. (2017). Primer reporte y distribución de *Delastria rosea* (Pezizales: Incertae sedis) en México. Acta Bot. Mex., 119: 139–144. https://doi.org/10.21829/abm119.2017.1236

Gucin, F. (1990). Macrofungi found surroundings of Elazg. Tur. J. Bot., 14: 171–177.

Gutiérrez, A., Morte, A. and Honrubia, M. (2003). Morphological characterization of the mycorrhiza formed by *Helianthemum almeriense* Pau with *Terfezia claveryi* Chatin and *Picoa lefebvrei* (Pat.) Maire. Mycorrhiza, 13(6) : 299–307. https://doi.org/10.1007/s00572-003-0236-7

Harrington, F.A. and Potter, D. (1997). Phylogenetic relationships within *Sarcoscypha* based upon nucleotide sequences of the internal transcribed spacer of nuclear ribosomal DNA. Mycologia, 89(2), 258–267. https://doi.org/10.1080/00275514.1997.12026779

Harrington, F.A., Pfister, D.H., Potter, D. and Donoghue, M. (1999). Phylogenetic studies within the Pezizales. I. 18S rRNA sequence data and classification. Mycologia, 91(1): 41–50. https://doi.org/10.1080/00275514.1999.12060992

Hashem, A.R. and AJ-Obaid, A.M. (1996). Mineral composition of soil and wild desert truffles in Saudi Arabia. J. King Saudi Univ. Sci., 8: 5–10.

Honrubia, M., Cano, A. and Molina-Ninirola, C. (1992). Hypogeous fungi from Southern Spanish semiarid lands. Persoonia, 14(4): 647–653.

Janakat, S.M., Al-Fakhiri, S.M. and Sallal, A.K. (2005). Evaluation of antibacterial activity of aqueous and methanolic extracts of the truffle *Terfezia claveryi* against *Pseudomonas aeruginosa*. Saudi Med. J., 26(6): 952–955.

Khabar, L., Najim, L., Janex-Favre, M.C. and Parguey-Leduc, A. (2001). Contribution a` l'étude de la flore mycologique du Maroc: les truffes marocaines, discomycètes. Bull. Soc. Mycol. Fr., 117(3): 213–229.

Khabar, L., Slama, A. and Neffati, M. (2005). Terfess common to Morocco and Tunisia. Actes of IWEMM, 27 novembre au 2 décembre 2005, Murcia, Espagne

Kovács, G.M. and Trappe, J.M. (2014). Nomenclatural history and genealogies of desert truffles. pp. 21–37. In: Kagan-Zur, V., Roth-Bejerano, N., Sitrit, Y. and Morte, A. (eds.). Desert Truffles:

Phylogeny, Physiology Distribution and Domestication. Springer International Publishing, Switzerland.

Kovács, G.M., Trappe, J.M., Alsheikh, A.M., Hansen, K., Healy, R.A. and Vági, P. (2011). *Terfezia* disappears from the American truffle mycota as two new genera and *Mattirolomyces* species emerge. Mycologia, 103(4): 831–840. https://doi.org/10.3852/10-273

Læssøe, T. and Hansen, K. (2007). Truffle trouble: What happened to the Tuberales? Mycol. Res., 111(9): 1075–1099. https://doi.org/10.1016/j.mycres.2007.08.004

Landvik, S., Egger, K.N. and Schumacher, T. (1997). Towards a subordinal classification of the Pezizales (Ascomycota): Phylogenetic analyses of SSU rDNA sequences. Nordic J. Bot., 17(4): 403–418. https://doi.org/10.1111/j.1756-1051.1997.tb00337.x

Ławrynowicz, M., Markowić, M., Milenković, M. and Ivanćević, B. (1997). *Terfezia terfezioides* - a new hypogeous fungus for Balkan Peninsula. Acta Mycol., 32(2): 233–238. https://doi.org/10.5586/am.1997.019

Ławrynowicz, M., Markowić, M., Milenković, M. and Ivanćević, B. (2014). *Terfezia terfezioides* - A new hypogeous fungus for Balkan Peninsula. Acta Mycol., 32(2): 233–238. https://doi.org/10.5586/am.1997.019

Li, C., Liang, D., Li, J., Duan, Y., Li, H. et al. (2012). Unravelling mitochondrial retrograde regulation in the abiotic stress induction of rice alternative oxydase 1 genes. Pl. Cell Environ., 36(4): 775–788. https://doi.org/10.1111/pce.12013

Liu, J.J., Sturrock, R. and Ekramoddoullah, A.K.M. (2010). The superfamily of thaumatin-like proteins: its origin, evolution, and expression towards biological function. Pl. Cell Rep., 29(5): 419–436. https://doi.org/10.1007/s00299-010-0826-8

Liu, Z., Li, Y., Wang, J., He, X. and Tian, C. (2015). Different respiration metabolism between mycorrhizal and non-mycorrhizal rice under low-temperature stress: A cry for help from the host. J. Agric. Sci., 153(4): 602–614. https://doi.org/10.1017/s0021859614000434

Lu, B., Zhang, F.M., Yu, F.Q. and Rinaldi, A.C. (2022). Ethnobiological notes and volatile profiles of two rare Chinese desert truffles. Mycology, 13(3) : 177–184. https://doi.org/10.1080/21501203.2022.2035005

Malençon, G. (1973). Champignons hypogés du Nord de l'Afrique. I. Ascomycètes. Persoonia, 7(2): 261–288.

Mandeel, Q.A. and Al-Laith, A.A.A. (2007). Ethnomycological aspects of the desert truffle among native Bahraini and non-Bahraini peoples of the Kingdom of Bahrain. J. Ethnopharmacol., 110(1): 118–129. https://doi.org/10.1016/j.jep.2006.09.014

Marjanović, A., Uehlein, N., Kaldenhoff, R., Zwiazek, J.J., Weiß, M. et al. (2005). Aquaporins in poplar: What a difference a symbiont makes! Planta, 222(2): 258–268. https://doi.org/10.1007/s00425-005-1539-z

Marqués-Gálvez, J.E., Miyauchi, S., Paolocci, F., Navarro-Ródenas, A., Arenas, F. et al. (2021). Desert truffle genomes reveal their reproductive modes and new insights into plant-fungal interaction and ectendomycorrhizal lifestyle. New Phytol., 229(5): 2917–2932. https://doi.org/10.1111/nph.17044

Marqués-Gálvez, J.E., Morte, A. and Navarro-Ródenas, A. (2020). Spring stomatal response to vapor pressure deficit as a marker for desert truffle fruiting. Mycorrhiza, 30(4) : 503–512. https://doi.org/10.1007/s00572-020-00966-8

Marqués-Gálvez, J.E., Morte, A., Navarro-Ródenas, A., García-Carmona, F. and Pérez-Gilabert, M. (2019). Purification and characterization of *Terfezia claveryi* TcCAT-1, a desert truffle catalase upregulated in mycorrhizal symbiosis. PLOS ONE, 14(7): e0219300. https://doi.org/10.1371/journal.pone.0219300

Maser, C., Trappe, J.M. and Nussbaum, R.A. (1978). Fungal-small mammal interrelationships with emphasis on Oregon coniferous forests. Ecology, 59(4) : 799–809. https://doi.org/10.2307/1938784

Maublanc, A. (1946) Les champignons de France. pp. 220–223. In: Les champignons de France, 3ème édition, Tome II, Beauvais, France.

Maublanc, A. (1952) Les champignons de France. pp. 232–233. In: Les champignons de France, 4ème édition, Tome II, Beauvais, France.

Montecchi, A. and Sarasini, M. (2000). Fungi ipogei d'Europa. Associazione Micologica Bresadola. Centro Studii Micologici, Vicenza, Italie, p. 714.

Moreno, G., Alvarado, P. and Manjón, J.L. (2014). Hypogeous desert fungi. pp. 3–20. In: Kagan-Zur, V., Roth-Bejerano, N., Sitrit, Y. and Morte, A. (eds.). Desert Truffles: Phylogeny, Physiology Distribution and Domestication. Springer International Publishing, Switzerland. https://doi.org/10.1007/978-3-642-40096-4_1

Moreno, G., Diez, J. and Manjon, J.L. (2002). *Terfezia boudieri* first records from Europe of a rare vernal hypogenous mycorrhizal fungus. Persoonia, 17: 637–641.

Moreno, G., Manjón, J.L. and Alvarado, P. (2019). A new *Terfezia* from Spain. Boletín de la Sociedad Micológica de Madrid, 43: 55–60.

Morte, A., Honrubia, M. and Gutierrez, A. (2008). Biotechnology and cultivation of desert truffles. pp. 467–483. In: Varma, A. (ed.). Mycorrhiza. Springer-Verlag, Berlin, Germany.

Morte, A., Lovisolo, C. and Schubert, A. (2000). Effect of drought stress on growth and water relations of the mycorrhizal association *Helianthemum almeriense-Terfezia claveryi*. Mycorrhiza, 10(3): 115–119. https://doi.org/10.1007/s005720000066

Morte, A., Navarro-Ródenas, A. and Nicolás, E. (2010). Physiological parameters of desert truffle mycorrhizal *Helianthemun almeriense* plants cultivated in orchards under water deficit conditions. Symbiosis, 52(2–3) : 133–139. https://doi.org/10.1007/s13199-010-0080-4

Morte, A., Pérez-Gilabert, M., Gutiérrez, A., Arenas, F., Marqués-Gálvez, J.E. et al. (2017). Basic and applied research for desert truffle cultivation. pp 23–42. In: Varma, A., Prasad, R. and Tuteja, N' (eds.). Mycorrhiza - Ecophysiology, Secondary Metabolites, Nanomaterials. Springer, Berlin. https://doi.org/10.1007/978-3-319-57849-1_2

Murcia, M.A., Maggi, L., Hussain, G., Jiménez-Monreal, A.M., Martínez-Tomé, M. and Alruqaie, I.M. (2014). Preservation of truffles. Soil Biology Desert Truffles, pp. 299–321. https://doi.org/10.1007/978-3-642-40096-4_19

Navarro-Ródenas, A., Bárzana, G., Nicolás, E., Carra, A., Schubert, A. and Morte, A. (2013). Expression analysis of aquaporins from desert truffle mycorrhiza. In symbiosis reveals a fine-tuned regulation under drought. Mol. Pl.-Microbe Interact., 26, 1068–1078. http://doi.org/10.1094/MPMI-07-12-0178-R

Navarro-Ródenas, A., Ruíz-Lozano, J.M., Kaldenhoff, R. and Morte, A. (2012). The aquaporin TcAQP1 of the desert truffle *Terfezia claveryi* is a membrane pore for water and CO_2 transport. Mol. Pl.-Microbe Interact., 25(2), 259–266. https://doi.org/10.1094/mpmi-07-11-0190

Norman, J.E. and Egger, K.N. (1996). Phylogeny of the genus *Plicaria* and Its relationship to *Peziza* inferred from ribosomal DNA sequence analysis. Mycologia, 88(6): 986–995. https://doi.org/10.2307/3761062

Norman, J.E. and Egger, K.N. (1999). Molecular phylogenetic analysis of *Peziza* and related genera. Mycologia, 91(5): 820–829. https://doi.org/10.1080/00275514.1999.12061087

O'Donnell, K., Cigelnik, E., Weber, N.S. and Trappe, J.M. (1997). Phylogenetic relationships among ascomycetous truffles and the true and false morels inferred from 18S and 28S ribosomal DNA sequence analysis. Mycologia, 89(1): 48–65. https://doi.org/10.1080/00275514.1997.12026754

Oder, N. (1988). Taxonomic investigations of important edible and poisonous mushrooms growing in the Konya center and some Districts of Konya. Selcuk Univ Fen Edeb Fak Derg, 8: 237–257

Omirou, M., Ioannides, I.M. and Ehaliotis, C. (2013). Mycorrhizal inoculation affects arbuscular mycorrhizal diversity in watermelon roots, but leads to improved colonization and plant response under water stress only. Appl. Soil Ecol., 63, 112–119. http://doi.org/10.1016/j.apsoil.2012.09.013

Owaid, M.N. (2016). Biodiversity and bioecology of Iraqi desert truffles (Pezizaceae) during season 2014. J. Aridland Agric., 2: 22–25. https://doi.org/10.19071/jaa.2016.v2.3046

Owaid, M.N. (2017). Bioecology and uses of desert truffles (Pezizales) in the Middle East. Walailak J. Sci. Technol., 15(3): 179–188. https://doi.org/10.48048/wjst.2018.3058

Pacioni, G. and El-kholy, H.K. (1994). Truffles from Egyptian desert. Micol Vegetazione Mediterr, 9(2): 69–84

Patouillard, M.N. (1894). Les Terfez de la Tunisie. J. Bot., 10: 153–156

Paz, A., Lavoise, C., Chautrand, P., Moreau, P.A. and Bellanger, J.M. (2018). The genus *Delastria* (Pezizaceae), a worldwide revision. Ascomycete.org, 10(6): 229–243. https://doi.org/10.25664/ART-0247

Percudani, R., Trevisi, A., Zambonelli, A. and Ottonello, S. (1999). Molecular phylogeny of truffles (Pezizales: Terfeziaceae, Tuberaceae) derived from nuclear rDNA sequence analysis. Mol. Phylog. Evol., 13(1): 169–180. https://doi.org/10.1006/mpev.1999.0638

Porcel, R., Aroca, R., Azcón, R. and Ruiz-Lozano, J.M. (2006). PIP aquaporin gene expression in arbuscular mycorrhizal *Glycine max* and *Lactuca sativa* plants in relation to drought stress tolerance. Pl. Mol. Biol., 60: 389–404. http://doi.org/10.1007/s11103-005-4210-y

Pozo, M.J., Verhage, A., García-Andrade, J., García, J.M. and Azcón-Aguilar, C. (2009). Priming plant defence against pathogens by arbuscular mycorrhizal fungi. 123–135. In: Azcón-Aguilar, C., Barea, J.M., Gianinazzi, S. and Gianinazzi-Pearson, V. (eds.). Mycorrhizas - Functional Processes and Ecological Impact. https://doi.org/10.1007/978-3-540-87978-7_9

Roux, C., Sejalon-Delmas, N., Martins, M., Parguey-Leduc, A., Dargent, R. and Bécard, G. (1999). Phylogenetic relationships between European and Chinese truffles based on parsimony and distance analysis of ITS sequences. FEMS Microbiol. Lett., 180(2): 147–155. https://doi.org/10.1111/j.1574-6968.1999.tb08789.x

Ruiz-Lozano, J.M. and Aroca, R. (2010b). Modulation of aquaporin genes by the arbuscular mycorrhizal symbiosis in relation to osmotic stress tolerance. pp. 357–374. In: Seckbach, J. and Grube, M. (eds.). Symbioses and Stress: Cellular Origin, Life in Extreme Habitats and Astrobiology, Springer, Dordrecht, Germany.

Ruiz-Lozano, J.M., Porcel, R. and Aroca, R. (2008). Evaluation of the possible participation of drought-induced genes in the enhanced tolerance of arbuscular mycorrhizal plants to water deficit. pp. 185–205. In: Varma, A. (ed.). Mycorrhiza: State of the Art, Genetics and Molecular Biology, Eco-Function, Biotechnology, Eco-Physiology, Structure and Systematics Springer, Berlin, Germany. http://doi.org10.1007/978-3-540-78826-3_10

Sawaya, W.N., Al-Shalhat, A., Al-Sogair, A. and Al-Mohammad, M. (1985). Chemical composition and nutritive value of truffles of Saudi Arabia. J. Food Sci., 50(2): 450–453. https://doi.org/10.1111/j.1365-2621.1985.tb13425.x

Seyidova, H. and Hüseyin, E. (2012). Macrofungi of Nakhchivan (Azerbaijan) autonomous republic. Tur. J. Bot., 36(6): 761–768. https://doi.org/10.3906/bot-1101-43

Slama, A., Fortas, Z., Neffati, M., Khabar, L. and Boudabous, A. (2006). Étude taxonomique de quelques Ascomycètes hypogés (Terfeziaceae) de la Tunisie meridionale. Bulletin de la Société Mycologique de France, 122(2–3): 187–195.

Spatafora, J.W. (1995). Ascomal evolution of filamentous ascomycetes: evidence from molecular data. Can. J. Bot., 73(S1): 811–815. https://doi.org/10.1139/b95-326

Taylor, F., Thamage, D., Baker, N., Roth-Bejerano, N. and Kagan-Zur, V. (1995). Notes on the Kalahari desert truffle, *Terfezia pfeilii*. Mycol. Res. 99(7): 874–878. https://doi.org/10.1016/s0953-7562(09)80744-2

Trappe, J.M. (1979). The order, families and genera of hypogeous Ascomycotina Truffles and their relatives. Mycotaxon, 9: 297–340.

Trappe, J.M. and Weber, N.S. (2001). North American desert truffles: the genus *Carbomyces* (Ascomycota, Carbomycetaceae). Harvard Pap. Bot., 6: 209–214.

Trappe, J.M., Claridge A.W. and Kagan-Zur, V. (2014). Ecology and distribution of desert truffles in the Kalahari of southern Africa. pp. 193–202. In: Kagan-Zur, V., Roth-Bejerano, N., Sitrit, Y. and Morte, A. (eds.). Desert Truffles: Phylogeny, Physiology Distribution and Domestication. Springer International Publishing, Switzerland.

Trappe, J.M., Claridge, A.W., Claridge, D.L. and Liddle, L. (2008b). Desert truffles of the Australian outback: Ecology, ethnomycology, and taxonomy. Econ. Bot., 62(3): 497–506. https://doi.org/10.1007/s12231-008-9041-8

Trappe, J.M., Claridge, A.W., Arora, D. and Smit, W.A. (2008a). Desert truffles of the African Kalahari: Ecology, ethnomycology, and taxonomy. Econ. Bot., 62(3): 521–529. https://doi.org/10.1007/s12231-008-9027-6

Trappe, J.M., Kovács, G.M. and Claridge, A.W. (2010a). Comparative taxonomy of desert truffles of the Australian outback and the African Kalahari. Mycol. Prog., 9(1): 131–143. https://doi.org/10.1007/s11557-009-0612-6

Trappe, J.M., Kovács, G.M. and Claridge, A.W. (2010b). Validation of the new combination *Mattirolomyces austroafricanus*. Mycol. Prog., 9(1): 145–145. https://doi.org/10.1007/s11557-009-0631-3

Turgeman, T., Lubinsky, O., Roth-Bejerano, N., Kagan-Zur, V., Kapulnik, Y. et al. (2015). The role of pre-symbiotic auxin signaling in ectendomycorrhiza formation between the desert truffle *Terfezia boudieri* and *Helianthemum sessiliflorum*. Mycorrhiza, 26(4): 287–297. https://doi.org/10.1007/s00572-015-0667-y

Türkoğlu, A. and Yağiz, D. (2012). Contributions to the macrofungal diversity of Uşak Province. Tur. J. Bot., 36: 580–589. https://doi.org/10.3906/bot-1103-2

Vanlerberghe G.C. (2013). Alternative oxidase: a mitochondrial respiratory pathway to maintain metabolic and signaling homeostasis during abiotic and biotic stress in plants. Int. J. Mol. Sci., 14(4): 6805–6847. https://doi.org/10.3390/ijms14046805

Veneault-Fourrey, C., Commun, C., Kohler, A., Morin, E., Balestrini, R. et al. (2014). Genomic and transcriptomic analysis of *Laccaria bicolor* CAZome reveals insights into polysaccharides remodelling during symbiosis establishment. Fungal Gen. Biol., 72: 168–181. https://doi.org/10.1016/j.fgb.2014.08.007

Wang, W., Shi, J., Xie, Q., Jiang, Y., Yu, N. and Wang, E. (2017). Nutrient exchange and regulation in arbuscular mycorrhizal symbiosis. Mol. Pl., 10(9): 1147–1158. https://doi.org/10.1016/j.molp.2017.07.012

Ward, D. (2008). The Biology of Deserts. Oxford University Press, Oxford, UK, p. 339.

Wu, Q.S., Srivastava, A.K. and Zou, Y.N. (2013). AMF-induced tolerance to drought stress in citrus: a review. Sci. Hortic., 164: 77–87. http://doi.org/10.1016/j.scienta.2013.09.010

Zaretsky, M., Kagan-Zur, V., Mills, D. and Roth-Bejerano, N. (2005). Analysis of mycorrhizal associations formed by *Cistus incanus* transformed root clones with *Terfezia boudieri* isolates. Plant Cell Reports, 25(1): 62–70. https://doi.org/10.1007/s00299-005-0035-z

Zitouni-Haouar, F.E.H., Bidartondo, M.I., Moreno, G., Carlavilla, J.R., Manjón, J.L. et al. (2023). Bioclimatic origin shapes phylogenetic structure of *Tirmania* (Pezizaceae): New species and new record from North Africa. J. Fungi, 9(5): 532. https://doi.org/10.3390/jof9050532

Zitouni-Haouar, F.E.H., Carlavilla, J.R., Moreno, G., Manjón, J.L. and Fortas, Z. (2018). Genetic diversity of the genus *Terfezia* (Pezizaceae, Pezizales): New species and new record from North Africa. Phytotaxa, 334(2): 183–194. https://doi.org/10.11646/phytotaxa.334.2.7

Chapter 4

Truffle Industry: An Overview

Hesham R. El-Seedi,[1,2,3] *Doaa M. Abo-Atya,*[4] *Shaden Khalifa,*[5] *Tiehua Zhang*[6] and *Jelena Popović-Djordjević*[7,*]

1. Introduction

Due to their chemical composition, nutritional quality, and health benefits, mushrooms have held a special place in many cultures since ancient times. Truffles, in particular, are noted for their unique characteristics (Wang and Marcone, 2011; Rosa-Gruszecka et al., 2017; Lee et al., 2020). Truffles are one of the oldest types of food. According to some documents, the first culinary use of truffles by people settled on the eastern coast of the Mediterranean Sea dates back to the Bronze Age (Rosa-Gruszecka et al., 2017). Due to their distinct aroma and delicious taste, they are very attractive to consumers (Wang and Marcone, 2011; Khalifa et al., 2019).

Generally, truffles are referred to two widespread categories: tuber and desert. Species of the genus Tuber are considered "true" truffles (order Pezizales). The genus Tuber includes over 180 species that have been reported so far. However, no more than 20 of these species have been found to be of commercial significance (Oliach et

[1] Department of Pharmaceutical Biosciences, Uppsala University, Biomedical Centre, Uppsala, Sweden
[2] International Research Center for Food Nutrition and Safety, Jiangsu University, Zhenjiang, China
[3] Chemistry Department, Faculty of Science, Islamic University of Madinah, Madinah, Saudi Arabia
[4] Department of Chemistry, Faculty of Science, Menoufia University, Shebin El-Kom, Egypt
[5] Department of Molecular Biosciences, The Wenner-Gren Institute (MBW), Stockholm University, Stockholm, Sweden
[6] College of Food Science and Engineering, Jilin University, Changchun , China
[7] University of Belgrade-Faculty of Agriculture, Department of Food Technology and Biochemistry, 11080 Belgrade, Serbia
* Corresponding author: jelenadj@agrif.bg.ac.rs

al., 2021). The most commercially significant species that naturally grow in Europe include *Tuber melanosporum* and *Tuber aestivum* (black truffles), as well as *Tuber borchii* and *Tuber magnatum* (white truffles) (Khalifa et al., 2019; Lee et al., 2020; Farag et al., 2021).

On the other hand, the most common «truffle-like» fungi, or desert truffles, that grow in semi-arid as well as arid geographical regions belong to the genera *Terfezia* (black truffles) and *Tirmania* (white truffles): *Terfezia metaxasi*, *Terfezia claveryi*, *Terfezia leonis*, *Terfezia boudieri*, and *Tirmania nivea* (Murcia et al., 2003; Khalifa et al., 2019; Lee et al., 2020; Farag et al., 2021). *Picoa* fungi also belong to dessert truffles but are less studied compared to *Terfezia* and *Tirmania* spp. (Murcia et al. 2003).

True truffles (*Tuber* sp.) are Ascomycetes fungi that establish ectomycorrhizal symbiosis with various tree species, primarily in the Mediterranean and temperate regions. Due to their unique and intense aroma, some European species are highly valued and can fetch high prices on the market, making them some of the most expensive foods (Patel et al., 2017; Popović-Djordjević et al., 2019; Oliach et al., 2021). For example, the prices (in Euros) of the most consumed truffle species are as follows: *T. magnatum* up to €5000 per kg, *T. melanosporum* €700–1200 per kg (the most famous and valued truffles), *T. aestivum* (the most common species in Europe, but less appreciated) from €25–600 per kg (depending on the country), *T. brumale* (less cherished among the winter truffles in Europe) €60–120 per kg, and *T. borchii*, up to €250 per kg (when other *Tuber* species are poor) (Segelke et al., 2020; Farag et al., 2021; Kappacher et al., 2022). *T. indicum* is an Asian truffle that morphologically resembles *T. melanosporum*. The aroma of *T. indicum* is less pronounced and contains a lower variety of sulfur compounds compared to the more expensive *T. melanosporum*. However, it is often found as an adulterant in different food products prepared from *T. melanosporum* (Farag et al., 2021). Among traditional edible mushrooms, *T. latisporum*, *T. pseudohimalayense*, and *T. subglobosum* are produced on a large scale in China (Yan et al., 2017).

In North Africa and the Middle East, desert truffles have long been used as a substitute for meat in the diet (Murcia et al., 2003). *Tirmania* and *Terfezia* spp. are known to be valuable food sources for the Saharan tribes (Rosa-Gruszecka et al., 2017). Nowadays, *Terfezia* spp. are especially prized in Arab countries and the Mediterranean region (Murcia et al., 2003).

2. Nutritional Potential of Truffles

Truffles contain various nutritionally valuable compounds, such as amino acids, carbohydrates, dietary fibers, proteins, unsaturated fatty acids, minerals, vitamins, and volatile organic compounds (Hamza et al., 2016a; Kivrak and Kivrak, 2018; Lee et al., 2020; Shimokawa et al., 2020; Ljubojević et al., 2023). The content of macronutrients (carbohydrates, proteins, fats, and fatty acids) and micronutrients (essential elements) in truffles varies between species, and within the same species from different locations, as presented in Table 1 (Wang and Marcone, 2011; Vahdani et al., 2017; Kivrak and Kivrak, 2018; Khalifa et al., 2019, Popović-Djordjević et al.,

2019; Lee at al., 2020; Shimokawa et al., 2020; Henkrar and Khabar, 2022; Freiberg et al., 2023; Ljubojević et al., 2023).

Polyunsaturated fatty acids (PUFA) and monounsaturated fatty acids (MUFA) are the dominant FAs in truffles. Among individual FAs, oleic, linoleic, linolenic, palmitic, and steric acids are the most abundant (Beuchat et al., 1993; Murcia et al., 2003; Stojković et al., 2013; Yan et al., 2017; Dahham et al., 2018; Kivrak and Kivrak, 2018; Henkrar and Khabar, 2022; Ljubojević et al., 2023). Free sugars such as glucose, psicopyranose, trehalose, turanose, rhamnose, galactose, and sugar alcohols like sorbitol, mannitol, inositol, and myo-inositol are predominant in various truffles (Yan et al., 2017; Li et al., 2020; Farag et al., 2021). Brassicasterol and ergosterol are the major sterols quantified in *Tuber* spp. ascocarps (Tejedor-Calvo et al., 2022), whereas in *Terfezia* spp. several sterols, including ergosterol, brassicasterol, ergosta7.22-dienol, campestanol, stigmasterol, β-sitosterol, and lupeol, have been identified (Dahham et al., 2018; Tejedor-Calvo et al., 2022). Among nutritionally valuable elements, K, Ca, Mg, P, S are the most abundant in most truffle ascocarps (Sawaya et al., 1985; Kagan-Zur and Roth-Bejerano, 2008; Kruzselyi and Vetter, 2014; Hamza et al., 2016a; Hamza et al., 2016b; Popović-Djordjević et al., 2019; Shimokawa et al., 2020; Ljubojević et al., 2023). It should be noted that the distinct chemical composition of the same species of truffles growing in different regions may be linked to various environmental factors, including soil types, the amount of precipitation, among others (Dahham et al., 2018).

Truffles can be preserved and stored for some time, with the preservation method largely depending on the truffle's aroma profile. Selecting the proper method is crucial for maintaining the quality of truffles (Lee et al., 2020; Chauhan et al., 2021). Several methods of preservation have been proposed: drying in an oven at 110°C; combination of gamma radiation, acid, alkali treatment, and refrigeration; blanching in 4% boiling natrium-chloride, spraying with 5% vinegar, freezing at -18°C; combined micro-perforated films with electron beam irradiation; modified atmosphere packaging (MAP) with micro-perforated films; refrigeration. During refrigeration, it has been noted that white truffles (*T. magnatum* and *T. borchii*) are more susceptible to deterioration compared to black truffles (*T. melanosporum* and *T. aestivum*) (Chauhan et al., 2021). Industrial processes such as freezing and canning truffles can decrease protein, lipid and ash content (Murcia et al., 2003).

Nowadays, a wide array of truffle products from *Tuber* spp. is available on the market: truffle butter, truffle honey, truffle pesto, truffle salt, truffle spread, black truffle rub, white truffle mayonnaise, white truffle porcini, and white truffle sauce (Fig. 1). Other gourmet products include: balsamic vinegar with white truffles, goat cheese with truffles, olive oil with white or black truffles, pralines with truffles, and plum brandy with truffles (https://www.gourmetfoodstore.com/truffles-and-mushrooms/; https://www.bsfproduct.com/proizvodi.html).

3. Bioactive Metabolites

Metabolomics is a novel method used to systematically study the peculiar chemical fingerprint of small-molecule metabolites that are produced in the cell processes of some organisms. In an ultra-high-performance liquid chromatography-tandem

Fig. 1. Truffle products found on the market: A) balsamic vinegar with white truffles and olive oil with black truffles, B) paprika spread ('ajvar') with black truffles, C) sliced truffles, D) olive oil with extract of black truffles, E) sea salt with truffle and apricot black truffle, F) mayonnaise with black truffle and ketchup with black truffle, G) pralines with truffles, and H) olive oil with white truffle flavor (Photos credit: J. Popović-Djordjević).

mass spectrometry (UHPLC-MS/MS)-based metabolomic study of the five most economically important tuber species from China, a total of 2376 metabolites were identified (*T. indicum, T. melanosporum, T. panzhihuanense*, *T. pseudoexcavatum,* and *T. sinoaestivum*). The metabolites, predominantly found in the studied species, were amino acids, alkaloids, carnitines, flavonoids, nucleosides and their analogs, organic acids, saccharides, and sulfur compounds. Some of these compounds in truffle extracts are associated with anti-inflammatory, antioxidant, and antitumor immunomodulatory activities (Li et al., 2019a). Antioxidant activities of other compounds from truffles, like vitamins, carotenoids, phenolic compounds, and polysaccharides, are also known (Akyüz et al., 2021).

Sterols have a crucial role in the organization and function of the cell plasma membrane in members of the fungi kingdom. Ergosterol is the distinctive sterol of truffle and 'truffle-like' fungi, and its role is comparable to that of cholesterol in vertebrates and of phytosterols in plants (Dupont et al., 2021). Among other sterols, brassicasterol, ergosta7,22-dienol, lanosterol, campestanol, stigmasterol, β-sitosterol, and lupeol are commonly found in *Tuber* and *Terfezia* fungi (Dahham et al., 2018; Patel et al., 2017; Henkrar and Khabar, 2022; Tejedor-Calvo et al., 2022). Carotenoids, β-carotene and lycopene, are also detected in some truffles (Yan et al., 2017).

Pacioni et al. (2015) identified anandamide, an important constituent of the endocannabinoid system (ECS), which is in charge of melanin synthesis in normal human epidermal melanocytes in the black truffle (*T. melanosporum*). Anandamide (*N*-arachidonoylethanolamine) is a fatty acid neurotransmitter, obtained from the

Table 1. Composition of macronutrients and micronutrients in *Tuber, Terfezia, Tirmania* and *Picoa* truffles.

Genus/Species	Country	Nutritional composition	Reference
Tuber			
T. aestivum Vittad.	Bosnia and Herzegovina	Fresh weight: water (75.5%); Dry weight (24.5%), in %: proteins-6.52, carbohydrates-16.43, fat-0.69, ash-0.87; Fatty acids (FA), in %: *saturated* (*SFA*): butyric (4:0)-0.42, caproic (6:0)-0.39, capriylic (8:0)-0.5, capric (10:0)-0.7, lauric (12:0)-2.49, tridecylic (13:0)-0.1, myristic (14:0)- 3.88, pentadecylic (15:0)-0.7, palmitic (16:0)-25.89, margaric (17:0)-0.53, stearic (18:0)-10.04, arachidic (20:0)-11.51, lignoceric (24:0)-1.07; *unsaturated* (*UFA*): myristoleic (C14:1)-0.25, pentadecenoic (C15:1,*cis*-10)-0.56, palmitoleic (C16:1)-0.82, heptadecanoic (C17:1, *cis*-10)-0.14, oleic (C18:1n9c)-15.9, elaidic (C18:1n9c)-28.17, linoleic (C18:2)-4.99, α-linolenic (C18:3n3)-4.21, γ-linolenic (C18:3n6)-0.54, *cis,cis*-11,14-eicosadienoic (C20:2)-0.76, timnodoic (C20:5n3)-1.22; Elements, in mg/kg: K-26409, Na- 49.9, Ca-3388, Mg-749, Fe-501.2, P-4112 and Zn-338.5, Cu-47.3, Cr-1.1 and Pb-0.45	Ljubojević et al., 2023
T. aestivum	Spain	Dry weight, in g/100g: carbohydrates-35.83, proteins-11.92, β-glucans-21.71, chitin-10.20; Sterols, in mg/g: ergosterol-2.27, brassicasterol-1.71, ergosta7,22-dienol-1.09 and stigmasterol-0.65	Tejedor-Calvo et al., 2022
T. aestivum	Serbia	Fresh matter: proteins (11-12.9%), carbohydrates (5.65%), minerals (K, Ca, P, Fe), sulfur containing amino acids and fatty acids (linoleic, palmitic, oleic)	Beara et al., 2014
T. aestivum	Hungary	Dry matter, in %: crude protein-19.11, crude fat-2.27, crude fiber-22.03, crude ash-7.55, chitin-7.55, carbohydrates-48.9; Soluble oligosaccharides and polysaccharides-9.0 and 49.92 mg/g, respectively; Protein fractions in mg/g: albumins-57.8, globulins-8.56, prolamines-3.22, prolamine-like substances-2.47, glutelines-25.7 and gluteline-like substances-2.09; Elements, in mg/kg: Al-166.7, B-19.2, Ba-8.48, Ca-3331, Cd-4.11, Cr-1.29, Cu-49.3, Fe-230.3, K-25647, Mg-1060, Mn-17.4, Na-148.2, Ni-1.68, P-7879, Sr-10.58, Ti-4.64 and Zn-160.5	Kruzselyi and Vetter, 2014
T. aestivum	Italy	Dry matter, in g/100g: soluble protein-11.0, total carbohydrate-5.65 (monosaccharide-0.42, polysaccharide-5.23)	Saltarelli et al., 2008
T. borchii Vittard.	Italy	Dry matter, in g/100g: soluble protein-13.1, total carbohydrate-3.59 (monosaccharide- 0.38, polysaccharide-3.21)	Saltarelli et al., 2008

T. brumale Vittard.	Serbia	Fresh matter: Essential major elements, in mg/kg: K -8640.3, Na-67.5, Ca-1358.4, Mg-395.4, Fe-232.3, P-2155.0, S-1318.6 and Zn-72.7; Essential trace elements (in mg/kg): Cu-16.7 and Mn-6.9	Popović-Đorđević et al., 2019
T. floridanum Grupe, Sulzbacher & ME Smith	Brazil	Dry weight, in %: carbohydrates-71, proteins-19, fat-5 and ash-5	Freiberg et al., 2023
T. japonicum Hir. Sasaki, A. Kinosh. & Nara	Japan	Fresh matter: moisture (76.7-78.8%); Dry weight (DW), in g/100g: crude protein (29.9-31.1), crude fat (1.12-1.34), crude fiber (12.2-14.4), and ash (6.95-8.32); Elements, in mg/100g DW: Na (45.0-61.8), P (171.0-223.0), Ca (3.54-4.1), K (650.1-683.9), Mg (14.4-16.8), Fe (1.04-1-93),Cu (1.58-1.71), Zn (5.16-7.82) and Mn (0.06-0.07); Free amino acids (FAA): aspartic acid, threonine, serine, asparagine, glutamic acid, glutamine, glycine, alanine, α-amino butyric acid, valine, cystathionine, isoleucine, lysine, arginine (content of FAA varied from 0.34 for arginine to 90.3 for glutamine, expressed as μmol/g DW)	Shimokawa et al., 2020
T. latisporum Juan Chen & P.G. Liu	China	Dry matter, in %: protein-14.64, fat-2.4, carbohydrate-74.63, ash- 8.33, total sugars -50.10 g/100g; Free sugars in g/100g: glucose-1.25, trehalose-13.84, sorbitol-33.82 and inositol-1.18; total carotenoids-170.6 mg of β-carotene/100 g extract and lycopene-235.0 g/100 g extract; Fatty acid (FA) composition, in mg/100 g: SFA- 121.2, MUFA- 265.2, PUFA- 360.6; Individual FA in mg/100 g: palmitic (C16:0)-72.36, palmitoleic (C16:1)-1.06, stearic (C18:0)-47.56, oleic (C18:1n9c)-250.97, linoleic (C18:2n6c)-347.47, α-linolenic (C18:3n3)-9.52, arachidic (C20:0)-1.01, *cis*-11-eicosenoic (C20:1c)-13.15, *cis*-11,14-eicosadienoic (C20:2c)-2.99 and arachidonic (C20:4n6)- 0.64	Yan et al., 2017
T. magnatum Picco (Piedmont truffle)	Italy	Fresh matter: moisture (75.1%); Dry weight (DW), in g/100g: crude protein (38.5), crude fat (1.13), crude fiber (11.5), and ash (9.41); Elements, in mg/100g DW: Na (11.0), P (216.1), Ca (26.9), K (902.1), Mg (21.0), Fe (1.50),Cu (1.82), Zn (11.3) and Mn (0.11); Free amino acids (FAA): aspartic acid, threonine, serine, asparagine, glutamic acid, glutamine, glycine, alanine, α-amino butyric acid, valine, cystathionine, isoleucine, lysine, arginine (content of FAA varied from 5.92 for asparagine to 256 for alanine, expressed as μmol/g DW)	Shimokawa et al., 2020
T. magnatum	Serbia	Fresh matter: Essential major elements, in mg/kg: K-10760.9, Na- 118.3, Ca-557.9, Mg-300.1, Fe-349.2, P-2769.6, S-1183.1 and Zn-168.1; Essential trace elements (in mg/kg): Cu-34.5 and Mn-7.1	Popović-Djordjević et al., 2019

Table 1. contd. ...

Table 1. *contd.* ...

Genus/Species	**Country**	**Nutritional composition**	**Reference**
T. magnatum	Serbia	Fresh matter: proteins (20.5-24%), carbohydrates (2.23%), minerals (P, K, Ca, Fe), sulphur containing amino acids and fatty acids (linoleic, palmitic, and oleic)	Beara et al., 2014
T. magnatum	Italy	Dry matter, in g/100g: soluble protein-24.0, total carbohydrate-2.23 (monosaccharide-0.53, polysaccharide-1.70)	Saltarelli et al., 2008
T. melanosporum Vittad.	France	Fresh matter: moisture (73.3%); Dry weight in g/100g: crude protein-29.1, crude fat-0.95, crude fiber-20.9 and ash-7.02; Elements, in mg/100g: Na-6.7, P-267.8, Ca-81.7, K-735.6, Mg-24.1, Fe- 1.20, Cu-1.75, Zn-3.6 and Mn-0.11	Shimokawa et al., 2020
T. melanosporum	China	Dry weight, in mg/g: total carbohydrate-64.6, reducing sugar-14.72 and water-soluble polysaccharides-30.0	Zhang et al., 2018
T. melanosporum	Italy	Dry matter, in g/100g: soluble protein-8.7, total carbohydrate-1.77 (monosaccharide-0.25, polysaccharide-1.52)	Saltarelli et al., 2008
T. nitidum Vittad.	Turkey	Fresh weight: moisture (82.06%); Dry weight, in g/100g: ash-14.87, carbohydrates-49.21, proteins-35.27, fat-3.24; Fatty acids (FA), in %: palmitic (C16:0)-12.38, palmitoleic (C16:1)-0.44, heptadecanoic (C17:0)-0.26, stearic (C17:0)-2.54, oleic (C18:1n9c)-6.38, linoleic (C18:2n6c)-76.79, linolenic (C18:3n3)-0.22, *cis*-11-eicosenoic acid (C20:1)-0.25, *cis*-11,14-eicosadienoic acid (C20:2)-0.58; Free amino acids (FAA), in mg/100g: *essential*: valine-61.29, threonine-10.72, leucine-60.38, isoleucine-48.06, lysine-12.20, methionine-34.32, phenylalanine-52.09, tryptophan-32.05; *nonessential*: glycine-1.37, alanine-5.46, serine-5.07, proline 9.87, 4-hydroxyproline-5.44, asparagines-10.11, aspartic acid-5.81, glutamine-21.13, glutamic acid-19.08, histidine-1.27, arginine-29.34, tyrosine-19.64	Kivrak and Kivrak, 2018
T. panzhihuanense X.J. Deng & Y. Wang	China	Ethanol extract of dried tubers, in %: proteins-4.15, carbohydrates 67.40; monosaccharide composition (mol%) of polysaccharide fractions: rhamnose-23.9, mannose-62.3, galactose-4.7 and glucose- 9.2	Li et al., 2020
T. pseudoexcavatum Y. Wang, G. Moreno, Riousset, Manjón & G. Riousset	China	Ethanol extract of dried tubers, in %: protein 3.79, carbohydrates- 57.90; monosaccharide composition (mol%) of polysaccharide fractions: rhamnose-19.8, mannose- 63.4, galactose-2.4 and glucose-14.4	Li et al., 2020

T. pseudohimalayense G. Moreno, Manjón, J. Diez & Garcia-Mont.	China	Dry matter, in %: proteins-14.28, fat-2.55, carbohydrate-74.40, ash- 8.77, total sugars-23.89 g/100g; Free sugars in g/100g: glucose-1.22, trehalose-4.22, sorbitol-17.55 and inositol-0.90; total carotenoids-271.3 mg of β-carotene/100 g extract and lycopene-355.2 g/100 g extract; Fatty acids (FA) composition in mg/100 g: SFA- 117.4, MUFA-223.9, PUFA-368; Individual FA, in mg/100 g: palmitic (C16:0)- 80.93, palmitoleic (C16:1)-3.32, heptadecanoic (C17:0)-0,72, stearic (C18:0)-0,72, oleic (C18:1n9c)- 219.63, linoleic (C18:2n6c)-365.24, arachidic (C20:0)-0.86, (C20:1c)-1.00, *cis*-11-eicosenoic (C20:2c)-1.41, arachidonic (C20:4n6)-0.81, eicosapentaenoic (C20:5n3)-0.53, behenic (C22:0)-0.97 and lignoceric (C24:0)- 0.78	Yan et al., 2017
T. subglobosum L. Fan & C.L. Hou	China	Dry matter, in %: proteins-10.96, fat-2.23, carbohydrate-78.68, ash-8.13, total sugars-29.98 g/100g; Free sugars in g/100g: glucose-0.92, trehalose-4.14, sorbitol-22.65 and inositol-2.00; total carotenoids-144.3 mg of β-carotene/100 g extract and lycopene-151.6 g/100 g extract; Fatty acid (FA) composition in mg/100 g: SFA- 70.4, MUFA-96.5, PUFA-249.4; Individual FA in mg/100g: myristic (C14:0)-0.59, pentadecanoic (C15:0)-0.44, palmitic (C16:0)-41.96, palmitoleic (C16:1)-1.63, heptadecanoic (C17:0)-0,58, stearic (C18:0)-26.06, oleic (C18:1n9c)-94.04, linoleic (C18:2n6c)-246.81, arachidic (C20:0)-0.75, (C20:1c)-0,78, *cis*-11-eicosenoic (C20:2c)-1.41, arachidonic (C20:4n6)-0.63 and eicosapentaenoic (C20:5n3)-0.54	Yan et al., 2017
T. texense Heimsch (Pecan truffle)	USA (Georgia)	Dry matter (DM), in %: protein-16.1, lipid-2.8, carbohydrate-75, ash-6.1; Lipids, in %: oleic -45.9, linoleic-38, palmitic-11.4; Elements, in %: potassium-2.49 phosphorus-1.29, and sulfur-0.68	Beuchat et al., 1993
Terfezia			
T. arenaria I(Moris) Trappe	Portugal	Fresh weight: moisture -77%; Dry weight (DW), in g/100g: carbohydrates-67, proteins-14, lipids-2.2, fibers-10 and ash-7.3	Ferreira et al., 2023
T. arenaria	Morocco	Dry weight, in %: (C16:0)-21.47, (C16:1)-1.07, (C18:0)-3.25, (C18:1)-27.02, (C18:2)-38.80, (C20:0)-8.56, (C22:0)-0.90: Sterols, in %: brassicasterols-96.83 and lanosterol-2.00	Henkrar and Khabar, 2022
T. boudieri Chatin	Morocco	Dry weight, in %: (C16:0) -20.82, (C16:1)- 0.83, (C18:0)-2.43, (C18:1)-28.33, (C18:2)-42.77, (C20:0)-4.58, (C22:0)-0.71; Sterols, in %: brassicasterols-97.19 and lanosterol-3.11	Henkrar and Khabar, 2022

Table 1. contd. ...

Table 1. contd. ...

Genus/Species	Country	Nutritional composition	Reference
T. boudieri	Egypt	Sugars - 92.70 µg/ml of extract: (monosaccharides: xylose and glucose; disaccharides: trehalose and turanose and sugar alcohols: mannitol, were predominant); major free amino acids (FAA): alanine and glutamic acid; major organic acids (OA): succinic and citric; major fatty acid (FA): ω-9 oleic acid	Farag et al., 2021
T. boudieri	Tunisia	**Fresh weight**: Moisture-78.23 g/100 g; **Dry weight**, in g/100 g: proteins-26.12, fat 8.01, carbohydrates-62.03, ash-4.49, ascorbic acid-12.20 and carotenoids-1.43; FA concentration, in %: hexadecanoic acid, methyl ester (C16:0)-20.40, 9,12-octadecadienoic acid, methyl ester (C18:2n-6)-54.10, 9-octadecenoic acid, methyl ester (C18:1 n-9)-22.0, octadecanoic acid, methyl ester (C18:0)-3.40; Mineral content, in mg/100 g: K -1512.60, Ca- 474.28, Mg- 126.19, Fe- 170.29, Na-26.89, P-322.70, Mn- 3.01, Cu- 0.92	Hamza et al., 2016a
T. boudieri	Israel	Dry matter, in %: proteins-20-27, fat-3-7.5, crude fiber-7-13, carbohydrates ~60, ascorbic acid-2-5; high levels of K, P, and fair amounts of Fe	Kagan-Zur and Roth-Bejerano, 2008
T. claveryi Chatin	Spain	Dry weight, in g/100g: carbohydrates-38.44, proteins-8.92, β-glucans-27.96, chitin-8.53; Sterols, in mg/g: ergosterol-2.30, brassicasterol-1.40 and ergosta7,22-dienol -1.16	Tejedor-Calvo et al., 2022
T. claveryi	Egypt	Sugars-58.71 µg/ml of extract: (monosaccharides: psicopyranose; disaccharides: trehalose and turanose and sugar alcohols: myo-inositol, were predominant); major free amino acids (FAA): glycine and proline; major organic acids (OA): succinic and citric; major fatty acid (FA): ω-9 oleic acid	Farag et al., 2021
T. claveryi	Iraq	Fatty acids: palmitic (saturated FA), oleic and linoleic acid (unsaturated FA) and sterols (campestanol, stigmasterol, β-sitosterol, lupeol)	Dahham et al., 2018
T. claveryi	Algeria	Fresh matter, in %: humidity-73, proteins-12.87, fats-3.9, sugars -6.4 and ash-5.5	Wahiba et al., 2016
T. claveryi	Spain	Dry matter, in g/kg: moisture-730.9, proteins-159.5, carbohydrates-645.5, lipids-69.5, fiber-83.2 and ash-42.5	Murcia et al., 2003
T. claveryi	Saudi Arabia	Fresh weight: total moisture-78%; Dry weight, in %: protein-16, carbohydrate-28	Bokhary and Sarwat, 1993

T. claveryi	Saudi Arabia	Fresh weight: moisture-75.44%; Dry weight, in %: protein-24.96, fat- 4.2, crude fibre-7.02, ash-6.39; ascorbic acid- 5.10 mg/100 g; Mineral composition, in mg/100g: Ca-129, Mg-104, P-756, Na-199, K-1730, Fe-10.68, Mn-0.48, Cu-1.69, Zn-5.10: Amino acid (AA) composition, in mg/100g: methionine+cystine-5.85, lysine- 4.13 and tryptophan-1.2	Sawaya et al., 1985
T. claveryi	Iraq	Dry matter, in %: protein-6.2, fats-2.5, fatty acids: linoleic 57-58% and palmitic 18-21%	Al-Shabibi et al.,1982
T. leptoderma (Tul. & C. Tul.) Tul. & C. Tul.	Morocco	Dry weight, in %: (C16:0) 16.08-17.46, (C16:1) 0.91-1.73, (C18:0) 1.03-1.60, (C18:1) 26.11-27.53, (C18:2) 48.51-48.63, (C20:0) 4.57- 5.75, (C22:0) 0.79-0.80; Sterols, in %: brassicasterols 92.41-96.87 and ergosterols 3.02-8.15	Henkrar and Khabar, 2022
Tirmania			
T. nivea (Desf.) Trappe	Tunisia	Fresh mass: moisture-77.63 g/100 g; Dry mass, in g/100 g: fat-6.78, proteins-28.81, carbohydrates-57.83, ash-5.06, ascorbic acid-10.63 mg/100 g and total carotenoids-1.17 mg/100 g; Mineral content, in mg/100 g: K-1263.12, Ca-427.51, Mg-366.12, Fe-217.05, Na-33.01, P-287.06, Mn-3.20 and Cu-0.77	Hamza et al., 2016b
T. nivea	Algeria	Fresh matter, in %: humidity-68, protein-11.97, fats-2.15, sugars-5.01 and ash-5.5	Wahiba et al., 2016
T. nivea	Saudi Arabia	Fresh weight: moisture-75.27%; Dry weight, in %: protein-27.18, fat-7.42, crude fibre-13.02, ash-5.40, ascorbic acid-1.56 g/100 g; Mineral composition, in mg/100 g: Ca-62, Mg-101, P-644, Na-110, K-1734, Fe-4.35, Mn-0.49, Cu-11.54, Zn-5.04; Amino acid (AA) composition, in g/100g: methionine+cystine : 4.52, lysine-5.7 and tryptophan-1.44	Sawaya et al., 1985
T. pinoyi (Maire) Malençon	Morocco	Fresh matter: moisture-81.5 g/100 g; Dry matter, in g/100g: total carbohydrate-64.74, crude protein-26.96, crude fat-3.01, ash-5.28, insoluble ash-0.0047 and dietary fiber (pectin - 8.96, cellulose-4.34); Elements in mg/kg: Cu-65.3, Zn-38.1 and Se-35.3	Bouatia et al., 2018
T. pinoyi	Libya	Dry matter, in g/100 g: carbohydrates-82.60, protein-8.06, fatty acids-300.65; The main FAs, in %: oleic-32.29, linoleic-29.72, and palmitic-27.97	Stojković et al., 2013
Picoa			
P. juniperi Vittadini	Spain	Fresh matter: moisture-637.8 g/kg; Dry matter, in g/kg: protein-225.4, lipids-199.4, carbohydrates-366.6, fibre-130.4, ash-82.1; Fatty acids (most abundant), in %: linoleic-52, oleic-23.5, palmitic-8	Murcia et al., 2003

non-oxidative metabolism of arachidonic acid (essential ω-6 fatty acid) (Patel et al., 2017). The chemical structures of some bioactive metabolites detected in truffles are shown in Fig. 2.

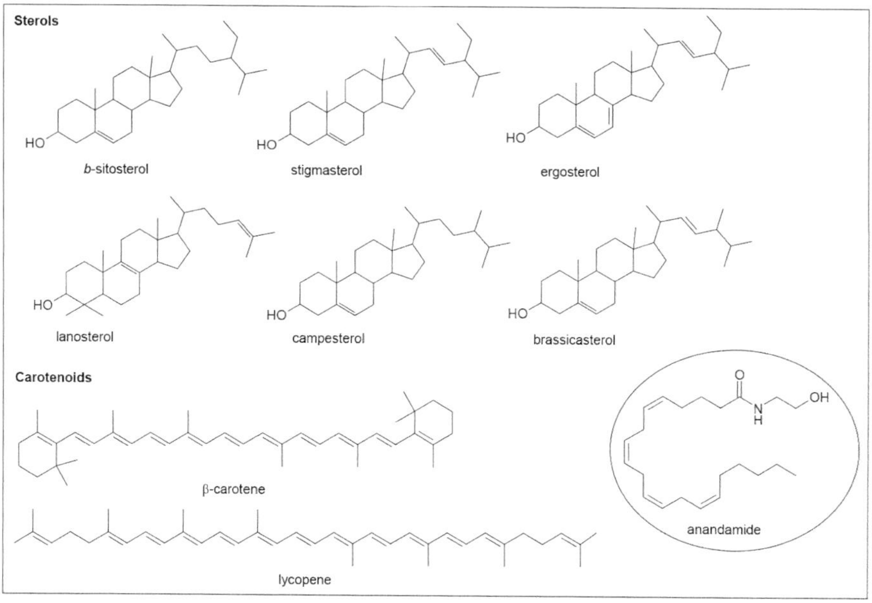

Fig. 2. Chemical structures of some bioactive metabolites in truffles.

3.1 Exopolysaccharides

One of the key ingredients of healthy food, which has emerged as a current research hotspot, is polysaccharide. Numerous biologically significant properties of the polysaccharides extracted from different truffle species have been demonstrated, including antioxidant (Zeng et al., 2020; Guan et al., 2022), antitumor (Zhao et al., 2014; Attia et al., 2018), hypoglycemic (Mudliyar et al., 2019), and immunomodulatory activity (Tejedor-Calvo et al., 2020). The biological function of polysaccharides is significantly influenced by their molecular weight (MW). High-molecular-weight (MW) polysaccharides may generally have better biological activity. High-MW polysaccharides are thought to have an effective binding affinity for immune cell carbohydrate receptors, which accounts for their increased activity. Low-MW polysaccharides have the capacity to enter immune cells and trigger an immune reaction (Gong et al., 2020). Moreover, the content of polysaccharides isolated from truffles cultivated in various locations differed. According to a study conducted by Li and collaborators, the highest polysaccharide content was found in Tibetan truffles, which were followed by those from Hebei, Liaoning, Gongshan, and Qujing (Li et al., 2019b). Microorganisms generate a wide variety of polysaccharides with various purposes, which could be intracellular polysaccharides (that accumulate inside of cells) or extracellular polysaccharides (exopolysaccharides) (Chaisuwan et al., 2020).

Exopolysaccharides (EPSs) could be defined as «extracellular macromolecules that microorganisms expel as a loosely attached slime layer or as a tightly bound capsule». They are particularly important in the fight against osmotic stress, antibiotics, toxic compounds, desiccation, phagocytosis, cell recognition, and phage attack (Angelin and Kavitha, 2020). The Tang et al. (2015) group has improved the submerged fermentation of truffles to promote not only truffle biomass but also the amount of extracellular and intracellular polysaccharides produced from truffles. A study was conducted to investigate the best culture conditions and mechanisms to enhance the production of EPS by *T. melanosporum* using *Dioscorea* saponin. The ideal parameters provided for the maximum production of EPS (9.853 g/l) include: inoculum ratio, 7 ml/100 ml medium; culture temperature, 20°C; rotary speed, 139 rpm. After the addition of *Dioscorea*, the EPS content in the basal fermentation medium increased to 2.4467 g/l, which was 38.49% higher compared to the control (1.7667 g/l) (Zhang et al., 2019).

In a recent study, in order to increase the production of EPSs and mycelial biomass, submerged fermentation was used to cultivate *T. borchii*. In order to maximize mycelial biomass and polysaccharides, the best concentrations of carbon and nitrogen sources were sucrose (80 g/l) and yeast extract (20 g/l), respectively. When the concentration of yeast extract was increased to 20 g/l, the molecular weight distribution of the EPSs, as determined by the Gel Permeation Chromatography (GPC) method, shifted to the high-molecular-weight range. On the 28th day of submerged fermentation, the EPS reached its highest average molecular weight of 92 kDa when it was extracted by NaOH at a yeast extract concentration of 20 g/l. The results exhibited high-molecular-weight EPSs, which were identified as β-(1-3)-glucan after elucidating the structure by Fourier-transform infrared spectroscopy (FTIR) (Chen et al., 2023). There haven›t been many studies done on exopolysaccharides from truffles; more research on extracting these polysaccharides and creating techniques to increase their production is strongly recommended.

3.2 Pigments

Many microorganisms, including bacteria, yeast, algae, actinomycetes, and fungi, offer a readily accessible source of pigments that frequently serve to imitate and accentuate the beauty of the related life form (Malik et al., 2012; Narsing Rao et al., 2017). Examples of these pigments include astaxanthin, canthaxanthin, zeaxanthin, indigoidine, prodigiosin, violacein, pyocyanin, xanthomonadin, β-carotene, lutein, lycopene, ankaflavin, anthraquinone, riboflavin, torularhodin, and prodigiosin, including rubrolone (Malik et al., 2012). Pigments possess a wide variety of biological activities, including anticancer (El-Naggar and El-Ewasy, 2017), antimicrobial (Vendruscolo et al., 2014), antioxidant (Correa-Llantén et al., 2012), anti-bacterial (Umadevi and Krishnaveni, 2013), and antifungal (Jayaseelan et al., 2014) properties. The pigments of the ectomycorrhizal fungi «truffles» may originate from the truffles themselves or from other microorganisms that exist in a symbiotic relationship with truffles. The bright orange appearance of the carrot truffle, *Stephanospora caroticolor*, could be easily noticed when soil has been partially removed. The mushroom interior part is likewise intensely orange in color. Two pigments were found using high-performance liquid chromatography (HPLC) in the yellow solution obtained from the methanol extraction of the truffle fruit bodies. Stephanosporin was identified as Compound 1 using NMR and LC-ESI/MS. Using GC/MS, Compound 2 was identified to be 2-chloro-4-nitrophenol.

Atomic absorption spectroscopy (AAS) and UV/Vis spectrometry analysis of the extract reveal that 2-chloro-4-nitrophenol primarily exists as its potassium salt, which accounts for the yellow color of the mixture (Lang et al., 2001). Melanocrocin is a polyene pigment that was extracted from the truffle-like, reddish-brown fungus *Melanogaster broomeianus*. It was found to be responsible for the brilliant orange color of the fungus extract (Aulinger et al., 2001). A study was done to look for the presence of bacteria that produce pigments in the reddish pigmentation that commonly appears in *T. magnatum* ascomata. It was assumed that carotenoid-producing species like *Microbacterium* and *Chryseobacterium* were most likely to be the source of the unusual red pigmentation. The LC-MS analysis of the pigmented part of *T. magnatum* carpophores displayed a m/z range of 380–680 in the ESI-MS spectrum. The MS spectrum revealed four peaks, each of which indicated the presence of one of the following four pigments: fucoxanthin ($[M+H]^+$ 659.48 m/z), β-carotene ($[M+H]^+$ 537.23 m/z), *cis*-1-glycosyl-apo-8›-lycopene ($[M+H]^+$ 581.25 m/z), and bixin ($[M+H]^+$ 396.31 m/z) (Amicucci et al., 2018). The pigment profile of the truffles needs more chemical and biological investigations. The chemical structures of some pigments found in truffles are shown in Fig. 3.

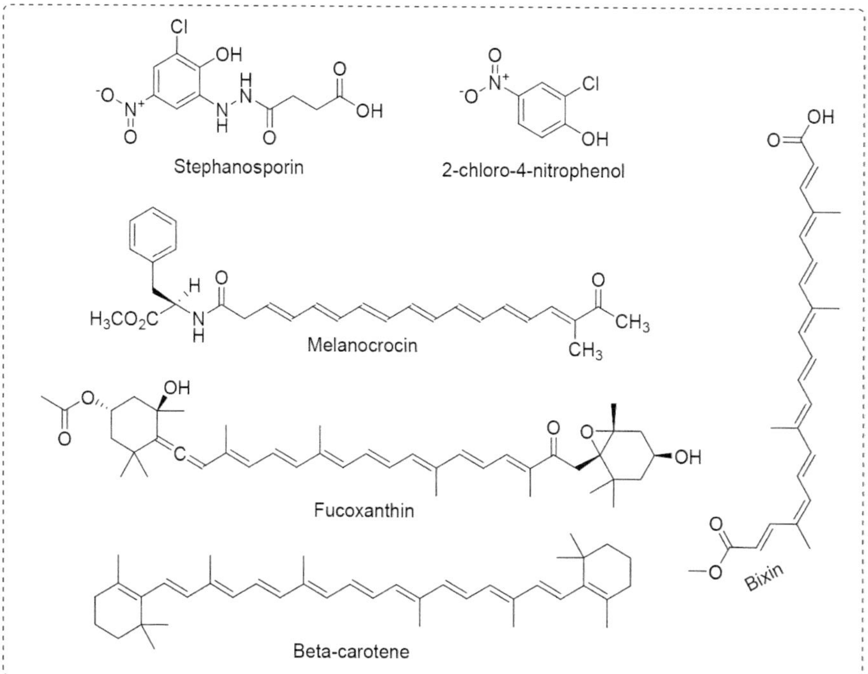

Fig. 3. Chemical structures of some pigments detected in truffles.

3.3 Aroma and Volatiles

Approximately 180 truffle species are thought to exist worldwide, of which roughly 30 are traded for a profit. They are among the priciest foods in the world due to their scarcity and distinctive aroma. The *T. magnatum* (Piedmont white truffle) and the *T. melanosporum* (Périgord black truffle) are examples of the most expensive truffle species, can cost up to €1,000/kg (Vahdatzadeh and Splivallo, 2018). The aroma of truffles can be subtle or strong and varies between different odors of cheese, pungent spices, earth, vanilla, garlic, dust, gasoline, and creamy odor (Khalifa et al., 2019; Patel et al., 2017). The complex aroma profile of truffles was found to be influenced by different environmental factors like climate and microbial populations of fruit bodies; genotypic variability; geographic origin, and maturity stage (Splivallo et al., 2011; Strojnik et al., 2020). Around the world, the truffle fruiting bodies of more than 200 different species have been found to contain volatile organic compounds (VOCs). As extraction methods become more sophisticated, this number is expected to rise (Choo et al., 2021). In a recent study, the aroma profiles of different truffle species were investigated and revealed that truffles belonging to the same species may differ in their aromatic profiles, but they all share similar volatiles that can be used as a species› fingerprint (Strojnik et al., 2020).

Table 2 contains a review of literature reports on numerous volatiles presented in different truffle species, including *Tuber aestivum*, *T. brumale*, *T. dicum*, *T. excavatum*, *T. macrosporum*, *T. magnatum*, *T. melanosprum*, *T. mesentericum*, *T. rufum*, and *T. uncinatum*. These truffles were from different origins, as analyzed using multiple tools, where GC-MS was the most dominant. The aroma of these volatiles varied between sulfurous, alliaceous, creamy, acidic, alcoholic, malty, fruity, narcissus, medicinal, garlic, buttery, ethereal, alcoholic, smoky, sweaty, mushroom-like, chocolate, roasty-like, metallic, popcorn, earthy, moldy, flower-like, and many other odors. The volatiles found in truffles belonged to different classes and had a wide range of chemical structures. The major volatile classes were hydrocarbons, sulfur compounds, phenols, aromatic hydrocarbons, and terpenes.

4. Truffle Production and Quality Requirements

In recent decades, there has been a global interest in the cultivation of truffles from the *Tuber* genus. Orchards are mostly established in southern and central Europe (Oliach et al., 2021; Lemmond et al., 2023), but also in non-traditional areas and countries (Australia, Chile, China, Greece, New Zealand, Morocco, South Africa, Turkey, and the United States) (Oliach et al., 2021; Rosa-Gruszecka et al., 2017). The species primarily cultivated in European orchards are *T. aestivum, T. borchii*, *T. magnatum*, and *T. melanosporum* due to their long-standing tradition in culinary use in these regions (Lemmond et al., 2023).

Truffles have become an emerging ‹global industry› in several countries, including France, Italy, Spain, and Australia, generating millions of Euros annually. In addition to economic benefits, the production of black truffles offers advantages, mainly truffle tourism, the production of inoculated seedlings, and a variety of

Table 2. Volatile organic compounds of different truffle species with their aroma and chemical structures: *Tuber aestivum* (TAE); *Tuber magnatum* (TMA); *Tuber brumale* (TBR); *Tuber melanosprum* (TME); *Tuber macrosporum* (TMC); *Tuber mesentericum* (TMS); *Tuber rufum* (TRF); *Tuber dicum* (TDI); *Tuber excavatum* (TEX); *Tuber uncinatum* (TUN); Headspace solid-phase microextraction gas chromatography-mass spectrometry (HS-SPME/GC-MS); Headspace-gas chromatography-mass spectrometry/ Aroma Dilution Analysis (HS-GC-MS)/ADA); Gas chromatography-olfactometry (GC-O); Proton-transfer-reaction time-of-flight mass spectrometer (PTR-TOF-MS).

Truffle species	Volatile compound	Aroma description	Identification method	Molecular weight (g/mol)	Chemical structure	Reference
TAE, TBR, TMA	Sulfide, allyl methyl-	Sulfurous	HS-SPME/GC-MS, PTR-TOF-MS	88.17	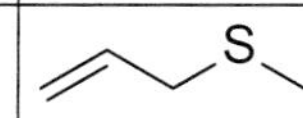	Strojnik et al., 2020; Vita et al., 2015
TAE, TBR, TMC	Propane, 1-(methylthio)-	Alliaceous creamy green leek	HS-SPME/GC-MS	90.19		Strojnik et al., 2020
TAE, TMA, TME	Anisole	Phenolic gasoline ethereal anise aromatic odor	HS-SPME/GC-MS, PTR-TOF-MS	108.14	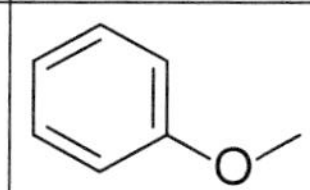	Strojnik et al., 2020; Vita et al., 2015
TAE, TME, TMA, TUN	Butanoic acid, 2-methyl-	Fruity sweet, sweaty	HS-SPME/GC-MS, HS-SPME/GC-MS, HS-GC-MS/ADA	102.13	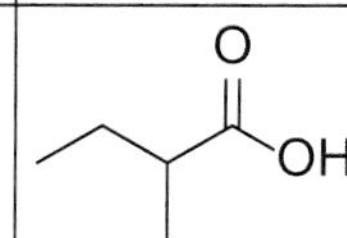	Strojnik et al., 2020; Schmidberger and Schieberle, 2017
TAE, TME, TMA	Acetic acid	Acidic, vinegar-like	HS-SPME/GC-MS, HS-GC-MS/ADA, SPME/GC-MS and GC-O	60.05	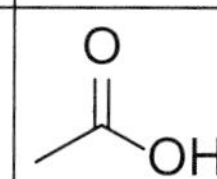	Strojnik et al., 2020; Schmidberger & Schie-berle, 2017; Piloni et al., 2005
TMA, TAE, TME	Ethanol	Alcoholic	HS-SPME/GC-MS, SPME/GC-MS and GC-O, PTR-TOF-MS	46.07	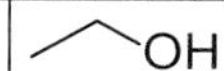	Strojnik et al., 2020; Piloni et al., 2005; Vita et al., 2015
TBR, TME, TMC	3-Octanol	Earthy mush-room herbal melon cit-rus woody spicy minty	HS-SPME/GC-MS	130.229	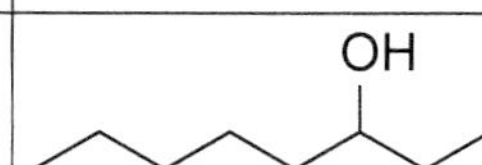	Strojnik et al., 2020

TBR	2-Nonanone	Fruity	HS-SPME/GC-MS	142.24	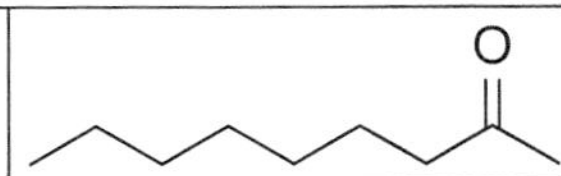	Strojnik et al., 2020
TMS, TMA, TBR, TEX	Benzene, 1,4-dimethoxy-	Sweet green hay newly mown hay fennel	HS-SPME/GC-MS	138.16	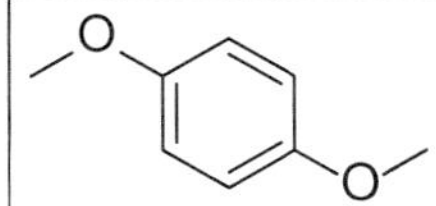	Strojnik et al., 202
TME, TMS	Benzene, 1,3-dimethoxy-	Medicinal	HS-SPME/GC-MS	138.16	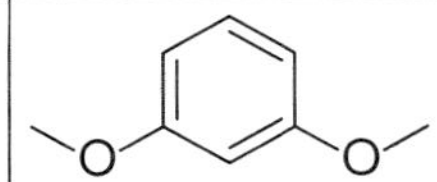	Strojnik et al., 2020
TME, TMA, TBR, TMC, TMS	Benzene, 1-methoxy-3-methyl-	Narcissus	HS-SPME/GC-MS, SPME/GC-MS and GC-O	122.16	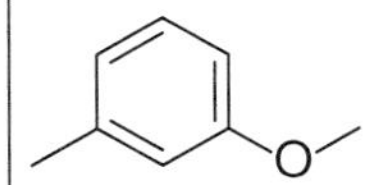	Strojnik et al., 2020; Piloni et al., 2005
TBR, TMS	Benzene, 1,4-dimethoxy-2-methyl-	–	HS-SPME/GC-MS	152.19	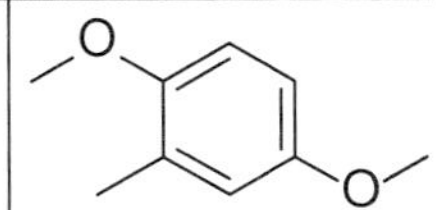	Strojnik et al., 2020
TBR	3-Ethylphenol, methyl ether	–	HS-SPME/GC-MS	136.19	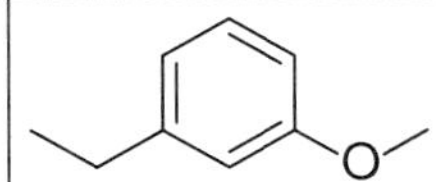	Strojnik et al., 2020
TBR, TME, TDI	Formic acid, 1-methylpropyl ester	–	HS-SPME/GC-MS	102.13	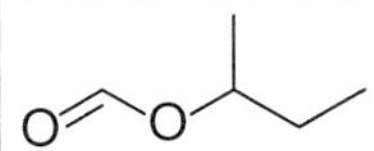	Strojnik et al., 2020
TBR	Formic acid, 1-methylethyl ester	Sweet-ethereal	HS-SPME/GC-MS	88.11	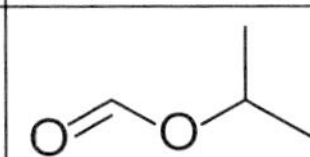	Strojnik et al., 2020

Table 2. contd. ...

Table 2. contd. ...

Truffle species	Volatile compound	Aroma description	Identification method	Molecular weight (g/mol)	Chemical structure	Reference
TBR, TMS, TMA, TAE, TUN	1-Octen-3-ol	Earthy, mushroom-like	HS-SPME/GC-MS, HS-GC-MS/ADA	128.21	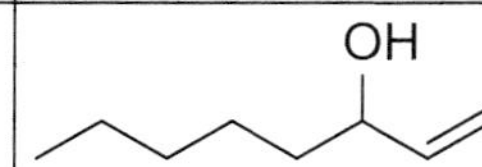	Strojnik et al., 2020; Schmidberger and Schieberle, 2017
TBR, TMA, TAE, TMC	3-Octanone	Fresh herbal lavender sweet mushroom	HS-SPME/GC-MS	128.21	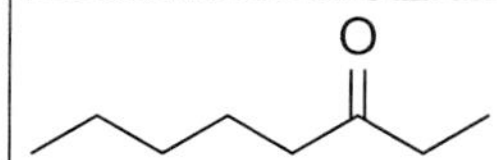	Strojnik et al., 2020
TAE, TME, TMA	2-Butanone	Acetone ethereal fruity camphoraceous	HS-SPME/GC-MS, PTR-TOF-MS	72.11	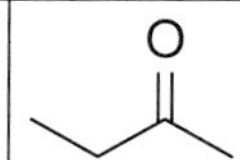	Strojnik et al., 2020; Vita et al., 2015
TMC, TEX	1-Propene, 1-(methylthio)-, (*E*)-	Acrid strong garlic-like odour	HS-SPME/GC-MS	88.17	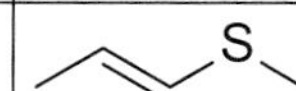	Strojnik et al., 2020
TEX	1-Propene, 1-(methylthio)-, (*Z*)-	Acrid strong garlic-like odour	HS-SPME/GC-MS	88.17	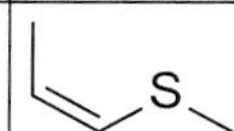	Strojnik et al., 2020
TMA, TME, TUN	Butanal, 2-methyl-	Chocolate, malty	HS-SPME/GC-MS, HS-GC-MS/ADA, PTR-TOF-MS	86.13	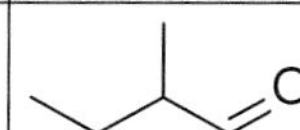	Strojnik et al., 2020; Schmidberger & Schieberle, 2017; Vita et al., 2015
TMA	Benzaldehyde	Fruity	HS-SPME/GC-MS	106.12	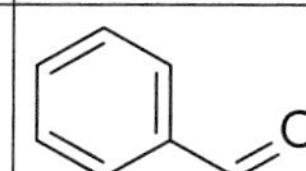	Strojnik et al., 2020
TMA, TME, TUN	1-Butanol, 2-methyl-	Roasted winey onion fruity fusel alcoholic whiskey, malty	HS-SPME/GC-MS, HS-SPME/GC-MS, HS-GC-MS/ADA	88.15	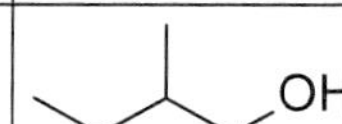	Strojnik et al., 2020; Schmidberger and Schieberle, 2017; Piloni et al., 2005
TUN	3-methylbutanol	Malty	HS-SPME/GC-MS, HS-GC-MS/ADA	88.15	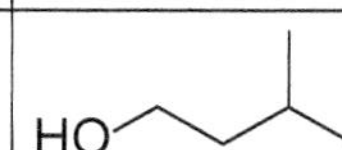	Schmidberger and Schieberle, 2017

TMA, TBR, TAE, TEX, TUN	Dimethyl sulfide	Sulfurous onion sweet corn vegetable cabbage tomato green radish, cabbage-like, broccoli, cauliflower, truffle	HS-SPME/GC-MS, HS-GC-MS/ADA, SPME/GC-MS and GC-O, PTR-TOF-MS	62.14	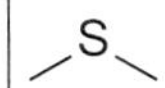	Strojnik et al., 2020; Schmidberger and Schieberle, 2017; Piloni et al., 2005; Vita et al., 2015
TME, TMA, TUN	Butanal, 3-methyl-	Ethereal aldehydic chocolate peach fatty, malty	HS-SPME/GC-MS, HS-GC-MS/ADA	86.13	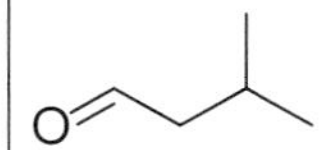	Strojnik et al., 2020; Schmidberger & Schie-berle, 2017
TME	1-Hexanol	Ethereal fusel oily fruity alcoholic sweet green	HS-SPME/GC-MS	102.17		Strojnik et al., 2020
TME, TUN, TMA	1-Propanol, 2-methyl-	Ethereal, malty	HS-SPME/GC-MS, HS-GC-MS/ADA, SPME/GC-MS and GC-O	74.12	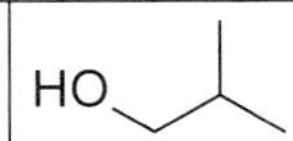	Strojnik et al., 2020; Schmidberger and Schieberle, 2017; Piloni et al., 2005
TUN, TMA	3-hydroxy-2-butanone/ acetoin	Buttery	HS-SPME/GC-MS, HS-GC-MS/ADA, PTR-TOF-MS	88.11	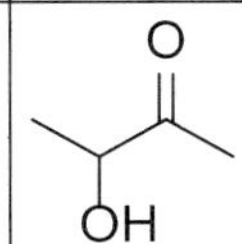	Schmidberger and-Schieberle, 2017; Vita et al., 2015
TME, TRF, TMA	Acetone	Solvent ethereal apple pear	HS-SPME/GC-MS, PTR-TOF-MS	58.08	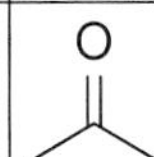	Strojnik et al., 2020; Vita et al., 2015
TME	1-butanol, 2-methyl acetate	Fruity	HS-SPME/GC-MS	130.18	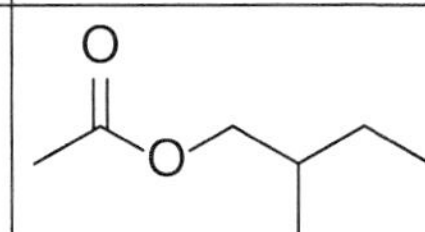	Strojnik et al., 2020

Table 2. contd. ...

Table 2. contd. ...

Truffle species	Volatile compound	Aroma description	Identification method	Molecular weight (g/mol)	Chemical structure	Reference
TAE, TME, TMA, TUN	Propanal, 2-methyl	Fresh aldehydic floral pungent, malty	HS-SPME/GC-MS, HS-GC-MS/ADA	72.11	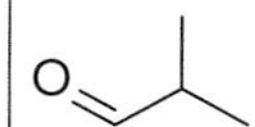	Strojnik et al., 2020; Schmidberger & Schieberle, 2017
TMA	2,4-Dithiapentane/ bis(methylthio)methane	Alliaceous sulfurous fresh onion garlic vegetable cabbage spicy mustard horseradish, garlic-like, sulfur, very strong	HS-SPME/GC-MS, HS-GC-MS/ADA, SPME/GC-MS and GC-O	108.23	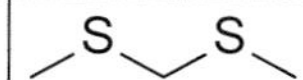	Strojnik et al., 2020; Schmidberger and Schieberle, 2017; Piloni et al., 2005
TRF	Butanenitrile, 2-methyl-	–	HS-SPME/GC-MS	83.13	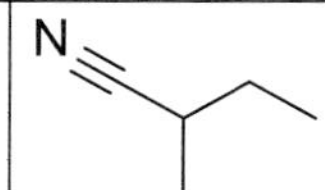	Strojnik et al., 2020
TRF	Pentane, 2-nitro-	-	HS-SPME/GC-MS	117.15	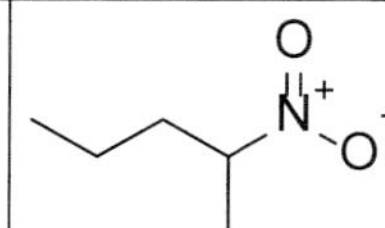	Strojnik et al., 2020
TDI	Formic acid, 2- methyl-propyl ester	-	HS-SPME/GC-MS	102.13	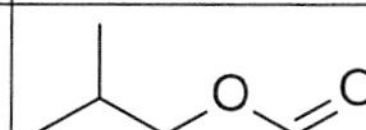	Strojnik et al., 2020
TMA, TUN	2,3-butanedione	Buttery	HS-GC-MS/ADA	86.09	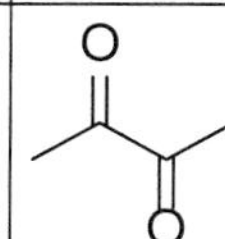	Schmidberger and Schieberle, 2017
TMA, TUN	3,4-dihydro-2*H*-pyrrole	Amine-like, sperm-like	HS-GC-MS/ADA	69.11	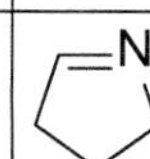	Schmidberger and Schieberle, 2017

TUN, TM	2,3-pentanedione	Buttery	HS-GC-MS/ADA, PTR-TOF-MS	100.12	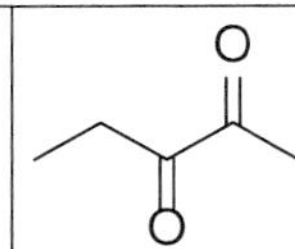	Schmidberger and Schieberle, 2017; Vita et al., 2015
TMA, TUN	1-octen-3-one	Mushroom-like	HS-GC-MS/ADA	126.20	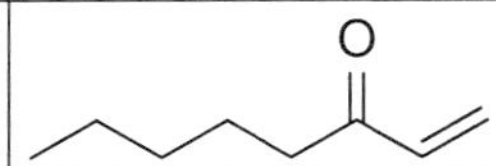	Schmidberger and Schieberle, 2017
TMA, TUN	2-acetyl-1-pyrroline	Popcorn, roasty-like	HS-GC-MS/ADA	111.14	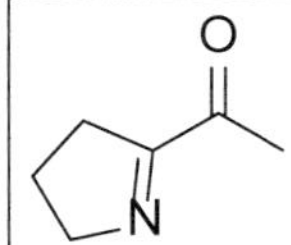	Schmidberger and Schieberle, 2017
TMA, TUN	Dimethyl trisulfide	Cabbage-like, rotten, cooked turnip	HS-GC-MS/ADA, SPME/GC-MS and GC-O	126.25	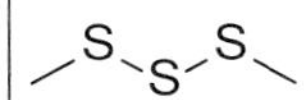	Schmidberger and Schieberle, 2017; Piloni et al., 2005
TMA	(*Z*)-1,5-octadien-3-one	Geranium-like, metallic	HS-GC-MS/ADA	124.18	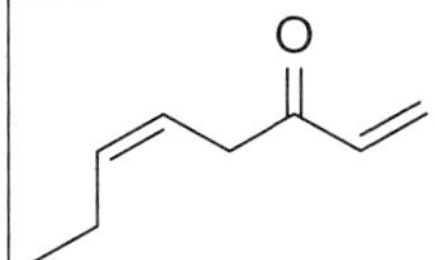	Schmidberger and Schieberle, 2017
TMA	2-propionyl-1-pyrroline	Popcorn, roasty-like	HS-GC-MS/ADA	125.17	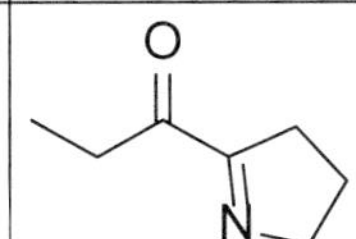	Schmidberger and Schieberle, 2017
TMA, TUN	(*E*)-2-octenal	fatty, nutty	HS-GC-MS/ADA	126.20	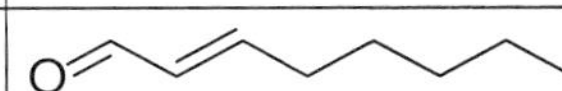	Schmidberger and Schieberle, 2017
TMA, TUN	3-(methylthio)propanal	Cooked potato-like	HS-GC-MS/ADA	104.17		Schmidberger and Schieberle, 2017

Table 2. contd. ...

Table 2. contd. ...

Truffle species	Volatile compound	Aroma description	Identification method	Molecular weight (g/mol)	Chemical structure	Reference
TMA	2,3-diethyl-5-methylpyrazine	Earthy	HS-GC-MS/ADA	150.22	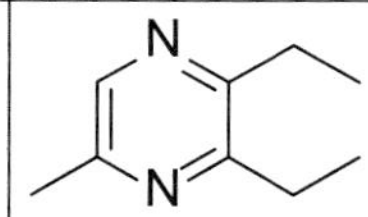	Schmidberger and Schieberle, 2017
TMA, TUN	Propanoic acid	Sweaty	HS-GC-MS/ADA	74.08	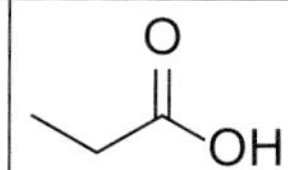	Schmidberger and Schieberle, 2017
TUN	(*E*)-2-nonenal	Fatty, green	HS-GC-MS/ADA	104.23		Schmidberger and Schieberle, 2017
TUN	3,5-dimethyl-2-ethenylpyrazine	Earthy	HS-GC-MS/ADA	134.18	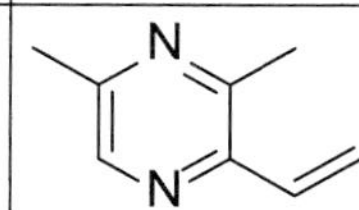	Schmidberger and Schieberle, 2017
TUN	2-methylpropanoic acid	Sweaty	HS-GC-MS/ADA	88.11	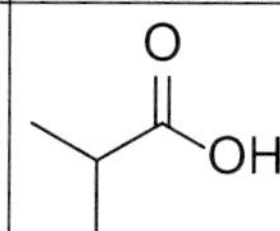	Schmidberger and Schieberle, 2017
TUN	2-ethenyl-3-ethyl-5-methylpyrazine	Moldy, earthy	HS-GC-MS/ADA	148.21	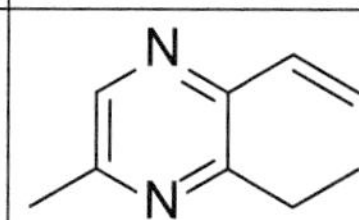	Schmidberger and Schieberle, 2017
TUN	Phenylacetaldehyde	Flower-like	HS-GC-MS/ADA	120.15	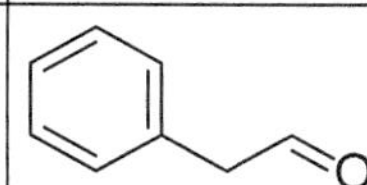	Schmidberger aand Schieberle, 2017

TMA, TUN	3-methylbutanoic acid	Sweaty	HS-GC-MS/ADA	102.13	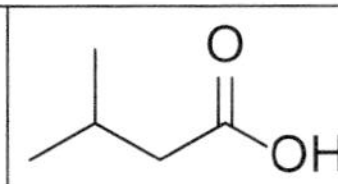	Schmidberger and Schieberle, 2017
TMA, TUN	(*E*,*E*)-2,4-nonadienal	Fatty, green	HS-GC-MS/ADA	138.21	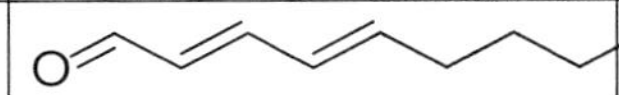	Schmidberger and Schieberle, 2017
TMA	3-(methylthio)propanol	Cooked potato-like	HS-GC-MS/ADA	106.18		Schmidberger and Schieberle, 2017
TMA	1,2,4-trithiolane	Sulfury, onion-like	HS-GC-MS/ADA	124.23	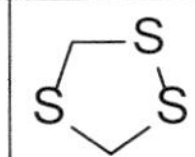	Schmidberger and Schieberle, 2017
TMA	3-methylnonan-2,4-di-one	Anise-like, hay-like, fishy	HS-GC-MS/ADA	170.25	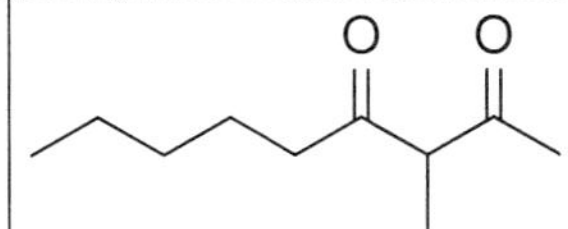	Schmidberger and Schieberle, 2017
TUN	Pentanoic acid	Sweaty	HS-GC-MS/ADA	102.13	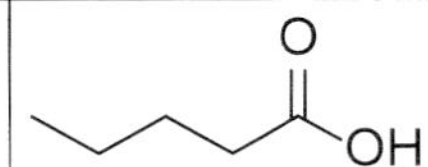	Schmidberger and Schieberle, 2017
TUN	(*E*,*E*)-2,4-decadienal	Fatty, fried	HS-GC-MS/ADA	152.24	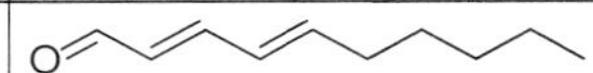	Schmidberger and Schieberle, 2017
TUN	Geosmine	Beet-like, moldy	HS-GC-MS/ADA	182.31	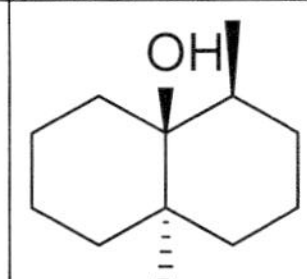	Schmidberger and Schieberle, 2017

Table 2. contd. ...

Table 2. contd. ...

Truffle species	Volatile compound	Aroma description	Identification method	Molecular weight (g/mol)	Chemical structure	Reference
TUN	2-methoxyphenol	Smoky	HS-GC-MS/ADA	124.14	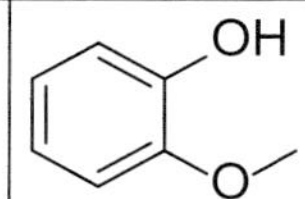	Schmidberger and Schieberle, 2017
TMA, TUN	4-methyl-2-methoxy-phenol	Vanilla-like, clove-like, smoky	HS-GC-MS/ADA	138.17	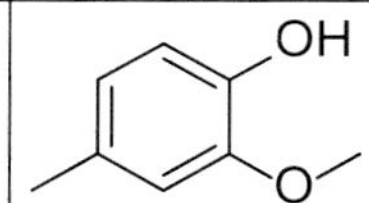	Schmidberger and Schieberle, 2017
TMA, TUN	*trans*-4,5-epoxy-(*E*)-2-decenal	Metallic	HS-GC-MS/ADA	168.24	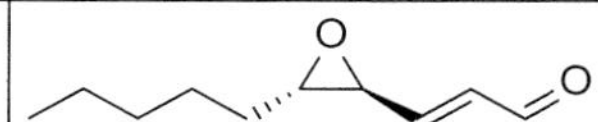	Schmidberger and Schieberle, 2017
TUN	Gamma-nonalactone	Coconut-like	HS-GC-MS/ADA	156.23	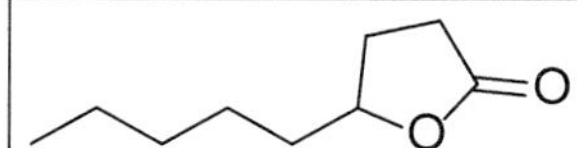	Schmidberger and Schieberle, 2017
TMA, TUN	4-hydroxy-2,5-dimeth-yl-3(2*H*)-furanone	Caramel-like	HS-GC-MS/ADA	128.13	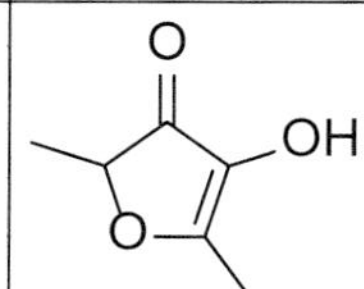	Schmidberger and Schieberle, 2017
TMA, TUN	Delta-nonalactone	Coconut-like	HS-GC-MS/ADA	156.23	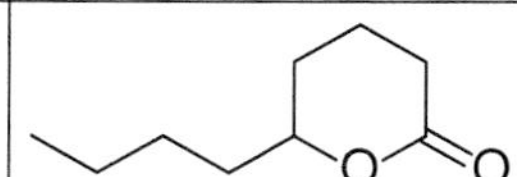	Schmidberger and Schieberle, 2017
TMA	2-propyl-4-methoxy-phenol	Clove-like, phenolic	HS-GC-MS/ADA	166.22	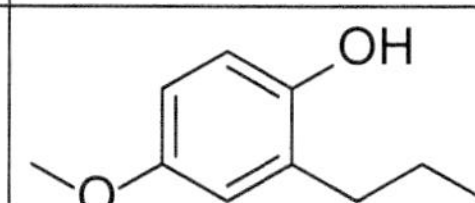	Schmidberger and Schieberle, 2017

TMA, TUN	3-hydroxy-4,5-dimeth-yl-2(5*H*)-furanone	Seasoning-like	HS-GC-MS/ADA	128.13	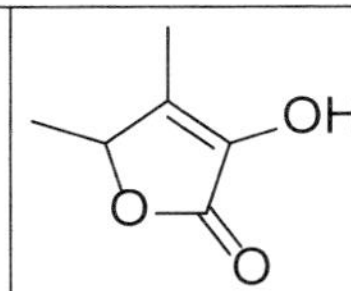	Schmidberger and Schieberle, 2017
TMA	3-propylphenol	Phenolic, leather-like	HS-GC-MS/ADA	136.19	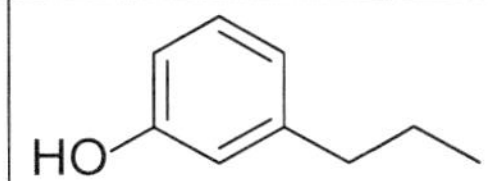	Schmidberger and Schieberle, 2017
TUN	2,6-dimethoxyphenol	Smoky, sweet, clove-like 2	HS-GC-MS/ADA	154.17	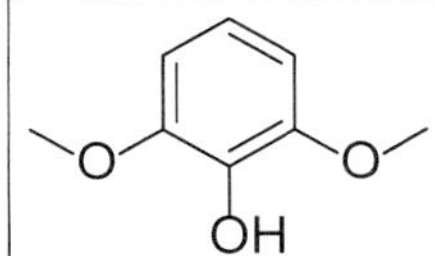	Schmidberger and Schieberle, 2017
TMA, TUN	Delta-dodecalactone	Peach-like	HS-GC-MS/ADA	198.31	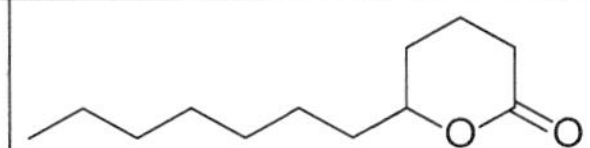	Schmidberger and Schieberle, 2017
TUN	4-vinylphenol	Phenolic	HS-GC-MS/ADA	120.15	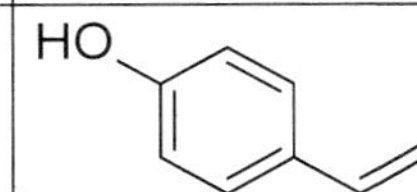	Schmidberger and Schieberle, 2017
TMA, TUN	Phenylacetic acid	Honey-like	HS-GC-MS/ADA	136.15	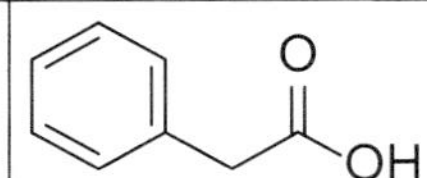	Schmidberger and Schieberle, 2017
TMA, TUN	Vanillin	Vanilla-like	HS-GC-MS/ADA	152.15	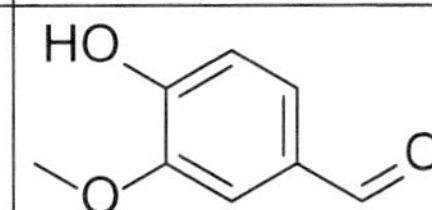	Schmidberger and Schieberle, 2017

Table 2. contd. ...

Table 2. contd. ...

Truffle species	Volatile compound	Aroma description	Identification method	Molecular weight (g/mol)	Chemical structure	Reference
TMA, TUN	Hydrogen sulfide	Rotten	HS-GC-MS/ADA, PTR-TOF-MS	34.08	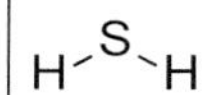	Schmidberger and Schieberle, 2017; Vita et al., 2015
TMA, TUN	Acetaldehyde	Green, fruity	HS-GC-MS/ADA, PTR-TOF-MS	44.05	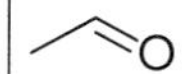	Schmidberger and Schieberle, 2017; Vita et al., 2015
TMA, TUN	Methanethiol	Cabbage-like	HS-GC-MS/ADA, PTR-TOF-MS	48.10	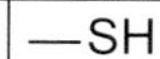	Schmidberger and Schieberle, 2017; Vita et al., 2015
TMA, TUN	Ethyl 2-methylpropa-noate	Fruity	HS-GC-MS/ADA	116.16	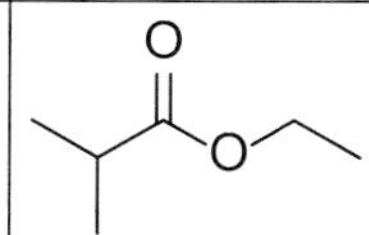	Schmidberger and Schieberle, 2017
TMA, TUN	Methyl 2-methylbuta-noate	Fruity	HS-GC-MS/ADA	116.16	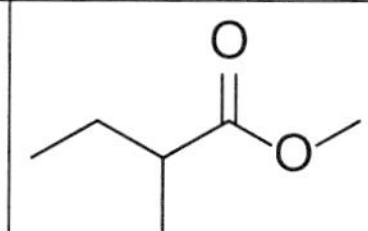	Schmidberger and Schieberle, 2017
TMA, TUN	Hexanal	Grassy, green	HS-GC-MS/ADA	100.16	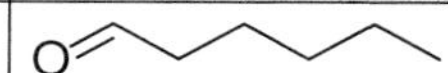	Schmidberger and Schieberle, 2017
TMA, TUN	Ethyl 3-methylbutanoate	Fruity	HS-GC-MS/ADA	130.19	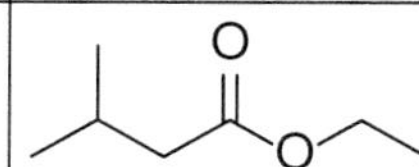	Schmidberger and Schieberle, 2017
TMA	Dimethyl disulphide	-	SPME/GC-MS and GC-O	94.19	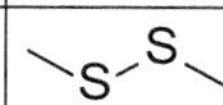	Piloni et al., 2005

TMA	Limonene	-	SPME/GC-MS and GC-O	136.24	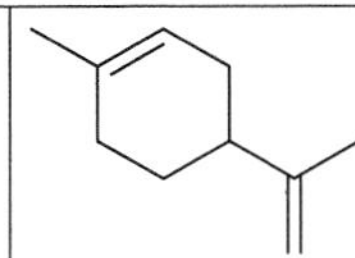	Piloni et al., 2005
TMA	2-acetyl-5-methylfuran	-	SPME/GC-MS and GC-O	124.14	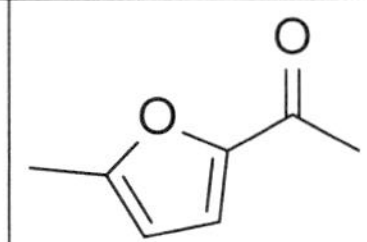	Piloni et al., 2005
TMA	Linalol	-	SPME/GC-MS and GC-O	154.25	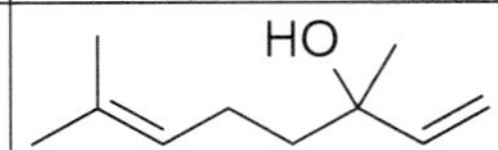	Piloni et al., 2005
TMA	Dimethyl sulfoxide	-	SPME/GC-MS and GC-O, PTR-TOF-MS	78.13	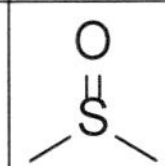	Piloni et al., 2005; Vita et al., 2015
TMA	3-ethyl-4-methyl-3-pent-en-2-one	-	SPME/GC-MS and GC-O	126.20	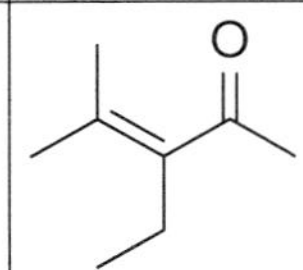	Piloni et al., 2005
TMA	3,4-dimethyl-3-hexen-2-one	-	SPME/GC-MS and GC-O	126.20	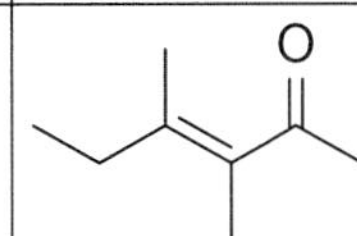	Piloni et al., 2005
TMA	Methyl(methylthio) methyl disulfide	-	SPME/GC-MS and GC-O, PTR-TOF-MS	140.28	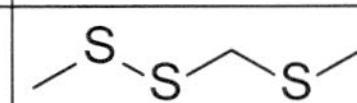	Piloni et al., 2005; Vita et al., 2015

Table 2. contd. ...

Table 2. contd. ...

Truffle species	Volatile compound	Aroma description	Identification method	Molecular weight (g/mol)	Chemical structure	Reference
TMA	Tris(methylthio)methane	Sulfur, garlic	SPME/GC-MS and GC-O	154.30	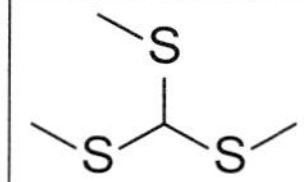	Piloni et al., 2005
TMA	Dimethyl sulfone	-	SPME/GC-MS and GC-O, PTR-TOF-MS	94.13	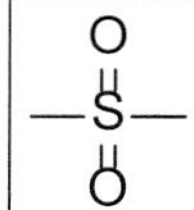	Piloni et al., 2005; Vita et al., 2015
TMA	Phenylethyl alcohol	-	SPME/GC-MS and GC-O	122.17	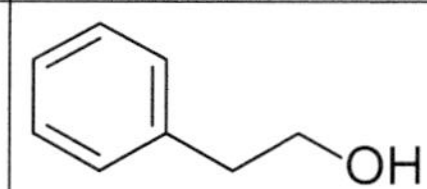	Piloni et al., 2005
TMA	(methylsulfinyl)(methylthio)methane	Sulfur	SPME/GC-MS and GC-O, PTR-TOF-MS	124.22	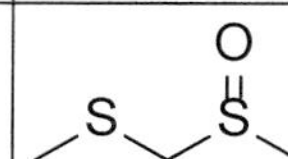	Piloni et al., 2005; Vita et al., 2015
TMA	Acetylene	-	PTR-TOF-MS	26.04	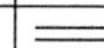	Vita et al., 2015
TMA	Formaldehyde	-	PTR-TOF-MS	30.03		Vita et al., 2015
TMA	Methanol	-	PTR-TOF-MS	32.04		Vita et al., 2015
TMA	Acetonitrile	-	PTR-TOF-MS	41.05	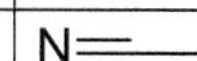	Vita et al., 2015
TMA	Formic acid	-	PTR-TOF-MS	46.03	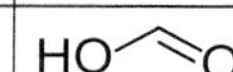	Vita et al., 2015
TMA	Cyclobutadiene	-	PTR-TOF-MS	52.08	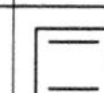	Vita et al., 2015
TMA	Methanethial	-	PTR-TOF-MS	46.09		Vita et al., 2015
TMA	2-Propenal	-	PTR-TOF-MS	56.06	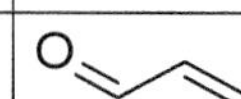	Vita et al., 2015

TMA	1-Butene	-	PTR-TOF-MS	56.11	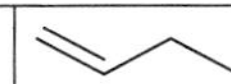	Vita et al., 2015
TMA	Methanetriol	-	PTR-TOF-MS	64.04	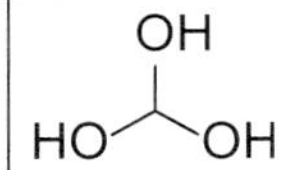	Vita et al., 2015
TMA	3-Penten-1-yne	-	PTR-TOF-MS	66.10	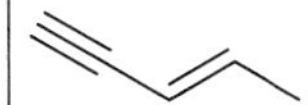	Vita et al., 2015
TMA	Pyrrole	-	PTR-TOF-MS	67.09	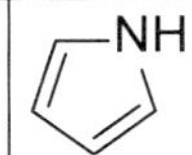	Vita et al., 2015
TMA	Furan	-	PTR-TOF-MS	68.08	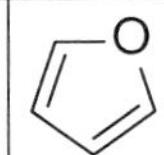	Vita et al., 2015
TMA	2-Methyl-1,3- butadiene/ isoprene	-	PTR-TOF-MS	68.12	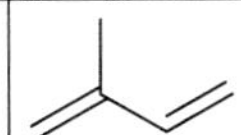	Vita et al., 2015
TMA	Acrylic Acid	-	PTR-TOF-MS	72.06	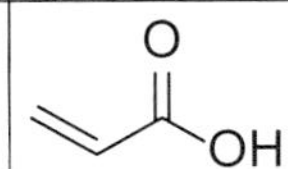	Vita et al., 2015
TMA	Dimethylformamide	-	PTR-TOF-MS	73.10	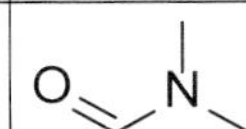	Vita et al., 2015
TMA	2-Methyl-1-propanol/ isobutanol	-	PTR-TOF-MS	74.12	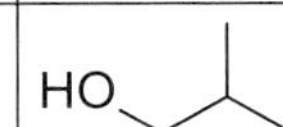	Vita et al., 2015
TMA	1-Propanethiol	-	PTR-TOF-MS	76.16	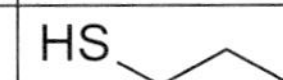	Vita et al., 2015

Table 2. contd. ...

Table 2. contd. ...

Truffle species	Volatile compound	Aroma description	Identification method	Molecular weight (g/mol)	Chemical structure	Reference
TMA	Pyridine	-	PTR-TOF-MS	79.10		Vita et al., 2015
TMA	Bis(methylthio) methane	-	PTR-TOF-MS	108.22	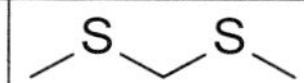	Vita et al., 2015
TMA	Cyclopentenyl carbe-nium	-	PTR-TOF-MS	81.14	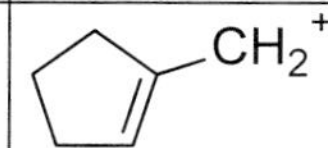	Vita et al., 2015
TMA	2-Methylfuran	-	PTR-TOF-MS	82.10	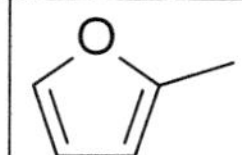	Vita et al., 2015
TMA	Thiophenium	-	PTR-TOF-MS	85.14	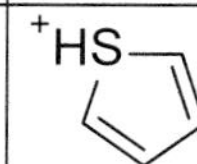	Vita et al., 2015
TMA	Butan-4-olide	-	PTR-TOF-MS	86.09	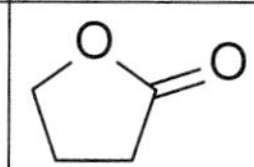	Vita et al., 2015
TMA	3,4-Dihydro-2Hthio-phene	-	PTR-TOF-MS	88.17	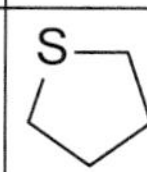	Vita et al., 2015
TMA	2,3-Butanediol	-	PTR-TOF-MS	90.12	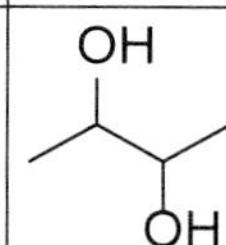	Vita et al., 2015
TMA	2-Methylmercaptoetha-nol/ toluene	-	PTR-TOF-MS	92.16	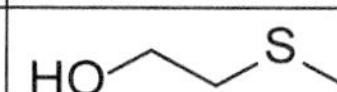	Vita et al., 2015

TMA	Furfural/ 2-Furancarbox-aldehyde	-	PTR-TOF-MS	96.09	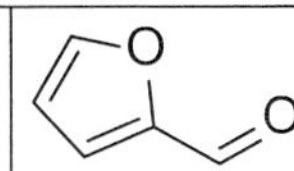	Vita et al., 2015
TMA	2,4-Dimethylfuran	-	PTR-TOF-MS	96.13	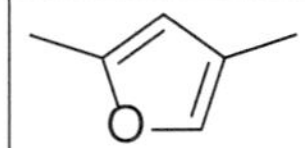	Vita et al., 2015
TMA	2-Methyl-4,5-dihydrothiophene	-	PTR-TOF-MS	100.18	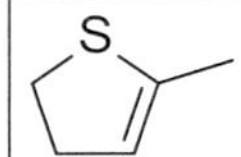	Vita et al., 2015
TMA	Ethynylbenzene	-	PTR-TOF-MS	102.14	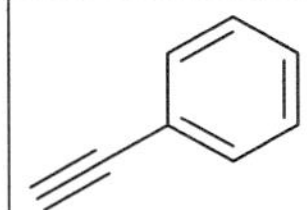	Vita et al., 2015
TMA	4-Hydroxy-3-methyl-2-butanone	-	PTR-TOF-MS	102.13	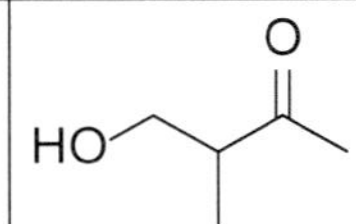	Vita et al., 2015
TMA	Methional	-	PTR-TOF-MS	104.17	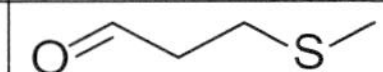	Vita et al., 2015
TMA	4-Hydroxybutanoic acid	-	PTR-TOF-MS	104.11	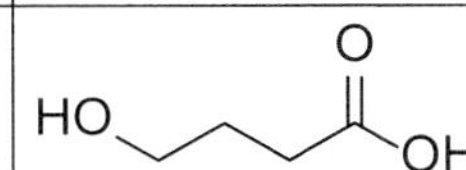	Vita et al., 2015
TMA	2-Methylthioacetic acid	-	PTR-TOF-MS	90.14	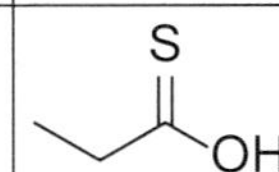	Vita et al., 2015

Table 2. contd. ...

Table 2. contd. ...

Truffle species	Volatile compound	Aroma description	Identification method	Molecular weight (g/mol)	Chemical structure	Reference
TMA	2-Ethynylthiophene	-	PTR-TOF-MS	108.16	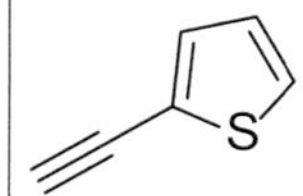	Vita et al., 2015
TMA	2-Acetylfuran	-	PTR-TOF-MS	110.11	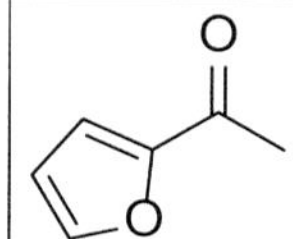	Vita et al., 2015
TMA	4-Methyl-1,3-heptadiene	-	PTR-TOF-MS	110.20	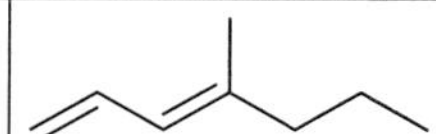	Vita et al., 2015
TMA	2,5-Dimethylthiophene	-	PTR-TOF-MS	112.19	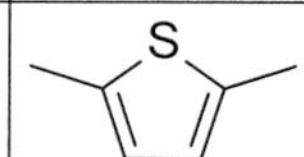	Vita et al., 2015
TMA	2-Heptenal	-	PTR-TOF-MS	112.17	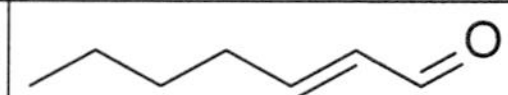	Vita et al., 2015
TMA	2-Methyl-3-furanthiol	-	PTR-TOF-MS	114.16	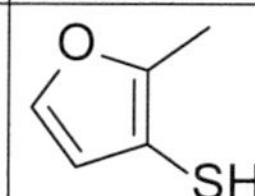	Vita et al., 2015
TMA	Methylsulfanyl cyclopentane/ cyclopen-tyl-1-thiaethane	-	PTR-TOF-MS	116.22	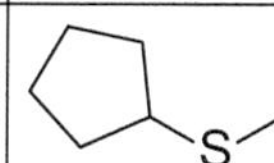	Vita et al., 2015
TMA	1-(Methylthio)pentane	-	PTR-TOF-MS	118.24	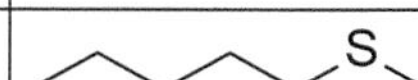	Vita et al., 2015
TMA	2-Hydroxyethyl propyl sulfide	-	PTR-TOF-MS	120.21	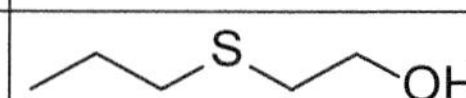	Vita et al., 2015

TMA	1,2,4-Trimethylbenzene	-	PTR-TOF-MS	120.20	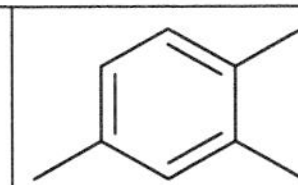	Vita et al., 2015
TMA	1-Methoxy-3-methylbenzene	-	PTR-TOF-MS	122.17	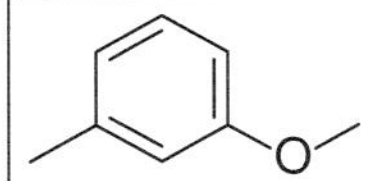	Vita et al., 2015
TMA	2-(2-hydroxyethylsulfanyl)ethanol	-	PTR-TOF-MS	122.18	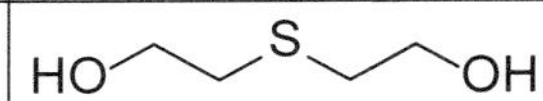	Vita et al., 2015
TMA	2-Butylfuran	-	PTR-TOF-MS	124.18	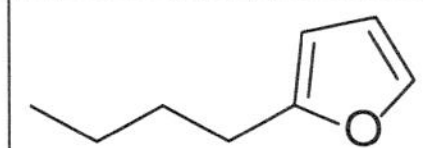	Vita et al., 2015
TMA	Methyl 2-Furancarboxylate	-	PTR-TOF-MS	126.11	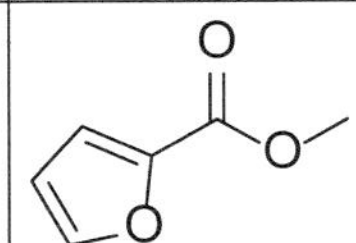	Vita et al., 2015
TMA	Naphthalene	-	PTR-TOF-MS	128.17	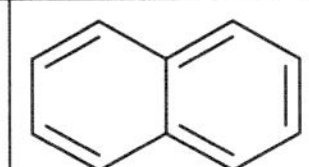	Vita et al., 2015
TMA	Methylsulfanylcyclohexane	-	PTR-TOF-MS	130.25	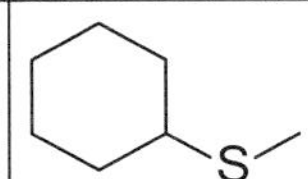	Vita et al., 2015
TMA	Ethyl 2- methylbutanoate/ Ethyl 2-methylbutyrate	-	PTR-TOF-MS	130.19	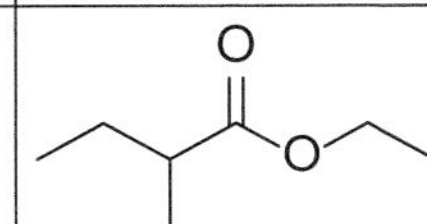	Vita et al., 2015

Table 2. contd. ...

Table 2. contd. ...

Truffle species	Volatile compound	Aroma description	Identification method	Molecular weight (g/mol)	Chemical structure	Reference
TMA	*P*-Cymenene	-	PTR-TOF-MS	132.21	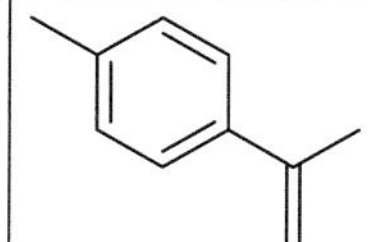	Vita et al., 2015
TMA	P-Cymene	-	PTR-TOF-MS	134.22	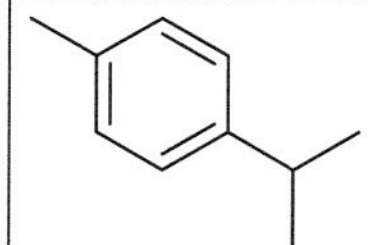	Vita et al., 2015
TMA	2-Pentylfuran	-	PTR-TOF-MS	138.21	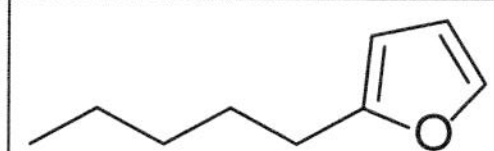	Vita et al., 2015
TMA	Nonanal	-	PTR-TOF-MS	142.24		Vita et al., 2015
TMA	2,3-Octanedione	-	PTR-TOF-MS	142.20	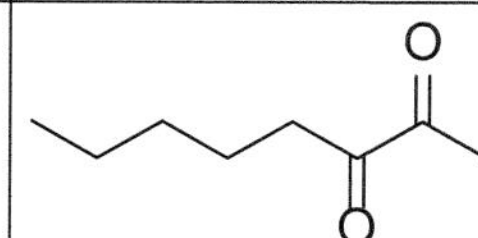	Vita et al., 2015
TMA	3-Phenyl-furan	-	PTR-TOF-MS	144.17	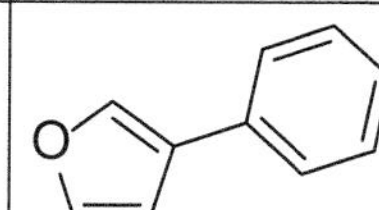	Vita et al., 2015
TMA	2-Ethyl-1-hexanethiol	-	PTR-TOF-MS	146.29	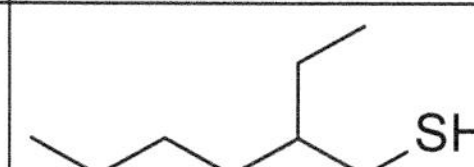	Vita et al., 2015

TMA	(1-Ethylpropyl)benzene	-	PTR-TOF-MS	148.25	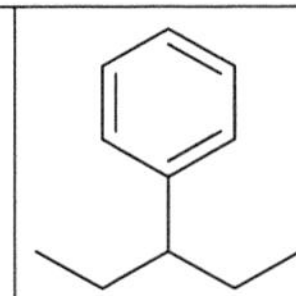	Vita et al., 2015
TMA	Methyl pentyl disulfide	-	PTR-TOF-MS	150.30	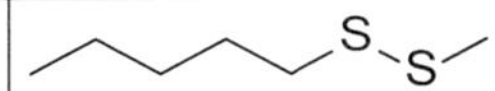	Vita et al., 2015
TMA	2-Methyl-5-(1-methylethenyl)-2-cyclohexen-1-ol (Carveol)	-	PTR-TOF-MS	152.24	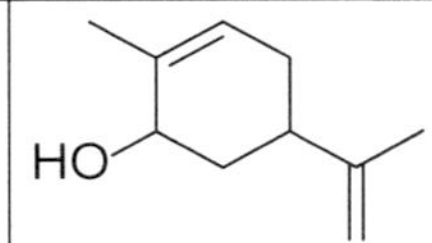	Vita et al., 2015
TMA	Diethyl trisulfide	-	PTR-TOF-MS	154.30	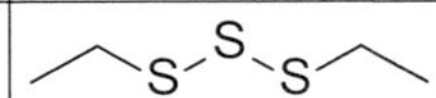	Vita et al., 2015
TMA	Decanal	-	PTR-TOF-MS	156.27	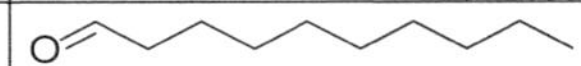	Vita et al., 2015
TMA	2-Methylbutyl-2-methyl-propanoate	-	PTR-TOF-MS	158.24	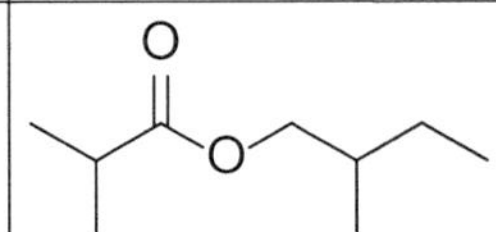	Vita et al., 2015
TMA	1,9-Nonanediol	-	PTR-TOF-MS	160.26	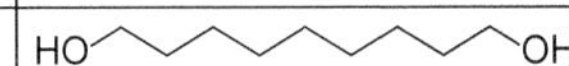	Vita et al., 2015
TMA	3-Methyl-2-(penta-2,4-dienyl)cyclo-pent-2-enone	-	PTR-TOF-MS	162.23	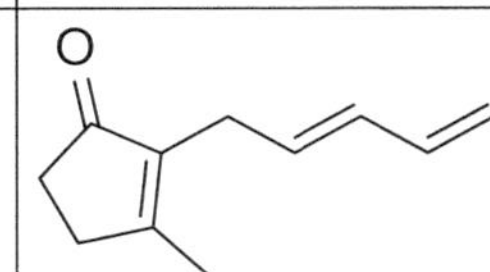	Vita et al., 2015
TMA	2-Heptylfuran	-	PTR-TOF-MS	166.26	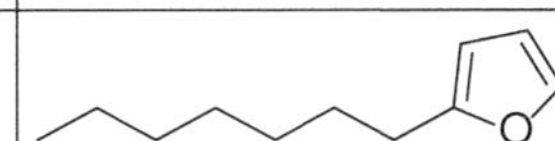	Vita et al., 2015

Table 2. contd. ...

Table 2. contd. ...

Truffle species	Volatile compound	Aroma description	Identification method	Molecular weight (g/mol)	Chemical structure	Reference
TMA	1,2,4-Trimethoxyben-zene	-	PTR-TOF-MS	168.19	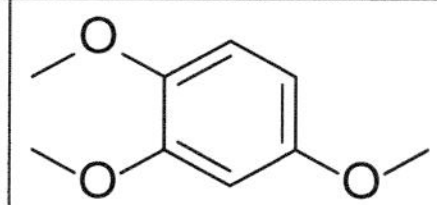	Vita et al., 2015
TMA	2-Undecanone	-	PTR-TOF-MS	170.30	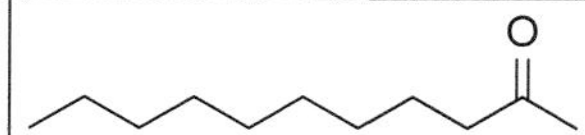	Vita et al., 2015
TMA	4-Hydroxybenzenesul-fonic acid	-	PTR-TOF-MS	174.17	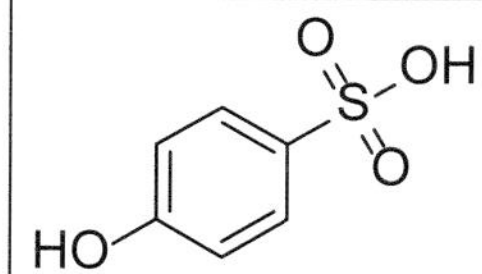	Vita et al., 2015
TMA	2-Hydroxy-4-isopropylnaphthalene	-	PTR-TOF-MS	186.25	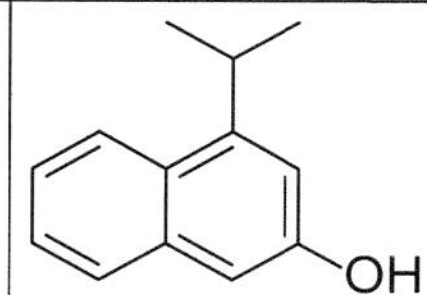	Vita et al., 2015
TMA	6,10-Dimethyl-5,9-undecadien-2-one/ gera-nylacetone	-	PTR-TOF-MS	194.32	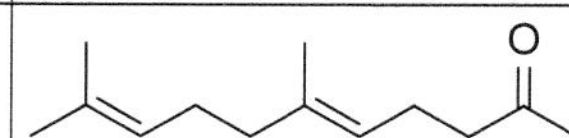	Vita et al., 2015
TMA	Cedrol	-	PTR-TOF-MS	222.37	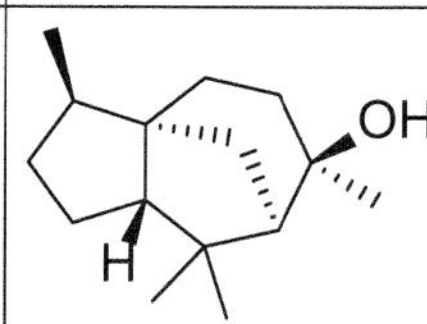	Vita et al., 2015

truffle-based products. Furthermore, it contributes to increasing the land value in rural regions (Oliach et al., 2021).

An analysis of the current situation within the truffle sector in the leading producing countries of the Mediterranean region (Croatia, France, Greece, Italy, and Spain) has identified several actions to support truffle production. These actions include strengthening the connection between truffles, gastronomy, and tourism, promoting truffle consumption among consumers, fostering appreciation of truffle production and evolving truffle culture and marketing, as well as developing tourism seminars for truffle growers (Oliach et al., 2021).

Standards for the quality of truffle trading are established by the Codex Alimentarius (Chauhan et al., 2021), while standards relevant to the marketing and trading quality control of truffles are developed by the United Nations Economic Commission for Europe (UNECE) (UNECE standard FFV-53, 2017). These standards define quality requirements for truffles post-preparation and packaging, including minimum and maturity requirements, classification, sizing, quality and size tolerance, presentation uniformity and packaging, and marking specifications such as identification, nature, and origin of produce. The standards specifically apply to fresh fruiting bodies of truffles (genus *Tuber*), excluding industrially processed truffles (Chauhan et al., 2021).

Due to the fact that some truffle species are very expensive (*T. magnatum* and *T. melanosporum*) and morphologically resemble the less expensive ones (e.g., *T. borchii, T. himalayense*, *T. indicum*, and *T. sinense*), they could be the subject of fraud. So, there is a need for quick and reliable fraud detection methods. In that respect, spectroscopic methods such as Fourier transform infrared (FT-IR) and near-infrared (NIR) can be applied for the analysis and authentication of truffles and truffle-based products as simple and cost-effective methods. Using FTIR and NIR, truffles could be discriminated not only by species but also by the area of production (Kappacher et al., 2022; Segelke et al., 2020; Zhao et al., 2006).

5. Industrial Significance of Truffles

The industrial application of truffles could be summarized in the food industry, cosmetics and perfume industries, and enzyme production (Nadim et al., 2015; Thomas et al., 2021). Due to their distinctive organoleptic properties and nutritional composition, truffles are mostly used in the culinary and food industry, shaping local lifestyle recognition. *T. melanosporum* (black truffle), often called the "black diamond" of gastronomy, is particularly valued in international high-class cuisine (Mustafa et al., 2020; Chauhan et al., 2021; Oliach et al., 2021; Thomas et al., 2021). Additionally, truffles have been studied for their antioxidant, antimicrobial, and anti-inflammatory properties (Stojković et al., 2013; Hamza et al., 2016a; Patel et al., 2017; Zhang et al., 2018; Khaled et al., 2021; Thomas et al., 2021; Owaid, 2022a). Although extracts of various truffle species express notable antimicrobial activity, no extract is still used for commercial purposes (Thomas et al., 2021). The use of dessert truffles in the production of *T. nivea* fungus biomass at a laboratory scale has been reported by Khaled et al. (2021). The biomass has potential as a natural industrial renewable supply for a variety of bioactive compounds, which could act as antimicrobial, anticancer, anti-inflammatory, antioxidant, and antitrypanosomal agents (Khaled et al., 2021).

Bio-nanotechnology, as a new branch of nanotechnology, incorporates biological materials to create nanoparticles for particular purposes. Lately, researchers have paid special attention to the green biosynthesis of silver-nanoparticles utilizing the extracts of bacteria, yeast, fungi, and algae. Dessert truffles (*Tirmania* and *Picoa* spp.) have been employed for the biosynthesis of silver-nanoparticles (AgNPs), which have shown promising potential in inhibiting several pathogenic bacteria and fungi strains (Alshammari and Abd El Aty, 2022; Owaid, 2022a). Owaid (2022b) synthesized green AgNPs using *Tirmania* sp., demonstrating that truffle-synthesized AgNPs exhibit fair antibacterial activity and notable synergistic effects with gentamicin against human pathogenic bacteria such as *Klebsiella* spp., *Pseudomonas aeruginosa* Migula, and *Staphylococcus aureus* Rosenbach. On the other hand, Picoa-AgNPs have shown modest antifungal activity toward *Pythium* sp., and *Aspergillus niger* van Tieghem (Owaid, 2022a).

Regarding the cosmetic and perfume industries, truffle-derived compounds serve as inspiration for various products. Although some skin care products containing truffle extracts and perfumes with a 'truffle' note are advertised, there is concern about whether they contain any truffle oil at all. This is more likely as none of the natural aroma compounds from truffles have been identified in such products, whereas a synthetically-derived compound such as 2,4-dithiapentane was reported (Thomas et al., 2021).

Enzymes are one of the most crucial industrial products, with a wide application in food, textile, and leather production. Fungi cells resemble small factories for enzyme production, capable of various catalytic tasks. Moreover, several advantages of fungi-derived enzymes, such as higher yield and efficient production using economical growth mediums, make fungi a promising source of enzymes. In this regard, some tuber species have been explored for their ability to produce amylase, cellulase, catalase, xylanase, laccase, lipase, and peroxidase (*T. maculatum*) as well as amylase, catalase, and peroxidase (*T. aestivum*) (Nadim et al., 2015). However, extensive research is still needed to fully justify the potential of truffles as enzyme producers.

6. Conclusion

Tuber, *Terfezia*, *Tirmania*, and *Picoa* fungi are primarily rich in carbohydrates and proteins, followed by ascorbic acid, dietary fibers, fats, and minerals. These fungi are abundant in functional bioactive compounds with potential therapeutic properties. Truffles are mostly used in the food industry and culinary applications, with limited data available on their use in other industries. Nowadays, a wide array of truffle products are available on the food market. The growing truffle industry has become important in countries including France, Italy, Spain, and Australia, resulting in millions of Euros of annual income. However, other uses of truffles still need exploration, such as their potential to serve as a renewable industrial source of enzymes and bioactive compounds, as well as a natural supply for the green synthesis of silver nanoparticles with promising antimicrobial properties.

Acknowledgements

The author (J.P.D.) acknowledges the Ministry of Science, Technological Development, and Innovation of the Republic of Serbia for financial support (Grant # 451-03-65/2024-03/200116), and D. Popović Beogračić for the design of Fig. 1.

References

Akyüz, M., Karagöz, I.D., İnci, S., Kılıç, İ.H. Kırbağ, S. et al. (2021). Cytotoxicity and DNA protective effects of the *Terfezia* and *Picoa* species from the eastern region of Turkey. Istanbul J. Pharm., 51(2): 198–203. DOI:10.26650/IstanbulJPharm.2020.0064

Alshammari, S.O. and Abd, El, Aty, A.A. (2022). Statically controlled mycogenic-synthesis of novel biologically active silver-nanoparticles using Hafr Al-Batin desert truffles and its antimicrobial efficacy against pathogens. Saudi J. Biol. Sci., 29: 103334. https://doi.org/10.1016/j.sjbs.2022.103334

Al-Shabibi, M.M.A., Toma, S.J. and Haddad, B.A. (1982). Studies on Iraqi truffles. I. Proximate analysis and characterization of lipids. Can. Inst. Food Technol. J., 15: 200–202.

Amicucci, A., Barbieri, E., Sparvoli, V., Gioacchini, A.M., Calcabrini, C. et al. (2018). Microbial and pigment profile of the reddish patch occurring within *Tuber magnatum* ascomata. Fungal Biol., 122(12): 1134–1141. https://doi.org/https://doi.org/10.1016/j.funbio.2018.07.007

Angelin, J. and Kavitha, M. (2020). Exopolysaccharides from probiotic bacteria and their health potential. Int. J. Biol. Macromol., 162: 853–865. https://doi.org/10.1016/j.ijbiomac.2020.06.190

Attia, W.Y., El-Naggar, R.E., Bawadekji, A. and Al Ali, M. (2018). Evaluation of some in vitro anti-carcinogenic activities of polysaccharides extracted from Ascomata of the desert truffle *Terfezia claveryi* Chatin. J. Appl. Environ. Biol. Sci., 8: 152–159.

Aulinger, K., Besl, H., Spiteller, P., Spiteller, M. and Steglich, W. et al. (2001). Melanocrocin, a polyene pigment from *Melanogaster broomeianus* (Basidiomycetes). Zeitschrift Für Naturforschung C, 56(7–8): 495-498.

Beara, I.N., Lesjak, M.M., Četojević-Simin, D.D., Marjanović, Ž.S., Ristić, J.D. et al. (2014). Phenolic profile, antioxidant, anti-inflammatory and cytotoxic activities of black (*Tuber aestivum* Vittad.) and white (*Tuber magnatum* Pico) truffles. Food Chem., 165: 460–466.

Beuchat, L.R., Brenneman, T.B. and Dove, C.R. (1993). Composition of the pecan truffle (*Tuber texense*). Food Chem., 46: 189–192.

Bokhary, H.A. and Sarwat, P. (1993). Chemical composition of desert truffles *Terfezia clavery*. J. Food Compos. Anal., 6(3): 285–293.

Bouatia, M., Touré, H.A., Cheikh, A., Eljaoudi, R., Rahali, Y. et al. (2018). Analysis of nutrient and antinutrient content of the truffle (*Tirmania pinoyi*) from Morocco. Int. Food Res. J., 25: 174–178. http://ifrj.upm.edu.my/25%20(01)%202018/(22).pdf

BSF PRODUCT, https://www.bsfproduct.com/proizvodi.html (Accessed on June 22, 2024)

Chaisuwan, W., Jantanasakulwong, K., Wangtueai, S., Phimolsiripol, Y., Chaiyaso, T. et al. (2020). Microbial exopolysaccharides for immune enhancement: Fermentation, modifications and bioactivities. Food Biosci., 35: 100564. https://doi.org/https://doi.org/10.1016/j.fbio.2020.100564

Chauhan, O.P., Vijay, V., Pandey, A.K. and Semwal, A.D. (2021). Biochemical and health properties of truffles. Def. Life Sci. J., 6(3): 251–258. DOI: 10.14429/dlsj.6.15659

Chen, C.-C., Nargotra, P., Kuo, C.-H. and Liu, Y.-C. (2023). High-molecular-weight exopolysaccharides production from *Tuber borchii* cultivated by submerged fermentation. Int. J. Mol. Sci., 24(5). https://doi.org/10.3390/ijms24054875

Choo, K.S.O., Bollen, M., Dykes, G.A. and Coorey, R. (2021). Aroma-volatile profile and its changes in Australian grown black Périgord truffle (*Tuber melanosporum*) during storage. Int. J. Food Sci. Tech., 56(11): 5762–5776. https://doi.org/https://doi.org/10.1111/ijfs.15171

Correa-Llantén, D.N., Amenábar, M.J. and Blamey, J.M. (2012). Antioxidant capacity of novel pigments from an Antarctic bacterium. J. Microbiol., 50(3). 374–379.

Dahham, S.S., Al-Rawi, S.S., Ibrahim, A.H., Majid, A.S.A. and Majid, A.M.S.A. et al. (2018). Antioxidant, anticancer, apoptosis properties and chemical composition of black truffle *Terfezia claveryi*. Saudi J. Biol. Sci., 25: 1524–1534. http://dx.doi.org/10.1016/j.sjbs.2016.01.031

Dupont, S., Fleurat-Lessard, P., Cruz, R.G., Lafarge, C., Grangeteau C. et al. (2021). Antioxidant properties of ergosterol and its role in yeast resistance to oxidation. Antioxidants, 10(7): 1024. https://doi.org/10.3390/antiox10071024

El-Naggar, N.E.-A. and El-Ewasy, S.M. (2017). Bioproduction, characterization, anticancer and antioxidant activities of extracellular melanin pigment produced by newly isolated microbial cell factories *Streptomyces glaucescens* NEAE-H. Sci. Rep-UK., 7(1): 42129.

Farag, M.A., Fathi, D., Shamma, S., Shawkat, M.S.A., Shalabi, S.M. et al. (2021). Comparative metabolome classification of desert truffles *Terfezia claveryi* and *Terfezia boudieri* via its aroma and nutrients profile. LWT, 142: 111046. https://doi.org/10.1016/j.lwt.2021.111046

Freiberg, J.A., Grebenc, T., Strojnik, L., Meireles, L.A., Jiménez, M.S.E. et al. (2023). Production and marketing of *Tuber floridanum* – ecology and gastronomic value of a recently described truffle species. Sci. Agric., 80: e20220102. https://doi.org/10.1590/1678-992X-2022-0102

Ferreira, I., Dias, T., Mouazen, A.M. and Cruz, C. (2023). Using science and technology to unveil the hidden delicacy *Terfezia arenaria*, a desert truffle. Foods, 12(19): 3527. https://doi.org/10.3390/foods12193527

Gong, P., Wang, S., Liu, M., Chen, F., Yang, W. et al. (2020). Extraction methods, chemical characterizations and biological activities of mushroom polysaccharides: A mini-review. Carbohydr. Res., 494: 108037. https://doi.org/https://doi.org/10.1016/j.carres.2020.108037

Guan, T., Wei, X., Xu, P., Chen, K., Zou, Y. et al. (2022). Comparison of structural and antioxidant activity of polysaccharide extracted from truffles. J. Food Sci., 87(7): 2999-3012. https://doi.org/https://doi.org/10.1111/1750-3841.16207

Gourmet Food Store. https://www.gourmetfoodstore.com/truffles-and-mushrooms/ (Accessed on June 22, 2024)

Hamza, A., Zouari, N., Zouari, S., Jdir, H., Zaidi, S. et al. (2016a). Nutraceutical potential, antioxidant and antibacterial activities of *Terfezia boudieri* Chatin, a wild edible desert truffle from Tunisia arid zone. Arab. J. Chem., 9: 383–389.

Hamza, A., Jdir, H. and Zouari, N. (2016b). Nutritional, antioxidant and antibacterial properties of *Tirmania nivea*, a wild edible desert truffle from Tunisia arid zone. Med Aromat. Plants, 5: 4. DOI: 10.4172/2167-0412.1000258

Henkrar, F. and Khabar, L. (2022). Proximate analysis of lipid composition in Moroccan truffles and desert truffles. Ital. J. Food Sci., 34(2): 67–73. DOI 10.15586/ijfs.v34i2.2202

Jayaseelan, S., Ramaswamy, D. and Dharmaraj, S. (2014). Pyocyanin: production, applications, challenges and new insights. World J. Microb. Biot., 30(4): 1159–1168. https://doi.org/10.1007/s11274-013-1552-5

Kagan-Zur, V. and Roth-Bejerano, N. (2008). Studying the brown desert truffles of Israel. Isr. J. Plant Sci., 56: 309–314.

Kappacher, C., Trübenbacher, B., Losso, K., Rainer, M., Bonn, G.K. et al. (2022). Portable vs. benchtop NIR-sensor technology for classification and quality evaluation of black truffle. Molecules, 27(3): 589. https://doi.org/10.3390/molecules27030589

Khaled, J.M., Alharbi, N.S., Mothana, R.A., Kadaikunnan, S. and Alobaidi, A.S. et al. (2021). Biochemical profile by GC–MS of fungal biomass produced from the ascospores of *Tirmania nivea* as a natural renewable resource. J. Fungi, 7(12): 1083. https://doi.org/10.3390/jof7121083

Khalifa, S.A.M., Farag, M.A., Yosri, N., Sabir, J.S.M., Saeed, A. et al. (2019). Truffles: From Islamic culture to chemistry, pharmacology, and food trends in recent times. Tr. Food Sci. Technol., 91. https://doi.org/10.1016/j.tifs.2019.07.008

Kivrak, Ş. and Kivrak, İ. (2018). Investigation of chemical composition and nutritional value of truffle mushroom (*Tuber nitidum* Vittad.). Süleyman Demirel Univ. J. Nat. Appl. Sci., DOI: 10.19113/sdufbed.22646

Kruzselyi, D. and Vetter J. (2014). Complex chemical evaluation of the summer truffle (*Tuber aestivum* Vittadini) fruit bodies. J. Appl. Bot. Food Qual., 87: 291–295, DOI:10.5073/JABFQ.2014.087.041

Lang, M., Spiteller, P., Hellwig, V. and Steglich, W. (2001). Stephanosporin, a "Traceless" Precursor of 2-Chloro-4-nitrophenol in the *Gasteromycete Stephanospora* caroticolor. Angew. Chem. Int. Edit.,

40(9): 1704–1705. https://doi.org/https://doi.org/10.1002/1521-3773(20010504)40:9<1704::AID-ANIE17040>3.0.CO;2-L

Lee, H., Nam, K., Zahra, Z. and Farooqi, M.Q.U. (2020). Potentials of truffles in nutritional and medicinal applications: a review. Fungal Biol. Biotechnol., 7: 9 https://doi.org/10.1186/s40694-020-00097-x

Lemmond, B., Sow, A., Bonito, G. and Smith, M.E. (2023). Accidental cultivation of the European truffle *Tuber brumale* in North American truffle orchards. Mycorrhiza, 33: 221-228 https://doi.org/10.1007/s00572-023-01114-8

Li, X., Zhang, X., Ye, L., Kang, Z., Jia, D. et al. (2019a). LC-MS-based metabolomic approach revealed the significantly different metabolic profiles of five commercial truffle species. Front. Microbiol., 10: 2227. doi: 10.3389/fmicb.2019.02227

Li, J.-M., Liang, H.-Q., Qiao, P., Su, K.-M., Liu, P.-G. et al. (2019b). Chemical composition and antioxidant activity of *Tuber indicum* from different geographical regions of China. Chem. Biodivers., 16(3): e1800609. https://doi.org/https://doi.org/10.1002/cbdv.201800609

Li, X., Zhu, Z., Ye, L., Kang, Z., Zhang, X. et al. (2020). Comparison of the partial structure and antioxidant activity of polysaccharides from two species of Chinese truffles. Molecules, 25(18): 4345. https://doi.org/10.3390/molecules25184345

Ljubojević, S., Vasilišin, L., Vučić, G. and Velemir, A. (2023). Chemical composition of summer truffle (*Tuber aestivum* Vittad.) from Bosnia and Herzegovina. EUREKA: Life Sciences, (6): 20–27. https://doi.org/10.21303/2504-5695.2023.003191

Malik, K., Tokkas, J. and Goyal, S. (2012). Microbial pigments: a review. Int. J. Microbial Res. Technol., 1(4): 361–365.

Mudliyar, D.S., Wallenius, J.H., Bedade, D.K., Singhal, R.S., Madi, N. et al. (2019). Ultrasound assisted extraction of the polysaccharide from *Tuber aestivum* and its *in vitro* anti-hyperglycemic activity. Bioact. Carbohydr. Diet. Fiber, 20: 100198. https://doi.org/https://doi.org/10.1016/j.bcdf.2019.100198

Murcia, M.A., Martínez-Tomé, M., Vera, A., Morte, A., Gutierrez, A. et al. (2003). Effect of industrial processing on desert truffles *Terfezia claveryi* Chatin and *Picoa juniperi* Vittadini): proximate composition and fatty acids. J. Sci. Food Agric., 83: 535–541. DOI: 10.1002/jsfa.1397

Mustafa, A.M., Angeloni, S., Nzekoue, F.K., Abouelenein, D., Sagratini, G. et al. (2020). An overview on truffle aroma and main volatile compounds. Molecules, 25(24): 5948. https://doi.org/10.3390/molecules25245948

Nadim, M., Deshaware, S., Saidi, N., Abd-Elhakeem, M.A., Ojamo, H. et al. (2015). Extracellular enzymatic activity of *Tuber maculatum* and *Tuber aestivum* mycelia. Adv. Microbiol., 5: 523–530. http://dx.doi.org/10.4236/aim.2015.57054

Narsing Rao, M.P., Xiao, M. and Li, W-J. (2017). Fungal and bacterial pigments: secondary metabolites with wide applications. Front. Microbiol., 8:1113. doi: 10.3389/fmicb.2017.01113

Oliach, D., Vidale, E., Brenko, A., Marois, O., Andrighetto, N. et al. (2021). Truffle market evolution: an application of the Delphi method. Forests, 12(9): 1174. https://doi.org/10.3390/f12091174

Owaid, M.N., Rabeea, M.A., Abdul A.A., Jameel, M.S. and Dheyab, M.A. et al. (2022a). Mycogenic fabrication of silver nanoparticles using *picoa*, Pezizales, characterization and their antifungal activity. Environ. Nanotechnol. Monit. Manag., 17: 100612. doi: 10.1016/j.enmm.2021.10061

Owaid, M.N. (2022b). Biosynthesis of silver nanoparticles from truffles and mushrooms and their applications as nanodrugs. Curr. Appl. Sci. Technol., 22(5): 1–11. DOI: 10.55003/cast.2022.05.22.009

Pacioni, G., Rapino, G., Zarivi, O., Falconi, A., Leonardi, M. et al. (2015). Truffles contain endocannabinoid metabolic enzymes and anandamide. Phytochemistry, 110: 104–110. doi: 10.1016/j.phytochem.2014.11.012

Patel, S., Rauf, A., Khan, H., Khalid, S. and Mubarak, M.S. et al. (2017). Potential health benefits of natural products derived from truffles: A review. Trends Food Sci. Tech., 70: 1–8. https://doi.org/https://doi.org/10.1016/j.tifs.2017.09.009

Piloni, M., Tat, L., Tonizzo, A. and Battistutta, F. (2005). Aroma characterisation of white truffle by GC-MS and GC-O. Ital. J. Food Sci., 17(4).

Popović-Djordjević, J., Marjanović, S.Ž., Gršić, N., Adžić, T., Popović, B. et al. (2019). Essential elements may be distinguishing factors between mycorrhizal potentials of *Tuber magnatum* Pico and *Tuber brumale* Vitt. as cohabiting truffle species in riparian forest habitat in Serbia. Chem. Biodivers., 16: e1800693; https://doi.org/10.1002/cbdv.201800693

Rosa-Gruszecka A., Hilszczańska, D., Gil, W. and Koselet, B. (2017). Truffle renaissance in Poland – history, present and prospects. J. Ethnobiol. Ethnomed., 13: 36. DOI 10.1186/s13002-017-0163-x

Saltarelli, R., Ceccaroli, P., Cesari, P., Barbieri, E. and Stocchi, V. et al. (2008). Effect of storage on biochemical and microbiological parameters of edible truffle species. Food Chem., 109: 8–16

Sawaya, W.N., Al-Shalhat, A., Al-Sogair, A. and Mohammad, M. (1985). Chemical composition and nutritive value of truffles of Saudi Arabia. J. Food Sci., 50: 450–453. https://doi.org/10.1111/j.1365-2621.1985.tb13425.x

Schmidberger, P.C. and Schieberle, P. (2017). Characterization of the key aroma compounds in white Alba truffle (*Tuber magnatum* pico) and Burgundy truffle (*Tuber uncinatum*) by means of the sensomics approach. J. Agr. Food Chem., 65(42): 9287–9296. https://doi.org/10.1021/acs.jafc.7b04073

Segelke, T., Schelm, S., Ahlers, C. and Fischer, M. (2020). Food authentication: truffle (*Tuber* spp.) species differentiation by FT-NIR and chemometrics. Foods, 9(7): 922. https://doi.org/10.3390/foods9070922

Shimokawa, T., Kinoshita, A., Kusumoto, N., Nakano, S., Nakamura, N. et al. (2020). Component features, odor-active volatiles, and acute oral toxicity of novel white-colored truffle *Tuber japonicum* native to Japan. Food Sci. Nutr., 8: 410–418. 10.1002/fsn3.1325

Splivallo, R., Ottonello, S., Mello, A. and Karlovsky, P. (2011). Truffle volatiles: from chemical ecology to aroma biosynthesis. New Phytol., 189(3): 688–699. https://doi.org/https://doi.org/10.1111/j.1469-8137.2010.03523.x

Stojković, D., Reis, F.S., Ferreira, I.C., Barros, L. and Glamočlija, J. (2013). *Tirmania pinoyi*: Chemical composition, *in vitro* antioxidant and antibacterial activities and *in situ* control of *Staphylococcus aureus* in chicken soup. Food Res. Int., 53(1): 56–62.

Strojnik, L., Grebenc, T. and Ogrinc, N. (2020). Species and geographic variability in truffle aromas. Food Chem. Toxicol., 142: 111434. https://doi.org/https://doi.org/10.1016/j.fct.2020.111434

Tang, Y.-J., Liu, R.-S. and Li, H.-M. (2015). Current progress on truffle submerged fermentation: a promising alternative to its fruiting bodies. Appl. Microbiol. Biot., 99(5): 2041–2053. https://doi.org/10.1007/s00253-015-6379-6

Tejedor-Calvo, E., Morales, D., Marco, P., Sánchez, S., Garcia-Barreda, S. et al. (2020). Screening of bioactive compounds in truffles and evaluation of pressurized liquid extractions (PLE) to obtain fractions with biological activities. Food Res. Int., 132: 109054. https://doi.org/https://doi.org/10.1016/j.foodres.2020.109054

Tejedor-Calvo, E., García-Barreda, S., Sánchez, S., Morte, A., Siles-Sánchez, MdlN. et al. (2022). Application of pressurized liquid extractions to obtain bioactive compounds from *Tuber aestivum* and *Terfezia claveryi*. Foods, 11(3): 298. https://doi.org/10.3390/foods11030298

Thomas, P., Elkhateeb, W. and Daba, G. (2021). Industrial applications of truffles and truffle-like fungi. pp. 82–88. In: Sridhar, K.R. and Deshmukh, S.K. (eds.). Advances in Macrofungi Industrial Avenues and Prospects. CRC Press, Boca Raton, Florida, USA.

Umadevi, K. and Krishnaveni, M. (2013). Antibacterial activity of pigment produced from Micrococcus luteus KF532949. Int. J. Anal. Chem., 4(3): 149–152.

UNECE STANDARD FFV-53UNECE STANDARD FFV-53, UNITED NATIONS New York and Geneva, 2017. https://unece.org/fileadmin/DAM/trade/agr/standard/standard/fresh/FFV-Std/English/53_Truffles.pdf (accessed on June 20, 2024).

Vahdani, M., Rastegar, S., Rahimzadeh, M., Ahmadi, M. and Karmostaji, A. et al. (2017). Physicochemical characteristics, phenolic profile, mineral and carbohydrate contents of two truffle species. J. Agr. Sci. Tech., 19(5): 1091–1101. http://jast.modares.ac.ir/article-23-11818-en.html

Vahdatzadeh, M. and Splivallo, R. (2018). Improving truffle mycelium flavour through strain selection targeting volatiles of the Ehrlich pathway. Sci. Rep.-UK, 8(1): 9304. https://doi.org/10.1038/s41598-018-27620-w

Vendruscolo, F., Tosin, I., Giachini, A.J., Schmidell, W. and Ninow, J.L. et al. (2014). Antimicrobial activity of Monascus pigments produced in submerged fermentation. J. Food Process. Pres., 38(4): 1860–1865.

Vita, F., Taiti, C., Pompeiano, A., Bazihizina, N., Lucarotti, V. et al. (2015). Volatile organic compounds in truffle (*Tuber magnatum* Pico): comparison of samples from different regions of Italy and from different seasons. Sci. Rep.-UK, 5(1): 12629. https://doi.org/10.1038/srep12629

Wahiba, B., Tehami, W., Kherraf, A., Abbouni, B. and Benali, M. et al. (2016). Nutritional and antioxidant profile of red truffles (*Terfezia claveryi*) and white truffle (*Tirmania nivea)* from Southwestern of Algeria. Der Pharmacia Letter, 8: 134–141.

Wang, S. and Marcone, M.F. (2011). The biochemistry and biological properties of the world›s most expensive underground edible mushroom: truffles. Food Res. Int., 44: 2567–2581.

Yan, X., Wang, Y., Sang, X. and Fan, L. (2017). Nutritional value, chemical composition and antioxidant activity of three *Tuber* species from China. AMB Expr., 7: 136. https://doi.org/10.1186/s13568-017-0431-0

Zeng, C., Chen, X., Jiang, W., Liu, Y. and Fang, C. et al. (2020). Isolation, purification and antioxidant activity of the polysaccharides from Chinese truffle *Tuber sinense*. Iran. J. Pharm. Res., 19(1): 436–447. https://doi.org/10.22037/ijpr.2020.1100954

Zhang, T., Jayachandran, M., Ganesan. K. and Xu, B. (2018). Black truffle aqueous extract attenuates oxidative stress and inflammation in STZ-induced hyperglycemic rats via Nrf2 and NF-κB pathways. Front. Pharmacol, 9: 1257. doi: 10.3389/fphar.2018.01257

Zhang, Z., Wang, J., Liu, L., Shun, Q., Shi, W. et al. (2019). The optimum conditions and mechanism for increasing exo-polysaccharide production of Truffles melanosporum by *Dioscorea* saponins. LWT, 107: 331–339. https://doi.org/https://doi.org/10.1016/j.lwt.2019.03.011

Zhao, D., Liu, G., Song, D., Liu, J-h., Zhou, Y. et al. (2006). Fourier transform infrared spectroscopic study of truffles, Proc. SPIE 6026, ICO20: Biomedical Optics, 60260H. doi:10.1117/12.667133

Zhao, W., Wang, X.-H., Li, H.-M., Wang, S.-H., Chen, T. et al. (2014). Isolation and characterization of polysaccharides with the antitumor activity from Tuber fruiting bodies and fermentation system. Appl. Microbiol. Biotechnol., 98(5): 1991–2002. https://doi.org/10.1007/s00253-013-5379-7

Diversity and Ecology

Chapter 5

Diversity, Distribution and Ecology of Truffles in Morocco

Hakkou, S.,[1,*] *Sabir, M.*[2] and *Machouri, N.*[1]

1. Introduction

The fungal world is home to a diversity of species of ecological, nutritional, pharmaceutical, and commercial interest. Among 140,000 mushrooms, only 2,000 are edible, and about 700 species possess valuable pharmaceutical properties (Wasser, 2002; Badalyan, 2012). Among these mushrooms are truffles, which are hypogeous fungi living entirely underground and belonging to the Ascomycetes (Morte et al., 2008; Trappe et al., 2008). They come in two types: true truffles, which belong to the Tuber genus and are associated with trees, and truffles, which are highly prized for their intense, exclusive aroma. The best-known truffles are *Tuber aestivum*, *T. borchii*, *T. brumale*, *T. indicum*, *T. macrosporum*, *T. magnatum*, *T. melanosporum*, *T. mesentericum*, and *T. uncinatum* (Callot, 1999; Rodriguez, 2008; Olivier et al., 2018). In addition, we also have the desert truffles known as "*Terfès*", which grow in areas with typically arid and semi-arid climates. They mainly refer to species of the *Terfezia, Tirmania*, and *Tuber* genera, but also to lesser-known species of *Picoa*, *Mattirolomyces*, and *Loculotuber* genera (Alsheikh, 1994; Morte et al, 2008; Rodriguez, 2008; Trappe et al., 2008; Zitouni-Haouar, 2016; Boufeldja, 2017). They are an important component of the mycological flora of arid and semi-arid forests and rangelands (Alsheikh, 1994; Bradai, 2014). Desert truffles are widely found

[1] Faculty of Letters and Human Sciences, Rabat, Morocco.
[2] National School of Forest Engineers, Sale, Morocco.
* Corresponding author: soukaina.hakkou@gmail.com

around the Mediterranean, in northern Africa, and in the Middle East. Some have been found in South Africa and China (Boufeldja, 2017; Bouatia et al., 2018).

The growth and development of truffles rely on the interplay of three key factors: soil conditions, climate, and the presence of a suitable host plant. The quantity and seasonal distribution of rainfall influence the yield and abundance of truffles. As for the soil, its physico-chemical properties play a major role in their growth. Through a symbiotic association, the host plant provides truffles with the nutrients (water, and mineral elements) they need for their development (Grente and Chevalier, 1980; Dexheimer et al., 1985; Fortas and Chevalier, 1992; Laessoe and Hansen, 2007; Fortas, 2009; Bouziani et al., 2010; Bradai et al., 2013; Hakkou et al., 2023). The distribution and variability of truffle species primarily stem from variations in these factors across different regions (De Laplane, 2010). In Morocco, desert truffles are produced in four main areas: the Oriental (eastern region), the Maâmora forest, the Abda-Doukkala Sahel, and the Moroccan Sahara. Morocco also produces true truffles of two types, spontaneous and cultivated (Abourouh, 2011; Khabar, 2016, Abourouh, 2020; Laqbaqbi, 2020; Hakkou et al., 2022). Particular attention has been paid to truffles, with the aim of understanding this fungus at different levels: ecological, botanical, physiological, and pharmaceutical. This study was carried out through field surveys (collection of samples of truffles, host plants, and soils) and surveys and workshops with the various players in the sector. It is within this framework that this article aims to improve and clarify our knowledge of truffles and their geographical distribution in the Kingdom of Morocco, study the factors involved in the production of Moroccan truffles, and qualify the production potential of truffles through the evaluation and spatialization of their productivity on a national scale.

2. Geographic Distribution of Truffles

Morocco is best known for its production of desert truffles, known as "*Terfès*", which are found over large areas of the country. They belong to the *Terfezia*, *Tirmania*, *Tuber*, *Delastria*, and *Picoa* genera (Fig. 1). Other species, true truffles, also exist, but are not collected due to the difficulty of detecting them. Indeed, they are not even known to the local population. Collectors often distinguish species by color. For example, they call truffles of the *Tirmania* genus "*Terfes labyed*" because they are whitish, and *Terfezia claveryi* "*Terfes Lahmer*" because they are reddish. They are also named for the host plant with which they are associated, such as *Tuber oligospermum*, called "*terfes taida*" because it occurs under pine, known in Arabic as "*taida*". Another is called by a derivative of its scientific name, such as *Picoa juneperi*, known in some areas as "*jobber*". Four major Moroccan zones are the most productive for truffles: Maamora Forest, Doukkala-Abda Sahel, the North-East region, and the Moroccan Sahara. The number of species varies from one zone to another. Some species can be found in several zones, while others are limited to a specific area.

2.1 Maamora Forest

There are mainly five well-known truffle species, three of which are associated with cork oak (*Quercus suber*) in connection with the host plant *Helianthemum guttatum*.

These are *Terfezia arenaria*, *T. leptoderma*, and *Tuber asa.* While two species associated with maritime pine (*Pinus pinaster* var. *atlantica*), *Tuber oligospermum* and *Delastria rosea* (Chatin, 1896; Malençon, 1973; Alsheikh, 1994; Khabar, 2002, 2016; Abourouh, 2011, 2020; Hakkou et al., 2022). In addition to these species, two others have been mentioned in the forest by Malençon (1973), Khabar (1988), and Abourouh (2020). These are *Tuber gennadii* associated with *H. guttatum* and *T. Borchii* Var. *sphaerosperma* associated with Aleppo pine (*Pinus halepensis*) and pinion pine (*Pinus pinea*).

2.2 Doukkala-Abda Sahel

Our field observations in February 2021 revealed the existence of two truffle species, *Terfezia boudieri* and *Tirmania pinoyi.* Collectors confirmed the presence of a third species, *Tuber oligospermum*. However, Moroccan literature only mentions a single *T. boudieri* species associated with *H. lipii* var. *sessiliflorum* and *H. ledifolium* in the "*Had Hrara*" region located to the east of the town of Safi (Abourouh, 2011; Khabar, 2016).

2.3 North-East Region

Five species are known to exist in the region. Three species are the most widespread and the most widely collected. These are *Tirmania pinoyi* and *T. nivea* associated with *H. hirtum*, and *Terfezia claveryi* associated with *H. lipii*. The other two species are rare and occur at specific sites rather than throughout the zone. They are *Terfezia boudieri* associated with *H. lipii* var. *sessiliflorum*, and *H. ledifolium*, and *Picoa juniperi* associated with *H. lipii.* These species have also been indicated by some previous studies (Malençon, 1973; Alsheikh, 1994; Serrhini et al., 1995; El Aji, 1999; Bouziani, 2009; Khabar, 2016; Hakkou et al., 2022). In addition, two other species, less known in the region, were encountered in 2013 by other researchers. These were *T. olbiensis* in the Beni Yaala Forest in Jerada under *Pinus halepensis* and *T. leptoderma* collected in Ain Beni Mathar associated with *Helianthemum* sp. (El Akil et al., 2016). Since 2000, the region has also been home to a *Tuber melanosporum* black truffle farm associated with holm oak planted by Doctor Abdelaziz Laqbaqbi in Tifezouine, Debdou region. The holm oak plants were produced and inoculated in France and introduced to Morocco.

2.4 Moroccan Sahara

The area is known for the production of two species: *T. pinoyi* and *T. nivea*, associated with *H. hirtum*, *H. lipii*, *H. ledifolium*, and *H. apertum*. Our field observations have shown that two other species are also present in the south of the country: *T. claveryi* and *Picoa juniperi*, associated with *H. lipii*.

On an irregular basis and of little quantitative importance to collectors, other forest areas are home to truffles. The holm oak forests of the Middle Atlas are home to truffles of the *Tuber* genus. The existing species are *T. excavatum*, *T. brumale*, *T. rufum*, and *T. uncinatum/aestivum*, harvested by Malençon in November 1960

Fig. 1. Some truffles collected in the fields of Morocco: *Terfezia arenaria* (a), *Terfezia boudieri* (b), *Terfezia claveryi* (c), *Tirmania nivea* (d), *Tirmania pinoyi* (e), and *Tuber asa* (f); Cultivated black truffle (*Tuber melanosporum*) in Morocco (© Hakkou).

(Malençon, 1973) and in May 2014 by Laqbaqbi (2020). The latter has set up two holm oak farms to grow *T. melanosporum* truffles in the Middle Atlas, not far from the town of Imouzzer Kandar.

Our surveys confirm the observations of several authors that truffles also exist in the Tangérois (Tangier), Loukkous (Larache), Souss (Agadir), Drâa (Ouarzazate), and Green Belt plantations south of Rabat (Malençon, 1973; Serrhini et al., 1995; Chatin, 1896; Khabar, 2016; Hakkou, 2019).

3. Environmental Factors

Truffle production depends mainly on three environmental factors, rainfall, soils, and host plant species.

3.1 Precipitation

While desert truffles are well-adapted to arid climates, rainfall plays a pivotal role in triggering the germination of host plants, thereby influencing truffle production. The required amount of rainfall for optimal production varies across regions and truffle species. Typically, successful truffle production is associated with rainfall occurring in two distinct periods. Firstly, the summer/autumn rains, characterized by thunderstorms from late August to early November, facilitate seed germination and favorable growth conditions for host plants. Secondly, the winter rains, spanning from December to March, prompt truffle development as fungi, capitalizing on the symbiotic relationship between truffles and host plant roots. Enhanced precipitation during the latter period replenishes soil water reserves, leading to more abundant production. Additionally, higher soil moisture levels result in larger truffle sizes. Consequently, the truffle harvesting season extends, often lasting until April-June, contingent upon regional variations.

By referring to the averages recorded during the years deemed favorable for production by collectors, we can provide some insight into the expected quantities conducive to favorable conditions (Table 1). The years considered to have good productivity by region are:

North-East Region: The most recent year of significant production was 2018, when average rainfall varied from 402.95 mm to 776.7 mm. Notably, a rainfall level of approximately 195 mm has been associated with good production, as observed in Figuig in 1995.

Maamora Forest: Average rainfall ranged from 485 mm to 1236.4 mm.

Doukkala-Abda Sahel: Average rainfall ranged from 338.6 mm to 530.25 mm.

Moroccan Sahara: The average rainfall ranged from 50.2 mm to 122.3 mm. Notably, a rainfall level of around 37.6 mm has been linked to good production, as observed in Dakhla in 2017.

Our findings regarding climatic conditions for desert truffles corroborate Khabar›s (2016) assertion that the production of desert truffles is contingent upon favorable rainfall in early autumn for certain species, while others depend on rainfall during winter and spring. According to Khabar (2002), in the Maâmora forest, truffles necessitate substantial rainfall before March, with a threshold of 240 mm being adequate to ensure good production. In the North-East region, El Aji (1999) documented that the years 1990/1991 and 1993/1994 witnessed good production in Bouaarfa, with rainfall measuring 222.6 mm and 214 mm, respectively.

3.2 Truffle-Growing Soils

Apart from the four aforementioned regions, soil analyses were also conducted in the Jaaba forest in the Middle Atlas, renowned for its production of true truffles, and the Zouada forest in the northern part of the country, known for its production of Terfezia arenaria, a common species found in the Maâmora forest. Across most of the studied areas, the soils exhibited a predominant sand content ranging from 13.5% to 87.5%, a moderate silt content ranging from 7.5% to 66.5%, and a low clay content ranging from 5% to 20%. Consequently, the soils predominantly exhibited sandy, silty, or sandy-loam textures, rendering them moderately suitable for truffle production. These soils were characterized as loose to slightly dense in consistency (Table 2).

The chemical analyses conducted on these soils (see Table 3) reveal slightly acidic to neutral to slightly alkaline pH values. Notably, the North-Eastern region and the Moroccan Sahara exhibit almost identical pH values, at 8.38 and 8.37, respectively.

Table 1. Rainfall (mm) in years of good truffle production in truffle-growing regions of Morocco (P, average annual precipitation; SD, standard deviation; CV, variation coefficient) (Source: Provincial Water and Forestry Departments, Morocco).

Year	Maâmora	North-East			Doukkala Abda Sahel		Moroccan Sahara	
	Rabat-Salé	Figuig	Jerada	Oujda	Sidi Msahel	El Jadida	Tan-Tan	Dakhla
1990	413.0	235	425	422.9				
1991	513.6	325	480	293.5				
1992	280.4	170	400	263.1				
1993	434.6	230	415	377.0				
1994	208.8	210	360	344.5				
1995	321.1	195	370	240.9			105.9	
1996	1091.7	275	445	393.7				
1997	545.3	215	315	352.3				
1998	261.4	120	340	324.1				
1999	418.8	190	385	248.1				
2000	504.2	130	265	225.8				
2001	349.0	85	375	243.3				
2002	683.4	215	395	252.9				
2003	602.5	220	530	305.6				
2004	619.0	205	480	310.4				
2005	404.9	225	320	217.1				
2006	505.2	275	400	235.0				
2007	514.9	225	485	346.7				
2008	848.9	510	645	358.4				
2009	651.5	285	435	249.9	385.0	475		
2010	951.8	210	540	322.1	283.0	394.2	283.0	
2011	424.4	370	495	300.6	438.0	449.7		
2012	414.2	300	505	305.4	143.0	474.3		
2013	402.9	160	480	202.2	82.0	348.6	71.4	0
2014	576.3	405	405	306.3	535.0	525.5	153.5	52.8
2015	243.3	235	400	295.1	156.0	272.5	53.9	46.5
2016	485.0	225	455	240.5	444.0	496.4	197.1	47.5
2017	423.2	176	380	207.0	305.0	274.1	84.4	37.6
2018	1236.4	294	498	325.0	274.0	577.7	19.8	15.7
2019	397.4	236	440	266.4	232.5	251.4	40.1	5.3
2020	348.7	211	415	241.5	347.0	320.7	42.7	16.8
2021	505.3	123	327	153.7	74.5	315.8	48.7	0.8
P value	518.2	233.9	425.2	286.6	283.9	397.7	100.0	24.8
SD	234.0	85.1	77.4	61.2	144.5	108.2	80.6	21.4
VC	45.2	36.4	18.2	21.4	50.9	27.2	80.5	86.2

Table 2. Soil texture and bulk density in truffle-growing areas of Morocco.

Area	Clay (%)	Loam (%)	Sand (%)	Texture	Bulk Density (g/cm^3)
North-East region	13.6	37.19	49.19	Loamy	1.38
Maamora forest	5.0	7.5	87.50	Sandy	1.47
Doukkala-Abda Sahel	7.0	17.4	75.60	Sandy-loam	1.58
Moroccan Sahara	11.37	30.93	57.70	Loamy-sand	1.67
Jaâba Forest, Middle Atlas	20.0	66.5	13.50	Loamy	0.75
Zouada Forest, Larache	10.0	7.8	82.20	Sandy-loam	1.76

Table 3. Chemical characteristics of soils in truffle-growing areas of Morocco.

Parameter North-East region		Area					
		Maâmora forest	Doukkala-Abda Sahel	Moroccan Sahara	Jaaba forest, Middle Atlas	Zouada forest, Larache	
pH	H_2O	8.38	6.20	7.6	8.37	7.56	5.85
	KCl	7.7	5.72	7.18	7.90	6.96	5.57
$CaCO_3$ (%)		5.76	0	4.13	4.58	0	0
% H_2O (%)		1.14	0.23	1.08	1.66	1	0.3
Organic matter (%)		1.68	1.95	2.65	0.55	12.36	0.72
Inorganic carbon (%)		0.97	1.13	1.53	0.32	7.17	0.42
Nitrogen (%)		0.22	0.13	0.12	0.13	0.17	0.03
C/N ratio		3.88	9.97	13.7	2.52	40.96	14.84
P_2O_5 (mg/kg)		15.45	6.54	60.3	14	43.47	3.51
K_2O (mg/kg)		397.65	57.57	62.77	306.4	1018.23	57.24
K^+ (meq/100 g)		0.83	0.12	0.13	0.64	2.14	0.12
Na^+ (meq/100 g)		0.37	0.26	0.32	1.03	0.39	0.20
Ca^{2+} (meq/100g)		23.91	2.53	13.66	22.06	18	1.75
Mg^{2+} (meq/100g		2.69	1.19	2.33	4.68	5	1.25

The pH value for the Jaaba forest closely resembles that of Sahel Doukkala-Abda, measuring 7.56 and 7.60, respectively. Their limestone content ($CaCO_3$) varies significantly from zone to zone, ranging from non-calcareous to slightly calcareous. Generally, these soils exhibit low organic matter content in the surface horizons (less than 2%), except in the Middle Atlas under dense holm oak, where it boasts a rich content of 12.36%. Assimilable phosphorus (P_2O_5) levels range from very low (Maâmora, Moroccan Sahara, and Zouada Forest) to low (North-East), well-supplied (Jaaba), and high (Doukkala-Abda Sahel). Exchangeable potassium levels exhibit considerable variation across zones, ranging from very low (Maâmora and Zouada forests) and low (Sahel de Doukkala-Abda) to very high levels in the northeast, the Moroccan Sahara, and the Jaaba forest, which records the highest value at 1018.23 mg/kg.

The C/N ratios are low in all arid zones, indicating rapid mineralization of organic matter. Conversely, in the humid Middle Atlas, C/N ratios are relatively high, suggesting an accumulation of surface organic matter derived from holm oak litter. Regarding exchangeable bases (Na^+, Ca_2^+, and Mg_2^+), values exhibit considerable variability between zones, yet indicate that these soils are not saturated with bases. The soils generally boast a high Ca_2^+ content, except in the Maâmora and Zouada forests, which are predominantly occupied by cork oak, a species known to be calcifuge.

Overall, soils in truffle-producing regions of Morocco are characterized as sandy to silty, with low organic matter content, relatively calcareous, and exhibiting a neutral pH. Moreover, these soils are not saturated with exchangeable bases. These findings align closely with those reported for truffle soils by various authors across North Africa, including El Aji (1999), Bouziani (2009), Khabar (2016), and Abourouh (2020) for Morocco, as well as Aïbeche (2008), Zitouni-Haouar (2010), and Bradai et al. (2013) for Algeria.

3.3 Truffle Host Plants

In Morocco, the plants with which desert truffles live in symbiotic association are generally *Helianthemum* (Fig. 2). These plants have several vernacular names depending on the region of origin, but the common names throughout the country are "*Tegcis*" and "*Yarguig*". Some truffle species in the Maâmora forest are linked to pine reforestation.

North-east region: *Helianthemum ledifolium*, known locally as "*Tegcis*" or "*Tzaoua*", *Helianthemum pergamaceum*, known locally as "*Serd Amalou*" or "*Izefzaf*", *Helianthemum hirtum*, known locally as "*Idoudi*" and *Helianthemum lipii* known locally as "*Reguig*".

Maamora forest: *Helianthemum guttatum*, known locally as "*Tegcis*" and *Pinus pinaster* var. *atlantica*, known locally as "*Taida*".

Doukkala-Abda Sahel: *Helianthemum guttatum* and *Helianthemum salicifolium*, both locally called "*Tegcis*" or "*laqsid*".

Moroccan Sahara: the most common host plant is *Helianthemum lipii*, known locally as "*Yarguig*" or "*Izzi*". As for true truffles, desert truffles are collected in the vicinity of holm oaks.

The host plant species identified in the studied areas (Fig. 3) align with those mentioned by several authors. For instance, Tahiri (1997) and Khabar (1988, 2016) have reported that Moroccan truffles are often found in association with herbaceous plants of the genus *Helianthemum*; *H. apenninum, H. apertum, H. croceum, H. glaucum, H. guttatum, H. hirtum, H. ledifolium*, and *H. lipii*; *Cistus*; *Cistus halimifolius, C. ladaniferus, C. salicifolius, C. monspeliensis*, and *C. salvifolius*; Pinus: *Pinus halepensis*, and *P. pinaster* var. *atlantica*.

Fig. 2. Host plant species of Moroccan truffles: *Helianthemum guttatum* (a), *Helianthemum hirtum* (b), *Helianthemum ledifolium* (c), *Helianthemum lipii* (d), *Helianthemum pergamaceum* (e), *Helianthemum salicifolium* (f), *Pinus pinaster var. atlantica* (g), *Quercus ilex* (h), (© Hakkou).

4. Truffle Productivity

Field observations, surveys, and participatory workshops with resource persons have shown that truffle productivity varies within the same region and from place to place. Some areas are known by collectors to be highly productive, others are moderately productive, and the remaining are not very productive. The country's highly productive areas are the Maâmora forest, the northern part of the northeastern region, the northern part of the Doukkala-Abda Sahel, and the central Moroccan

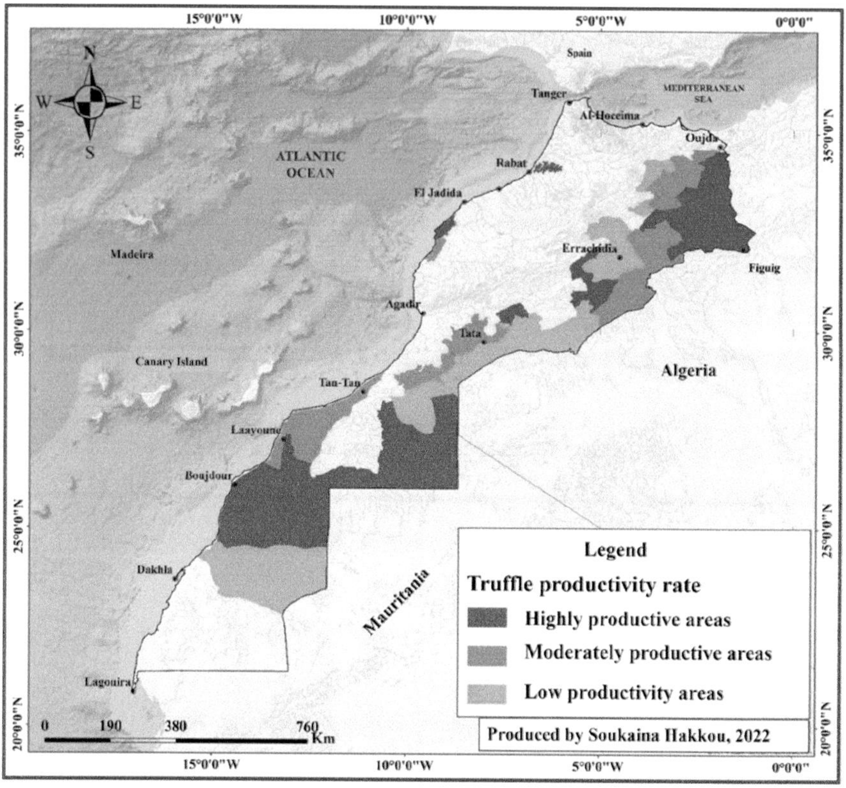

Fig. 3. Locations of truffle production in Morocco.

Sahara. Moderately productive areas are the southern part of the Sahel Doukkala-Abda, the center and extreme south of the northeastern region, and the northwestern part of the Sahara. Productivity becomes increasingly low towards the west in the north-eastern region and towards the east and extreme south of the Moroccan Sahara.

Surveys with wholesalers who buy truffles in all the country›s truffle-growing regions show that production differs in terms of productivity and season. Chronologically, production starts first in January in the Doukkala-Abda Sahel, then a few days to a few weeks later in the Maâmora forest, and finally in the same period in the Moroccan Sahara and Oriental regions. When rainfall conditions allow, the Bouaarfa white truffle is early, starting as early as November. Rainfall determines productivity and the length of the production season. The latter generally lasts until March for the Doukkala-Abda Sahel, until April for the Maâmora, Oriental, and Moroccan Saharas, and until June for the Upper Moulouya (Midelt). Productivity is highest in the Bouaarfa, Boujdour, and Smara areas.

In years of good productivity, the rate rises to 6 tonnes or more per day in the Sahara at Boujdour and Smara and in the northeastern region at Bouaarfa, where red truffles predominate. In the Sahel Doukkala-Abda region and the Maâmora forest, sales can reach 2 to 3 tonnes per weekly souk. Analysis of the data collected shows that truffle production in the Kingdom of Morocco has fallen sharply in recent years due to a succession of drought years, especially in the North-East and Sahara regions. This shows that this resource is highly vulnerable to climate change and it follows inter- and intra-annual variations in rainfall in truffle-growing areas.

5. Conclusion

Morocco is one of the world›s best-known producers of desert truffles. There are around ten species, distinguished by color, shape, host plant, climatic and edaphic conditions, and production zone. Natural truffle production takes place in the Kingdom›s four main truffle-growing zones: the North-East region, the Maâmora forest, the Doukkala-Abda Sahel, and the Moroccan Sahara. Other species of the *Tuber* genus, known as true truffles, also exist, but their production is not regular, and they are not abundant. They are found in the holm oak forests of the Middle Atlas. Morocco has also been growing black truffles, *Tuber melanosporum*, on three sites: one at Tifezouine (Debdou in the North-East region) since 2000 and two others at Imouzzer Kandar since 2008 (Middle Atlas). Truffle production in the studied areas necessitates specific conditions that are congruent with some factors while divergent with others. These truffles typically require rainfall during two distinct periods: the first in summer or autumn and the second in winter. The soils in these regions are predominantly sandy to silty, characterized by low organic matter content, generally deficient in limestone ($CaCO_3$), and feature a pH close to neutral, while not being saturated with exchangeable bases. The prevalent host plants in these areas are Helianthemum species, except for certain species in the Maâmora forest and true truffles associated with tree species such as holm oak and maritime pine. Knowledge of the truffle species existing in Morocco, their precise geographical distribution, the conditions favorable to their production, and their productivity can contribute to their development.

References

Abourouh, M. (2011). Truffes du désert du Maroc: diversité et modes d'exploitation. 6ème Rencontre de Micosylva , Mértola (ADPM), Portugal, Mars 15–18, p. 23.

Abourouh, M. (2020). Terfess et truffes de la Maâmora : importances biologique, écologiques et socio-économique. Consultation nationale Restauration de la forêt de chêne-liège pour le développement et la valorisation des truffes dans la forêt de la Maâmora, FAO, p. 78.

Aïbèche, C. (2008). Caractérisation écologiques et mycologiques d'une espèce de terfez du littoral Ouest algérien. Essai de mycorhization contrôlée avec sa plante-hôte naturelle *Helianthemum guttatum*. Mémoire de Magister, Univ d'Oran, Es-Sénia, p. 85.

Alsheikh, A.M. (1994). Taxonomy and mycorrhizal ecology of the desert truffles in the Genus *Terfezia*. Ph.D. dissertation. Oregon State University, Corvallis, p. 239.

Badalyan, S. (2012). Medicinal Aspects of Edible Ectomycorrhizal Mushrooms. pp. 317–334. In: Zambonelli, A., Bonito, G.M. (eds.). Edible Ectomycorrhizal Mushrooms, Soil Biology 34, Springer-Verlag Berlin Heidelberg.

Bouatia, M., Touré, H.A., Cheikh, A., Eljaoudi, R., Rahali, Y. et al. (2018). Analysis of nutrient and antinutrient content of the truffle (*Tirmania pinoyi*) from Morocco. Int. Food Res. J., 25(1): 174–178.

Boufeldja, W. (2017). Evaluation du potentiel nutritionnel et antioxydant de quelques variétés de Truffes du Sud-ouest algérien. Effets antimicrobien et antiinflammatoire. Thèse de doctorat. Université Djillali Liabes, p. 112.

Bouziani, N. (2009). Contribution à l'étude et à la mise en valeur du potentiel truffier de la région orientale du Maroc. Thèse. Université Mohammed Premier, Faculté des Sciences Oujda, p. 177.

Bouziani, N., Boukka, M., Belabed, A. and Tahri, E.H. (2010). Le Terfess au Maroc Oriental: Isolement et culture *in vitro*. Rev. Microbiol. Ind. San et Environn., 4(2): 110–120.

Bradai, L. (2014). Inventaire et bioecologie des terfez dans quelques régions du Sahara septentrional (oued M'ya, oued Righ et Ghardaïa). Thèse de doctorat ès science. Université Kasdi Merbah – Ouargla, p. 124.

Bradai, L., Bissati, S. and Chenchouni, H. (2013). Étude mycologique et bioécologique de la truffe blanche du désert (*Tirmania nivea* Desf. Trappe 1971) dans la région de Oued M'ya (Ouargla, Sahara Algérien). Revue des BioRessources, 3(1): 6–14.

Callot, G. (1999). La truffe, la terre, la vie. Ed. INRA., Paris, p. 209.

Chatin, A. (1896). Un terfas d'Espagne et trois nouveaux terfas du Maroc. Bull. Soc. Bot., 43: 397–399.

De Laplane, D. (2010). Des truffes sauvages aux truffes cultivées en Loudunais. Mémoires de l'Académie des Sciences, Arts et Belles-Lettres de Touraine, 23: 215–241.

Dexheimer, J., Gerard, J., Leduc, J.P. and Chevalier, G. (1985). Comparative ultrastructurale study of symbiotic mycorrhizal associations between *Helianthemum salicifolium-Terfezia claveryi* and *Helianthemum salicifolium-Terfezia leptoderma*. Can. J. Bot., 63: 582–591.

El Aji, A. (1999). Contribution à l'étude des Terfez des parcours de l'oriental marocain : germination des ascospores in vitro et mycorhization de l'*Helianthemum ledifolium*. Mémoire de troisième cycle en agronomie, École nationale d'agriculture, Meknès, p. 84.

El Akil, M., Outcoumit, A., Ouazzani Touhami, A., Benkirane, R. and Douira, A. (2016). Study of eastern Morocco desert truffles. Int. J. Curr. Res., 8: 33922–33929.

Fortas, Z. and Chevalier, G. (1992). Effet des conditions de culture sur la mycorhization de l'*Helianthemum guttatum* par trois espèces de terfez des genres *Terfezia* et *Tirmania* d'Algérie. Can. J. Bot., 70: 2453–2460.

Fortas, Z. (2009). Diversité des espèces de terfez (truffes des sables) des zones arides algériennes. Séminaire International: Protection et préservation des écosystèmes sahariens, Ouargla, p. 51.

Grente, J. and Chevalier, G. (1980). La trufficulture, un or noir, pour la Région Méditerranéenne. Forêt Méditerranéennes, 1(2): 151–1 62.

Hakkou, S. (2019). Valorisation écotouristique des plantes aromatiques et médicinales dans la région de Rabat-Salé-Kénitra-. Mémoire de master. FLSH, Université Mohamed V, Rabat, p. 251.

Hakkou, S., Sabir, M. and Machouri, N. (2022). Le Maroc truffier : la répartition géographique des truffes et truffes du désert et leur productivité. Rev. Mar. Sci. Agron. Vét . 10(3): 395–407.

Hakkou, S., Sabir, M. and Machouri, N. (2023). Contribution to the study of the production factors of Moroccan truffles (Soil, Climate, Vegetation). J. Mater. Environ. Sci., 14(1): 13–30.

Khabar, L. (1988). Le genre *Terfezia* Tull. (Terfass) de la Forêt de la Maâmora (Région de Salé): étude systématique, écologique, morphologique, cytologique et ultrastructurale. Université Mohammed V, Rabat. Thèse de troisième cycle, p. 150.

Khabar, L. (2002). Études pluridisciplinaires des truffes du Maroc et perspectives pour l'amélioration de production des Terfess de la forêt de la Maâmora. Doctorat d›État Es- Sciences, Université Mohamed V, Faculté des Sciences, Rabat, Marocco, p. 167.

Khabar, L. (2016). Les Terfess et les truffes du Maroc, biodiversité et valorisation. Ed: Univ. Européennes, p. 276.

Laessoe, T. and Hansen, K. (2007). Truffle trouble: what happened to the Tuberales. Mycol. Res., 111: 1075–1099.

Laqbaqbi, A. (2020). La truffe entre France et Maroc ou d'un coté à l'autre de la Méditerranée. Imprimerie GRAPHO, 12 Aveyron, p. 80.

Malençon, G. (1973). Champignons hypogés du Nord de l'Afrique. I. Ascomycètes. Persoonia, 7: 261–288.

Morte, A., Honrubia, M. and Gutierrez, A. (2008). Biotechnology and cultivation of desert truffles. In: Varma, A. (ed.). Mycorrhiza: state of the art genetics and molecular biology, eco-function, biotechnology, eco-physiology, structure, and systematics. Springer Verlag, Berlin, Heidelberg, pp. 467–483.

Olivier, J.-M, Savignac, J.-C. and Sourzat, P. (2018). Truffe et trufficulture. Périgueux, Fanlac, p. 352.

Rodriguez, A. (2008). Desert truffles. Sur. http://www.trufamania.com/desert-truffles.htm

Serrhini, M., El Youssoufi, K. and Allali, K. (1995). Une ressource naturelle exceptionnelle des milieux pastoraux de l'Oriental et du centre Sud Marocain : le terfèze, situation actuelle et perspectives d'amélioration. École Nationale d'agriculture, Meknès, p. 33.

Tahiri, A. (1997). Relation mycorhizienne entre *Helianthemum ledifolium* (L) Mill, et des espèces de Terfez du Maroc. Mémoire de troisième cycle en agronomie, Ecole nationale d'agriculture, Meknès, p. 80.

Trappe, J.M., Claridge, A.W., Arora, D. and Smit, W.A. (2008). Desert truffles of the African Kalahari: Ecology, ethnomycology, and taxonomy. Econ. Bot., 62(3): 521–529.

Wasser, S.P. (2002). Medicinal mushrooms, as a source of antitumor and immunomodulating polysaccharides. Appl. Microbiol. Biotechnol., 60: 256–274.

Zitouni-Haouar, F.H. (2010). Etude des associations mycorhiziennes entre quatre espèces de terfez et diverses plantes Cistacées et ligneuse en conditions contrôlées. Mém. Magister, Univ, Oran, p. 262.

Zitouni-Haouar, F.H. (2016). Etude de la diversité des truffes du désert et de leurs associations mycorhiziennes. Thèse de doctorat. Université d'Oran 1 Ahmed Ben Bella, p. 95.

Chapter 6

Diversity, Distribution and Significance of Truffles in Türkiye

Mehmet Akyüz[1,*] and *Sevda Kırbağ*[2]

1. Introduction

The fungi kingdom is the second largest biological community after insects (Purvis and Hector, 2000). They evolved about 1.8 billion years ago, making them an ancient, diverse and heterogeneous group of organisms (Hawksworth, 1991; Kirk et al., 2008). Fungi play a crucial role in ecosystems as endophytes, pathogens and saprobes. They exhibit considerable variation in morphology, ranging from unicellular to filamentous to sporocarps, and possess diverse life cycles. Many fungi form specialized associations with other organisms, especially plants. Fungi displays an enormous diversity of taxa, with their broad ecologies, different life cycle strategies, and morphologies ranging from single-celled water chytrids to molds, yeasts, dermatophytes, and macrofungi (Hawksworth and Lucking, 2017).

The estimated numbers for fungi have been an intriguing topic, with numerous papers discussing them over time (Hawksworth, 1991, 2001; O'Brien et al., 2005; Blackwell, 2011; Hawksworth and Lücking, 2017; Willis, 2017; Wu et al., 2019; Hyde et al., 2020; Lücking et al., 2021; Baldrian et al., 2022; Phukhamsakda et al., 2022). However, as a result of recent advances in molecular genetics, the integration of DNA analysis into taxonomy and the use of reliable statistical phylogenetic methods, it is still too early to estimate the exact number of fungi. Present assessments of fungal diversity exhibit considerable uncertainty, with estimates ranging from 1.5

[1] Department of Biology, Faculty of Arts and Science, Bitlis Eren University, Bitlis, Türkiye.
[2] Department of Biology, Faculty of Science, Fırat University, Elazığ, Türkiye.
* Corresponding author: makyuz@beu.edu.tr

to 12 million species. However, only approximately 150.000 species have been formally named and classified thus far (Hyde, 2022; Bhunjun et al., 2022). The phylum Ascomycota and Basidiomycota contain the majority of described fungal species with 92.724 in Ascomycota and 50.385 in Basidiomycota (Catalogue of Life et al., 2021). Among the species identified, there are approximately 14.000 to 22.000 varieties of mushrooms known, with over 3.000 being edible, and around 700 of these types utilized for medicinal applications (Hawksworth, 2001; Mueller and Schmit, 2007; Kirk et al., 2008; Wasser, 2011; Cerletti et al., 2021).

The human relationship with mushrooms is ancient and fascinating, and they are believed to be endowed with supernatural powers. Mushrooms were believed by the Greeks to provide strength to soldiers in wars, whilst the Romans deemed them the "food of the gods", and the Chinese considered them as the "elixir of life" (Manzi et al., 1999; Valverde et al., 2015). Macrofungi have served as both food and medicinal agents throughout history, and remain crucial to human existence today supplying nourishment, vital compounds, medical applications and economic income (Hyde et al., 2020).

A mushroom is classified as a macrofungus possessing a distinctive fruiting body, observable either above ground (epigeous) or below ground (hypogeous), and is sufficiently sizable to be visible without magnification and to be harvested manually (Chang, 2001). Edible fungi can be categorized into two groups: "epigeous" carpophores, commonly referred to as mushrooms, and "hypogeous" species that thrive underground, known as truffles (Lee et al., 2020). Among edible mushrooms, hypogeous mushrooms show some distinctive characteristics, and recently received more attention than other epigeous mushrooms. Various terms have been used to describe potato-shaped fungi that grow under or close to the soil surface, including "truffle, true truffle, forest truffle, desert truffle, false truffle, truffle-like fungi, sequestrate, and hypogeous" fungi. These terms refer to a specific type of truffles known as hypogeous fungi. Hypogeous fungi are defined as species with closed or ‹sequestrated› globose fruiting bodies growing completely below or partially covered by the soil surface (Kagan-Zur et al., 2014; Zambonelli et al., 2016). Truffles comprise a diverse range of mycorrhizal macrofungi that grow beneath the surface of the soil, forming a mutually beneficial association with the roots of certain trees, shrubs, and herbs. When all these concepts are taken into consideration, it would be more accurate to evaluate the hypogeous species that grow under the soil or close to the surface in two separate groups as "true" and "false" truffles. In this chapter, the distribution of true truffles of commercial and economic importance in Türkiye is discussed. In addition, truffle-like (false truffles), which are important in terms of biodiversity, are also mentioned and their records and distribution in Türkiye are given.

1.1 False (Truffle-Like) and True Truffles

Within the Ascomycota and Basidiomycota phylum, there exist both true and false hypogeous species. Despite differing ontogenic processes, these species often exhibit a similar external appearance and ecological behavior (Moreno et al., 2014). Hypogeous fungi within the Basidiomycota are occasionally referred to as "false truffles". Initially, this term was employed to differentiate these truffles from those within the Ascomycota. Nevertheless, false truffles or truffle-like species are also present in the Ascomycota (Kagan-Zur et al., 2014; Bonito and Smith, 2016). It would therefore be incorrect to say that false truffles belong exclusively to the Basidiomycota. Numerous fungal species generate their fruiting bodies underground, commonly referred to as "truffles" However, strictly speaking, true truffles pertain to the hypogeous fruit bodies of the genus *Tuber* (Ascomycota, Pezizales). Other similar structures are more accurately labeled as false truffles, truffle-like, or desert truffles (Leonardi et al., 2021). While the terms ‹truffle› and ‹true truffle› may be suitable for hypogeous species in the Ascomycota that grow in forests or arid, semi-arid or desert regions, it is more accurate to use ‹truffle-like›, ‹semi-hypogeous›, ‹sequestrate›, or ‹false truffle› for hypogeous species within the Basidiomycota.

The hypogeous fungi were classified into two separate groups: True truffle and truffle-like or false truffle. While the former belong to the Ascomycota, the latter can be found in both the Ascomycota and Basidiomycota. From this point onward in the study, only true and truffle-like truffles within the Ascomycota phylum have been considered. Thus, the term 'truffle-like' is specifically applied to species within the Ascomycota. While false truffles are important in terms of fungal biodiversity, true truffles are more important in terms of their economic level, commercial potential, gastronomic properties (such as nutritional, flavor, organoleptic properties) and medicinal properties rather than biodiversity. In general, it would be helpful to categorize true truffles, which possess both commercial and economic value as well as nutritional, culinary, gastronomic, and medicinal benefits, into two distinct groups: forest truffles (*Tuber* species) and semi-arid, arid or desert truffles (*Terfezia*, *Tirmania* and *Picoa* species).

While the true truffles (*Tuber*) possess unique flavors, aromas, organoleptic and gastronomic properties, as well as nutritional and medicinal effects, they are often expensive and rare (Zambonelli et al., 2016). Therefore, it would be unwise to overlook the commercial, economic, nutritional, gastronomic, and medicinal benefits of desert truffles (Kagan-Zur et al., 2014). For the purposes of this book, we will refer to certain Ascomycota as 'false truffles', and use the term 'true truffles' to describe forest truffles (*Tuber*) and desert truffles (*Terfezia*, *Tirmania* and *Picoa*) in the Ascomycota. True truffles are consumable hypogeous fruit bodies generated by various genera of fungi within the Ascomycota. Among these, *Terfezia* and *Tirmania* (Pezizaceae Dumort.), *Picoa* (Pyronemataceae Corda), and *Tuber* (Tuberaceae Dumort.) are classified into different orders within Pezizales (Moreno et al., 2014).

2. Diversity and Distribution of False Truffles

Türkiye is a region of highly diverse biodiversity located in the Northern Hemisphere. This diversity is due to varying elevations, climatic conditions and a combination of three phytogeographic regions: Euro-Siberian, Mediterranean, and Iranian-Turanian. Plant diversity in temperate and subtropical regions is significant for the habitat of various fungal species. In Türkiye, the most common tree families are mainly ectomycorrhizal; namely Betulaceae, Fagaceae, Pinaceae, and Salicaceae. These tree species serve as habitats for *Tuber* and other truffle-like or false truffle species, specifically (Gryndler, 2016; Bonito and Smith, 2016). These mutualistic ectomycorrhizal relationships create very beneficial interactions. Cistaceae, in the form of shrubs, dominate semi-arid, arid or desert regions establish ectomycorrhiza-like mutualisms with desert truffles (*Terfezia*, *Tirmania* and *Picoa*) (Díez et al., 2002; Castellano and Türkoğlu, 2012; Kagan-Zur et al., 2014; Bradai et al., 2015ab; Zambonelli et al., 2016; Bonito and Smith, 2016; Leonardi et al., 2021). The distribution of truffle species is closely linked to climatic and edaphic conditions.

The results of extensive research on macrofungi in other countries (Hawksworth, 1991, 2001; O'Brien et al., 2005; Laessoe and Hansen, 2007; Mueller and Schmit, 2007; Kirk et al., 2008; Blackwell, 2011; Kagan-Zur et al., 2014; Bonito and Smith, 2016; Zambonelli et al., 2016; Hawksworth and Lücking, 2017; Wu et al., 2019; Hyde et al., 2020; Lee et al., 2020; Leonardi et al., 2021; Bhunjun et al., 2022) have significantly contributed to the identification of different fungal species in Türkiye. Türkiye harbors a significant diversity of fungi within its diverse ecosystem. The use of molecular techniques has facilitated the documentation of a large number of fungal discoveries, so that a large number of fungal species have been meticulously documented and classified by mycologists in Türkiye in recent years. This comprehensive survey covers all taxa described in books (Barutçiyan, 2012, 2021; Sesli et al., 2020), checklists (Sesli and Denchev, 2014; Şen et al., 2016; Akata et al., 2022; Atila and Kazankaya, 2022; Uzun, 2023), theses and academic papers (see references in Table 1, 2) on the subject and provides valuable insights.

This research provides information on 27 taxa associated with truffle-like or false truffles belonging to Ascomycota, categorized into 5 families (see Table 1). All the aforementioned taxa and their taxonomies, locations, habitats, host plant, collection season and references are shown in Table 1. Those references presented in Table 1 provide comprehensive species-specific information on the ecology, distribution, hosts, soil properties, microscopic and macroscopic characteristics of various taxa, as well as the molecular identification of specific species. Table 1 presents concise information regarding the species detected in Türkiye to date, including their taxonomy, hosts, appearing period, and reference records. These false or truffle-like species live in mycorrhizal association with high tree plants (Hoeksema and Thompson, 2007; Trappe and Claridge, 2010; Bonito and Smith, 2016). They can be found in various seasons, depending on the species, environment, soil characteristics, and favorable seasonal conditions (Laessoe and Hansen, 2007).

While some species are consumed by local people, most do not possess the gastronomic characteristics, unique musky aroma and delicious flavor content of true truffles. Nonetheless, their significance in terms of biodiversity remains of utmost importance. These species are not widely acknowledged as they lack economic and commercial value. They are known to people living in the areas where they are found, but are relatively unknown in most areas. Most of these false truffle or truffle-like species are not commonly eaten due to their distinct taste, flavor, cooking properties, nutritional composition, and organoleptic properties in comparison to forest and desert truffles. Rather, these species comprise the food source for indigenous wildlife in the area. Thus, numerous false truffle species are not consumed and retain significance exclusively for biodiversity. Many of the common beliefs about false or truffle-like species mentioned earlier have already been documented by various researchers (Table 1).

Table 1. Diversity and distribution of false truffles (truffle-like) in Türkiye.

Species	**Locality**	**Host**	**Season**	**Reference**
ASCOMYCOTA				
Elaphomycetaceae Tul. ex Paol.				
Elaphomyces anthracinus Vittad.	Trabzon	under *Fagus orientalis* Lipsky, *Rhododendron ponticum* L. and *Quercus* sp.	November	Uzun (2021)
Elaphomyces citrinus Vittad.	Trabzon	mixed forest, under *Castanea sativa* Mill., *Corylus avellana* L. and *F. orientalis*	September	Uzun and Kaya (2020)
Elaphomyces cyanosporus Tul. & C. Tul.	Trabzon	under *F. orientalis* and *Carpinus betulus* L	July	Uzun and Kaya (2020)
	Trabzon	under mixed forest and under *Fagus* L. sp.	April, September, November	Uzun and Kaya (2022)
	Rize	under *Fagus orientalis* and *Rhododendron ponticum*	August	Uzun and Kaya (2020)
Elaphomyces decipiens Vittad.	Trabzon	under *Castanea* sp., *Fagus orientalis* and *Rhododendron ponticum*	February	Uzun and Kaya (2021)
Elaphomyces granulatus Fr.	Giresun	under *Fagus orientalis* and *Rhododendron ponticum*	February	Uzun and Kaya (2019a)
	Rize	under *Fagus orientalis*, *Castanea sativa*, *Alnus glutinosa* and *Rhododendron ponticum* mixed forest	February	
	Trabzon	under *Fagus orientalis*, *Castanea sativa*, *Picea orientalis* and *Rhododendron ponticum* mixed forest	August, November	
	Muğla	-	November	Allı (2022)

Elaphomyces leucocarpus Vittad.	Artvin	mixed stand of *Alnus glutinosa*, *Picea orientalis*, *Salix coprea*, *Rhododendron ponticum*, and *Fagus orientalis*	May	Türkoğlu et al. (2015)
	Trabzon	mixed stand of *Quercus* sp., *Corylus* sp., *Pinus sylvestris*, *Rhododendron ponticum*, and *Fagus orientalis*	October	
	Tekirdağ	under *Pinus nigra*	Januray, March	
Elaphomyces muricatus Fr.	Artvin	mixed stand of *Alnus glutinosa*, *Picea orientalis*, *Salix coprea*, *Rhododendron ponticum*, and *Fagus orientalis*	May	Türkoğlu et al. (2015)
	Karabük	mixed stand of *Quercus* sp., *Pinus sylvestris*, and *Fagus orientalis*	November	
	Trabzon	under mixed forest	April, May	Uzun and Kaya (2022)
Elaphomyces septatus Vittad.	İstanbul	*Fagus*, *Quercus*, *Pinus* L., and *Tilia* L. mixed forest, around *Fagus* tree, under soil and plant debris	March	Uzun and Kaya (2019b)
Helvellaceae Fr.				
Balsamia vulgaris Vittad.	Muğla	under *Quercus robur* L.	March	Allı and Doğan (2019)
Barssia hellenica Kaounas, Agnello, P. Alvarado & Slavova	Gaziantep	in soil, under *Abies cilicica* (Ant. et Kotschy) Carr. subsp. *Cilicica* and mixed forest	April	Uzun et al. (2018)
Brassia gunerii (H.H. Doğan, Bozok & Taşkın) K. Hansen & X.H. Wang	Osmaniye	under *C. libani*	June	Doğan et al. (2018)
Pezizaceae Dumort.				

Table 1. contd. ...

Table 1. contd. ...

Species	Locality	Host	Season	Reference
Delastria rosea Tul. & C. Tul.	Edirne	pine forest, under pine (*Pinus* sp.)	March, December	Akata et al. (2023)
Pezizaceae Dumort.				
Hydnobolites cerebriformis Tul. & C. Tul.	Rize	mixed forest, under *Corylus*	August	Uzun and Kaya (2018)
	Trabzon	mixed forest, under *Fagus orientalis*	November	
Pachyphlodes citrina (Berk. & Broome) Doweld	Trabzon	under *Quercus*	May, June	Uzun and Kaya (2018)
	Rize	under *Corylus*	July	
Pachyphlodes conglomerata (Berk. & Broome) Doweld	Giresun	under *Fagus* and *Carpinus*	September	Uzun and Kaya (2018)
	Trabzon	under *Alnus*, *Castanea*, and *Rhododendron*	September	
Pyronemataceae Corda				
Genea hispidula Berk. ex Tul. & C. Tul.	Trabzon	in *Fagus orientalis* Lipsky forest	October	Uzun and Kaya (2019c)
Genea klotzschii Berk. & Broome	Samsun	*Corylus* sp.,	November	Türkoğlu and Castellano (2014)
Genea lobulata (Mor.-Arr., J. Gómez & Calonge) P. Alvarado & Mor.-Arr.	Niğde	under *Quercus* sp.,	April	Berber et al. (2022)
Genea sphaerica Tul. & C. Tul.	İzmir	under mixed stand of *Pinus nigra* and Quercus cerris	June	Türkoğlu et al. (2015)
Genea verrucosa Vittad.	Aydın	under *Pinus brutia*	November	Türkoğlu and Castellano (2014)
Geopora clausa (Tul. & C. Tul.) Burds.	Karaman and Trabzon	in soil under *Quercus* sp. and *Pinus pinaster*	May and November	İleri et al. (2020), Uzun and Kaya (2019d)

Geopora cooperi Harkn	Bolu	-	November	Türkoğlu et al. (2015)
	Denizli	-	May and November	
	Burdur	-	November, April	
	Muğla	-	April	
	İzmir, Manisa	under pine forest	January	Solak et al. (2003)
Hydnocystis piligera Tul. & C. Tul.	Aydın	under *Pinus brutia*	November	Kaygusuz et al. (2018)
Stephensia bombycina (Vittad.) Tul. & C.Tul.	Samsun	*Corylus* sp.	November	Türkoğlu and Castellano (2014)
Tuberaceae Dumort.				
Choiromyces meandriformis Vittad.	Uşak	-	April	Türkoğlu and Castellano (2014)
	Bolu	-	September	
	Samsun	-	November	
Reddellomyces parvulosporus Trappe, Castellano & Malajczuk	Muğla	collected from *Eucalyptus* forests	December to March	Ünal et al. (2016)
Reddellomyces westraliensis Trappe, Castellano & Malajczuk	Muğla	collected from *Eucalyptus* forests	December to March	Ünal et al. (2016)

Table 2. Diversity and distribution of true truffles with commercial and economic importance in Türkiye.

Species	**Locality**	**Host**	**Season**	**Reference**
ASCOMYCOTA				
Pyronemataceae Corda				
Picoa juniperi Vittad	Uşak	associated with *Geranium*, *Helianthemum* and *Cerastium* sp.	April	Türkoğlu and Yağız (2012)
	Denizli	*Helianthemum* sp.	April, July, May	Türkoğlu and Castellano (2014)
	Nevşehir	-	May	
	Konya	-	March, April	
	Elazığ	-	April, June	
	Afyon, Antalya	-	April	
	Kayseri	-	June	Türkoğlu et al. (2015)
	Muğla	near the *Helianthemum salicifolium* (L.) Mill.	April	Şahin et al. (2021)
	Niğde	in soil among grasses	April	Berber et al. (2022)
	Elazığ, Malatya	near the *H. salicifolium* and *H. ledifolium*	from March to May	Akyüz et al. (2015)
	Konya	in soil areas	April	Çevik et al. (2021)
	Konya	in soil among grasses	April	Çelik et al. (2020)
	Karaman	in soil among *Helianthemum* sp.	May	Çetinkaya et al. (2021)
Picoa lefebvrei (Pat.) Maire	Denizli, Konya, Aksaray	-	April	Türkoğlu et al. (2015)
	Karaman	in soil among *Helianthemum* sp.	April, May	Çetinkaya et al. (2021)
	Karaman	in soil among *Helianthemum* sp.	April, May	İleri et al. (2020)
	Elazığ, Şanlıurfa	-	April, May	Gücin et al. (2010)
	Konya	in steppe, in soil	April	Çevik et al. (2021)
	Konya	near the *Helianthemum* sp.	April, May	Şahin et al. (2020)

	Niğde	in soil among grasses	April	Berber et al. (2022)
	Elazığ, Malatya	near the *H. salicifolium* and *H. ledifolium*	from March to May	Akyüz et al. (2015)
	Konya	in soil among grasses	April	Çelik et al. (2020)
	Şanlıurfa	in soil	May	Kaya (2015)
Pezizaceae Dumort.				
Terfezia albida Ant. Rodr. Muñ.-Moh. & Bordallo	Karaman	associated with *Helianthemum* spp.	April, May	Uzun et al. (2016)
	Karaman	in soil among *Helianthemum* sp.,	April, May	Çetinkaya et al. (2021)
Terfezia arenaria (Moris) Trappe	Aydın	under *Asphodelus aestivus*	May	Türkoğlu et al. (2015)
	Konya			Öder (1988); Kaşık et al. (1998)
	Malatya			Işıloğlu and Öder (1995)
	Isparta	-	-	Afyon (1996)
Terfezia boudieri Chatin	Uşak	in grass	April	Türkoğlu and Yağız (2012)
	Karaman	with *Helianthemum* sp.,	April, May	Doğan and Öztürk (2006)
	Batman		April	Demir et al. (2007)
	Denizli	near the *H. salicifolium*	May	Gezer et al. (2014)
	Şanlıurfa	in soil	April, May	Kaya (2015)
	Gaziantep	in soil	April	Kaya et al. (2012)
	Gaziantep	in soil among *Helianthemum* sp.	March	Kaya et al. (2019)
	Elazığ, Malatya	near the *H. salicifolium* and *H. ledifolium*	from March to May	Akyüz et al. (2015)
	Konya, Karaman, Ankara, Aksaray, Kırşehir, Eskişehir, Kütahya, Mardin, Şanlıurfa, Elazığ, Diyarbakır, Malatya, Mardin, Gaziantep	near the *Helianthemum* spp.	from March to May	Akyüz et al. (2012) Kagan-Zur and Akyüz (2014)

Table 2. contd. ...

Table 2. contd. ...

Species	**Locality**	**Host**	**Season**	**Reference**
Pezizaceae Dumort.				
	Elazığ	near the *H. salicifolium*	March-May	Gücin (1990)
	Diyarbakır	near the *H. salicifolium*	March, April	Yıldız and Ertekin (1997)
	Şanlıurfa, Mardin	on *H. salicifolium*	April-May	Dündar et al. (2012)
	Karaman Niğde	In soil among *Helianthemum* sp. -	April, May April-May	İleri et al. (2020) Kaşık et al. (2001)
	Karaman	in soil among *Helianthemum* sp.	April- May	Çetinkaya et al. (2021)
Terfezia cistophila Ant. Rodr., Bordallo, Kaounas & A.Morte	Trabzon	under soil with *C. creticus*	March	Uzun and Kaya (2019c)
Terfezia claveryi Chatin	Denizli, Konya, Şanlıurfa, Yozgat, Aksaray, Diyarbakır, Karaman	near the *Helianthemum* spp.	April	Türkoğlu et al. (2015)
	Konya	in soil, among *Helianthemum* sp.	April	Çevik et al. (2021)
	Elazığ, Malatya	near the *Helianthemum* spp.	March-May	Akyüz et al. (2015)
	Konya	near the *Helianthemum* spp.	April, May	Şahin et al. (2020)
	Niğde	among *Helianthemum* spp.	April	Berber et al. (2022)
	Adana	under *Helianthemum* spp.	-	Doğan and Kurt (2016)
	Karaman	in soil among *Helianthemum* sp.	April, May	İleri et al. (2020)
	Konya	in soil among *Helianthemum* sp.	April	Çelik et al. (2020)

Terfezia leptoderma Tulasne & C.Tulasne	Uşak	in association with *Geranium*, *Helianthemum* and *Cerastium* sp.	April	Castellano and Türkoğlu (2012)
	Uşak, Denizli		April	Türkoğlu and Castellano (2014)
Terfezia olbiensis Tulasne & C. Tulasne	Uşak, Konya, Nevşehir	-	April	Türkoğlu and Castellano (2014)
	Konya	near the *Helianthemum* sp.	April, May	Şahin et al. (2020)
	Elazığ, Malatya	near the *H. salicifolium* and *H. ledifolium*	from March to May	Akyüz et al. (2015)
	Gaziantep	mixed forest, under soil	May	Uzun et al. (2015)
Tirmania pinoyi (Maire) Malençon	İzmir	in pasture	January	Ersel and Solak (2004)
Tuberaceae Dumort				
Tuber aestivum Vittad.	Denizli	-	June, July, May	Türkoğlu et al. (2015)
	Muğla	-	January to May	
	İzmir	-	April	
	Hatay, Burdur, Düzce, Osmaniye	-	June	
	Antalya	-	June, July, May	
	Ordu	-	October	
	İstanbul	-	August	
	Bolu	-	November	
	Kırklareli	-	September	
	Artvin	-	December	
	Konya	under *Quercus pubescens* Willd. and *Q. cerris* L.	May	Alkan et al. (2018)

Table 2. contd. ...

Table 2. contd. ...

Species	**Locality**	**Host**	**Season**	**Reference**
Pezizaceae Dumort.				
	Muğla	in *Pinus brutia* Ten. forest	March	Korkmaz and Duru (2023)
	Kırklareli	oak and hornbeam mix forest	July	
	Denizli	under *Pinus nigra*	March-June	Gezer et al. (2014)
	Denizli, Aydın, Muğla	from *Pinus brutia* (Ten.) and various *Quercus* forests	in spring and early summer	Özderin et al. (2018)
	Muğla	in campus areas	April	Allı (2022)
Tuber borchii Vittad	Kahramanmaraş	under *Populus* sp.	May	Kaya (2009)
	Tekirdağ	under mixed stands of *Pinus nigra*, *Quercus* sp., and *Carpinus betulus*	November	Elliott et al. (2016)
	Muğla	under mixed stands of *Quercus cerris*, *Q. pubescens* and *Pinus brutia*	March	
	Aydın	under mixed stands of *Quercus ilex* and *Pinus brutia*	March	
	Samsun	under *Corylus* spp,	December	
	Denizli	under *Pinus brutia* and *P. nigra*	May	Gezer et al. (2014)
	Muğla	in *Pinus brutia* Ten. forest	April	Korkmaz and Duru (2023)
	Kırklareli	oak and hornbeam mix forest	March	
	Muğla	in campus areas, in forest	April	Allı (2022)
	Konya	in soil under *Quercus* sp.	April	Çelik et al. (2020)
	Gaziantep	in soil under *Quercus* sp.	October	Kaya et al. (2019)
	Konya	in soil under *Quercus* sp.	June, July	Çevik et al. (2021)

Tuber brumale Vittad.	Niğde	-		Öztürk et al. (1997)
	Samsun	under *Corylus* sp.	November, October	Türkoğlu and Castellano (2014)
	Osmaniye	under *Pinus brutia*	January	
	Denizli	under *Pinus nigra*	March	Gezer et al. (2014)
Tuber excavatum Vittad.	Denizli	under mixed *Quercus* spp. and *Pinus* spp.	June	Türkoğlu and Castellano (2014)
	Konya	in soil under *Quercus* sp.	April	Çelik et al. (2020)
	Trabzon	under mixed stants of *Carpinus betulus*, *Corylus* sp. and *Picea orientalis*	September	Uzun and Yakar (2018)
Tuber ferrugineum Vittad.	Aydın	under mixed stands of *Quercus ilex* and *Pinus brutia*	March	Elliott et al. (2016)
	Muğla	under mixed stands of *Pinus brutia* and *Quercus coccifera*	April	
	Denizli	under mixed stands of *Pinus brutia*, *Quercus coccifera* and *Q. ithaburensis*	May	
	Antalya	under *Tilia* sp.	May	
	Konya	in soil under *Quercus* sp.	April	Çelik et al. (2020)
Tuber fulgens Quél.	Kırklareli	in oak and hornbeam mixed forest	August	Akata et al. (2020)
Tuber macrosporum Vittad.	Tekirdağ	in oak forest, under *Quercus* sp.	September	Doğan (2021)
	Edirne	in oak forest, under *Quercus* sp.	November	
Tuber mesentericum Vittad.	Denizli	under *Pinus nigra*	May, June	Castellano and Türkoğlu (2012)
	Denizli	-	June	Türkoğlu and Castellano (2014)
Tuber nitidum Vittad.	Denizli	under *Juniperus excelsa*, *J. foetidissima*, and *Quercus pubescens*	May	Castellano and Türkoğlu (2012)

Table 2. contd. ...

Table 2. contd. ...

Species	Locality	Host	Season	Reference
Pezizaceae Dumort.				
	Osmaniye	under mixed *Quercus* spp. and *Pinus* spp.	April, May	Türkoğlu and Castellano (2014)
	Kastamonu	under *Abies nordmanniana*, *Pinus sylvestris*	May	
	Burdur	-	July	
	Burdur	-	November	Türkoğlu et al. (2015)
	Denizli	-	March	
	Konya	in soil under *Quercus* sp., in soil under *Populus* and *Salix* sp.	April	Çelik et al. (2020)
	Karaman	in soil under *Quercus* sp.	May	İleri et al. (2020)
	Mersin	in soil under *Pinus* sp.	March	Kaplan et al. (2023)
Tuber oligospermum (Tul. & C. Tul.) Trappe	Şanlıurfa	under pine	April	Akata et al. (2022)
Tuber puberulum Berk. & Broome	Denizli	under mixed stands of *Pinus brutia*, *Quercus cerris* and *Q. ilex*		Elliott et al. (2016)
		under mixed stands of *Pinus brutia*, *Quercus coccifera* and *Q. ithaburensis*		
		under *Quercus coccifera* and *Q. trojana*		
		under *Pinus* spp. and *Quercus* spp.		
	Muğla	under *Quercus cerris*	March	
		under mixed stands of *Pinus brutia* and *Quercus coccifera*	April	
	Osmaniye	*Pinus* spp. and *Quercus* spp.	April	

	Aydın	under mixed stands of *Quercus cerris* and *Q. ithaburensis*	March	
	Artvin	under different mixed stands of *Fagus orientalis*, *Castanea sativa*, *Rhododendron ponticum*, *Alnus glutinosa* and *Corylus* sp.	November	Uzun and Yakar (2018)
	Trabzon	under different mixed stands of *Fagus orientalis*, *Castanea sativa*, *Rhododendron ponticum*, *Alnus glutinosa* and *Corylus* sp.	August, September	Uzun and Yakar (2018)
	Trabzon	In soil under broadleaved forest	September	Uzun and Kaya (2022)
Tuber rufum Pico	Denizli,	-	May, October	Türkoğlu and Castellano (2014)
	Muğla, Konya	-	April	
	Kastamonu, Antalya	-	June	
	Burdur	-	July, November, April	Türkoğlu et al. (2015)
	Bolu	-	November	
	Muğla	-	March	
	Aydın	-	March	
	Osmaniye	under mixed stand of *Quercus cerris*, *Pinus nigra*, and *Pinus brutia*	May	
	Denizli	-	May	

3. Diversity and Distribution of True Truffles

In the introduction section, we described the true truffles divided into two groups: forest truffles and arid, semi-arid, or desert truffles. A total of 22 different taxa were identified within three families belonging to Ascomycota (see Table 2). A total of twenty-two taxa from three true truffle families, namely Tuberaceae (*Tuber*), Pezizaceae (*Terfezia* and *Tirmania*), and Pyronemataceae (*Picoa*), including forest (*Tuber* species) and desert (*Terfezia*, *Tirmania* and *Picoa* species) truffles, were identified in different provinces of Türkiye. Researchers have identified twelve *Tuber* taxa from the Tuberaceae family in different regions of Türkiye. Twelve types of forest truffle belonging to the Tuberaceae were identified in various Turkish provinces, including *Tuber aestivum*, *T. brumale*, *T. excavatum*, *T. nitidum*, *T. rufum*, *T. mesentericum*, *T. borchii*, *T. ferrugineum*, *T. puberulum*, *T. macrosporum*, *T. fulgens*, and *T. oligospermum*. These findings have been reported by multiple researchers and detailed in Table 2.

3.1 Forest Truffles

Basic information has been provided by many scientists in the past about forest truffles (Wang and Marcone, 2011; Zambonelli et al., 2016; Thomas et al., 2019; Lee et al., 2020), considered as one of the world›s natural, delicious and expensive food sources. This information serves as the main source for us to carry out the present study. The *Tuber* genus comprises approximately 180 species, mainly located in the Northern Hemisphere. Although various *Tuber* species are locally consumed and appreciated, *Tuber magnatum*, *T. melanosporum*, *T. aestivum*, and *T. borchii* are considered the most economically significant fungi (Zambonelli et al., 2016). Numerous truffle species fulfill a crucial role in forest ecosystems, especially in disrupted forests, where they frequently act as ectomycorrhizal symbionts. Beyond their economic and gastronomic significance, they serve as indispensable ecological agents. Observations have revealed these species forming mycorrhizal structures with various tree species across forest areas with diverse climatic profiles, encompassing regions such as the Mediterranean, Iranian-Turanian, and Siberian zones.

The formation of truffles depends on various factors, such as the quantity and timing of precipitation during the rains, the nature and characteristics of the soil, the availability of water, the prevailing weather conditions, host and vegetation factors, and appropriate habitat (Wang et al., 2014; Jaillard et al., 2016; Le Tacon, 2016; Zambonelli et al., 2016). The *Tuber* genus exhibits a broad distribution across Asia, Europe, and North America, yet it has been largely absent in the Southern Hemisphere up to this point (Chen et al., 2016). Edible fungi belonging to the genus *Tuber*, characterized by their mycorrhizal Ascomycota nature and subterranean fruiting bodies, are highly appreciated for their unique flavor profile and considerable economic importance. Truffle fungi, caviar, and champagne are widely considered the most luxurious culinary choices among indulgent gastronomic preferences. Truffles, celebrated for their unique scent, are known for their exorbitant prices and rarity (Splivallo and Culleré, 2016). In Türkiye, forest truffles (known as Domalan) are typically found in specific forested areas and are considered rare. Their highly prized flavor and aroma make them particularly valuable and expensive, leading to their collection for use in luxury restaurants and by gourmet enthusiasts for commercial and economic purposes rather than for regular consumption by locals.

While experienced and skilled individuals may sometimes come across truffles near the surface, trained dogs are typically relied upon to detect these prized fungi. The primary reason for the utilization of truffles as a delicacy in various esteemed cuisines, such as French, Italian, and Spanish, is their distinctive aroma (Beara et al., 2014). Furthermore, they rank as the most expensive mushrooms in global food markets due to their rarity and scarcity, a fact particularly evident during unfavorable harvesting seasons. Forest truffle species, known as true truffles-*Tuber*, are recognized in their regions of origin and are obtained in limited quantities. Since the availability of these valuable species is ensured by specially trained dogs, they are normally rarely obtained by local individuals (Table 2). True truffles (e.g., *Tuber melanosporum*, *Tuber magnatum*, and related species) are delicious and expensive food in several countries (Bonito and Smith, 2016). They have an important place in the favorite dishes of the elite and gourmets, especially those which are important to gastronomy. Therefore, these truffles are economically valuable species with gastronomic and nutritional properties, particularly due to their taste, flavor, aroma, and organoleptic properties, and also medicinal characteristics. Perhaps, true truffles are among the most priciestand most expensive and exclusive gourmet food items globally because of their distinct taste and limited availability during specific seasons.

3.2 Desert Truffles

Ten desert truffles taxa were identified in various Turkish provinces, belonging to two families. Eight of these taxa were from Pezizaceae, specifically *Terfezia boudieri*, *T. claveryi*, *T. olbiensis*, *T. leptoderma*, *T. albida*, *T. arenaria*, *T. cistophila*, and *Tirmania pinoyi*. The remaining two taxa were from Pyronemataceae, namely *Picoa lefebvrei* and *P. juniperi*. These findings have been reported by multiple researchers and are outlined in detail in Table 2. Many researchers from past to present have provided valuable information about these species. These sources have allowed us to learn more about them (Diez et al., 2002; Mandeel and Al-Laith, 2007; Kagan-Zur and Roth-Bejerano, 2008; Trappe et al., 2008, 2014; Al-Laith, 2010; Loizides et al., 2012; Shavit, 2013; Kagan-Zur et al., 2014; Bradai et al., 2014, 2015ab; Thomas et al., 2019; Morte et al., 2020; Henkrar et al., 2022). Desert truffles, also known as arid or semi-arid truffles, are a type of mycorrhizal fungi that develop in association with the root systems of *Helianthemum* spp. plants, alongside fungi species such as *Terfezia*, *Tirmania*, or *Picoa*. (Bordallo and Rodriguez, 2014; Roth-Bejerano et al., 2014). They have a natural distribution around the Mediterranean, Europe, America, North Africa, China, Australia, and the Middle East (Loizides et al., 2012; Kagan-Zur et al., 2014; Morte et al., 2021).

Climatic factors, precipitation, soil properties, habitat, and host plant are crucial elements in the formation of desert truffle (Mandeel and Al-Laith, 2007; Kagan-Zur and Roth-Bejerano, 2008; Bonifacio and Morte, 2014; Bradai et al., 2015b). Desert truffles are believed to be popular due to their nutritional value and unique flavor, texture and delicious taste. Although less strongly flavored than forest truffles, they are considered a valuable type for their distinctive musky aroma and flavor. Arid, semi-arid or desert truffles are a popular gastronomic delicacy in various nations (Khalifa et al., 2019). They are seasonal and socio-economically significant, and typically appear after the rainy season from March to April, in Türkiye. Specific hosts within the Cistaceae family, soil properties, rainy seasons, and climatic factors have a significant influence on the prevalence of desert truffles (Table 2). Similar studies (Kagan-Zur et al., 2014) in different chapters of this book have demonstrated that truffles are in high demand due to their natural origins as a food source and their limited availability during specific times of the year in certain regions.

The growth of truffles is influenced by factors such as rain, soil type, temperature, and host plant; however, the preferred truffle varieties, as well as their preparation and preservation methods, and edibility can vary depending on the mushroom type and cultural differences in different regions. Desert truffles are widespread in certain areas of Türkiye where *Helianthemum* host plants grow naturally (Fig. 1a), as observed by Gürhan (2016), Akyüz et al. (2012, 2015) and other researchers (Table 2). The local people in these areas refer to these truffles as Keme, Kumi, Kemi, Topalak, Dobelon, Dobalan, Dolaman, Dümbelek, Duvberon, Domalan, and Trüf. Typically, desert truffles emerge during a brief span in spring, lasting around 1 to 1.5 months. The soil type conducive to their growth has been identified as sand-clay-slimy. These fungi exhibit distinct flavors, aromas, and valuable organoleptic properties.

Truffles are believed to be popular due to their nutritional value, delectable taste, and purported aphrodisiac qualities. Since they occur close to the surface, they can be easily found by all individuals and especially by experienced collectors. Notably, *T. boudieri* is the dominant species and holds commercial significance. It is a seasonal and socioeconomically important mushroom (Fig. 1b). In the region,

Fig. 1. Host plant and desert truffles in Türkiye: a, Host plant, *Helianthemum salicifolium*; b, *Terfezia boudieri*; c, *Terfezia claveryi*; d, other desert truffles (1: *Terfezia olbiensis*; 2, *Picoa juniperi*; 3: *Picoa lefebvrei*).

T. claveryi (Fig. 1c) is also very rare and is confused with *T. boudieri* by the local people. As *Picoa* species are notably small and other *Terfezia* species remain unrecognized, they are exclusively consumed by the local population on a daily basis. As a result, *T. boudieri* is favored for commercial purposes, while other types such as *P. lefebvrei*, *P. juniperi*, *T. olbiensis*, and similar species are primarily utilized for everyday consumption as they do not yield significant financial gains (Fig. 1d). They feature extensively in dishes enjoyed by diverse demographic groups, from indigenous populations to urban, middle-class and affluent families alike. These truffles are an essential element in many recipes and are highly valued. The distinctive characteristics of these species enhance their economic value. They are exclusively found in specific locations for a short period during the spring and serve as a natural food source that presents potential economic prospects. These mushrooms are a fundamental component in various recipes appreciated by a diverse group of consumers, including indigenous communities and urban, middle-class, and affluent families. The popularity of truffles is attributed to their nutritional value and distinctive flavor, texture, and delicious taste. This underscores their significance, as supported by previous research mentioned earlier.

4. Conclusion

Until now, Türkiye has identified a total of 49 truffle taxa from 8 families within the Ascomycota phylum. Among these, 27 taxa associated with false or truffle-like truffles have been identified, representing 5 families, while 22 taxa associated with true truffles have been classified into 3 families. Many false truffles are not suitable for human consumption as a food source due to their taste, flavor, aroma, nutritional content, and organoleptic characteristics. Therefore, they do not provide any economic or commercial benefits. False truffles have important ecological functions with their host plants, but they are also a food source for local mammalian wildlife. For this reason, many of these species are not consumed as food in Türkiye, but they are very important in terms of mushroom biodiversity.

In addition, a total of twenty-two taxa from three true truffle families, twelve *Tuber* taxa from the Tuberaceae family, seven *Terfezia* and one *Tirmania* taxa from Pezizaceae, and two *Picoa* taxa from Pyronemataceae, were identified in different provinces of Türkiye. Certain species of forest truffles (*Tuber*) are highly valued in the global food market, revered by gourmets and often served as the main course in high-end restaurants. These twelve taxa are found in some regions of Türkiye. They remain mediocre among the local population because they are found deep in the soil, are rare and in small quantities, are detected by specially trained dogs, and appeal to the wealthy upper class. They are very expensive due to their taste and flavor content, are obtained in small quantities and are usually sold commercially instead of being consumed daily. Although not as prevalent as desert truffle species and not recognized by all sections of the population, they hold significant value among particular segments.

In Türkiye, desert truffles are widespread in specific regions, typically appearing during the spring season for a short period, usually around 1 to 1.5 months. It is common to observe multiple taxa together. These truffles are easily recognized by all segments of the public and can be readily found. *T. boudieri* stands out as the primary commercial product, recognized by all sectors of society for its flavor, aroma, and nutritional value. As an inexpensive delicacy, it appeals to consumers from all the layers of society. Moreover, these desert truffles are more well-known to the public than forest truffles. It has been observed that *T. boudieri* is the only product with economic value, generating profit. In contrast, *P. lefebvrei*, *T. olbiensis*, *P. juniperi* and also other species have no economic value as they are used for daily consumption and are more popular among locals. The first product is too small, the second degrades quickly, and the third is rare and hard to find. *T. claveryi* is also rare in the area and can be mistaken for *T. boudieri* by the locals. Other desert truffle species are predominantly utilized for daily consumption as they do not yield significant financial profit. In our opinion, desert truffles are more popular in Turkish culture, even though they are not as tasty and expensive as forest truffles. The keme mushroom (*T. boudieri* or desert truffles) is an integral part of traditional society. It holds great importance for the local community, appealing to all segments of society and featuring heavily in food culture. This food is also affordable and offers a range of delicious cooking options such as Keme kebab or others.

References

Afyon, A. (1996). Some macrofungi of Isparta (Turkey). Turk. J. Bot., 20: 161–165.

Akata, I., Sevindik, M. and Şahin, E. (2020). *Tuber fulgens* Quél. A new record for Turkish truffles. Turk. J. Agric. Food. Sci. Technol., 8: 2472–2475. 10.24925/turjaf.v8i11.2472-2475.3884

Akata, I., Şen, İ., Sevindik, M. and Kabaktepe, Ş. (2022). Truffle checklist of Turkey II with a new record. Turk. J. Agric. Food. Sci. Technol., 10: 1913–1920. 10.24925/turjaf.v10i10.1913-1920.5482

Akata, I., Şen, İ., Şahin, E., Çöl, B. and Keskin, E et al. (2023). *Delastria*, a new genus of hypogeous fungi record for the Turkish mycobiota. Trakya Univ. J. Nat. Sci., 24: 85–90. 10.23902/trkjnat.1331537

Akyüz, M., Kirbag, S. and Kursat, M. (2012). Ecological aspects of the arid and semi-arid truffle in Turkey: evaluation of soil characteristics, morphology, distribution, and mycorrhizal relationships. Turk. J. Bot., 36: 386–391. 10.3906/bot-1103-4

Akyüz, M., Kırbağ, S., Bircan, B. and Gürhan, Y. (2015). Diversity and distribution of arid-semi arid truffle (*Terfezia* and *Picoa*) in Elazığ-Malatya Region of Turkey. Mycosphere, 6: 766–783. 10.5943/mycosphere/6/6/11

Alkan, S., Aktaş, S. and Kaşık, G. (2018). *Tuber* species in Turkey and a new locality for *Tuber aestivum*. Selçuk Univ. J. Sci. Fac., 44: 25–29.

Al-Laith, A.A.A. (2010). Antioxidant components and antioxidant/antiradical activities of desert truffle (*Tirmania nivea*) from various Middle Eastern origins. J. Food. Compost. Anal. 23: 15–22. 10.1016/j.jfca.2009.07.005

Allı, H. and Doğan, H.H. (2019). A new genus (*Balsamia*) addition for Turkish mycota. Mantar Dergisi, 10: 23–25. 10.30708mantar.482166

Allı, H. (2022). Macrofungi growing in Mugla Sıtkı Kocman University Campus. Mantar Dergisi, 13: 96–104. 10.30708.mantar.1118349

Atila, F. and Kazankaya, A. (2022). Türkiye'de doğal yayılış gösteren *Terfezia* türleri ve ekolojileri. Mantar Dergisi, 13: 119–130. 10.30708.mantar.1185592

Baldrian, P., Větrovský, T., Lepinay, C. and Kohout, P. (2022). High-throughput sequencing view on the magnitude of global fungal diversity. Fungal Divers., 114: 539–547. 10.1007/s13225-021-00472-y

Barutçiyan, J. (2012). Türkiye›nin Mantarları - I. Oğlak Yayıncılık, Türkiye.

Barutçiyan, J. (2021). Makromantarlar. Ginko Bilim, Türkiye.

Beara, I.N., Lesjak, M.M., Četojević-Simin, D.D., Marjanović, Ž.S., Ristić, J.D. et al. (2014). Phenolic profile, antioxidant, anti-inflammatory and cytotoxic activities of black (*Tuber aestivum*

Vittad.) and white (*Tuber magnatum* Pico) truffles. Food Chem., 165: 460–466. 10.1016/j.foodchem.2014.05.116

Berber, O., Uzun, Y. and Kaya, A. (2022). Macrofungi determined in Ulukışla (Niğde-Turkey) district. KSU J. Agric. Nat., 25: 1007–1015. 10.18016/ksutarimdoga.vi.960925

Bhunjun, C.S., Niskanen, T., Suwannarach, N., Wannathes, N., Chen, Y.-J. et al. (2022). The numbers of fungi: are the most speciose genera truly diverse? Fungal Divers., 114: 387–462. 10.1007/s13225-022-00501-4

Blackwell, M. (2011). The Fungi: 1, 2, 3 ... 5.1 million species? Am. J. Bot., 98: 426–438. 10.3732/ajb.1000298

Bonifacio, E. and Morte, A. (2014). Soil properties. pp. 57–67. *In*: Kagan-Zur, V., Roth-Bejerano, N., Sitrit, Y. and Morte, A (eds.). Desert truffles: Phylogeny, physiology, distribution and domestication. Soil Biology, Springer, Berlin, Heidelberg. 10.1007/978-3-642-40096-4_4

Bonito, G.M. and Smith, M.E. (2016). General systematic position of the truffles: Evolutionary theories. pp. 3–18. *In*: Zambonelli, A., Iotti, M. and Murat, C. (eds.). True truffle (*Tuber* spp.) in the World. Soil ecology, systematics and biochemistry. Soil Biology, Springer, Cham. 10.1007/978-3-319-31436-5_1

Bordallo, J.J. and Rodríguez, A. (2014). Cryptic and new species. pp. 39–53. *In*: Kagan-Zur, V., Roth-Bejerano, N., Sitrit, Y. and Morte, A. (eds.). Desert truffles: Phylogeny, physiology, distribution and domestication. Soil Biology, Springer, Berlin, Heidelberg. 10.1007/978-3-642-40096-4_3

Bradai, L., Bissati, S. and Chenchouni, H. (2014). Desert truffles of the North Algerian Sahara: diversity and bioecology. Emir. J. Food. Agric., 26: 425–435. 10.9755/ejfa.v26i5.16520

Bradai, L., Neffar, S., Amrani, K., Bissati, S. and Chenchouni, H. et al. (2015a). Ethnomycological survey of traditional usage and indigenous knowledge on desert truffles among the native Sahara Desert people of Algeria. J. Ethnopharmacol., 162: 31–38. 10.1016/j.jep.2014.12.031

Bradai, L., Bissati, S., Chenchouni, H. and Amrani, K. (2015b). Effects of climate on the productivity of desert truffles beneath hyper-arid conditions. Inter. J. Biometeorol., 59: 907–915. 10.1007/s00484-014-0891-8

Castellano, M.A. and Türkoğlu, A. (2012). New records of truffle taxa in *Tuber* and *Terfezia* from Turkey. Turk. J. Bot., 36: 295–298. 10.3906/bot-1106-10

Catalogue of Life (2021). Species 2000 & ITIS Catalogue of Life, 21-04-21. Digital resource at www.catalogueofife.org. Species 2000: Naturalis, Leiden, Netherlands

Cerletti, C., Esposito, S. and Iacoviello, L. (2021). Edible mushrooms and beta-glucans: Impact on human health. Nutrients, 13: 2195. 10.3390/nu13072195

Chang, S.T. (2001). Mushrooms and mushroom cultivation. Encyclopedia of life sciences. John Wiley & Sons, Ltd. 10.1038/npg.els.0000370

Chen, J., Murat, C., Oviatt, P., Wang, Y. and Le Tacon, F. et al. (2016). The black truffles *Tuber melanosporum* and *Tuber indicum*. pp. 19–32. *In*: Zambonelli, A., Iotti, M. and Murat, C. (eds.). True truffle (*Tuber* spp.) in the World. Soil Biology, Springer, Cham. 10.1007/978-3-319-31436-5_2

Çelik, A., Uzun, Y. and Kaya, A. (2020). Macrofungal biodiversity of Güneysınır district (Konya-Turkey). Mantar Dergisi, 11: 75–83. 10.30708.mantar.662015

Çetinkaya, A., Uzun, Y. and Kaya, A. (2021). Macrofungi determined in Ayrancı and Yeşildere (Karaman) districts. Mantar Dergisi, 12: 42–49. 10.30708.mantar. 838640

Çevik, F.T., Uzun, Y. and Kaya, A. (2021). Macrofungi determined in Ereğli (Konya) district. Mantar Dergisi, 12: 138–147. 10.30708.mantar960666

Demir, S., Demirel, K. and Uzun, Y. (2007). Macrofungi of Batman province. Ekoloji, 16: 37–42.

Diez, J., Manjon, J.L. and Martin, F. (2002). Molecular phylogeny of the mycorrhizal desert truffles (*Terfezia* and *Tirmania*), host specificity and edaphic tolerance. Mycologia, 94: 247–259. 10.1080/15572536.2003.11833230

Doğan, H.H. and Öztürk, C. (2006). Macrofungi and their distribution in Karaman province, Turkey. Turk. J. Bot., 30: 193–207.

Doğan, H.H. and Kurt, F. (2016). New macrofungi records from Turkey and macrofungal diversity of Pozantı-Adana. Turk. J. Bot., 40: 209–217. 10.3906/bot-1501-22

Doğan, H.H., Bozok, F. and Taşkın, H. (2018). A new species of *Barssia* (Ascomycota, Helvellaceae) from Turkey. Turk. J. Bot., 42, 636–643. 10.3906/bot-1801-33

Doğan, H.H. (2021). A new truffle species addition, *Tuber macrosporum* Vittad., to Turkish mycota. Trakya Univ. J. Nat. Sci., 22: 139–146. 10.23902/trkjnat.873651

Dündar, A., Yeşil, O.F., Acay, H., Okumuş, V., Özdemir, S. et al. (2012). Antioxidant properties, chemical composition, and nutritional value of *Terfezia boudieri* (Chatin) from Turkey. Food. Sci. Technol. Inter., 18: 317–328. 10.1177/1082013211427954

Elliott, T.F., Türkoğlu, A., Trappe, J.M. and Güngör, M.Y. (2016). Turkish truffles 2: eight new records from Anatolia. Mycotaxon, 131: 439–453. 10.5248/131.439

Ersel, F.Y. and Solak, M.H. (2004). Contributions to the macrofungi of İzmir province. Turk. J. Bot., 28: 487–490.

Gezer, K., Kaygusuz, O., Çelik, A. and Işıloğlu, M. (2014). Ecological characteristics of truffles growing in Denizli province, Turkey. J. Food. Agric. Envir., 12: 1105–1109. 10.1234/4.2014.5296

Gryndler, M. (2016). True truffle host diversity. pp. 267–281. *In*: Zambonelli, A., Iotti, M. and Murat, C (eds). True Truffle (*Tuber* spp.) in the World: Soil ecology, systematics and biochemistry. Soil Biology, Springer, Cham. 10.1007/978-3-319-31436-5_16

Gücin, F. (1990). Elazığ çevresinde belirlenen makrofunguslar. Turk. J. Bot., 14: 171–177.

Gücin, F., Kaya, A., Soylu, M.K. and Uzun, Y. (2010). *Picoa* Vittad., A new truffle genus record for Turkey. Biol. Divers. Conserv., 3: 23–25. BioDiCon. 143-0410

Gürhan, Y. (2016). Determination of some ecological, morphological feature and nutritive contents of *Picoa lefebvrei* (Pat.) maire. Bitlis Eren University Graduate School of Natural and Applied Sciences, Department of Biology, Master thesis.

Hawksworth, D.L. (1991). The fungal dimension of biodiversity: magnitude, significance, and conservation. Mycol. Res., 95: 641–655. 10.1016/S0953-7562(09)80810-1

Hawksworth, D.L. (2001). The magnitude of fungal diversity: The 1.5 million species estimate revisited. Mycol. Res., 105: 1422–1432. 10.1017/S0953756201004725

Hawksworth, D.L. and Lücking, R. (2017). Fungal diversity revisited: 2.2 to 3.8 million species. Microbiol. Spectr., 5: 10–1128. 10.1128/microbiolspec.funk-0052-2016

Henkrar, F., Meyad, C., Sabaa, S. and Khabar, L. (2022). Desert truffles and truffles in Morocco: Biodiversity of promising fungi to combat desertification. Environ. Sci. Proc., 16: 35. 10.3390/environsciproc2022016035

Hoeksema, J.D. and Thompson, J.N. (2007). Geographic structure in a widespread plant–mycorrhizal interaction: pines and false truffles. J. Evol. Biol., 20: 1148–1163. 10.1111/j.1420-9101.2006.01287.x

Hyde, K.D. (2022). The numbers of fungi. Fungal Divers., 114: 1. 10.1007/s13225-022-00507-y

Hyde, K.D., Jeewon, R., Chen, Y.J., Bhunjun, C.S. (2020). The numbers of fungi: is the descriptive curve flattening? Fungal Divers., 103: 219–271. 10.1007/s13225-020-00458-2

Işıloğlu, M. and Öder, N. (1995). Malatya yöresinin makrofungusları. Turk. J. Bot., 19: 321–324.

İleri, R., Uzun, Y. and Kaya, A. (2020). Macromycetes of Karadağ (Karaman) and its environs. Mantar Dergisi, 11: 57–63. 10.30708.mantar.654111

Jaillard, B., Oliach, D., Sourza, P. and Colinas, C. (2016). Soil characteristics of *Tuber melanosporum* habitat. pp. 169–190. *In*: Zambonelli, A., Iotti, M. and Murat, C. (eds.). True truffle (*Tuber* spp.) in the World. Soil ecology, systematics and biochemistry. Soil Biology, Springer, Cham. 10.1007/978-3-319-31436-5_11

Kagan-Zur, V. and Roth-Bejerano, N. (2008). Desert truffles. Fungi, 1: 32–37.

Kagan-Zur, V., Roth-Bejerano, N., Sitrit, Y. and Morte, A. (2014). Desert truffles: Phylogeny, physiology, distribution and domestication. Soil Biology, Springer, Berlin, Heidelberg. 10.1007/978-3-642-40096-4

Kagan-Zur, V. and Akyüz, M. (2014). Asian Mediterranean desert truffles. pp. 159–171. *In*: Kagan-Zur, V., Roth-Bejerano, N., Sitrit, Y. and Morte, A. (eds.). Desert truffles: Phylogeny, physiology, distribution and domestication. Soil Biology, Springer, Berlin, Heidelberg. 10.1007/978-3-642-40096-4_11

Kaplan, D., Uzun, Y. and Kaya, A. (2023). Macromycetes determined in Silifke (Mersin) district. KSU J. Agric. Nat., 26: 11–21. 10.18016/ksutarimdoga.vi.1060627

Kaşık, G., Öztürk, C., Akköz, C. and Doğan, H.H. (1998). SÜ Alaaddin Keykubat kampusünde belirlenen bazı makrofunguslar. Selçuk Üniv. Fen-Edb. Fak. Derg., 15: 87–99.

Kaşık, G., Öztürk, C. and Toprak, E. (2001). Macrofungi of Niğde province (Turkey). Herb. J. Syst. Bot., 8: 137–142.

Kaya, A. (2009). Macromycetes of Kahramanmaraş province (Turkey). Mycotaxon, 108: 31–34. 10.5248/108.31

Kaya, A., Demirel, K. and Uzun, Y. (2012). Macrofungal diversity of Araban (Gaziantep/Turkey) district. Biol. Divers. Conserv., 5: 162–166. BioDiCon. 258-0712

Kaya, A. (2015). Contributions to the macrofungal diversity of Atatürk Dam Lake basin. Turk. J. Bot., 39: 162–172. 10.3906/bot-1404-70

Kaya, A., Uzun, Y., Karacan, H.İ. and Yakar, S. (2019). Contributions to the macromycota of Gaziantep province. Kastamonu Univ. J. For. Fac., 19: 329–341. 10.17475/kastorman.662722

Kaygusuz, O., Çolak, Ö.F., Matočec, N. and Kušan, I. (2018). New data on Turkish hypogeous fungi. Nat. Croat., 27: 257–269. 10.20302/NC.2018.27.16

Khalifa, S.A., Farag, M.A., Yosri, N., Sabir, J.S.M., Saeed, A. et al. (2019). Truffles: From islamic culture to chemistry, pharmacology, and food trends in recent times. Trends Food. Sci. Technol., 91: 193–218. 10.1016/j.tifs.2019.07.008

Kirk, P., Cannon, P., Minter, D. and Stalpers, J. (2008). Dictionary of the fungi, 10th edn. CABI, Wallingford, UK.

Korkmaz, C. and Duru, M.E. (2023). Investigation of aroma components of *Tuber aestivum* and *Tuber borchii* collected from different location in Türkiye. Mantar Dergisi, 14: 82–91. 10.30708. mantar.1299554

Laessoe, T. and Hansen, K. (2007). Truffle trouble: what happened to the Tuberales? Mycol. Res., 111: 1075–1099. 10.1016/j.mycres.2007.08.004

Le Tacon, F. (2016). Influence of climate on natural distribution of *Tuber* species and truffle production. pp. 153–167. *In*: Zambonelli, A., Iotti, M. and Murat, C. (eds.). True truffle (*Tuber* spp.) in the World. Soil ecology, systematics and biochemistry. Soil Biology, Springer, Cham. 10.1007/978-3-319-31436-5_10

Lee, H., Nam, K., Zahra, Z. and Farooqi, M.Q.U. (2020). Potentials of truffles in nutritional and medicinal applications: A review. Fungal. Biol. Biotechnol., 7: 1–17. 10.1186/s40694-020-00097-x

Leonardi, M., Iotti, M., Pacioni, G., Hall, I.R. and Zambonelli, A et al. (2021). Truffles: Biodiversity, ecological significances, and biotechnological applications. pp. 107–146. *In*: Abdel-Azeem, A.M., Yadav, A.N., Yadav, N. and Usmani, Z. (eds.). Industrially important fungi for sustainable development. Fungal Biology, Springer, Cham. 10.1007/978-3-030-67561-5_4

Loizides, M., Hobart, C., Konstandinides, G. and Yiangou, Y. (2012). Desert truffles: The mysterious jewels of antiquity. Field. Mycol., 13: 17–21. 10.1016/j.fldmyc.2011.12.004

Lücking, R., Aime, M.C., Robbertse, B., Miller, A.N., Aoki, T. et al. (2021). Fungal taxonomy and sequence-based nomenclature. Nat. Microbiol., 6: 540–548. 10.1038/s41564-021-00888-x

Mandeel, Q.A. and Al-Laith, A.A.A. (2007). Ethnomycological aspects of the desert truffle among native Bahraini and non-Bahraini peoples of the Kingdom of Bahrain. J. Ethnopharmacol., 110: 118–129. 10.1016/j.jep.2006.09.014

Manzi, P., Gambelli, L., Marconi, S., Vivanti, V. and Pizzoferrato, L. et al. (1999). Nutrients in edible mushrooms: an interspecies comparative study. Food Chem., 65: 477–482. 10.1016/S0308-8146(98)00212-X

Moreno, G., Alvarado, P. and Manjón, J.L. (2014). Hypogeous desert fungi. pp. 3–20. *In*: Kagan-Zur, V., Roth-Bejerano, N., Sitrit, Y. and Morte, A. (eds.). Desert truffles. Soil Biology, Springer, Berlin, Heidelberg. 10.1007/978-3-642-40096-4_1

Morte, A., Gutiérrez, A. and Ródenas, A.N. (2020). Advances in desert truffle mycorrhization and cultivation. pp. 205–219. *In*: Pérez-Moreno, J., Guerin-Laguette, A., Arzú, R.F. and Yu, F.Q. (eds.). Mushrooms, humans and nature in a changing world: Perspectives from ecological, agricultural and social sciences. Springer, Cham. 10.1007/978-3-030-37378-8_7

Morte, A., Kagan-Zur, V., Navarro-Ródenas, A. and Sitrit, Y. (2021). Cultivation of desert truffles–A crop suitable for arid and semi-arid zones. Agronomy, 11: 1462. 10.3390/agronomy11081462.

Mueller, G.M. and Schmit, J.P. (2007). Fungal biodiversity: what do we know? What can we predict? Biodivers. Conserv., 16: 1–5. 10.1007/s10531-006-9117-7

O'Brien, H.E., Parrent, J.L., Jackson, J.A., Moncalvo, J.-M. and Vilgalys, R. et al. (2005). Fungal community analysis by large-scale sequencing of environmental samples. Appl. Environ. Microbiol., 71: 5544–5550. 10.1128/AEM.71.9.5544-5550.2005

Öder, N. (1988). Taxonomic investigations of important edible and poisonous mushrooms growing in the Konya center and some districts of Konya. Selçuk Univ. Fen. Edeb. Fak. Derg., 8: 237–257.

Özderin, S., Yılmaz, F. and Allı, H. (2018). Determining mycorrhiza rate in some oak species inoculated with *Tuber aestivum* Vittad. (summer truffle). Turk. J. For., 19: 226–232. 10.18182/tjf.435372

Öztürk, C., Kaşık, G. and Toprak, E. (1997). Ascomycetes makrofunguslarından Türkiye için iki yeni kayıt. Herb. J. Syst. Bot., 4: 53–56.

Phukhamsakda, C., Nilsson, R.H., Bhunjun, C.S., de Farias, A.R.G., Sun, Y.R. et al. (2022). The numbers of fungi: Contributions from traditional taxonomic studies and challenges of metabarcoding. Fungal Divers., 114: 327–386. 10.1007/s13225-022-00502-3

Purvis, A. and Hector, A. (2000). Getting the measure of biodiversity. Nature, 405: 212–219. 10.1038/35012221

Roth-Bejerano, N., Navarro-Ródenas, A. and Gutiérrez, A. (2014). Types of mycorrhizal association. pp. 69–80. *In*: Kagan-Zur, V., Roth-Bejerano, N., Sitrit, Y. and Morte, A. (eds.). Desert truffles: Phylogeny, physiology, distribution and domestication. Soil Biology, Springer, Berlin, Heidelberg. 10.1007/978-3-642-40096-4_5

Sesli, E. and Denchev, C.M. (2014). Checklists of the myxomycetes, larger ascomycetes, and larger Basidiomycetes in Turkey. 6th ed. Mycotaxon Checklists Online.

Sesli, E., Asan, A. and Selçuk, F. (2020) Türkiye mantarları listesi (The checklist of fungi of Turkey). Ali Nihat Gökyiğit Vakfı Yayını, İstanbul.

Shavit, E. (2014). The history of desert truffle use. pp. 217–241. *In*: Kagan-Zur, V., Roth-Bejerano, N., Sitrit, Y. and Morte, A. (eds.). Desert truffles: Phylogeny, physiology, distribution and domestication. Soil Biology, Springer, Berlin, Heidelberg. 10.1007/978-3-642-40096-4_15

Solak, M.H., Gücin, F., Işıloğlu, M. and Pacioni, G. (2003). A new record of *Geopora cooperi* f. *cooperi* from West Asia. Pak. J. Bot., 35: 473–475.

Splivallo, R. and Culleré, L. (2016). The smell of truffles: from aroma biosynthesis to product quality. pp. 393–407. *In*: Zambonelli, A., Iotti, M. and Murat, C. (eds.). True truffle (*Tuber* spp.) in the World. Soil ecology, systematics and biochemistry. Soil Biology, Springer, Cham. 10.1007/978-3-319-31436-5_23

Şahin, A., Kaşık, G., Alkan, S. and Özcan, M.M. (2020). Total phenol, antioxidant activity, fatty acid composition and mineral contents of the species of fungi known as domalan. J. Agroaliment. Processes. Technol., 26: 149–154.

Şahin, E., Keskin, E. and Akata, I. (2021). Novel and diverse mycoviruses co-inhabiting the hypogeous ectomycorrhizal fungus *Picoa juniperi*. Virology, 552: 10–19. 10.1016/j.virol.2020.09.009

Şen, İ., Allı, H. and Civelek, H.S. (2016). Checklist of Turkish truffles. Turk. J. Life. Sci., 1: 103–109.

Thomas, P.W., Elkhateeb, W.A. and Daba, G. (2019). Truffle and truffle-like fungi from continental Africa. Acta. Mycol., 54: 11–32. 10.5586/am.1132

Trappe, J.M., Claridge, A.W., Arora, D. and Smit, W.A. (2008). Desert truffles of the African Kalahari: Ecology, ethnomycology, and taxonomy. Econ. Bot., 62: 521–529. 10.1007/s12231-008-9027-6.

Trappe, J.M. and Claridge, A.W. (2010). The hidden life of truffles. Sci. Am., 302: 78–85. 10.1038/scientificamerican0410-78

Trappe, J.M., Claridge, A.W. and Kagan-Zur, V. (2014). Ecology and distribution of desert truffles in the Kalahari of Southern Africa. pp. 193–202. *In*: Kagan-Zur, V., Roth-Bejerano, N., Sitrit, Y. and Morte, A. (eds.). Desert truffles: Phylogeny, physiology, distribution and domestication. Soil Biology, Springer, Berlin, Heidelberg. 10.1007/978-3-642-40096-4_13

Türkoğlu, A. and Yağız, D. (2012). Contributions to the macrofungal diversity of Uşak province. Turk. J. Bot., 36: 580–589. 10.3906/bot-1103-2

Türkoğlu, A. and Castellano, M.A. (2014). New records of some Ascomycete truffle fungi from Turkey. Turk. J. Bot., 38: 406–416. 10.3906/bot-1303-24

Türkoğlu, A., Castellano, M.A., Trappe, J. and Güngör, M.Y. (2015). Turkish truffles I: 18 new records for Turkey. Turk. J. Bot., 39: 359–376. 10.3906/bot-1406-42

Uzun, Y., Kaya, A., Karacan, İ.H., Kaya, Ö.F. and Yakar, S. (2015). Macromycetes determined in Islahiye (Gaziantep/Turkey) district. Biol. Divers. Conserv. 8: 209–217.

Uzun, Y., Çetinkaya, A. and Kaya, A. (2016). Two new hypogeous species records for Turkish macromycota from Ayrancı and Yeşildere (Karaman) districts. 4th International Symposium on Development of Kop Region, Karaman–Turkey, October 21–23, pp 182–185.
Uzun, Y. and Kaya, A. (2018). First records of *Hydnobolites* and *Pachyphlodes* species from Turkey. Mycotaxon, 133: 415–421. 10.5248/133.415
Uzun, Y. and Yakar, S. (2018). New locality record for two *Tuber* species in Turkey. Ant. J. Bot., 2: 88–92. 10.30616/ajb.467302
Uzun, Y., Yakar, S., Karacan, I.H. and Kaya, A. (2018). New additions to the Turkish Pezizales. Turk. J. Bot., 42: 335–345. 10.3906/bot-1708-11
Uzun, Y. and Kaya, A. (2019a). *Elaphomyces granulatus*, a new hypogeous ascomycete record for Turkey. KSU J. Agric. Nat., 22: 85–88. 10.18016/ksutarimdoga.vi.466008
Uzun, Y. and Kaya, A. (2019b). A new *Elaphomyces* record for Turkey. Mantar Dergisi, 10: 40–43. 10.30708mantar.498766
Uzun, Y. and Kaya, A. (2019c). New additions to Turkish Pezizales from the Eastern Black Sea region. Turk. J. Bot., 43: 262–270. 10.3906/bot-1802-34
Uzun, Y. and Kaya, A. (2020). *Elaphomyces citrinus* and *E. cyanosporus*, new for Turkey. Mycotaxon, 135: 339–344. 10.5248/135.339
Uzun, Y. (2021). *Elaphomyces anthracinus*, a new hypogeous ascomycete record for Turkish mycota. Anatol. J. Bot., 5: 29–31. 10.30616/ajb.857752
Uzun, Y. and Kaya, A. (2021). First record of *Elaphomyces decipiens* for the mycobiota of Turkey. Mantar Dergisi, 12: 134–137. 10.30708.mantar955067
Uzun, Y. and Kaya, A. (2022). Macromycetes determined in Tonya (Trabzon) district. KSU J. Agric. Nat., 25: 66–77. 10.18016/ksutarimdoga.vi.857201
Uzun, Y. (2023). The checklist of Turkish Pezizales species. Anatol. J. Bot., 7: 1–20. 10.30616/ajb.1230337
Ünal, G., Türkoğlu, A. and Güngör, M.Y. (2016). Taxonomic studies on the macrofungi growing in *Eucalyptus* forests in Mugla province. Turkish J. Agric. Food. Sci. Technol., 4: 244–247. 10.24925/turjaf.v4i3.244-247.612
Valverde, M.E., Hernández-Pérez, T. and Paredes-López, O. (2015). Edible mushrooms: improving human health and promoting quality life. Inter. J. Microbiol., 2015: 376–387. 10.1155/2015/376387
Wang, S. and Marcone, M.F. (2011). The biochemistry and biological properties of the world›s most expensive underground edible mushroom: Truffles. Food. Res. Inter., 44: 2567–2581. 10.1016/j.foodres.2011.06.008
Wang, H.F., Zhao, Y.L. and Fan, Y.J. (2014). Ecological characters of truffles. pp. 2903–2908. *In*: Li, S., Jin, Q., Jiang, X. and Park, J. (eds.). Frontier and future development of information technology in medicine and education. Lecture Notes in Electrical Engineering, Springer, Dordrecht. 10.1007/978-94-007-7618-0_368
Wasser, S.P. (2011). Current findings, future trends, and unsolved problems in studies of medicinal mushrooms. Appl. Microbiol. Biotechnol., 89: 1323–1332. 10.1007/s00253-010-3067-4
Willis, K.J. (2017). State of the world's plants 2017—report. Royal Botanic Gardens, Kew.
Wu, B., Hussain, M., Zhang, W., Stadler, M., Liu, X. et al. (2019). Current insights into fungal species diversity and perspective on naming the environmental DNA sequences of fungi. Mycology, 10: 127–140. 10.1080/21501203.2019.1614106
Yıldız, A. and Ertekin, A.S. (1997). Contribution to the macrofungal flora of Diyarbakır. Turk. J. Bot., 21: 119–122.
Zambonelli, A., Iotti, M. and Murat, C. (2016). True trufle (*Tuber* spp.) in the world: Soil ecology, systematics and biochemistry. Springer International Publishing, Cham. 10.1007/978-3-319-31436-5

Chapter 7

Diversity, Ecology and Mycorrhizal Association of Desert Truffles in Morocco

Hicham Kaddouri,[1,2,*] *Lahcen Ouahmane*[1] and *Abdessamad Tounsi*[2]

1. Introduction

Several studies have been conducted on the global importance of truffles in terms of agriculture, medicine, and economics. Although some research has found that fungi are heterotrophic organisms, they acquire organic matter through three main modes of nutrition (symbiotic, parasitic and saprotrophic). While many species of saprophytic fungi have been successfully cultivated for a long time (Oei, 2003), for example, Judas ear (*Auricularia auricula*), cultivated in China since 600 BC, or *Agaricus bisporus*, dating back to the 17th century, are other types of fungi that are more difficult to cultivate (Hall and Zambonelli, 2012). The first to understand the symbiotic nature of chimeric associations between roots and fungi, but described long ago, he named mycorrhizae from the Greek Mukè, fungus, and rhizome root (Frank, 2005). Like all fungi, truffles are unable to synthesize their organic matter and use solar energy due to the absence of chlorophyll, a pigment responsible for photosynthesis.

They live in association with plants for their supply of organic substances and sugars that are important in exchange for minerals from their hosts. The reciprocal exchange between fungi and their hosts defines the relationship of symbiosis. They

[1] Cadi Ayyad University, Laboratory of Microbial Biotechnologies, Agrosciences and Environment (BioMAgE), Morocco.
[2] Faculty of Sciences and Techniques, University of Sultan Moulay Slimane Morocco, Avenue Mohamed, Beni Mellal, Morocco.
* Corresponding author: kaddouri888@gmail.com

are characterized by a mycelium that travels through the soil looking for food, penetrating the root system of the plant and causing disruption of its structure. The association between the host plant's secondary roots (rootlets) and the mycelium leads to the formation of a structure essential for exchange: the mycorrhiza. There are two types of symbiotic connections with the roots of plants belonging to the family *Cistaceae* (ectomycorrhizal and endomycorrhizal). These include annual and perennial species from the genera *Cistus*, *Helianthemum* and *Tuberaria* (Fortas and Chevalier, 1992; Slama et al., 2010; Zitouni-Haouar et al., 2014).

In this symbiotic relationship, the plant provides the fungus with carbon, which ensures its hydromineral nutrition. Additionally, both partners protect each other from the aggression of the soil. Therefore, in mycorrhizal species, cultivation of a single fungus is difficult, if not impossible, and is limited to the vegetative stage if the entire plant-fungus relationship is not understood. Hence, in light of this brief introduction, it is important to focus our research on magical resources. Truffles are ascomycetes that belong to the order Pezizales. Their hypogeous fructification is edible (Yao et al., 1995). They grow and develop in the Mediterranean basin and the Middle East, mostly in semi-arid and desert regions (Bradai et al., 2015), on calcareous soil rich in calcium (Bokhary and Parvez, 1992). Desert truffles are ascomycetes that only develop in arid and semi-arid regions. They have symbiotic relationships with the roots of trees like poplar, oak, and hazelnut (Harrison, 1997). Generally, the most common terfess host plants are herbaceous or perennial *Helianthemums* (Khabar, 2014).

The most essential qualities of truffles are their environmental and health benefits. They are full of proteins, vitamins, minerals, carbohydrates, and fats (Bokhary et al., 1987; Bokhary and Parvez, 1993). Some types of truffles, like *Tuber aestivum* (summer truffle), *Tuber magnatum* (white truffle), and *Tuber melanosporum* (black perigord truffle), are also appreciated for their aroma (Diaz et al., 2003). They have a symbiotic relationship with the roots of trees such as poplar, oak, and hazelnut (Harrison, 1997). Generally, the most common terfess host plants are herbaceous or perennial *Helianthemums* (Khabar, 2014). Desert truffles are popular for their nutraceutical benefits. A study aimed to produce fungal biomass from *Tirmania nivea* using an artificial growth medium. The fungal biomass was grown in potato dextrose broth and analyzed for antimicrobial activity. The crude extracts showed potential as antibacterial, anticancer, antifungal, anti-inflammatory, antioxidant, anti-trypanosomal, and antiviral agents. The study highlights the potential of desert truffles as a natural renewable resource (Khaled et al., 2021).

Mycorrhiza results from the association established between vascular plants and their symbiotic soil fungi. Mycorrhiza improves the ability of plants to resist environmental factors, including nutrient deficiency, water stress, and soil nuisance (Barea et al., 2011). In Morocco, research works related to truffles are not yet well developed; they are of taxonomic, floristic (Chatin, 1891), phytosociological, cytological, ultrastructural, and geographical orders (Harki et al., 2010; Khabar, 2014). It was found that mycorrhizas were closely related to desert truffles. In spite of the significance of mycorrhizal connections to desert truffles, there have been few investigations into this subject. As a result, the major goal of this study is to document recent studies of mycorrhiza-desert knowledgeabout mycorrhiza-desert truffle interactions, identify important knowledge gaps, and provide solutions to fill these gaps in future studies. For the objective of this study, a methodology based on the following main steps was chosen: Documentary research on knowledge of Moroccan truffles: Google Scholar, PubMed, ScienceDirect, SpringerLink, Web of Science and Scopus.

2. Desert Truffles and Their Classification

The *Carbomyces*, *Elderia*, *Eremiomyces*, *Kalaharituber*, *Mattirolomyces*, *Mycoclelandia*, *Picoa*, *Terfezia*, *Tirmania*, *Stouffera*, and *Ulurua* genera are the taxa that make up the family of desert truffles (Kovács and Trappe, 2013). The identification of truffles is based on taxonomic criteria: the shape of the spores, the size, the ornamentation, and the structure of the gleba and the peridium (Mello et al., 2006). Most of the morphological studies to differentiate truffle species have been done on the fruiting body, the ascocarp. Concerning the ascocarp (truffle itself), these morphological differences are observed at different levels: 1) macroscopic (size of the ascocarp, color of the gleba, aspect of the peridium); 2) microscopic (size of the gleba, aspect of the peridium); 3) microscopic (morphology of the asci and spores). The size of the fruiting body depends on the environment and the species.

Their chemical and molecular characteristics are another way to distinguish them. Indeed, the shape of the ascocarps is spherical (subglobose or piriform and glabrous), characterizing the species of *T. nivea* (Jamali and Banihashemi, 2012). Their skin is light brown or creamy white, and the pith, called gleba, is initially white or yellowish and becomes brown with age. Generally, the fruiting body is characterized by a spherical pear-shaped form of width 5–12 (20) cm, lobed or irregular. The outside is white, cream, ochre, rusty-ochre, or brown-ochre. In maturity, the gleba is compact, white, and elastic, with light pink or ochre spots and subtle white veins. It has a lovely odor. In contrast, the fruiting bodies of *T. arenaria* are 3–12 (15) cm wide, usually with a rudimentary and irregularly spherical base. The outer part is dry, beige, ochre-beige, or pink-beige and, in most cases, lobed, reddish-brown, grayish-blackish brown, or brown at maturity. The gleba is characterized by yellow colors and a slightly spermatic smell (Loizides et al., 2012).

The *T. claveryi* ascocarps collected in Morocco had the following shapes: spherical, ovoid, and cordiform, with fresh weights of between 17 and 50 g. The gleba exhibited a few millimeters of spotting, and the peridium was initially pinkish-white in hue before turning brownish-black. There were thin, pale veins dividing the fertile, spherical, pinkish (before turning light brown), and round nodules in the gleba. The gleba had a firm, slightly spongy texture. Hyaline, gray, spherical ascospores with an alveolus network were found inside the asci, which were ovoid in form. It is possible for *Tirmania* sp. ascocarps to change shape and size from sub-globular and pyriform to turbine-like, with lobes or irregular shapes (Khabar, 2016). The four main truffle-growing regions in Morocco are the Doukkala-Abda Sahel, Moroccan Sahara, Maâmora Forest, and the North-East. There are about ten species of desert truffles in these regions, which belong to the genera *Tuber*, *Delastria*, *Terfezia*, *Tirmania*, and *Picoa* (Hakkou et al., 2022a). Tables 1 and 2 encompass the biodiversity of desert truffles.

Table 1. Diversity of desert truffles.

Country	**Taxonomy**		**Biotic and abiotic factors**	**Harvest period**	**Reference**
Genus, *Terfezia*; Division, Ascomycota; Class, Pezizomycetes; Order: Pezizales; Family, Terfeziaceae					
Morocco	*Terfezia arenaria* (Moris)	*Tuber arenarium* Moris (1829)	Climate: Semi-arid Soil: acid, light sandy; Host plant: *Helianthemum guttatum*	Mid-February To early June	Trappe, 1971 ; Malençon, 1973, Abourouh, 2011; Khabar, 2016; Harir et al., 2019; Abourouh, 2020; Hakkou et al., 2022
		Choiromyces Leonis Tul. and C. Tul. (1845)			
		Rhizopogon Leonis Tul. and C. Tul (1850)			
		Terfezia leonis Tul. and C. Tul. (1851)			
	Terfezia boudieri Chatin (1891)	*Terfezia deflersii* Patouillard (1894)	Climate: Semi-arid and arid Soil: basic, gypseous and calcareous; Host plant: *Helianthemum Lippii* var. sessiliflorum and *Helianthemum ledifolium*	March and April	Chatin, 1891; Roth-Bejerano et al., 2004; Khabar, 2016; Abourouh, 2020
	Terfezia claveryi Chatin (1891)	*Terfezia hafizzi* Chatin (1892)	Climate: Semi-arid, Arid, pre-Saharan, and Saharan; Soil: basic, limestone or marl-gypsum soil; Host plant: *Helianthemum lipii* and *Helianthemum apertum*	Early March to early June	Chatin, 1891; Malençon, 1973; Alsheikh, 1994; Khabar, 2002; El Akil et al., 2016; Khabar, 2016
		Terfezia hanotauxi Chatin (1895)			
	Terfezia leptoderma Tulasne (1851)	*Choiromyces leptodermus* Tulasne and C. Tulasne (1845)	Climate: Semi-arid, Arid; Soil: sandy acid or more compact, clayey and siliceous; Host plant: *Helianthemum guttatum*	The third week of February and until May	Díez et al., 2002; Khabar, 2002; El Akil et al., 2016; Abourouh, 2020
	Terfezia olbiensis Tul. and C. Tul. (1845)	*Choiromyces olbiensis* Tul. and C. Tul. (1845)	Climate: Arid; Soil: sandy; Host plant: *Pinus halepensis*	December and January	Alsheikh, 1994; El Akil et al., 2016

Table 1. contd. ...

Table 1. contd. ...

Country	Taxonomy		Biotic and abiotic factors	Harvest period	Reference
Genus, *Tuber*; Division, Ascomycota; Class, Pezizomycetes; Order, Pezizales; Family, Tuberaceae					
Morocco	*Tuber asa* Tulasne (1851)	*Terfezia gennadii* Chatin (1896)	Climate: Semi-arid; Soil: acidic soils sandy or more compact, clayey and siliceous soils; Host plant: *Helianthemum guttatum*	End of the month of February until the end of April	Khabar, 2002, 2016; Abourouh, 2020
		Tuber lacunosus Mattirolo (1900)			
		Tuber gennadii (Chat) atouillard (1903)			
	Tuber oligospermum (Tul. and Tul.) Trape (1979)	*Terfezia oligosperma* Tul and Tul. (1851)	Climate: Semi-arid; Soil: sandy; Host plant: *Pinus pinaster* var. *atlantica*	December until April or May	Khabar, 2002, 2016; Abourouh, 2020
		Delastreopsis oligosperma (Tul.) Matt. (1904)			
	Tuber gennadii (Chatin) Patouillard (1903)	*Terfezia gennadii* Chatin (1896)	Climate: Semi-arid; Soil: sandy soils or more compact, clayey, and siliceous; Host plant: *Helianthemum guttatum*	March to June	Abourouh, 2020
		Tuber lacunosum Mattirolo (1900)			
		Loculotuber gennadii (Chatin) Trappe, (1992)			

Morocco	*Tirmania Nivea* (Desf.) Trappe, 1971	*Tuber niveum* Desf. (1823)	Climate: Arid, Pre-Saharan, and Saharan; Soil: sandy; Host plant: *Helianthemum hirtum* and *Helianthemum lipii*	From February through May	Malençon, 1973; El Aji, 1999; El Akil et al., 2016
		Terfezia ovalispora Patouillard (1890)			
		Tirmania ovalispora Pat. (1892)			
		Tirmania africana Chatin, (1892)			
	Tirmania pinoyi (Maire) Malençon (1973)	*Terfezia Pinoy* Maire (1906)	Climate: Arid, Pre-Saharan, and Saharan; Soil: sandy; Host plant: *Helianthemum hirtum* and *Helianthemum lipii*	From February through May	Malençon, 1973; El Aji, 1999; El Akil et al., 2016
Genus, *Delastria;* Division, Ascomycota; Class, Pezizomycetes; Order, Pezizales; Family, Terfeziaceae					
Morocco	*Delastria rosea* Tulasne (1843)	*Terfezia rosea* (Tul.) Torrend (1907)	Climate: semi-arid; Soil: sandy; Host plant: *Pinus pinaster* var. atlantica	From February through May	Khabar, 2002, 2016; Abourouh, 2020
Genus, *Picoa*; Division, Ascomycota; Class, Pezizomycetes; Order, Pezizales; Family, Helvellaceae					
Morocco	*Picoa juniperi* Vittadini (1831)	*Picoa juniperina* Tulasne (1851)	Climate: Arid, Pre-Saharan, and Saharan; Soil: calcareous; Host plant: *Helianthemum lipii*	February to April	El Aji, 1999; Khabar, 2016

3. Geographic Distribution of Truffles

Geographically, truffles are spread over arid and semi-arid regions around the Mediterranean Basin and the Middle East (Bradai et al., 2015). In Morocco, and according to Abourouh (2011) and Khabar (2016), four major areas have been identified for truffle growing: the Oriental, the Maâmora Forest, the Sahel of Abda-Doukkala, and the Moroccan Sahara. Ten species may be found in Morocco's semi-arid and arid areas, where yearly rainfall is low and average temperatures are high (Zniber et al., 2022). They are associated with *Cistus*, *Pinus*, and many other species, but they are more frequently found in *Helianthemum* spp. (Zniber et al., 2022).

The southern region of Morocco is home to a variety of desert truffles. The first is the Red Terfes of Tafilalet, also known as *T. claveryi* (*T. hafizi*). It is situated close to Ksar Es-Souk, Boudenib, and Figuig (Abou Lakhal and Bni Guil areas), in the regions of Tendrala, Ain Beni Mather, Erfoud, Bouarfa, and Bou Bernous (Khabar, 2016; Bermaki et al., 2017). Using *H. lipii* as a host plant species, it can be collected following the rains that occur from March to May. Additionally, *T. boudieri* can be found a few kilometers east of Safi in the Had Hrara and Abda plains in the Oualidia area. Additionally, it can be found in the areas around Ain Beni Mather and Erfoud. More significantly, it can be found in the March and April-harvested *Helianthemum* spp. (*H. ledifolium* and *H. lipii*) in the desert plains. *Ti. nivea* and *Ti. pinoyi* have been discovered as two additional species that fall under the umbrella of the genus *Tirmania*. Ain Beni Mather, Bouarfa, Erfoud, Figuig, Rissani and Tendrara are among the south-easterly areas where they are dispersed. The crop season lasts from early December to early March, when *H. hirtum* serves as the host plant.

Tunisia, Algeria, Egypt, Libya, Iraq, Saudi Arabia, Kuwait, Syria, Palestine, and Iran are among the countries where desert truffles are found. Additionally, they are found in the southern regions of France, Spain, Italy, Greece, and Portugal in the Mediterranean Basin (Khabar, 2016). *T. claveryi* is found in the African nations of Libya (Hashem et al., 2018), Tunisia, Algeria, Egypt (Khabar, 2016), and even in France and Spain (particularly in the southern semi-desert regions) (Khabar, 2016). *T. boudieri* can be found in Libya and Egypt, sometimes in Tunisia and Algeria, under the plant *Helianthemum lipii*, and occasionally under the plant *Rhatherium suaveolens*. In Tunisia, the desert truffles are mainly found in the center and south of the country.

In Algeria, truffles grow in the regions of Bechar (Taghit, Kenadssa, Tabelbala, Béni Abbes, Abadla), Ouargla (Oued Mya), Tindouf, Timimoun, Touggourt, Tamanrasset, El Golea, Ghardaïa, Biskra, Djelfa, Batna, Boussaâda, Saida, Naama, Mécherai, and Laghouat (Chevalier, 1984). In France, Spain, Portugal, Hungary, and northern Serbia, one finds *Terfezia arenaria T. boudieri*, *T. claveryi*, *T. leptoderma*, and *T. olbiensis* (Maire, 1906; Malençon, 1973; Dıaz et al., 2003; Culleré et al., 2010).

In addition, the following species are found in Turkey, Syria, Lebanon and the Middle East, *Picoa juniperi*, *P. lefebvrei*, *Terfezia arenaria*, *T. boudieri*, *T. claveryi, T. leptoderma*, *Tirmania nivea*, and *T. pinoyi*, were identified by Alsheikh and Trappe in 1983. Four truffle species were identified in non-Mediterranean Asian countries like Saudi Arabia, Bahrain, Qatar, and Kuwait during the census. The desert white truffle species *Tirmania nivea* is locally known as Zubaidi. Consumers prefer the latter species over the other. *Picoa lefebvrei*, *Terfezia claveryi*, and *T. boudieri* are the brown truffles known locally as Kholaissi (Al-Laith, 2010).

4. Biology, Development and Lifecycle of Truffles

4.1 Biology

A recent study carried out on desert truffles found that hypogeous fungi create a symbiotic relationship with *Cistaceae* family plants, and the reproductive methods and biochemical mechanisms behind the ectendomycorrhizal relationship of these fungi are unknown (Marqués-Gálvez et al., 2021). *Terfezia claveryi* and *Tirmania nivea*, two highly prized edible desert truffles, have had their genomes sequenced and compared to those of other *Pezizomycetes*. In the same context, Marqués-Gálvez et al. (2021) identified fungal genes involved in sexual reproduction in desert truffles as well as genomic and secretomic features distinct from other *Pezizomycetes* in desert truffles, such as the proliferation of a significant number of gene families with unknown Pfam domains and a number of small, species- or desert truffle-specific secreted proteins that were differentially managed in the symbiosis. The factors that induce sexual reproduction of the truffle are not known, and many burned sites where the vegetative presence of the truffle has been verified remain non-productive.

Laboratory studies on unicellular and filamentous Ascomycetes have highlighted the importance of factors such as substrate modification (including nutrient restriction), light, pH, and atmospheric conditions that induce either vegetative or sexual reproduction (Debuchy et al., 2010). Not many studies have looked into what causes fructification in macromycetes mushrooms, but it looks like sudden changes in temperature, the amount of nutrients available, and the amount of water available may all affect the growth of the mycelium and cause fructifications to occur (Kües and Liu, 2000; Oei, 2003). According to Pacioni et al. (2014), hydric stress encourages the commencement of young ascocarp in truffe, and the mycelium's strong activity (measured by CO_2 emission) precedes their creation. In general, it appears that moderate water stress is advantageous to the vegetative growth of the truffe (rate of mycorrhization). Olivera et al. (2014) indicate that careful irrigation is necessary to encourage the maintenance of truffles.

4.2 Ascocarp Development

As is typical for ascomycetes, the mycelium haplotype forms the ectomycorrhizae of truffe (Rubini et al., 2014). The genes present locally at the level of the ectomycorrhizae will act as the individual mother during reproduction, forming the gleba, the sterile chair of the ascocarp, in which they participate through nourishment. Researchers used an isotopic CO_2 measurement experiment to track the carbon produced by the host tree's photosynthetic process. This showed that the ascocarp gets its carbon-based food from the tree through the ectomycorrhizae, which are connected to the ascocarp in ways that were not known before (Le Tacon et al., 2013). While this method has not directly identified the source of other compounds (such as phosphate

Table 2. *Tuber* truffles in Morocco.

Country	Taxonomy		Biotic and abiotic factors	Harvest period	Reference
Morocco	*Tuber Melanosporum* Vittadini, 1831	*Tuber brumale* Pulpa obscura, odora P. Micheli (1729)	Climate: Semi-arid and subhumid to humid; Soil: well-drained and porous aerated limestone; Host plant: *Quercus ilex*	November through March	Laqbaqbi, 2020
		Tuber gulosorum (Scopoli) Pico (1788)			
		Tuber nigrum Bulliard, (1788)			
		Tuber cibarium nigrum Bulliard (1791)			
		Tuber cibarium Persoon, (1801)			
		Tuber cibarium Bulliard Fries (1823)			
		Tuber gulonum (Corda) Paoletti (1889)			
	Tuber Borchii var. sphaerosperma Malençon (1973)	*Tuber sphaerospermum* Malençon (2006)	Climate: Semiarid; Soil: sandy; Host plant: *Pinus halepensis* and *Pinus pinea*	April and May	Malençon, 1973; Abourouh, 2020
	Tuber rufum (Pico, 1788)	*Tuberum tertium* (Mattioli, 1544)	Climate: subhumid and humid; Soil: calcareous; Host plant: *Quercus ilex*	June to November	Malençon, 1973
		Tuber cinereum Tul et C. Tul. (1844)			
		Tuber lucidum Bonnet (1884)			
		Tuber caroli Bonnet (1885)			
		Tuber rutilum R Hesse, (1894)			

Morocco	*Tuber uncinatum* Chatin (1887)	*Tuber Aestivum* Vittadini (1831)	Climate: subhumid and humid; Soil: Acidic soils sandy; Host plant: *Quercus ilex*	September to January	Malençon, 1973
	Tuber brumale Vittadini (1831)	*Tuber brumale* Vittadini var. moschatum Ferry de la Bellone (1888)	Climate: humid; Soil: calcareous Host plant: *Quercus ilex*	November to March	Olivier et al., 2018
Algeria	*Tirmania*	*Tirmania pinoyi* (Maire)	Host plant: *Helianthemum lippii* (L.) Dum. Courset		Bradai et al., 2013
Tunisia	*Terfezia*	*Terfezia boudieri* (Chatin)	Host plant: *Helianthemum sessiliflorum* Desf.		Hamza et al., 2016
Egypt	*Terfezia*	*Terfezia arenaria* (Moris) Trappe	Host plant: *Helianthemum lippii* (L.) Dum. Courset		El-Kholy, 1989
		Terfezia claveryi (Chatin)			
	Tirmania	*Tirmania nivea* (Desf.) Trappe			
Bahrain	*Tirmania*	*Tirmania nivea* (Desf.) Trappe	Host plant: *Helianthemum kahiricum* Del.		Mandeel and Al-Laith, 2007
	Phaeangium	*Phaeangium lefebvrei* (Pat.)			
Kuwait	*Tirmania*	*Tirmania nivea* (Desf.) Trappe	Host plant: *Helianthemum kahiricum* Dei.		Alsheikh and Trappe, 1983
Jordan	*Terfezia*	*Terfezia claveryi* (Chatin)	Host plant: *Helianthemum* spp.		Janakat et al., 2004
Iraq	*Terfezia*	*Terfezia claveryi* (Chatin)	Host plant: *Helianthemum* spp.		Dahham et al., 2016
Saudi Arabia	*Terfezia*	*Terfezia claveryi* (Chatin)	Host plant: *Helianthemum vesicarium* Boiss. *Helianthemum kahiricum* Del. *Helianthemum lippii* (L.) Dum.		Sawaya et al., 1985; Bokhary et al., 1987; Hashem and Al-Obaid, 1996; Janakat et al., 2005; Schillaci et al., 2017
		Terfezia boudieri (Chatin)			
	Tirmania	*Tirmania nivea* (Desf.) Trappe			
		Tirmania pinoyi (Maire)			
	Picoa	*Picoa juniperi* Vittad.			
		Picoa lefebvrei (Pat.) Maire			
		Phaeangium lefebvrei (Pat.) Maire			

and azote) it is possible that truffle mycelium directly absorbs them from the soil (Barry et al., 1994). Additionally, bacterial involvement such as azote fixators from the Bradyrhizobiaceae family may play a role in nutrient availability for truffles particularly *T. magnatum* (Barbieri et al., 2010; Antony-Babu et al., 2014).

The gleba surrounds the ascocarp, which has thousands of brightly colored spores that make it hard for them to pass through a digestive tube (Piattoni et al., 2014). It has been established that these spores are the result of heterothallic sexual reproduction (Rubini et al., 2011), despite Truffe's long-held homothallic reputation. This idea was put forward because heterozygosis was not seen in one of the earlier studies that used codominant genetic markers (two markers for microsatellites that can find different alleles, or heterozygosis) (Bertault et al., 2001). Thus, the presence of heterozygotes would have increased the likelihood of a crossover between two individuals within this haploid species. However, it is likely that only the gleba's ADN was extracted and/or that the two microsatellite markers used were not sufficiently polymorphic (carried little genetic variability). A few years later, the sequencing of the truffe genome haploid strain Mel28 (Martin et al., 2010) showed the existence of a sexual allele (MAT1-2-1, abbreviated as MT2) that is indicative of heterothallism. The sexually antagonistic sequence MAT1-1 (abbreviated as MAT1) has subsequently been identified (Rubini et al., 2011). It is proven that the truffle mushroom is a heterothallic mushroom by extracting the ADN of the spores in a certain way that 1) concentrates the ascospores and 2) tears apart the ascus and spore walls. As a result, the genotype of the mother is determined by working with the gleba's ADN, and the genotype of the father can be determined by comparing the genotype of the mother to the zygote, which was obtained by working with the spores ADN. Because other species cannot directly achieve the parental genotypes, Truffe's particularity makes it an original biological function model.

4.3 Lifecycle

In the context of defining the symbiotic view cycle of truffles, it is necessary to talk about the emergence and functioning of the mycorrhizal association. The fungi colonize the surfaces and the cortex. Externally, tiny fungal hyphae efficiently explore the soil. 90% of all plants depend on this symbiotic relationship, which differs in the anatomical structures, fungal taxa, and plant families involved (Brundrett, 2002; van der Heijden et al., 2015). The fossil record and phylogenetic analyses show that mycorrhizal communities first appeared in the Ordovician period. At that time, they were very important in helping new land plants settle down (Le Tacon and Selosse, 1994; Remy et al., 1994). Indeed, the fossil structure found is very similar to that of the arbuscular endomycorrhiza (AM) formed. Today, there are only 244 known species of *glomeromycetes*, or about 80% of all plants (Smith and Read, 2008). More recently, 100–200 Ma ago, ectomycorrhizal associations emerged (van der Heijden, M. G. et al., 2015). This association concerns woody plants belonging to the families *Cistaceae*, *Pinaceae*, *Fagaceae*, *Tiliaceae*, and others. (Brundrett, 2002), which dominate the Mediterranean temperate and boreal forests (Smith and Read, 2008). The loss of genes is responsible for cellulose and lignin degradation (Plett and Martin, 2011). However, ectomycorrhizal fungi (ECMs) are endowed with a catalytic capacity to utilize soil organic and mineral matter (Rineau et al., 2012). It is crucial to note that the truffle can develop in pure culture using cellulose as a source of carbon (Mamoun and Olivier, 1991).

The truffles are ascomycetes with hypogeous fruiting that grow underground, from 5 to 10 cm deep. Throughout the world, more than hundreds of different species of truffles are known, and regularly, the discovery of new species is observed. Taxonomically, these mushrooms belong to the order Pezizales and the families *Tuberaceae* and *Pezizaceae* (El Enshasy et al., 2013). For centuries, truffles belonging to the family *Tuberaceae* were widely edible, so were the later variety of lineages, mainly truffle mushrooms (Laessoe and Hansen, 2007). Worldwide, the *Tuber* genus of the *Tuberaceae* family includes about 180 species (Bonito et al., 2010). The Tuber genus truffles that are found in Europe's temperate forests, New Zealand, Australia, America, and Asia qualify for culinary preparations (Kinoshita et al., 2011; Berch and Bonito, 2016). Generally, the desert truffles that grow in semi-arid and arid regions like Morocco, Egypt, Tunisia, Saudi Arabia, Syria, Kuwait, Iraq, and South Africa belong to the genus *Tirmania* and *Terfezia* (El Enshasy et al., 2013; Bradai et al., 2015). *Terfezia claveryi, Tuber aestivum*, *T. brumale*, *T. himalayense*, *T. indicum*, *T. melanosporum*, and *T. uncinatum* are species of black truffles. Whereas *Tuber aculatum, T. borchii*, *T. japonicum*, *T. latisporum*, *T. magnatum*, *T. oregonense* and *Tirmania nivea* are white truffles (Wang and Marcone, 2011). The Italian species (Eastern Europe), *T. magnatum*, has been declared the most expensive truffle (Mello et al., 2006). Therefore, it is important to point out that, in the Italian markets, the price of one kilogram of this species is 8000 dollars (Jeandroz et al., 2008). Truffles are characterized by a genome with a large molecular weight; *T. melanosporum*, for example, has a 125 Mb genome (Chen et al., 2014).

Truffles are ectomycorrhizal fungi that have symbiotic roots. They have a complicated life cycle, with the mycelium forming a symbiotic relationship with host species mainly through the roots of trees like oak, hazelnut, pine, and poplar (Mello et al., 2006; Patel, 2012). Since truffles grow their sexual fruiting bodies underground, insects and mammals are necessary for spore distribution (Laessoe and Hansen, 2007). After the ascospores are released, the haploid spore turns into haploid, free-living mycelium. This makes an ectomycorrhizal connection with the roots of host plants. Asci with an ascoma, which is the sexual fruiting body, are then made when hyphae stick together (Fig. 1) (Paolocci et al., 2006; Chen et al., 2019).

For example, the life cycle of *Tuber melanosporum* and the fruit of this species are localized in open woodlands. During the vegetative phase, the mycelium establishes ectomycorrhizal relationships in the soil with host plants (Figure 1). Its most well-known host species, mainly for inoculation are: *Quercus* spp., *Fagaceae* and *Corylus* spp., *Betulaceae* (Olivier et al., 2018). According to Comandini et al. (2006), *Cistus* are fewer known hosts for the presence of truffle species or the genus *Tuber*, or *arbutus* (a species of small tree shrubs of the family *Ericaceae*) (Lancellotti et al., 2014). It is, therefore, necessary to point out that the burned truffles are an important indicator of the presence of mycelium in the ground (Streiblová et al., 2012). It is most observable in the spring. In some places, few or no burns are visible, but the name always refers to the area where the truffle grows. When these burnt areas are productive, the fruiting of the truffle ascomycetes begins in spring, a period

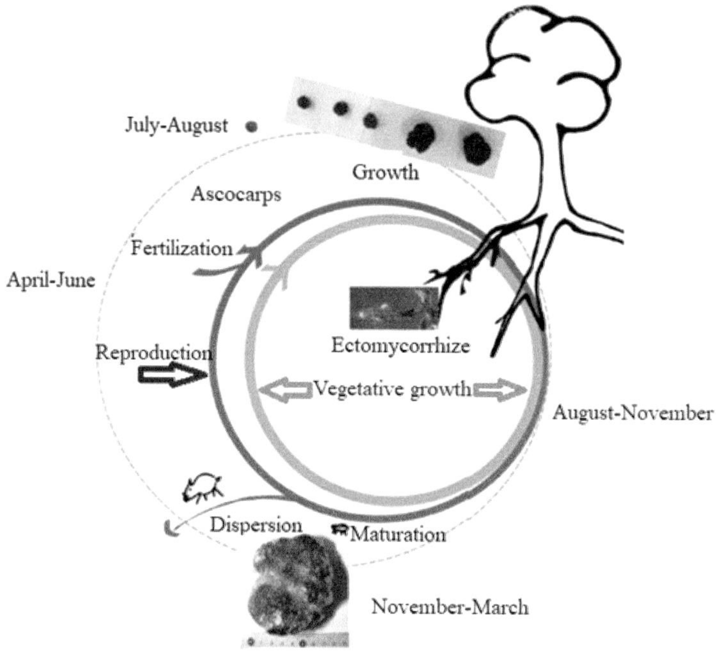

Fig. 1. The annual life cycle of the truffle (Taschen, 2015).

known as birth that can last until the end of July (Kulifaj, 1984). After a few months, the development and enlargement of the ascocarp occurs in the soil, releasing a particular aroma (sulfur compounds such as dimethyl sulfide and alcohols such as 2-methyl-1-propanol and 1-octen-3-ol) (Splivallo et al., 2011). From November until the first frosts in the soil, it reaches its sporal and aromatic maturity.

5. Ecology of Desert Truffles

In Morocco, in Great Sahara locations, desert truffles are proven to exist in soils of homogeneous texture, such as calcareous, aerated, and sandy soils; sometimes, they are located in soils of calcareous sandy loam nature (Khabar, 2016). The existence of desert truffles is rarely noted in acidic soils. It is interesting to point out that species of the genus *Helianthemum* are often symbiotic with desert truffles, such as *Helianthemum asperum*, *H. hirtum, H. ledifolium, H. lipii* and others. The organic matter concentration, fertility, and exchange capability of the soil must be low. Desert truffles need phosphate, nitrogen, potassium, calcium, and iron, yet it has been found that low levels of phosphate and iron enhance mycorrhizal potential. It should be noted that the mycorrhization potential is increases when phosphate and iron contents are low. The quantity and frequency of rainfall are significant limiting factors for desert truffles in terms of climatic conditions. In general, rain is needed in the winter or spring season, but for some species, rain is needed both in the winter and spring seasons. Desert truffles can be found in a variety of bioclimatic phases, including dry, semi-arid, sub-Saharan, and Saharan.

Despite being in a desert setting, the regions of Beni Guil and Abu Lakhal have the highest concentrations of desert truffles, with an average temperature range of 17.9–18.6°C. As a result, these two places experience precipitation that averages 170 mm per year. The significant amounts of rainfall in these areas throughout the night and early in the morning, as well as dewfall, thunder, and lightning, significantly aid the development of truffles. The great desert and a bulk of these characteristics are found in southern Morocco (Bermaki et al., 2017). Saudi Arabia is conducting ecological research on desert truffles in other parts of the globe. Hashem et al. (2018) listed precipitation as another element that is crucial for the arid truffle's growth. Reduced precipitation may change the host plant's physiological makeup and cause a decline in the production of spores and ascocarps. Additional factors include the weather, ambient temperature, and the physicochemical properties of the earth. The two main host plants on alkaline soils are *H. ledifolium* and *H. salicifolium*. The reaping season starts in early December, since it rains in October.

5.1 Edaphic Factors

Table 3 illustrates the range of characteristics observed in the majority of plantations established in Spain. Generally, the soils employed for plantation are of poor quality and characterized by a clay loamy texture, an alkaline pH (with a pH of 8.5), and a low level of electrical conductivity (at 123 µs/cm). Additionally, the soils exhibit low values of organic carbon (at 0.9–3.9%), as well as an imbalance in the carbon-to-nitrogen ratio, which is between 7 and 10, respectively. It has been observed that the characteristics, particularly those related to pH and texture, vary depending on the species of *Terfezia* that grows in acidic soils. While soil texture has been shown to influence the shape of ascomata, this factor is not regarded as a significant determinant in truffle formation, as *T. claveryi* ascomata have been produced in a range of soil types, including clay loam (in plantations in Murcia) and sandy soils (Lanzarote and Canary Islands plantations) (Malençon, 1973).

Analogous findings have been observed in *T. boudieria* ascocarps in plantations established in sand loamy and gypsy soils (Slama et al., 2010). Prior to this study, soil fertilization was not necessary. Fertilization has not been applied to the oldest plantation, which is currently 12 years old. Mycorrhizal *H. almeriense* plants continue to produce *T. claveryi* truffles, suggesting that the plantation has not yet required fertilization. In Spain, the cultivation of Terfezia is often carried out in conjunction with other crops, such as almonds and olives, in order to optimize land use and irrigation facilities. This approach is possible because arbuscular mycorrhizal fungi from almond and olive trees are not competing for the same host plants as the desert truffle species, allowing them to coexist.

5.2 Mycorrhizal Association

Desert truffles are culinary mushrooms that require plant roots as a living environment in order to thrive. In addition to the significant nutrients, they contain a high concentration of therapeutically beneficial secondary metabolites (Zniber et al., 2022). There are approximately 10 species that may be found in the semiarid and

Table 3. The range of soil physicochemical characteristics in which the majority of *Terfezia* plantations have been established in Spain.

Soil qualities	Range
pH (1:2.5 water v/v)	6.8–8.7
Electrical conductivity (1:5 µs/cm)	123.1–302
Sodium (meq/100 g)	0.15–1.20
Potassium (meq/100 g)	0.28–1.5
Calcium (meq/100 g)	4.94–23.38
Magnesium (meq/100 g)	0.95–3.7
Organic matter (%)	0.58–3.92
Total organic carbon (%)	0.34–2.28
Total nitrogen (%)	0.058–0.267
C/N ratio	2.8–10.15
Total carbonate (%)	5.1–80.1
Active lime (%)	3.4–24.76
Phosphorus (ppm)	7.52–66.4
Chloride (meq/100 g)	0.05–0.09
Sulphates (meq/100 g)	0.01–0.32
Iron (ppm)	1.79–79.5
Copper (ppm)	0.31–2.73
Manganese (ppm)	3.03–57.12
Zinc (ppm)	0.3–3.12

arid climate zones of Morocco, which are characterized by low annual rainfall and high average temperatures (Zniber et al., 2022). They may be related to the *Cistus* and *Pinus* species as well as a few other species, but they are most frequently found with the *Helianthemum* species (Zniber et al., 2022). In the same sense, a group of edible hypogeous fungi known as desert truffles create mycorrhizal symbiosis with perennial and annual plant shrubs from the plant family *Cistaceae*, which is adapted to arid and semi-arid regions (Kovács and Trappe, 2013; Roth-Bejerano et al., 2014). These areas are recognized by an aridity index of AI 0.5, which is the proportion between yearly precipitation and interest in evapotranspiration, possessing poor fertile soils with a sandy texture and low inputs of organic substances (Bonifacio and Morte, 2014).

Additionally, desert truffles interact with a mutualistic partner with mycorrhizal fungi and exist in symbiosis with plants. According to many studies, desert truffles in Algeria's Sahara region have symbiotic relationships with *Helianthemum lippii* (e.g., Díez et al., 2002; Mandeel and Al-Laith, 2007). The hypogeous ascomycetes develop mycorrhizal associations that share mutualistic relationships with several species of *Helianthemum* roots (Mandeel and Al-Laith, 2007; Kagan-Zur et al., 2014). Another study cites that mutualistic symbionts help their host plants develop and survive in certain environmental conditions (Morte et al., 2010). For instance, *Terfezia claveryi* mycorrhizas stimulate the assimilation of phosphorus, nitrogen, and potassium in *Helianthemum almeriense*, which increases the plant's survival during droughts (Morte et al., 2000). Because of this interaction, the expression of plant and fungal aquaporins as well as the amount of hydrogen peroxide in roots are finely regulated (Navarro-Ródenas et al., 2013; Marqués-Gálvez et al., 2020). Several *Terfezia* species (*T. boudieri* and *T. claveryi*) and *Tirmania* species (*T. nivea* and *T. pinoyi*) have been found to have mycorrhizal associations with *Helianthemum lippii*. They also form advantageous mycorrhizal connections with *Helianthemum* species (*H. ledifolium* and *H. salicifolium*) (Awameh and Alsheikh, 1980; Kovács and Trappe, 2013). *Helianthemum* species form both ectomycorrhizal and ectendodomycorrhizal connections (Claridge et al., 2013).

Although these species have been observed to lack a Hartig net and their mantle in the roots of the endomycorrhizal desert truffles, they still display undifferentiated hyphae, as seen, for instance, in associations between *Helianthemum* spp. and a number of species of *Terfezia* and *Tirmania* (Dexheimer et al., 1985). Thus, it is very important to mention that ectendomycorrhiza is the name for the kind of mycorrhiza created by these fungi (EEM). EEMs have a thin, disorganized fungal mantle covering the colonized roots, intercellular Hartig nets, intracellular hyphae that enter cortical cells and create coil-like structures, and all of these features (Morte et al., 1994; Yu et al., 2001). Intercellular or intracellular mycorrhizal structures can rely on a variety of factors. In field conditions, low auxin, low phosphate, and/or low water availability favor the intracellular mycorrhizal form, while in vitro, high auxin, phosphate, and/or water concentrations favor the intercellular mycorrhizal form (Gutiérrez et al., 2003; Zaretsky et al., 2006; Navarro-Ródenas et al., 2012). Due to this, it is possible to consider desert truffles as a type of intermediary connection between real ectomycorhizal and endomycorrhizal interactions. But the division between these two typical mycorrhizal connections is crystalline and fluid simultaneously. It should be noted that the traits and characteristics of mycorrhizal relationships are usually and frequently determined by both internal and external factors.

The ability of mycorrhizal associations to endure some degree of harsh environmental conditions is superior to that of plants in dry and semiarid areas of the world (Morte et al., 2009). A few varieties of truffle, including *Picoa lefebvrei* (*Phalangium lefebvrei*) and *Terfezia claveryi* are hypogeous ascomycetes that thrive in semiarid climates on marl and gypsum-rich soils. Additionally, these species establish mycorrhizal connections with the bulk of perennial and annual *Helianthemum* species (Honrubia et al., 1992). An ectomycorrhizal link exists between *Helianthemirhiza hirsuta* and *Helianthemum ovatum* (Kovács et al., 2011). This specific association has a simple mycorrhizal system with borders that are often cottony, straight or slightly curved, bending, or tortuous, and vary in color from ochre to brown. A thorough investigation into desert truffles and their interactions with different plants was carried out by Kovacs et al. (2002). The interaction with plants and ensuing behavior of both partners under aseptic conditions were investigated using modified Melin-Norkrans (MMN) substrates enriched with various phosphate concentrations. In order to assess the changes caused by *Terfezia* terfezioides interaction with *Helianthemum ovatum* and *Robinia pseudoacacia*, this task was carried out. Black locust colonization has frequently remained delicate and weaker than *Helianthemum*'s root system.

The *Helianthemum*do the intercellular hyphae usually create a Hartig net with such finger-like structures only in Helianthemum; without prior clear evidence, these kinds of contacts could not be classified as mycorrhizae. This is one of the few essential characteristics. They also observed that the RFLP profiles of the nuclear DNA ITS of 19 fungal fruiting bodies that had been concurrently collected from the same habitat showed no appreciable variation. They also showed that the three randomly selected specimens' ITS sequences shared similarities. The inconsistency between these types emphasizes the necessity for the creation and design of species-specific PCR primers, which could play a significant role in clearly identifying the host plants. A mycorrhizal relationship was created by *Helianthemum almeriense* with *Picoa lefebvrei* and *Terfezia claveryi*. In a thorough review of the mycorrhizal associations of desert truffles with the genus *Helianthemum*, Gutierrez et al. (2004) examined the morphological traits of this association. Additionally, they discussed the structure and ultrastructure of the established mycorrhizae of this plant species with *Picoa lefebvrei* and *Terfezia claveryi* and the morphology of the mycorrhizal associations, in which *Helianthemum almeriense* is one of the partners.

To further investigate the fungal inocula and mycorrhizal connections with plants and subsequent plantations, knowledge of the use of biotechnological tools and management methods has been developed and introduced for a select few *Terfezia* species. According to Morte and Honrubia (1997), agar or modified Melin-Norkrans (MMN) inocula can be used straight from the plates as inocula for the creation and synthesis of *in vitro* mycorrhization". Since they are mycorrhizal, true *Terfezia* species have been used in mycorrhizal bioassay tests with plants, mainly from the genus *Helianthemum* (Kovács et al., 2002). The truffle's anatomy and ultrastructure were thoroughly examined and subsequently *Terfezia* terfezioides' relationships with *Helianthemum ovatum* and *Robinia pseudoacacia* in vitro were investigated as well. Morte et al. (2008) carried out a comparitive study based on the techniques employed in the efficient synthesis of mycorrhiza between desert truffles and *Helianthemum* species. Agar or modified Melin-Norkrans (MMN) media can be used to make inocula, according to Morte et al. (1994) and Morte and Honrubia (1997). According to the source of the fungal inocula (whether it be from spores or mycelium), the source of the plant (whether it be seedlings or micropropagated plantlets), and the culture conditions, Morte et al. (2008) conducted a similar investigation into the processes involved in the efficient synthesis of mycorrhiza from *Helianthemum* species and desert truffles. Khanaqa (2006) suggested cultivating desert truffles in plantations with Olea europaea, which would have a dual beneficial effect. The tree will have increased water retention, lessen soil erosion, and enhance soil fertility.

According to Lopez-Nicolás et al. (2013), the compound cyclodextrin was effective at encouraging the mycelial growth of *T. claveryi*. It was discovered that the dominant sunflower species in Libya, *Helianthemum kahiricum* and *Helianthemum lippii*, developed mycorrhiza with arid truffles (Bouzadi et al., 2017). Mycorrhizal connections can be formed by the majority of desert truffles, but *Picoa lefebvrei* and *Terfezia claveryi* are the two most frequently cited candidates for this particular quality. To improve farming, ecological, and biotechnological practices in dry and desert landscapes, implementing mycorrhizal technology could prove to be a crucial component in Saudi Arabia's efforts to improve a variety of arid plants. This area needs careful and thorough exploration.

6. Biochemical and Physiological Considerations

Desert truffles have distinctive economic and medicinal significance and are extensively used throughout the globe. Polysaccharides have been found to be present in medicinal mushrooms, and these polysaccharides have been shown to have antibacterial and anticancer properties (Schillaci et al., 2017). Rare reports on the chemistry and makeup of truffles, which frequently represent their economic importance, are extremely rare. Truffles are a fantastic and plentiful supply of carbs, fiber, high-quality protein, and amino acids (Murcia et al., 2003; Mandeel and Al-Laith, 2007). According to estimates, proteins make up 20% of the dry bulk of desert truffles, which is more than the majority of plants and fungi. Desert truffles are also frequently suggested for their therapeutic benefits against a variety of diseases (Murcia et al., 2003; El Enshasy et al., 2013). Additionally, truffles contain a lot of chemical compounds that have important health benefits, including fighting inflammation, cancer, mutations, and bacteria (Murcia et al., 2002; Al-Laith, 2010; Dib-Bellahouel and Fortas, 2011). Truffles are also said to have anti-inflammatory and antioxidant qualities (Murcia et al., 2002; Kagan-Zur et al., 2013).

Truffles contain bioactive organic components like carotenoids, ergosterol, flavonoids, phenolic acids, and phenolic derivatives, which account for their antioxidant activity (Al-Laith, 2010; Villares et al., 2012). In a separate study, conducted by Bouzadi et al. (2017) in Libya two types of wild desert truffles *Terfezia* and *Tirmania* were identified and characterized.Their nutritional values based on dry weight were: protein (16.3% and 18.5%), fat (6.2 and 5.9%), and carbohydrate (67.2 and 65%). Other studies indicate that *Terfezia* and *Tirmania* ascocarps (fruiting bodies), which are found in desert truffles, are well-known delicacies that are heavily promoted primarily for their flavor and nutritive value (Kagan-Zur et al., 2014). Thus, *Terfezia boudieri* contains saturated and unsaturated fatty acids, such as behenic, stearic, margaric, pentadecanoic, arachidic, palmitoleic, palmitic, stearic, oleic, linoleic, and linolenic, according to Bokhary et al. (1989), and Bouzadi et al. (2017) also reported an abundance of potassium, iron, proteins, and phenolics in (black truffle) *Terfezia claveryi* and *Tirmania nivea* (white truffle). Numerous researchers have established the therapeutic qualities of desert truffles, including their antibacterial and antimicrobial properties (e.g., Chellal and Lukasova, 1995; Janakat et al., 2004, 2005). About two hundred different secondary metabolites have been found in truffles, including terpenoids, flavonoids, fatty acids, essential oils, phenolics, and aromatic chemicals (Splivallo, 2008). The dry biomass of Iraqi truffles (*Terfezia claveryi*, *Tirmania nivea* and *T. pinoyi*) contains 9.7 to 25.5% phosphorus, 8.1% to 13.8% protein, and 16.6% to 24.8% carbohydrates (Hussain and Al-Ruqaie, 1999). Also included in truffles are the important minerals K, Na, Ca, Al, Zn, Mn, and Mg (Wang and Marcone, 2011). Furthermore, up to 80% of truffle biomass is comprised of insoluble polysaccharides, hemicelluloses, chitin, and pectin (Cheung, 1997).

Desert truffles also contain bioactive substances such as phenols, anthocyanins, and carotenoids (Gouzi et al., 2013). In addition to noticing that the desert truffle fruiting bodies contain a significant quantity of sugar (15.4%), Slama et al. (2009) noted that only 2.02% of the various individual sugars determined are soluble and the remaining 10.5% are insoluble. Truffles from the desert are rich in calcium, potassium, phosphorus, magnesium, and iron. For instance, *Terfezia boudieri* provides adequate levels of potassium, calcium, phosphorus, and magnesium. Truffles found in the desert typically have a higher calcium content than black and white truffles (*Tuber melanosporum* and *Tuber magnatum*). To have a clear understanding of the desert truffles present in Morocco, a detailed study will be necessary. All of these facts, however, make desert truffles an intriguing topic for future researchers of this valuable fungus. On modified Melin-Norkrans agar medium and potato dextrose media, desert truffle mycelium grow best at neutral pH, which suggests that it could be used to make desert truffles more productive and useful for business purposes (Morte et al., 1994; Morte and Honrubia, 1995; Morte and Honrubia, 1997).

There are several stringent laboratory and physio-biochemical conditions for good desert growth and productivity. Pérez-Gilabert et al. (2001) described the characterization and biochemical properties of the inactive *Terfezia claveryi*. Due to the fact that desert truffles can grow under circumstances of water stress, several studies have examined the effects of polyethylene glycol-induced osmotic stress on two desert truffles (Mexal and Reid, 1973; Coleman et al., 1989; Navarro-Ródenas et al., 2011). According to Navarro-Ródenas et al. (2011), who discovered that it grows under dry conditions, alkaline phosphatase (ALP) activity is a major indicator of the metabolic activity of these fungi. They found that *Terfezia claveryi* strain TcS2 and *Picoa lefebvrei* strain OL2 are two species of desert truffles that have demonstrated the ability to withstand water stress at pressures lower than 1.07 MPa. Even the formation of mycelial inoculum in these arid truffles was found to be enhanced by water stress. According to Akyüz et al. (2012), *Terfezia boudieri* arid truffles have medicinal uses and contain phenolic compounds, antioxidants, and other minerals.

7. Market and Interest in Desert Truffles

The presence of truffles in desert environments has been documented since antiquity, with evidence of their exploitation by Mediterranean cultures dating back to the earliest periods of human habitation in the area. It is now evident that these truffles were traded by the Greeks and Romans, with the latter importing them from Libya for sale in several markets within their own empire (Honrubia et al., 2007). The products in question have continued to be distributed and consumed, and they remain closely associated with three distinct Mediterranean cultures (Christianity, Judaism and Islam) (Morte et al., 2009). A considerable quantity of various species of *Terfezia* is collected and sold in southern Europe, in North Africa, as well as in other countries bordering the Mediterranean Sea. It is notable, however, that over the past 50 years, natural areas of desert truffle have been on the decline.

During the Second World War, extensive areas of the coastal desert in Egypt and Libya were subjected to mining operations. It appears that the Gulf War has resulted in the deterioration of numerous truffle-inhabiting areas in Kuwait (Morte et al., 2008). The construction of numerous large-scale projects in Spain, Dubai, and Abu Dhabi over the past 15 years has contributed to the observed phenomenon. Furthermore, the phenomenon of climate change and global warming has resulted in a reduction in total rainfall, which has consequently led to a decline in the quantity of naturally produced desert truffles. The harvesting of desert truffles is a manual process, requiring the ability to identify the fissure created by the fungus in the soil in proximity to the host plant species. It is not necessary for a dog or another olfactory animal to be present when searching for desert truffles, provided that the individual undertaking the search is able to recognize the characteristics of the host plant and the specific type of soil crack that is typical of the area in question.

The interest in desert truffles is not solely culinary, gastronomic, or commercial; it is also nutritional, with truffles offering a high nutritional value (Murcia et al., 2003). Furthermore, some studies have indicated that it possesses antibiotic capacity (Janakat et al., 2005). Additionally, the compound demonstrated hepatoprotective properties against carbon tetrachloride (CCl4) (Janakat and Nassar, 2010). One of the most important characteristics of desert truffles, from a nutritional standpoint, is their antioxidant capacity, manifested in their ability to impede lipid oxidation processes (Murcia et al., 2002). It has been observed that the antioxidant properties of the substance in question are greater than those of several commonly utilized food antioxidants, even following industrial processes such as freezing and canning (Murcia et al., 2002). Besides, they are a priori "ecological" insofar as their production takes place within an ecosystem. Additionally, they have the potential to promote rural development, particularly in a sustainable farming context. This is because they may be established as an alternative or complementary crop (Morte et al., 2009).

The price of the desert truffles ranges from 20 to 250 Euros. However, it should be noted that the prices of these mushrooms are considerably lower compared to those of the *Tuber* species. One reason for this is that desert truffles exhibit a relatively weak flavor profile when compared to *Tuber* species. Nevertheless, marketing initiatives may facilitate the promotion of desert truffles and augment their commercial value (Morte et al., 2008). A considerable proportion of the *Terfezia* and *Tirmania* species gathered from the Egyptian and Moroccan regions are subsequently traded in markets situated in Abu Dhabi, Doha, Kuwait, and Riyadh. This could provide a potential explanation for the growing interest observed among some of these countries in the cultivation of desert truffles. The practice of cultivation has the potential to mitigate the overexploitation of natural desert truffle areas, thereby preventing the extinction of certain desert truffle species.

8. Conclusion

Desert truffles are both highly valuable commercially and highly beneficial medically. However, further study is still needed to improve our understanding of the taxonomy of desert truffles and to conduct in-depth analyses of their biochemical and medicinal properties. Both molecular and morphological methods should be used for this study. A potent technique, in vitro mycorrhization, also has positive effects on Morocco's conservation of truffle biodiversity. It might boost these indicators as well as the development, production, and overall caliber of desert truffles. The accurate taxonomic delimitation of truffles using conventional and molecular technologies, as well as their bioprospecting through biotechnological interventions, have become very important because truffles are very valuable on the market. Given the importance of truffles, the main objective of this scientific review is to optimize the cultivation conditions of some southeastern truffles in Morocco based on mycorrhization. To achieve this objective a brief bibliographic analysis of the geographical distribution of desert truffles throughout the world, particularly in Morocco is presented as a beginning. This includes their biology, ecology, host plants, chemical composition, mycorrhizal associations, characteristics of soil edaphic factors, market and interest in desert truffles.

References

Abourouh, M. (2011). Truffes du désert du Maroc: Diversité et modes d'exploitation. 6ème Rencontre de Mycosilva, Mértola (ADPM) Portugal. https://tinyurl.com/3afza3zz

Abourouh, M. (2020). Restauration de la forêt de chêne liège pour le développement et la valorisation des truffes dans la forêt de la Maâmora, FAO, Terfess et truffes de la Maâmora, importances biologique, écologiques et socio-économique. https://tinyurl.com/ymuz79t7

Akyüz, M., Kirbağ, S. and Kurşat, M. (2012). Ecological aspects of the arid and semi-arid truffle in Turkey: Evaluation of soil characteristics, morphology, distribution, and mycorrhizal relationships. Turk. J. Bot., 36: 386–391. https://doi.org/10.3906/bot-1103-4

Al-Laith, A.A.A. (2010). Antioxidant components and antioxidant, antiradical activities of desert truffle (*Tirmania nivea*) from various Middle Eastern origins. J. Food Comp. Anal., 23: 15–22. https://doi.org/10.1016/j.jfca.2009.07.005

Alsheikh, A. and Trappe, J. (1983). Desert truffles, the genus *Tirmania*. Trans. Br. Mycol. Soc., 81: 83–90. https://doi.org/10.1016/S0007-1536(83)80207-1

Al-Sheikh, A. and Trappe, J. (1990). Features and analysis of spore germination in the brown kame *Terfezia claveryi*. J. Mycol., 72: 495–499. https://doi.org/10.1080/00275514.1980.12021210

Alsheikh, A.M. (1994). Taxonomy and Mycorrhizal Ecology of the Desert truffles in the Genus *Terfezia*. Ph.D. Dissertation, Oregon State University, Corvallis, USA, p. 239. https://tinyurl.com/3stmvbh5

Antony-Babu, S., Deveau, A., Van Nostrand, J.D., Zhou, J., Le Tacon, F. et al. (2014). Black truffle associated bacterial communities during the development and maturation of *Tuber melanosporum* ascocarps and putative functional roles. J. Environ. Microbiol., 16: 2831–2847. https://doi.org/10.1111/1462-2920.12294

Awameh, M.S. and Alsheikh, A. (1980). Ascospore germination of black kame, *Terfezia boudieri*. J. Mycol., 72: 50–54. https://doi.org/10.1080/00275514.1980.12021154

Barbieri, E., Ceccaroli. P., Saltarelli, R., Guidi, C., Potenza, L. et al. (2010). New evidence for nitrogen fixation within the Italian white truffle *Tuber magnatum*. J. Fungal Biol., 114: 936–942. https://doi.org/10.1016/j.funbio.2010.09.001

Barea, J., Palenzuela, J., Cornejo, P., Sánchez-Castro, I., Navarro-Fernández, C. et al. (2011). Ecological and functional roles of mycorrhizas in semi-arid ecosystems of Southeast Spain. J. Arid Environ., 75: 1292–1301. https://doi.org/10.1016/j.jaridenv.2011.06.001

Barry, D., Staunton, S. and Callot, G. (1994). Mode of the absorption of water and nutrients by ascocarps of *Tuber melanosporum* and *Tuber aestivum* a radioactive tracer technique. Can, J. Bot., 72: 317–322. https://doi.org/10.1139/b94-04

Berch, S.M. and Bonito, G. (2016). Truffle diversity (*Tuber*, Tuberaceae) in British Columbia. Mycorrhiza, 26: 587–594. https://doi.org/10.1007/s00572-016-0695-2

Bermaki, F.Z., Khabar, L. and Ezzanega, A. (2017). Bioecology of desert truffles in the province of Figuig in Eastern Morocco. Journal Rev. Mex. Micol., 46: 29–36. https://doi.org/10.33885/sf.2017.46.1174

Bertault, G., Rousset, F., Fernandez, D., Berthomieu, A., Hochberg, M.E. et al. (2001). Population genetics and dynamics of the black truffle in a man-made truffle field. J. Hered., 86: 451–458. https://doi.org/10.1046/j.1365-2540.2001.00855.x

Bokhary, H. and Parvez, S. (1993). Chemical composition of desert truffles *Terfezia claveryi*. J. Food Comp. Anal., 6: 285–293. https://doi.org/10.1006/jfca.1993.1031

Bokhary, H. and Parvez, S. (1992). Soil mycoflora from truffle native areas of Saudi Arabia. Mycopathol., 118: 103–107. https://doi.org/10.1007/BF00442538

Bokhary, H., Suleiman, A. and Basalah, M. (1989). The fatty acid components of the desert truffle Al Kamah of Saudi Arabia. J. Food Prot., 52: 668–69. https://doi.org/10.4315/0362-028x-52.9.668

Bokhary, H., Suleiman, A., Basalah, M. and Parvez, S. (1987). Chemical composition of desert truffles from Saudi Arabia. Can. Inst. Food Sci. Technol. J., 20: 336–341. https://doi.org/10.1016/S0315-5463(87)71328-5

Bonifacio, E. and Morte, A. (2014). Soil properties. pp. 57–67. In: Kagan-Zur, V., Roth-Bejerano, N., Sitrit, Y. and Morte, A. (eds.). Desert truffles: Phylogeny, Physiology, Distribution, Domestication. Springer Berlin, Heidelberg. https://doi.org/10.1007/978-3-642-40096-4_4

Bonito, G.M., Gryganskyi, A.P., Trappe, J.M. and Vilgalys, R. (2010). A global meta-analysis of Tuber ITS rDNA sequences species diversity, host associations and long-distance dispersal. J. Mol. Ecol., 19: 4994–5008. https://doi.org/10.1111/j.1365-294X.2010.04855.x

Bouzadi, M., Grebenc, T., Turunen, O., Kraigher, H., Taib, H. et al. (2017). Characterization of natural habitats and diversity of Libyan desert truffles. J. Biotech., 7: 1–10. https://doi.org/10.1007/s13205-017-0949-5

Bradai, L., Bissati, S. and Chenchouni, H. (2013). Etude mycologique et bioécologiques de la truffe blanche du désert (*Tirmania nivea* Desf. Trappe 1971) dans la région de Oued M'ya (Ouargla, Sahara algérien). J. Rev. Bioresour., 3: 6–14. https://doi.org/10.12816/0008856

Bradai, L., Bissati, S., Chenchouni, H. and Amrani, K. (2015). Effects of climate on the productivity of desert truffles beneath hyper-arid conditions. Int. J. Biometeorol., 59: 907–915. https://doi.org/10.1007/s00484-014-0891-8

Brundrett, M.C. (2002). Coevolution of roots and mycorrhizas of land plants. New phytol., 154: 275–304. https://doi.org/10.1046/j.1469-8137.2002.00397.x

Chatin, M. (1891). Contribution à l'histoire botanique de la truffe: kamé de Damas (*Terfezia claveryi*). Journal Bulletin de la Société Botanique de France, 38: 332–35. https://doi.org/10.1080/00378941.1891.10828584

Chellal, A. and Lukasova, E. (1995). Evidence for antibiotics in two Algerien truffles *Terfezia* and *Tirmania*. Pharmazie, 50: 228–229. https://tinyurl.com/2wdf6e5n

Chen, J., Li, J.M., Tang, Y.J., Xing, Y.M., Qiao, P. et al. (2019). Chinese black truffle associated bacterial communities of Tuber indicum from different geographical regions with nitrogen fixing bioactivity. Front. Microbiol., 10: 2515. https://doi.org/10.3389/fmicb.2019.02515

Chen, P.-Y., Montanini, B., Liao, W.-W., Morselli, M., Jaroszewicz, A. et al. (2014). A comprehensive resource of genomic, epigenomic and transcriptomic sequencing data for the black truffle *Tuber melanosporum*. Gigascience, 3: 25. https://doi.org/10.1186/2047-217X-3-25

Cheung, P.C.K. (1997). Dietary fibre content and composition of some edible fungi determined by two methods of analysis. J. Sci. Food Agric., 73: 255–260. https://tinyurl.com/yckmfrrv

Chevalier, G. (1984). Une nouvelle méthode de production de plants mycorhizés par la truffe, l'inoculation en motte roulée Melfert. J. Agron., 4(2): p. 211. https://tinyurl.com/2bsxtjyr

Claridge, A.W., Trappe, J.M. and Paull, D.J. (2013). Ecology and distribution of desert truffles in the Australian outback. pp. 203–214. In: Kagan-Zur, V., Roth-Bejerano, N., Sitrit, Y. and Morte, A. (eds.), Desert truffles: Phylogeny, Physiology, Distribution, Domestication. Springer Berlin, Heidelberg. https://doi.org/10.1007/978-3-642-40096-4_14

Coleman, M.D., Bledsoe, C.S. and Lopushinsky, W. (1989). Pure culture response of ectomycorrhizal fungi to imposed water stress. Can. J. Bot., 67: 29–39. https://doi.org/10.1139/b89-005

Comandini, O., Contu, M. and Rinaldi, A. (2006). An overview of Cistus ectomycorrhizal fungi. Mycorrhiza, 16: 381–395. https://doi.org/10.1007/s00572-006-0047-8

Culleré, L., Ferreira, V., Chevret, B., Venturini, M.E., Sánchez-Gimeno, A.C. et al. (2010). Characterization of aroma active compounds in black truffles, *Tuber melanosporum* and summer truffles, *Tuber aestivum* by gas chromatography olfactometry. Food Chem., 122: 300–306. https://doi.org/10.1016/j.foodchem.2010.02.024

Dahham, S.S., Al-Rawi, S.S., Ibrahim, A.H., Majid, A.S.A. and Majid, A.M.S.A. et al. (2016). Antioxidant, anticancer, apoptosis properties and chemical composition of black truffle *Terfezia claveryi*. Saudi J. Biol. Sci., 25: 1524–1534. https://doi.org/10.1016/j.sjbs.2016.01.031

Debuchy, R., Berteaux-Lecellier, V. and Silar, P. (2010). Mating systems and sexual morphogenesis in ascomycetes. Cell. Mol. Biol. Filamentous Fungi, 499–535. https://doi.org/10.1128/9781555816636.ch33

Dexheimer, J., Gerard, J., Leduc, J.-P. and Chevalier, G. (1985). Étude ultrastructurale comparée des associations symbiotiques mycorhiziennes *Helianthemum salicifolium, Terfezia claveryi* et *Helianthemum salicifolium, Terfezia leptoderma*. Can. J. Bot., 63: 582–591. https://doi.org/10.1139/b85-073

Dıaz, P., Ibáñez, E., Senorans, F. and Reglero, G. (2003). Truffle aroma characterization by headspace solid phase microextraction. J. Chromat. A, 1017: 207–214. https://doi.org/10.1016/j.chroma.2003.08.016

Dib-Bellahouel, S. and Fortas, Z. (2011). Antibacterial activity of various fractions of ethyl acetate extract from the desert truffle, *Tirmania pinoyi*, preliminarily analyzed by gas chromatography mass spectrometry (GC-MS). Afr. J. Biotechnol., 10(47): 9694–9699.

Díez, J., Manjón, J.L. and Martin, F. (2002). Molecular phylogeny of the mycorrhizal desert truffles (*Terfezia* and *Tirmania*), host specificity and edaphic tolerance. Mycologia, 94: 247–259. https://doi.org/10.1080/15572536.2003.11833230

El Aji, A. (1999). Contribution à l'étude des Terfez des parcours de l'oriental marocain: Germination des ascospores in vitro et mycorhization de l'*Helianthemum ledifolium*. Mémoire de Troisième Cycle en Agronomie, École Nationale d'Agriculture, Meknès. https://tinyurl.com/43stvzm3

El Akil, M., Outcoumit, A., Touhami, A.O., Benkirane, R. and Douira, A. et al. (2016). Study of eastern Morocco desert truffles. Int. J. Curr. Res., 8: 33922–33929.

El Enshasy, H., Elsayed, E.A., Aziz, R. and Wadaan, M.A. (2013). Mushrooms and truffles: Historical bio-factories for complementary medicine in Africa and in the Middle East. Evid. Based Compl. Alt. Med., 2013: 620451. http://dx.doi.org/10.1155/2013/620451

El-Kholy, H. (1989). Genetic and Physiological Studies on Truffles. PhD. Thesis, Faculty of Agriculture, Cairo University, Egypt.

Fortas, Z. and Chevalier, G. (1992). Effet des conditions de culture sur la mycorhization de l'*Helianthemum guttatum* par trois espèces de terfez des genres *Terfezia* et *Tirmania* d'Algérie. Journal Can. J. Bot., 70: 2453–2460. https://doi.org/10.1139/b92-303

Frank, B. (2005). On the nutritional dependence of certain trees on root symbiosis with belowground fungi. J. Mycorrhiza, 15: 267–275. https://doi.org/10.1007/s00572-004-0329-y

Gouzi, H., Leboukh, M. and Bouchouka, E. (2013). Antioxidant and antiradical properties of methanolic extracts from Algerian wild edible desert truffles (*Terfezia* and *Tirmania*, Ascomycetes). Int. J. Med. Mushr., 15(5): 471–486. https://doi.org/10.1615/IntJMedMushr.v15.i5.50

Gutiérrez, A., Morte, A. and Honrubia, M. (2003). Morphological characterization of the mycorrhiza formed by *Helianthemum almeriense* Pau with *Terfezia claveryi* Chatin and *Picoa lefebvrei* (Pat.) Maire. Mycorrhiza, 13: 299–307. https://doi.org/10.1007/s00572-003-0236-7

Hakkou, S., Sabir, M. and Machouri, N. (2022a). Le Maroc truffier: la répartition géographique des truffes et truffes du désert et leur productivité. Moroccan J. Agron. Veter. Sci., 10: 395-407.

Hall, I.R. and Zambonelli, A. (2012). Laying the foundations in edible ectomycorrhizal mushrooms. pp. 3–16. In: Zambonelli, A. and Bonito, G.M. (eds.). Edible Ectomycorrhizal Mushrooms. Springer Berlin Heidelberg. https://doi.org/10.1007/978-3-642-33823-6_1

Hamza, A., Zouari, N., Zouari, S., Jdir, H., Zaidi, S. et al. (2016). Nutraceutical potential, antioxidant and antibacterial activities of *Terfezia boudieri* Chatin, a wild edible desert truffle from Tunisia arid zone. J. Arab. J. Chem., 9: 383–389. https://doi.org/10.1016/j.arabjc.2013.06.015

Harir, M., Bendif, H., Yahiaoui, M., Bellahcene, M., Zohra, F et al. (2019). Evaluation of antimicrobial activity of *Terfezia arenaria* extracts collected from Saharan desert against bacteria and filamentous fungi. 3 Biotech, 9: 281. https://doi.org/10.1007/s13205-019-1816-3

Harki, E., Farah, A. and Bouseta, A. (2010). Volatile compounds from four species of Moroccan truffles. Moroccan J. Biol., 12-2110/N 6-7: 10–15.

Harrison, M.J. (1997). The arbuscular mycorrhizal symbiosis in plant microbe interactions. pp. 1–34. In: Stacey, G., Keen, N.T. (eds.). Plant-Microbe Interactions. Springer, Boston, MA, USA. https://doi.org/10.1007/978-1-4615-6019-7_1

Hashem, A., Alqarawi, A.A., Shah, M.A., Wirth, S., Egamberdieva, D. et al. (2018). Desert truffles in Saudi Arabia: Diversity, ecology, and conservation. pp. 353–369. In: Sharma, S. and Varma, A. (eds.). Microbial Resource Conservation. Springer Cham. https://doi.org/10.1007/978-3-319-96971-8_13

Hashem, A.R. and Al-Obaid, A.M. (1996). Mineral composition of soil and wild desert truffles in Saudi Arabia. J. King Saud Univ., 8: 5–10.

Honrubia, M., Cano, A. and Molina-Ninirola, C. (1992). Hypogeous fungi from southern Spanish semi-arid lands. J. Pers. Mol. Phyl. Evol. Fungi, 14: 647–653.

Honrubia, M., Morte, A. and Gutiérrez, A. (2007). Las Terfezias. Un cultivo para el desarrollo rural en regiones áridas y semi-áridas. Journal Truficultura, Fundamentos y Técnicas, Capítulo, 10: 365–397.

Hussain, G. and Al-Ruqaie, I.M. (1999). Occurrence, chemical composition, and nutritional value of truffles: an overview. Pak. J. Biol. Sci., 2: 510–514. https://doi.org/10.3923/pjbs.1999.510.514

Jamali, S. and Banihashemi, Z. (2012). Fungi associated with ascocarps of desert truffles from different parts of Iran. J. Crop Prot., 1(1). 20.1001.1.22519041.2012.1.1.3.4

Janakat, S. and Nassar, M. (2010). Hepatoprotective activity of desert truffle (*Terfezia claveryi*) in comparison with the effect of Nigella sativa in the rat. J. Pak. J. Nutr., 9: 52–56.

Janakat, S., Al-Fakhiri, S. and Sallal, A.K. (2004). A promising peptide antibiotic from *Terfezia claveryi* aqueous extract against Staphylococcus aureus in vitro. Phytotherapy Research: An Int. J. Devoted Pharmacol. Toxicol. Eval. Nat. Prod. Deriv., 18: 810–813. https://doi.org/10.1002/ptr.1563

Janakat, S.M., Al-Fakhiri, S.M. and Sallal, A. (2005). Evaluation of antibacterial activity of aqueous and methanolic extracts of the truffle *Terfezia claveryi* against *Pseudomonas aeruginosa*. Saudi Med. J., 26: 952–955.

Jeandroz, S., Murat, C., Wang, Y., Bonfante, P. and Tacon, F.L. et al. (2008). Molecular phylogeny and historical biogeography of the genus *Tuber*, the true truffles. J. Biogeogr., 35: 815–829. https://doi.org/10.1111/j.1365-2699.2007.01851.x

Kagan-Zur, V., Roth-Bejerano, N., Sitrit, Y. and Morte, A. (2013). Desert truffles: Phylogeny, physiology, distribution and domestication. Springer Science and Business Media.

Kagan-Zur, V., Roth-Bejerano, N., Sitrit, Y. and Morte, A. (2014). Desert truffles: Phylogeny, physiology, distribution and domestication. Springer Berlin, Heidelberg, p. 397. https://doi.org/10.1007/978-3-642-40096-4

Khabar, L. (2014). Mediterranean Basin: North Africa. pp. 143–158. In: Kagan-Zur, V., Roth-Bejerano, N., Sitrit, Y. and Morte, A. (eds.). Desert truffles: Phylogeny, Physiology, Distribution, Domestication. Springer Berlin, Heidelberg. https://doi.org/10.1007/978-3-642-40096-4_10

Khabar, L. (2016). Les terfess et les truffes du Maroc: Biodiversité et valorisation. Éditions Universitaires Européennes (French Edition), ISBN-13 978-3639547900

Khaled, J.M., Alharbi, N.S., Mothana, R.A., Kadaikunnan, S. and Alobaidi, A.S et al. (2021). Biochemical profile by GC-MS of fungal biomass produced from the Ascospores of *Tirmania nivea* as a natural renewable resource. J. Fungi, 7: 1083. https://doi.org/10.3390/jof7121083

Khanaqa, A. (2006). Truffle production in the Kingdom of Saudi Arabia-potential and limitation. J. Appl. Bot. Food Qual., 80(1): 14–18.

Kinoshita, A., Sasaki, H. and Nara, K. (2011). Phylogeny and diversity of Japanese truffles (*Tuber* spp.) inferred from sequences of four nuclear loci. Mycologia, 103: 779–794. https://doi.org/10.3852/10-138

Kovács, G., Bagi, I., Vágvölgyi, C., Kottke, I. and Oberwinkler, F. et al. (2002). Studies on the root associations of the truffle *Terfezia terfezioides*. J. Acta Microbiol. Immunol. Hung., 49: 207–213. https://doi.org/10.1556/amicr.49.2002.2-3.6

Kovács, G.M. and Trappe, J.M. (2013). Nomenclatural history and genealogies of desert truffles. pp. 21–37. In: Kagan-Zur, V., Roth-Bejerano, N., Sitrit, Y. and Morte, A. (eds.), Desert truffles: Phylogeny, Physiology, Distribution, Domestication. Springer Berlin, Heidelberg. https://doi.org/10.1007/978-3-642-40096-4_2

Kovács, G.M., Balazs, T.K., Calonge, F.D. and Martin, M.P. (2011). The diversity of Terfezia desert truffles: new species and a highly variable species complex with intrasporocarpic nrDNA ITS heterogeneity. Mycologia, 103: 841–853. https://doi.org/10.3852/10-312

Kües, U. and Liu, Y. (2000). Fruiting body production in basidiomycetes. Appl. Microbiol. Biotechnol., 54: 141–152. https://doi.org/10.1007/s002530000396

Kulifaj, M. (1984). *Tuber melanosporum* Vitt.: Contribution à l'étude de la morphogenèse et de la physiologie de l'ascocarpe. Ph.D. Thesis, Inist-CNRS, Canada. https://tinyurl.com/2p86aut9

Laessoe, T. and Hansen, K. (2007). Truffle trouble: what happened to the Tuberales? Mycol. Res. J., 111: 1075–1099. https://doi.org/10.1016/j.mycres.2007.08.004

Lancellotti, E., Iotti, M., Zambonelli, A. and Franceschini, A. (2014). Characterization of *Tuber borchii* and *Arbutus unedo* mycorrhizas. Mycorrhiza, 24: 481–486. https://doi.org/10.1007/s00572-014-0564-9

Laqbaqbi, A. (2020). La truffe entre France et Maroc ou d'un coté à l'autre de la Méditerranée. Journal Imprimerie GRAPHO, 12. https://tinyurl.com/3r63trk6

Le Tacon, F. and Selosse, M.-A. (1994). La place des symbioses mycorhiziennes dans l'évolution et la colonisation des continents par la vie. Acta Botanica Gallica, 141: 405–419. https://doi.org/10.1080/12538078.1994.10515177

Le Tacon, F., Zeller, B., Plain, C., Hossann, C. and Bréchet, C. et al. (2013). Carbon transfer from the host to *Tuber melanosporum* mycorrhizas and ascocarps. Plos One, 8(5): e64626. https://doi.org/10.1371/journal.pone.0064626

Loizides, M., Hobart, C., Konstandinides, G. and Yiangou, Y. (2012). Desert truffles: The mysterious jewels of antiquity. Field Mycol., 13: 17–21. https://doi.org/10.1016/j.fldmyc.2011.12.004

López-Nicolás, J.M., Pérez-Gilabert, M., García-Carmona, F., Lozano-Carrillo, M.C. and Morte, A et al. (2013). Mycelium growth stimulation of the desert truffle *Terfezia claveryi* chatin by β-cyclodextrin. J. Biotechnol. Prog., 29: 1558–1564. https://doi.org/10.1002/btpr.1791

Maire, M. (1906). Contributions à l'étude de la flore mycologique de l'Afrique du Nord. Journal Bulletin de la Société Botanique de France 53: CLXXX–CCXV. https://doi.org/10.1080/00378941.1906.10831941

Malençon, G. (1973). Champignons hypogés du nord de l'Afrique I. Ascomycetes. J. Pers. Mol. Phyl. Evol. Fungi, 7: 261–279.

Mamoun, M. and Olivier, J. (1991). The effect of carbon source and form of mineral nitrogen on the development of *Tuber melanosporum* (Vitt) in pure culture. Application to production of mycelium biomass. Agronomy, 11: 521–527. https://doi.org/10.1051/agro:19910609

Mandeel, Q.A. and Al-Laith, A.A.A. (2007). Ethnomycological aspects of the desert truffle among native Bahraini and non-Bahraini peoples of the Kingdom of Bahrain. J. Ethnopharmacol., 110: 118–129. https://doi.org/10.1016/j.jep.2006.09.014

Marqués-Gálvez, J.E., Miyauchi, S., Paolocci, F., Navarro-Ródenas, A., Arenas, F. et al. (2021). Desert truffle genomes reveal their reproductive modes and new insights into plant, fungal interaction and ectendomycorrhizal lifestyle. New Phytol., 229: 2917–2932. https://doi.org/10.1111/nph.17044

Marqués-Gálvez, J.E., Morte, A. and Navarro-Ródenas, A. (2020). Spring stomatal response to vapor pressure deficit as a marker for desert truffle fruiting. Mycorrhiza, 30: 503–512. https://doi.org/10.1007/s00572-020-00966-8

Martin, F., Kohler, A., Murat, C., Balestrini, R., Coutinho, P.M. et al. (2010). Périgord black truffle genome uncovers evolutionary origins and mechanisms of symbiosis. J. Nature, 464: 1033–1038. https://doi.org/10.1038/nature08867

Mello, A., Murat, C. and Bonfante, P. (2006). Truffles: much more than a prized and local fungal delicacy. Fed. Eur. Microbiol. Soc. Microbiol. Lett., 260: 1-8. https://doi.org/10.1111/j.1574-6968.2006.00252.x

Mexal, J. and Reid, C. (1973). The growth of selected mycorrhizal fungi in response to induced water stress. Can. J. Bot., 51: 1579–1588. https://doi.org/10.1139/b73-20

Morte, A. and Honrubia, M. (1995). Improvement of mycorrhizal synthesis between micro propagated *Helianthemum almeriense* plantlets with *Terfezia claveryi* (desert truffle). pp. 863–868. In: Mushroom science XIV, Volume 2, Proceedings of the 14th international congress on the science and cultivation of edible fungi, 17-22 September, Oxford, UK.

Morte, A. and Honrubia, M. and Gutiérrez, A. (2008). Biotechnology and cultivation of desert truffles. pp. 467–483. In: Verma, A. (ed.). Mycorrhiza: state of the art, genetics molecular biology, eco-function, biotechnology, eco-physiology, structure and systematics. Springer Berlin, Heidelberg. https://doi.org/10.1007/978-3-540-78826-3

Morte, A., Lovisolo, C. and Schubert, A. (2000). Effect of drought stress on growth and water relations of the mycorrhizal association *Helianthemum almeriense* - *Terfezia claveryi*. Mycorrhiza, 10: 115–119.

Morte, A., Navarro-Ródenas, A. and Nicolás, E. (2010). Physiological parameters of desert truffle mycorrhizal *Helianthemun almeriense* plants cultivated in orchards under water deficit conditions. J. Symb., 52: 133–139. https://doi.org/10.1007/s13199-010-0080-4

Morte, A., Zamora, M., Gutiérrez, A. and Honrubia, M. (2009). Desert truffle cultivation in semiarid Mediterranean areas. Mycor. Func. Pro. Ecol. Imp., 221–233. https://doi.org/10.1007/978-3-540-87978-7_15

Morte, M. and Honrubia, M. (1997). Micropropagation of *Helianthemum almeriense*. pp. 163–177. In: Bajaj, Y.P.S. (eds.). High-Tech and Micropropagation VI. Biotechnol. Agric. For., 40. Springer, Berlin, Heidelberg. https://doi.org/10.1007/978-3-662-03354-8_12

Morte, M.A., Cano, A., Honrubia, M. and Torres, P. (1994). *In vitro* mycorrhization of micro propagated *Helianthemum almeriense* plantlets with *Terfezia claveryi* (desert truffle. Agric. Food Sci., 3: 309–314. https://doi.org/10.23986/afsci.72700

Murcia, M.A., Martínez-Tomé, M., Jiménez, A.M., Vera, A.M., Honrubia, M. et al. (2002). Antioxidant activity of edible fungi (truffles and mushrooms): losses during industrial processing. J. Food Prot., 65: 1614–1622.

Murcia, M.A., Martínez-Tomé, M., Vera, A., Morte, A., Gutierrez, A. et al. (2003). Effect of industrial processing on desert truffles Terfezia claveryi Chatin and *Picoa juniperi* Vittadini: proximate composition and fatty acids. J. Sci. Food Agric., 83: 535–541. https://doi.org/10.1002/jsfa.1397

Navarro-Ródenas, A., Bárzana, G., Nicolás, E., Carra, A., Schubert, A. et al. (2013). Expression analysis of aquaporins from desert truffle mycorrhizal symbiosis reveals a fine-tuned regulation under drought. Mol. Pl. Micr. Interac., 26: 1068–1078. http://dx.doi.org/10.1094/MPMI-07-12-0178-R

Navarro-Ródenas, A., Lozano-Carrillo, M.C., Pérez-Gilabert, M. and Morte, A. (2011). Effect of water stress on in vitro mycelium cultures of two mycorrhizal desert truffles. Mycorrhiza, 21: 247–253. https://doi.org/10.1007/s00572-010-0329-z

Navarro-Ródenas, A., Pérez-Gilabert, M., Torrente, P. and Morte, A. (2012). The role of phosphorus in the ectendomycorrhiza continuum of desert truffle mycorrhizal plants. Mycorrhiza, 22: 565–575. https://doi.org/10.1007/s00572-012-0434-2

Oei, P. (2003). Mushroom Cultivation: Appropriate Technology for Mushroom Growers. Backhuys Publishers, 3rd Edition, Leiden, The Netherlands, p. 429.

Olivera, A., Bonet, J.A., Oliach, D. and Colinas, C. (2014). Time and dose of irrigation impact *Tuber melanosporum* ectomycorrhiza proliferation and growth of *Quercus ilex* seedling hosts in young black truffle orchards. Mycorrhiza, 24: 73–78. https://doi.org/10.1007/s00572-013-0545-4

Olivier, J., Savignac, J. and Sourzat, P. (2018). Truffe et trufficulture. Fanlac, 4th Edition, p. 343.

Pacioni, G., Leonardi, M., Di Carlo, P., Ranalli, D., Zinni, A et al. (2014). Instrumental monitoring of the birth and development of truffles in a *Tuber melanosporum* orchard. J. Mycor., 24: 65–72. https://doi.org/10.1007/s00572-014-0561-z

Paolocci, F., Rubini, A., Riccioni, C. and Arcioni, S. (2006). Reevaluation of the life cycle of Tuber magnatum. Appl. Environ. Microbiol., 72: 2390–2393. https://doi.org/10.1128/AEM.72.4.2390-2393.2006

Patel, S. (2012). Food, health and agricultural importance of truffles. Curr. Tr. Biotechnol. Pharm., 6(1): 15–27.

Pérez-Gilabert, M., Morte, A., Honrubia, M. and García-Carmona, F. (2001). Partial purification, characterization, and histochemical localization of fully latent desert truffle, *Terfezia claveryi* Chatin, polyphenol oxidase. J. Agric. Food Chem., 49: 1922–1927. https://doi.org/10.1021/jf001009n

Piattoni, F., Amicucci, A., Iotti, M., Ori, F., Stocchi, V. et al. (2014). Viability and morphology of *Tuber aestivum* spores after passage through the gut of *Sus scrofa*. J. Fungal Ecol., 9: 52–60. https://doi.org/10.1016/j.funeco.2014.03.002

Plett, J.M. and Martin, F.J. (2011). Blurred boundaries: life style lessons from ectomycorrhizal fungal genomes. Tr. Gen., 27: 14–22. https://doi.org/10.1016/j.tig.2010.10.005

Remy, W., Taylor, T.N., Hass, H. and Kerp, H. (1994). Four hundred-million-year-old vesicular arbuscular mycorrhizae. Proc. Nat. Acad. Sci., 91: 11841–11843. https://doi.org/10.1073/pnas.91.25.1184

Rineau, F., Roth, D., Shah, F., Smits, M., Johansson, T. et al. (2012). The ectomycorrhizal fungus *Paxillus involutus* converts organic matter in plant litter using a trimmed brown rot mechanism involving

Fenton chemistry. J. Environ. Microbiol., 14: 1477–1487. https://doi.org/10.1111/j.1462-2920.2012.02736.x

Roth-Bejerano, N., Li, Y.-F. and Kagan-Zur, V. (2004). Homokaryotic and heterokaryotic hyphae in *Terfezia*. Antonie Van Leeuwenhoek, 85: 165–168. https://doi.org/10.1023/b:anto.0000020283.99376.55

Roth-Bejerano, N., Navarro-Ródenas, A. and Gutiérrez, A. (2014). Types of mycorrhizal association. Journal Desert Truffles: Phylogeny, physiology, distribution, domestication 69–80. https://doi.org/10.1007/978-3-642-40096-4_5

Rubini, A., Belfiori, B., Riccioni, C., Tisserant, E., Arcioni, S. et al. (2011). Isolation and characterization of MAT genes in the symbiotic ascomycete *Tuber melanosporum*. New Phytol., 189: 710–722. https://doi.org/10.1111/j.1469-8137.2010.03492.x

Rubini, A., Riccioni, C., Belfiori, B. and Paolocci, F. (2014). Impact of the competition between mating types on the cultivation of *Tuber melanosporum*: Romeo and Juliet and the matter of space and time. J. Mycor., 24: 19–27. https://doi.org/10.1007/s00572-013-0551-6

Sawaya, W., Al-Shalhat, A., Al-Sogair, A. and Al-Mohammad, M. (1985). Chemical composition and nutritive value of truffles of Saudi Arabia. J. Food Sci., 50: 450–453. https://doi.org/10.1111/j.1365-2621.1985.tb13425.x

Schillaci, D., Cusimano, M.G., Cascioferro, S.M., Di Stefano, V., Arizza, V. et al. (2017). Antibacterial activity of desert truffles from Saudi Arabia against *Staphylococcus aureus* and *Pseudomonas aeruginosa*. Int. J. Med. Mushr., 19. https://doi.org/10.1615/intjmedmushrooms.v19.i2.30

Slama, A., Fortas, Z., Boudabous, A. and Neffati, M. (2010). Cultivation of an edible desert truffle *Terfezia boudieri* Chatin. J. Afr. J. Microbiol. Res., 4: 2350–2356.

Slama, A., Neffati, M. and Boudabous, A. (2009). International Symposium on Medicinal and Aromatic Plants-SIPAM2009 8532009: 285–290.

Smith, S. and Read, D. (2008). Introduction. pp. 1–9. In: Smith, S. and Read, D. (eds.). Mycorrhizal Symbiosis, 3rd Edition, Academic Press.

Splivallo. R. (2008). Biological Significance of truffle secondary metabolites. pp. 141–165. In: Karlovsky, P. (ed.). Secondary Metabolites in Soil Ecology Soil Biology, Springer Berlin Heidelberg. https://doi.org/10.1007/978-3-540-74543-3_8

Splivallo, R., Ottonello, S., Mello, A. and Karlovsky, P. (2011). Truffle volatiles: from chemical ecology to aroma biosynthesis. New Phytol., 189: 688–699. https://doi.org/10.1111/j.1469-8137.2010.03523.x

Streiblová, E., Gryndlerova, H. and Gryndler, M. (2012). Truffle brûlé: an efficient fungal life strategy. FEMS Microbiol. Ecol., 80: 1–8. https://doi.org/10.1111/j.1574-6941.2011.01283.x

Taschen, E. (2015). Interactions biotiques et biologie reproductive de la truffe noire, *Tuber melanosporum* (Vittad.): des truffières spontanées aux plantations. Université Montpellier 2, Sciences et Techniques. https://tinyurl.com/3new6er3

Trappe, J.M. (1971). A synopsis of the Carbomycetaceae and Terfeziaceae (Tuberales). Trans. Br. Mycol. Soc., 57: 85–92. https://doi.org/10.1016/S0007-1536(71)80083-9

van der Heijden, M.G.A., Martin, F.M., Selosse, M.A. and Sanders, I.R. (2015). Mycorrhizal ecology and evolution: the past, the present, and the future. New Phytol., 205: 1406–1423. https://doi.org/10.1111/nph.13288

Villares, A., García-Lafuente, A., Guillamon, E. and Ramos, A. (2012). Identification and quantification of ergosterol and phenolic compounds occurring in Tuber spp. truffles. J. Food Comp. Anal., 26: 177–182. https://doi.org/10.1016/j.jfca.2011.12.003

Wang, S. and Marcone, M.F. (2011). The biochemistry and biological properties of the world's most expensive underground edible mushroom: Truffles. J. Food Res. Int., 44: 2567–2581. https://doi.org/10.1016/j.foodres.2011.06.008

Yao Y, Spooner B. and Hawksworth, D. (1995). Author citation of the generic name *Peziza* (Pezizales, Pezizazeae). Syst. Ascomyc., 14: 17–24.

Yu, T.E., Egger, K.N. and Peterson, L.R. (2001). Ectendomycorrhizal associations characteristics and functions. Mycorrhiza, 11: 167–177. https://doi.org/10.1007/s005720100110

Zaretsky, M., Kagan-Zur, V., Mills, D. and Roth-Bejerano, N. (2006). Analysis of mycorrhizal associations formed by *Cistus incanus* transformed root clones with *Terfezia boudieri* isolates. Pl. Cell Rep., 25: 62–70. https://doi.org/10.1007/s00299-005-0035-z

Zitouni-Haouar, F.E.-H., Fortas, Z. and Chevalier, G. (2014). Morphological characterization of mycorrhizae formed between three *Terfezia* species; desert truffles and several Cistaceae and Aleppo pine. Mycorrhiza, 24: 397–403. https://doi.org/10.1007/s00572-013-0550-7

Zniber, I., Boukcim, H., Khabar, L., Ducousso, M., Henkrar F. et al. (2022). Characterization of desert truffles in the Great Moroccan Sahara: A Review. Environ. Sci. Proc., 16: 55. https://doi.org/10.3390/environsciproc2022016055

Chapter 8

Black Truffles (*Tuber melanosporum*) in Holm Oak Ecosystems in Morocco

Hakkou, S.,[1,*] *Sabir, M.*[2] and *Machouri, N.*[1]

1. Introduction

The decline in natural truffle production due to climate challenges, conflicts, and increased demand has prompted producers around the world to embrace trufficulture. The transition from maintaining and enhancing natural truffle production in forests, to purposeful cultivation (trufficulture) has been significant (Oliver et al., 2018). The earliest documented success occurred around 1740 in Montbard,Cote-d'Or, when truffle ascocarps were introduced under hornbeams. Subsequent adequate fruiting validated this approach. Trufficulture has since become a vital source of these prized fungi (Zambonelli et al., 2015; Le Tacon, 2017).

Other methods were explored in the pursuit of truffle cultivation. One such approach involved planting harvested acorns beneath oak trees to produce truffles. The pioneering effort by Jean-Pierre-François Ripert de Monclar in 1750, a public prosecutor in the Aix-en-Provence parliament, marked the beginning of this fascinating journey . He sowed seeds of various species, including acorns, on his Bourgane property in Saint-Saturnin-les-Apt in the Vaucluse to harvest large quantities years later. Trials then followed until 1800 in France, when Joseph Talon, considered in most documents as the first truffle grower, had acorns sown without intending to grow truffles in the Loudunais, and 8 or 10 years later, while walking with his sow, he detected a truffle. Thus, he added other plantations and managed to obtain

[1] Facuty of Letters and Human Sciences, Rabat, Morocco.
[2] National School of Forest Engineers, Sale, Morocco.
* Corresponding author: soukaina.hakkou@gmail.com

a hundred kg per hectare in years of good production. This method has been very successful in France, especially in regions where truffle spores already exist in the soil (Callot, 1999; Zambonelli et al., 2015; Le Tacon, 2017; Oliver et al., 2018). The black truffle, *Tuber melanosporum*, because of its virtues and its monetary value, is one of the most prized truffles. It is the most cultivated truffle around the world, even in non-truffle-base countries. Thus, in the 1950s, truffle growing became common and popular in several countries seeking to improve their annual production. Thus, in France, Italy, and Spain, natural production only constitutes a small percentage of total production (Berch and Bonito, 2014; Reyna and Garcia-Barreda, 2014).

Morocco is the first country in North Africa to have introduced the *Tuber melanosporum* culture in 2000 by Doctor Abdelaziz Laqbaqbi, a doctor trained in France. He was impressed by this local product during his assignment as an orthopedic surgeon at the Cahors Hospital in 1994. After a feasibility study on the introduction of black truffles in Debdou (North-East region), his family village in the middle of the forest of holm oak developed on limestone, he introduced and planted holm oak seedlings produced and inoculated in France (Laqbaqbi, 2020). Six years later, in 2006, the first cultivated black truffles were harvested in Morocco. The success of this experiment encouraged the doctor to set up two other truffle farms in Imouzzer Kandar, in the Middle Atlas Causse. It is worth mentioning that in Morocco, there is a natural production of certain species of true truffles in the forests of holm oaks in the Middle Atlas. These are *T. excavatum, T. brumale*, *T. rufum*, and *T. uncinatum*/*aestivum* (Malençon, 1973). In addition to the production of truffles for food needs, trufficulture makes it possible to rehabilitate the forest stands of holm oak. The objective of this study is to describe truffle farming in Morocco as a success story to be disseminated in rural areas for innovative agriculture that reconciles the socioeconomic development of populations and the preservation of the environment.

2. Truffle Farms

2.1 Tifezouine Farm in Debdou

In 2000, it was the first farm to receive holm oak plants produced and inoculated in France (Fig. 1). It is located in the commune of Sidi Ali Belkacem in the province of Taourirt in the Oriental region. Having an area of 6 ha, it is installed in a forest enclave in the form of a basin belonging to the family of Laqbaqbi (2020). It overlooks the town of Debdou upstream of the valley of Oued Sellawit, a tributary of Oued Debdou. The soil is calcareous brown in nature, deep, and rich in organic matter. The climate is semi-arid with a cool variant and ranks among the continental climates. It is a moderate-altitude climate (900–1000 m) and rainy. The total annual rainfall is 400 to 550 mm in normal years, and the average temperature is 18.1°C (Mokhtari, 2016; Seghir et al., 2019; Laqbaqbi, 2020).

The farm, most of which is covered with truffle oak, is also known for the cultivation of certain fruit trees, such as the pear tree (*Pyrus communis*), and market gardening. It has a water source, which facilitates the irrigation of the plantations, especially during the initial years. Around the farm, there are vast expanses of natural holm oak forests.

2.2 Ain Chiffaa Farm in Imouzzer Kandar

Not far from the second farm, a 10-hectare plot belonging to the Ministry of "Habous" was rented in 2015 by Laqbaqbi (2020) to install the third farm (Fig. 1). The plot has the same climatic and edaphic conditions as the previous ones and is also located in the rural commune of Ain Sebaa Lajrouf. The planted truffle holm oaks are of Moroccan origin and precisely from Debdou. Their mycorrhization was done in France. The farm is surrounded by a mixed hedge made up of olive trees and a low stone wall resulting from the stoning of the land. It contains a few dwarf palms (*Chamaerops humilis*) and a few pomegranates (*Punica granatum*). A plot of land on the farm is intended for the cultivation of saffron (*Crocus sativus*).

2.3 Ain Jarrah Farm in Imouzzer Kandar

The farm is located in the rural town of Ain Sebaa Lajrouf, not far from the municipality of Imouzzer du Kandar in the province of Sefrou in the region of Fez-Meknes (Fig. 1). This is the second farm set up after the success of the truffle plantations in Tifezouine. The 12-hectare land belongs to the water and forest administration, and it has been granted to the doctor Laqbaqbi in the form of a temporary occupation since 2008. The climatic and edaphic conditions close to those of Debdou have made the region an area of choice. Thus, rainfall is 690 mm/year, and average annual temperatures vary from 27 to 35°C (Adallal, 2019). The land has undergone a rock removal operation to evacuate large limestone rocks. The farm, which contains only planted Moroccan and French truffle oak trees, is surrounded by a windbreak made from common cypress (*Cupressus sempervirens*).

3. Cultivation of *Tuber melanosporum*

In 2000, Morocco experienced for the first time the introduction of the black truffle, *Tuber melanosporum*. This innovation is due to the doctor Abdelaziz Laqbaqbi, who had the courage to take the risk and embark on this adventure, which subsequently succeeded. The success of this experiment required passion for this product, a mastery of the techniques of inoculation of holm oak plants, plenty of negotiations with the forest authorities of Morocco, and significant investments relating to land and farm management (Fig. 2a). Cooperation with scientists from the national institute of agricultural research (INRA) of France contributed to the success of this experiment through the mastery of plant inoculation. Thus, the first Debdou farm favored the installation of the other two.

The land chosen for truffle growing has no cultural history. Stone removal and levelling off the ground are necessary before cultivating. The large stones removed are used to build farm fence walls. Holm oak seedlings are planted in 50 × 50 × 50 cm holes in rows 6 × 7 m apart. For the first truffle farm in Debdou, the holm oak truffle plants were produced and inoculated in France and imported to Morocco. Currently, they are produced in Debdou, Morocco, and will continue to be mycorrhized in France. When planting, the stones are removed, and the fine soil is gathered around the holm oak seedlings. A basin 1 m in diameter, surrounded by a bead of 20 cm

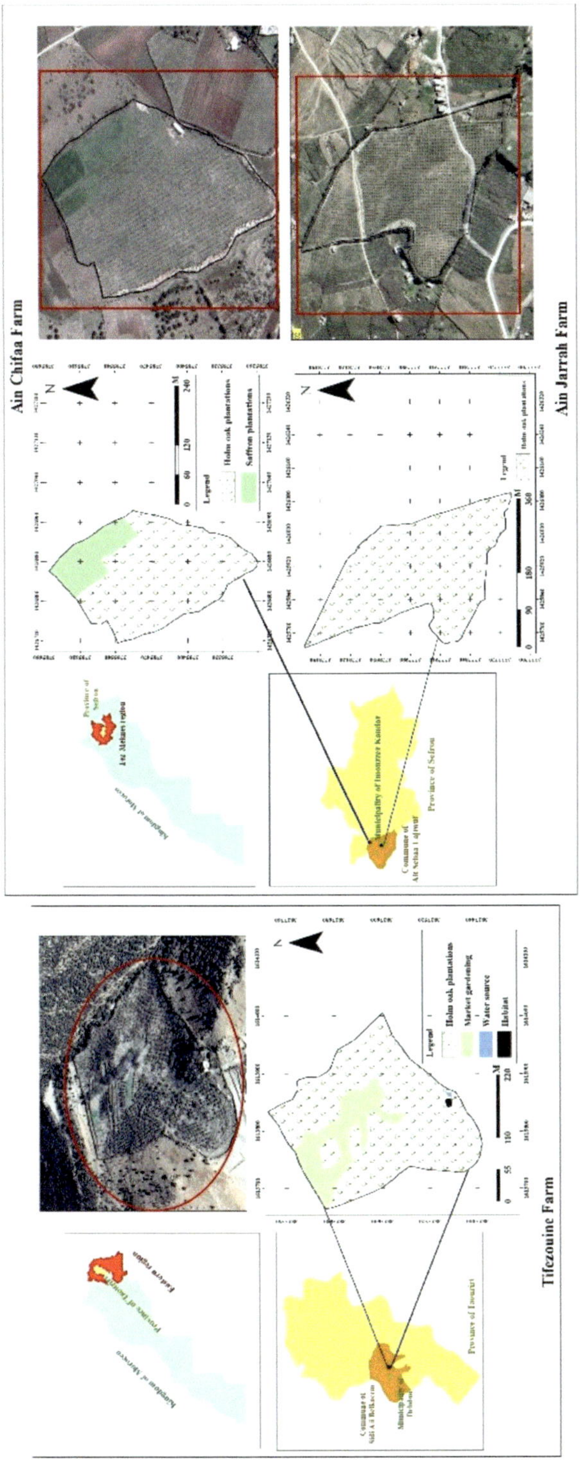

Fig. 1. Three Moroccan black truffle (*Tuber melanosporum*) farms (Tifezouine, Ain Chifaa, and Ain Jarrah farms).

height, is created around each tree to receive irrigation and rainwater. Young plants do not receive mineral fertilizers.

During the summer season (June, July, August, and September), the truffle farm requires intensive irrigation. Each tree receives 60 liters of water every two weeks. The "drip" irrigation system that was installed was with drippers for the first three years, and then it was changed to micro-sprinklers the following years. Weeding between rows of trees is necessary to avoid competition from weeds and is carried out systematically each year in the spring. Manual weeding of tree depressions is carried out once a month.

Maintenance was carried out from spring, after the end of the harvest season, until the end of summer. They begin with hoeing with a sap to a depth of 10 cm. Then, holes 20 cm in diameter and deep, called "truffle traps", are installed to ensure the seeding of the truffle trees. A mixture of substrates, amended with calcium carbonate, which includes spores of the black truffle, prepared by Laqbaqbi (2020), is put in these holes. The seeding of truffles is done annually. The holm oak plants grown in this way reach an average height of one meter in the third year. At this age, branch pruning is practiced. After each truffle harvest season, the holm oak trees are pruned (dead branches are removed). Laqbaqbi (2020) estimates that the truffle farms he

Fig. 2. The black truffle oaks of Imouzzer du Kandar (a), One of the dogs detected a black truffle (b) and a fresh black truffle dugout from the ground (c) (© Hakkou).

set up in 2000 required a significant investment of around 60000 MAD/ha without irrigation and 120000 MAD/ha with the installation of a drip irrigation system (1 US$ = 10.20 MAD). Generally, the work required is hoeing in the spring, ploughing between the rows of trees, and regular irrigation from June to September. The truffle field would have a lifespan of 30 years on average.

4. Harvesting *Tuber melanosporum*

Truffle oaks come into production in the 3rd or 4th year after planting. It is through the burn that forms in the basin under the tree and makes the weeds disappear, that we can be reassured that the mycorrhization is successful and that production has begun. At first, the truffles occur only near the trunk. Afterwards, they spread a little further, but remain in the basin around the tree. Exposed creeping roots do not give truffles, as do closed and shaded areas that block the sun's rays.

Natural truffles ripen in November and last until April. At the level of the truffle farms in Morocco, the harvest begins in the beginning of December and lasts until the end of March. Harvesting, also known as digging, is done weekly by Laqbaqbi (2020) every weekend, accompanied by his three trained dogs, to find the truffle by its strong smell. It is the dogs that detect the existence of the fungus, which is 40 cm deep most of the time (Fig. 2b). Once the dog begins to dig, it is stopped so that it does not eat the truffle or dig into it with its claws. Dogs are rarely wrong. They are rewarded each time a truffle is found; they are hugged and fed with pieces of meat . Then comes the role of the man. He takes a handful of soil and smells it. If the smell is strong and indicates that the truffle is there, he gently digs it up with an iron stake (Fig. 2c). The harvested truffles are kept in a ventilated bag until the end of the digging operation. Each tree can yield 300 g, which is the equivalent of 1–2 pieces. The weight of the truffles is 200 g on average, but some can reach up to 1 kg. Generally, one hectare produces up to 5 kg of truffles per year.

Towards the end of the day, it comes time to prepare the harvested truffles before sending them to their final destination. The truffles are brushed under running water using a more or less rigid brush that the rough peridium of the truffle can withstand without being damaged (Fig. 3a–d). They are then wiped with paper tissues and placed in the sun for a few minutes for drying and preventing them from rotting. The entire harvest is cleaned once by removing part of the peridium to ensure that it is ripe and not rotten. The removed parts and the rotten ones serve as seeds. They get buried in the ground in the depressions around the holm oak trees.

Sometimes during the harvest, some truffles are already broken under the ground or break when trying to get them out, especially when the place of existence is not well defined. They are called broken pieces and are sold to restaurants at half price to make mashed truffles. The truffles are then weighed before being put in fabric bags to be sent to buyers. In restaurants, they can be stored in the fridge at –18° for up to three years.

Moroccan *Tuber melanosporum* has a strong special aroma. Such black truffles are used in culinary preparations due to their distinguished taste and aroma (e.g., in omelets and along with shrimp) (Fig. 3e, f).

Fig. 3. Different shapes of the black truffles (a and b), black truffle flesh (c), and sorted black truffles (left, whole truffles; right, broken truffles), dishes prepared from black truffle: Omelet (e) (© Hakkou); Shrimps with black truffle (f) (© Laqbaqbi, 2020).

5. Economic Importance of Trufficulture

The black truffle sells on average at 12 MAD/g (US$1.2) or 12000 MAD/kg (US$1,200). The entire harvest from Laqbaqbi's (2020) farms is sold in Morocco (restaurants, hotels, and private individuals). It is very popular, and the demand far exceeds the supply. Based on this experience, truffle farming could contribute positively to sustainable environmental development through the reforestation of land with holm oak, which is a local forest tree adapted to the climatic and edaphic conditions of the country. Also, this operation makes it possible to reconstitute the natural ecosystem.

6. Conclusion

Given its strategic geographical location and favorable climatic and edaphic conditions, Morocco presents an enticing opportunity for agricultural investors interested in truffle cultivation. The natural abundance of holm oak forests and wild truffles makes this endeavor promising. By expanding truffle cultivation to available land withinMoroccan forests, we can simultaneously boost income generation and safeguard the environment, thereby alleviating poverty and reducing human impact on forest ecosystems.

References

Adallal, R. (2019). Variabilité environnementale des lacs du Moyen Atlas marocain: fonctionnement hydrogéochimique, hydrologique et réponse au changement climatique. Thèse de doctorat. Université d'Aix-Marseille (AMU) et Université Cadi Ayyad de Marrakech (UCAM), Morocco, p. 212.

Berch, S.M. and Bonito, G. (2014). Cultivation of mediterranean species of Tuber (Tuberaceae) in British Columbia, Canada. Mycorrhiza, 24: 473–479. 10.1007/s00572-014-0562-y

Callot, G. (1999). La truffe, la terre, la vie. Institut National de la Recherche Agronomique (INRA) Editiona, Paris, France, p. 209.

Laqbaqbi, A. (2020). La truffe entre France et Maroc ou d'un coté à l'autre de la Méditerranée. Imprimerie GRAPHO 12 Aveyron, France, p. 80.

Le Tacon, F. (2017). La domestication des truffes. Académie de Stanislas, 415–442.

Malençon, G. (1973). Champignons hypogés du Nord de l'Afrique. I. Ascomycètes. Persoonia, 7: 261–288.

Mokhtari, N. (2016). Etude de la dynamique de la désertification dans le bassin versant de la Moulouya en intégrant les données issues de la télédétection et les données socio-économiques. Thèse de doctorat. Université de Liège - Gembloux Agro-Bio Tech - ULiège, Belgium, p. 266.

Olivier, J.-M., Savignac, J.-C. and Sourzat, P. (2018). Truffe et trufficulture. Périgueux: Editions Fanlac, France, p. 352.

Reyna, S. and Garcia-Barreda, S. (2014). Black Truffle cultivation: a global reality. Forest Systems 23: 317–328.

Seghir, A., Mazoz, L. and Idrissi, A.J. (2019). Cartographie de sol dans la zone méridional de la plaine de Tafrata au Maroc Centro-Oriental et évaluation de leur sensibilité à la désertification. Rev. Mar. Sci. Agron. Vét., 7(2): 371–378.

Zambonelli, A., Iotti, M. and Hall, I.R. (2015). Current status of truffle cultivation: recent results and future perspectives. Micol. Ital., 44: 31–40.

CHAPTER 9

Ecological-Vegetational Characteristics of Truffle Habitats as an Important Basis for Choosing the Optimal Method of Their Cultivation

Srdjan Ljubojević,[1,*] *Ladislav Vasilišin,*[2] *Goran Vučić*[2] and *Tihomir Radovac*[3]

1. Introduction

In 2023, the estimated turnover from truffles reached US$378.7 million, showing an upward trend compared to previous years. Detailed studies predict substantial growth in turnover over the next decade, with an expected value of US$906.3 million by 2023, assuming a compound annual growth rate of 8.9%. (Anonymous, 2023). Notably, more than 50% of global truffle production now comes from controlled cultivation efforts (Mello et al., 2006). Truffle cultivation hinges on selecting the right truffle that aligns with local ecological conditions. Objective methods are

[1] Faculty of Ecology and Institute for Scientific Research Independent, University of Banja Luka, Bosnia and Herzegovina.
[2] Faculty of Technology, University of Banja Luka, Bosnia and Herzegovina.
[3] Society for the Environment and Sustainability on Earth, Banja Luka, Bosnia and Herzegovina.
* Corresponding author: ljuboje@blic.net

applied to maximize yield. Factors considered include parent material, soil quality, geographical location, terrain orography, climate, and the existing plant composition. Certain truffle species exhibit a symbiotic relationship with specific host plants, emphasizing the importance of understanding these ecological dynamics.

This chapter discusses two truffle species: *Tuber aestivum* (summer truffle or Burgundy truffle) and *Tuber excavatum* (hollowed truffle). The Dinarides, a large mountain massif in Southeastern Europe, spans eight countries: Slovenia, Croatia, Bosnia and Herzegovina (B & H), Montenegro, Serbia, Kosovo, Albania and North Macedonia. The name originates from Mount Dinara, situated on the Croatia- B & H border. Although not the highest peak the Dinara massif is centrally located within the Dinarides.The mountain range runs parallel to the Adriatic Sea in a northwest-southeast direction (Fig. 1a). B & H's territory is categorized into regions, areas, and districts. The research focused on the Interior Dinarides region, specifically the West Bosnian limestone-dolomite area and the Koprivnica district (Fig. 1b). Koprivnica district predominantly features dolomite terrain along the left side of the Vrbas River. Elevations range from 500 to over 1700 m. Common soil types include rendzina on dolomite as well as complexes of rendzina and calcomelansol on dolomite, and calcomelanosol and calcocambisol on limestone (Stefanović et al., 1983).

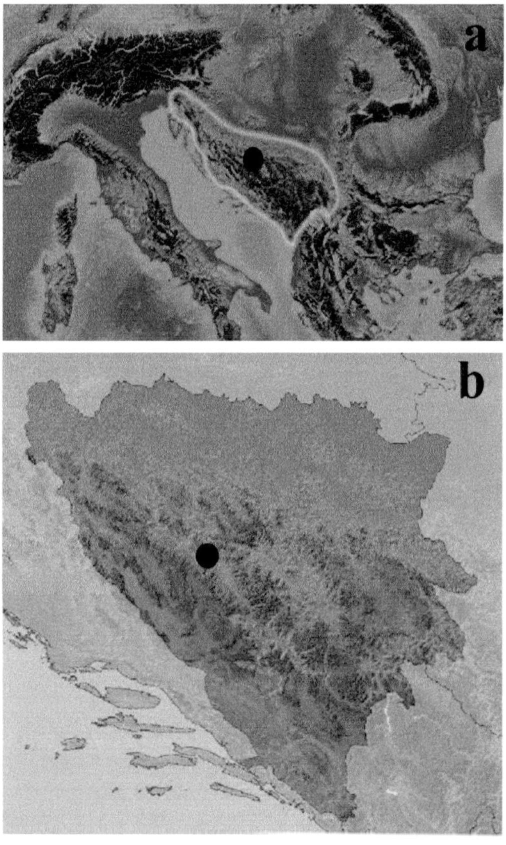

Fig. 1. Location of the research area: a, within the Dinarides; b, within B & H.

2. Ecological Characteristics of the Truffle Grounds

2.1 Parent Material and Soils

Based on geological maps of B & H and the former SFR of Yugoslavia, it is evident that the parent substrate in the research area originated in the Triassic period (Anonymous, 1989; Vujnović et al., 1981; Čičić, 2002). This substrate consists of a uniform layer of Upper Triassic dolomites. Among these dolomites, banded and layered varieties are prevalent, while massive gray and light gray colored dolomites occur less frequently. Additionally, smaller limestone inclusions are present within the dolomite package. Differential-thermal analysis reveals that the majority of the dolomite comprises the mineral dolomite (85%), followed by calcite (around 10%). The remaining portion (approximately 5%) consists of clay material and other components. Notably, the thickness of this dolomite package reaches up to 800 m (Vujnović et al., 1981).

By observing soil profiles, the rendzina soil type was detected. Between the classic A and C horizons, a transitional AC horizon was observed, the depth of which varies depending on the degree of weathering of the parent rock. The field findings agree with the Pedological Map of B & H (Stefanović et al., 1983) and with the land classification within the Forestry Master Plan for the observed area (Anonymous, 1998). According to the land classification of the former SFR Yugoslavia, the rendzina land type belongs to the humus-accumulative land class (Škorić et al., 1985). According to the International Society of Soil Science, rendzina is defined as Rendzic Leptosol (Spaargaren et al., 1994). Different forest associations are formed on rendzinas, but grass formations also appear on them, since they are favored by the accumulation of humus, characteristic of this type of land.

2.2 Climatic Characteristics

Average monthly air temperatures (°C) and mean monthly precipitation (mm) for the investigated locality were taken from the online climate database CRU TS 3.24.01., produced by the Climatic Research Unit at the University of East Anglia during 1901–2015 (Harris and Jones, 2017). This database was supplemented with data from the local meteorological station for the period 2016–2022 (Anonymous, 2010–2020; Anonymous, 2023/1). The duration of the vegetation period is expressed according to the average number of days with an air temperature higher than 10°C. The reading was taken from the map of B & H created by the Department of Forest Ecology at the Faculty of Forestry, University of Sarajevo (Stefanović et al., 1983). Potential evapotranspiration (PET) according to Thornthwaite, is expressed as an average of mm PET for the period April–September. The reading was taken from the map of B & H made by the Department of Forest Ecology at the Faculty of Forestry, University of Sarajevo (Stefanović et al., 1983), where PET is the amount of rainwater that would evaporate from the soil and plant cover provided that the soil maintains its optimal moisture content throughout the considered months.

The character of the climate of the researched area was evaluated by Lang and Milosavljević (1980) and Gračanin and Ilijanić (1977). Climate classification according to Lang is based on the rain factor (Rf), which is calculated using the formula: Rf = P/t

(Where, P denotes annual rate of precipitation in mm; t represents the mean annual air temperature in °C). If Rf is <40, the climate is considered arid; if Rf = 40–60, the climate is considered semiarid; if Rf = 61–100, the climate is characterized as semi-humid; if Rf = 101–160, the climate is characterized as humid; if Rf >160, the climate is considered perhumid.

Heat codes of climate, according to Gračanin and Ilijanić (1977) are determined based on the average monthly temperature (AMT). If AMT is < 0.5°C the climate is considered nival (NC), if AMT is 0.5–4°C the climate is considered cold (CC), if AMT is 4–8°C the climate is considered moderately cold (MCC), if AMT is 8–12°C the climate is considered moderately warm (MWC), if AMT is 12–20°C, the climate is considered warm (WC); if AMT values exceed 20°C, the climate is considered hot (HCC). The highest average monthly rainfall is in the month of November and amounts to 132.5 mm (Table 1). Total annual precipitation did not change significantly in the two observed cycles. For the period 1901–1959 the annual average was 1128.6 mm and for the period 1960–2021 the annual average was 1115.8 mm. If we look at the vegetation period (IV–IX months), the average precipitation for the period 1901–1959 was 531.7 mm and for the period 1960–2021 it was 539.8 mm. The seasonal distribution of precipitation is uniform: spring, 252.8 mm; summer, 256.3 mm; autumn, 331.8 mm; winter, 258.9 mm. The PET according to Tohornthwaite for the period April–September amounts to 400–500 mm (Stefanović et al., 1983). If we take into account the difference in seasonal rainfall as an input variable and potential evapotranspiration as an output variable, we get a surplus of water in the ecosystem to the tune of 39.8 mm. Lang's rain factor is 136.74, which gives the climate of this region a humid character (interval 100–160). The average annual air humidity, in the period 2009–2019, ranges within relatively narrow limits of 78–82% (Table 2). In the corresponding period, the number of days with snow varies between 40 and 79, with frost between 95 and 132 and with rain between 125 and 195 days.

2.3 Anthropogenically-Influenced Regions

Anthropogenic disturbances visible to the naked eye were recorded on all three truffle grounds. In the immediate vicinity of truffle ground # 1, rubble (construction waste) and garbage were deposited. The appearance of a ruderal group of species supports the conclusion that the waste is disposed of over a long period of time. The nearest summer truffle ascocarp was found 68 m away from this illegal dump site. The truffle ground # 2 was "crossed" by ruts from off-road vehicles, and the nearest ascocarp found was only 9.5 m away from them.

The truffle ground # 3 is the furthest away from human settlements and permanent communication networks. However, this does not mean that anthropogenic influences are excluded. On the contrary, several of them were recorded. Approximately 120 m from the found ascocarp, there is an abandoned quarry. At 14 m from the found ascocarp, there is a stump of a white pine and a part of the trunk with a logging top next to it, without proper residue treatment. It can be assumed that this is an illegal activity (timber theft). At the same location, only 12 m from the extracted truffle, a

Table 1. Average monthly temperature (AMT, °C), average monthly precipitation (AMP, mm) and heat codes of climate (HCC) (NC, nival climate; CC, cold climate; MCC, moderately cold climate; MWC, moderately warm climate; WC, warm climate).

Parameter	I	II	III	IV	V	VI	VII	VIII	IX	X	XI	XII
AMT 1901–1959	−2.9	−1.5	2.3	7.0	11.7	14.9	17.2	16.8	13.3	8.1	3.4	-1.0
AMT 1960–2021	−1.8	−0.3	3.1	7.7	12.2	15.7	17.7	17.7	13.7	8.9	4.2	-0.9
AMP 1901–1959	70.7	72.4	67.5	85.9	97	106	83	68.6	91.2	126	135.4	124.9
AMP 1960–2021	72.5	76.9	70.2	86.9	95.7	99.4	86.7	76.2	94.9	104.4	132.5	119.5
HCC	NC	NC	CC	MCC	MWC	WC	WC	WC	WC	MWC	MCC	NC

Table 2. Year-wise humidity with the number of days of snow, frost and rain.

Month	Characteristics	Unit	Year							
			2009	2010	2011	2012	2016	2017	2018	2019
I-XII	Humidity	%	78	81	78	78	79	79	82	80
I-XII	Snow	Number of days	40	41	46	41	79	56	74	41
I-XII	Frost		100	97	127	132	115	99	102	95
I-XII	Rain		175	195	131	154	167	125	146	145

hearth (lamb skewer) was observed. The described anthropogenic disturbances also support the view that truffles have the capability to persist in disturbed habitats, which are stressful to other fungi. Owing to their direct involvement in the nutrition of their host, the addition of nutrients is usually not needed (or conducive) for the production of truffles (Benucci et al., 2012).

3. Vegetation of the Truffle Grounds

In the Koprivnica region, we came across three natural truffle grounds: 1) Mountain beech forest (*Fagetum montanum*); 2) Forest-grassland mosaic; 3) Forest of spruce and white pine (*Piceo - Pinetum sens. lat.*). The first of the mentioned natural truffle-grounds has a characteristic "truffle landscape", which would certainly attract the attention of any serious truffle lover, which is not the case with the other two locations. To be honest, if we had not read the work of Stobbe et al. (2013), we would hardly have decided to search the spruce forest. To make the coincidence complete, our site is exactly located on the indicative map of the area given in the aforementioned work.

3.1 Truffle Ground # 1 - Mountain Beech Forest

The mountain beech forest with soils rich in bases (*Fagetum montanum*), *Fagion illyricum* alliance, adjoins the climatogenic forest with a sessile oak and common hornbeam (*Querco-Carpinetum*) and Carpinion *betuli* alliance. The community is observed within an altitude range of 806 to 950 m asl. Approximately in the middle, the community is crossed by a regional highway. In the part above the highway (883–950 m) the line of the main slope is NE (Fig. 2a), and the oak occurs as single saplings, while in the part of the forest below the road (806–883 m) the line of the main slope is N, and the oak also occurs in the canopy layer (Fig. 2b). Within the altitude belt 883–950 m we have found 12 woody species: *Acer campestre*, *A. obtusatum*, *A. pseudoplatanus*, *Carpinus betulus*, *Cerasus avium*, *Corylus avellana*, *Crataegus oxyacantha*, *Fagus sylvatica*, *Ligustrum vulgare*, *Rosa canina*, *Quercus petraea* and *Pyrus pyraster*, of which only four are able to establish EcM with truffles; these are: *Carpinus betulus*, *Corylus avellana*, *Fagus sylvatica* and *Quercus petraea*. *Carpinus betulus*, *Fagus sylvatica* and *Pyrus pyraster* occur in the tree floor and also in the form of new growth. *Corylus avellana* occurs in the upper part of the shrub layer, while *Quercus petraea* occurs only in the form of new growth.

Fig. 2. Truffle grounds: a, Truffle ground # 1 (883–950 m asl); b, Truffle ground # 1 (806–883 m asl); c, Truffle ground # 2 (a forest-grassland mosaic); d, Truffle ground # 3 (a forest of spruce and white pine) (photo: S. Ljubojević).

The ground flora consists of 20 plant species: *Achillea myllefolium*, *Aposeris foetida*, *Asarum europaeum*, *Asperula odorata*, *Athyrium filix-femina*, *Bupleurum rotundifolium*, *Fragaria vesca*, *Glechoma hirsuta*, *Heleborus odorus*, *Hepatica nobilis*, *Medicago lupina*, *Mercurialis perennis*, *Ornitogalum sphaerocorpum*, *Primula vulgaris*, *Polygonatum odoratum*, *Rannunculus aconitifolius*, *Sanicula europaea*, *Sedum acre*, *Valeriana officinalis* and *Veratrum album*. The most abundant species are: *Aposeris foetida*, *Asarum europaeum*, *Asperula odorata*, *Hepatica nobilis*, *Mercurialis perennis*, *Athyrium filix-femina* and *Sanicula europaea*. They are characteristic for mountain beech forests (*Fagetum montanum*), but they also occur in the zones below and above these forests - in the zone of oak forests (below) and in the zone of mixed forests of beech and fir, with spruce (above). In the peripheral parts of the stand, with a greater influx of direct solar radiation, the following species are encountered: *Glechoma hirsuta*, *Heleborus odorus*, *Ornitogalum sphaerocorpum*, *Polygonatum odoratum*. On somewhat wetter positions of the forest edges, the following species can be found: *Bupleurum rotundifolium*, *Medicago lupina*, *Rannunculus aconitifolius*, *Valeriana officinalis*, *Veratrum album*. On patches of the open canopy, where blocks of stone emerge on the surface of the land, *Fragaria vesca* and *Primula vulgaris occur*, more or less evenly distributed over the entire observed area.

Two mushrooms were found: *Boletus aestivalis* and *Russula cyanoxantha*, both characteristic of forest habitats. *Boletus aestivalis* was found in the marginal part of the stand, while *Russula cyanoxantha* was found under the canopy. Within the altitude belt 806–883 m we have found nine woody species: *Acer campestre*, *Carpinus betulus*, *Corylus avellana*, *Crataegus oxyacantha*, *Fagus sylvatica*, *Helianthemum numularium*, *Ligustrum vulgare*, *Quercus petraea* and *Rosa canina*, four of which are able to establish EcM with truffles: *Carpinus betulus*, *Corylus avellana*, *Fagus syvatica* and *Quercus petraea*. *Acer campestre*, *Carpinus betulus*, *Fagus sylvatica* and *Quercus petraea* occur in the tree layer, and the other species in the shrub layer. Sessile oak forms a fairly dense new growth in the ground cover.

The low chamaephyte shrub, *Helianthemum numularium*, can be cautiously considered as a potential host plant given the fact that it has been found to form an effective ectomycorrhiza with the desert truffle *Picoa lefebvrei* in France (Riousset et al., 1989, 1996) and Spain (Calonge et al., 1995; Moreno et al., 2000), especially since representatives of the genus *Helianthemum* are faithful companions of desert truffles from the genera: *Terfezia*, *Tirmania* and *Picoa* (Jamali and Banihashemi, 2012). The ground flora consists of 17 plant species: *Achillea myllefolium*, *Ajuga reptans*, *Asarum europaeum*, *Dianthus armeria*, *Filipendula vulgaris*, *Galium sylvaticum*, *Geranium sanguineum*, *Lotus corniculatus*, *Melampyrum nemorosum*, *Mercurialis perennis*, *Pedicularis beachyodonata*, *Polygala comosa*, *Scabiosa leucophylla*, *Trifolium montanum*, *Thymus serpillum*, *Veronica chamaedrys* and *Vicia cracca*. In the heart of the forest, where it is densest, we encounter the species: *Asarum europaeum*, *Galium sylvaticum*, *Geranium sanguineum* and *Mercurialis perennis*. Also in the forest, but under a slightly more open canopy, the species found are: *Ajuga reptans*, *Filipendula vulgaris*, *Melampyrum nemorosum* and *Veronica chamaedrys*. In the marginal parts of the forest, in the contact zone with the mesophilic meadows of the order *Arrhenatheretalia*, the following species occur: *Achillea myllefolium*, *Lotus corniculatus*, *Pedicularis beachyodonata* and *Vicia cracca*. On the slightly more xerothermic marginal positions of the open canopy, the following species occur: *Dianthus armeria*, *Polygala comosa*, *Scabiosa leucophylla*, *Trifolium montanum* and *Thymus serpillum*. Three mushrooms were found. *Cantharellus cibarius* and *Russula virescens* are characteristic of forest habitats, while *Macrolepiota procera* is found on the forest-meadow border.

3.2 Truffle Ground # 2 - Forest-Grassland Mosaic

This mosaic is about vegetation edges (ecotones) between grassland and trees, which were created in the process of reforestation, as a result of the reduction or complete cessation of traditional land use (Fig. 2c). So far, communities with a similar floristic composition have not been described (or identified) in B & H, so we are not in a position to identify this vegetation unit with specific syntaxa. Within the height zone of 851–900 m asl, and on the terrain whose line of the main slope is N-S oriented, we found six species of trees: *Carpinus betulus*, *Fagus sylvatica*, *Pinus sylvestris*, *Populus tremula*, *Quercus petraea* and *Sorbus aria*, of which *Quercus petraea* occurs only in the form of a new growth. *Carpinus betulus*, *Fagus sylvatica*, *Pinus sylvestris* and *Quercus petraea* are known to establish EcM with species of the genus *Tuber*. For *Populus tremula* it is stated that it can form both ecto- and endo-mycorrhiza (Harley and Harley, 1987), and that it can be a host species of the genus *Tuber*.

On this truffle-ground, we do not have a classic understory layer, because here the woody species grow next to each other, not above or below each other. Instead of an understory layer, we are talking about woody species with a scrubby habit, with the exception of *Crataegus oxyacantha*, whose growth habit is precisely a low tree, with a single stem. A total of five representatives of this category were recorded: *Crataegus oxyacantha*, *Juniperus communis*, *Rhamnus cathartica*, *Viburnum lantana* and *Ononis spinosa*, none of which are able to enter into an ectomycorrhizal relationship with species from the genus *Tuber*. The ground flora consists of 15 plant species: *Anthyllis vulneraria*, *Asperula cynanchia*, *Centaurea rupestris*, *Centaurea scabiosa*, *Cirsium acaule*, *Colchicum autumnale*, *Dorycnium herbaceum*, *Fragaria vesca*, *Gentiana cruciata*, *Petrorhagia saxifraga*, *Plantago* spp., *Polygala comosa*, *Thymus serpillum*, *Tragopogon pratensis* and *Trifolium montanum*. Most of these species occur in grassland communities belonging to the class *Festuco-Brometea*, where they find the optimum conditions for their development in xerothermic and xeromesophilic meadows and pastures, on shallow carbonate soils. Some of these species (*Asperula cynanchica* and *Dorycnium herbaceum*) can also be found in light pine forests of the order *Pinetalia heldreichii-nigrae*, and in deciduous thickets of the order *Ostryo-Carpinetalia orientalis*. *Colchicum autumnale* can also be found in hygromesophilic and mesophilic grassland vegetation of the class *Molinio-Arrhenatheretea*. *Polygala comosa* also inhabits light oak and beech forests and thickets, while *Thymus serpillum* is also found in subalpine mountain pastures on silicate substrates (class: *Juncetea trifidi*, Syn.: *Caricetea curvulae*). Two mushrooms were found: *Calvatia utriformis* and *Hygrocybe conica*, both characteristic for meadow habitats.

3.3 Truffle Ground # 3 - Forest of Spruce and White Pine

The forest of spruce and white pine (Pinus - *Pinetum sens*) has a disjunct range within the belt of the zonal (climate-regional) community of beech and fir (Abieti - *Fagetum illyricum*) (Fig. 2d). It is always found on plateaus, and plains in sunny positions (Stefanović et al., 1983). Abundant new growth of fir within the stand and beech in the peripheral parts of the stand clearly indicate the tendency of this association to develop towards a zonal community. Within the height zone of 1000–1050 m, and on the terrain whose main slope is SW, we found four species of trees: *Picea abies*, *Pinus sylvestris*, *Abies alba* and *Fagus sylvatica*. Norway spruce and Scots pine are in the form of mature trees, while European beech and silver fir are in the form of new growth. Trees of Scots pine and the new growth of European beech appear only along the edge of the truffle ground. Each of the listed species is capable of establishing EcM with species from the genus *Tuber*. In the understory, we also found four species: *Corylus avellana*, *Hedera helix*, *Ligustrum vulgare* and *Sambucus nigra*, of which only Common hazel is capable of establishing EcM with species from the genus Tuber. The other three species are endomycorrhizal species. Mainly due to the dense crown canopy, the shrub layer is poorly expressed. Common hazel occurs singly, on small clearings of the canopy, where a greater influx of light is enabled, while the presence of common privet is limited exclusively to the marginal zone of the stand. Only one specimen of black elder was seen, and that was right next to a weak spring. The presence of common ivy reminds us of Gryndler et al. (2014). They found that *Tuber aestivum* can establish certain symbiotic relationships with

Hedera helix, as a so-called non-host species. Unfortunately, no new truffles will be born from this marriage.

The ground cover consists of six plant species: *Aremonia agrimonoides*, *Carlina acaulis*, *Fragaria vesca*, *Glechoma hederacea*, *Knautia arvensis* and *Sanguisorba minor*. *Aremonia agrimonoides* occurs abundantly. It is a characteristic species of the Illyrian association of beech forests (*Fagion ilyricum*), (Stefanonović et al., 1983). *Glechoma hederacea* is also well represented as a characteristic species of the order *Populetalia albae* (Šilić, 1977). It indicates that the habitat is not xerothermic and that soil moisture is not a limiting factor. The presence of the forest strawberry, a species of wide ecological amplitude, without a pronounced indicator value, is also evident. The remaining three species occur exclusively at the edge of the forest, in bright and sunny places. Four types of mushrooms were found: *Calocera viscosa*, *Lycoperdon perlatum*, *Lycoperdon pratense* and *Lycoperdon umbrinum*. Jelly antler fungus occurs on fallen spruce branches and stumps. *Lycoperdon perlatum* and *L. umbrinum* are companions of spruce trees, while meadow pufball occurs only in the marginal zone, where it arrives from nearby meadows.

4. Found Truffles and their Spatial Arrangement in Relation to Host Trees

Two types of truffles were found: *Tuber aestivum*, the summer truffle and *Tuber excavatum*, the hollowed truffle. According to the commonly known categorization, *Tuber aestivum* belongs to the group of black truffles, characterized by the black and warty peridium, while *T. excavatum* belongs to the group of white truffles, with a whitish and smooth peridium. Both species have a broad host range (Molina et al., 1992; Benucci et al., 2012), which is confirmed by our review of the literature. In addition to sharing similar host vegetation, the observed species also share similar terrain and climatic conditions (Stobbe et al., 2012). From the aspect of market value, the ascocarps of the summer truffle is considered economically important or marketable non-wood forest products (NWFP), while the ascocarps of the hollowed truffle belong to the non-marketable NWFP category (Benucci et al., 2012). For the purposes of this work, a total of 12 well-developed carpophores were extracted. The actual abundance of ascocarps *in situ* is much higher, however, we do not engage in professional truffle farming. We collect small quantities (both in number and mass) exclusively for research purposes. After analysis, we use the material for reproductive needs in the same or similar habitats. We do not sell (traffic) collected material.

In the truffle-grounds described above, at the beginning of November 2020, we found 12 normally developed fruiting bodies, ten of which belong to the species *T. aestivum* and two to the species *T. excavatum*. *T aestivum* was found in each of the three observed truffle grounds, while *T. excavatum* was found only in the mountain beech forest. Several sources state that *T. excavatum* is a faithful companion of *T. aestivum* (Lawrynowicz, 1990; Riousset et al., 2001; Bratek et al., 2010; Stobbe et al., 2012; Gryndler et al., 2017), which did not fully come true in our case. The host trees for *T. aestivum* were: common beech, common hornbeam, sessile oak and Norway spruce, while the hosts of *T. excavatum* were common beech and common hornbeam. *T. aestivum* was found in positions with a wide range of canopy openness, from 0.8 to 0.1cm, which supports the findings that the summer truffle is a species of wide ecological valence. At the positions from which the hollowed truffle was extracted, the closure of the canopy was 0.6 and 0.4. The thinnest host-tree next to which *T. aestivum* was found had a breast-height diameter of 20 cm and the thickest of 65 cm. With *T. excavatum*, that range is much narrower and amounts to 40 and 42 cm. The distance between the ascocarp and trunk of the host-tree is also a variable feature. In the case of *T. aestivum*, the shortest distance was registered in relation to the oak - only 30 cm, while the largest distance was 250 cm in the case of beech. With *T. excavatum*, these distances were 120 and 220 cm. Ascocarps were usually found in the soil immediately below the litter layer. The depth of the ascocarp of *T. aestivum* varied between 5 and 6 cm, despite the fact that in 70% of cases it was 5 cm. Ascocarps of *T. excavatum* were found somewhat deeper, at depths of 6 and 7 cm. At the same time, the soil depth was in the range of 17–25 cm. The slopes of the terrain where the truffles were found varied from 0–21%, with all exposures represented except the northern one. The stoniness and rockiness of the terrain ranged from 0–10%. The soil is pH correlated with the associated vegetation formation. The lowest pH was registered in the spruce forest (pH, 6.8), slightly higher in the meadow (pH, 7.3), and the highest in the beech forest (pH, 7.7–7.8). On the day when the truffles were collected, the humidity in the soil ranged from 38–39% in the meadow, to 74–75% in the dense spruce forest. On June 15 of the following year, the soil moisture ranged from 44% in the meadow to 78–79% in the spruce forest with a dense structure (Tables 3 and 4).

4.1 Methodology of Data Collection

Data collection in the field was done on several occasions. In the first field trip, at the end of October 2020, truffles were searched for and found with trained Lagotto Romagnolo dogs. Each individual fruiting body collection site was simultaneously registered using two GPS devices, "Garmin" and "Magelan", and a marker was placed exactly 20 cm north of the truffle hole. In this way, interference in communication between the dog and its handler during the search was avoided. Ascocarps were extracted using a trapezoidal shovel, in such a way as to disturb the existing condition as little as possible. Immediately after extraction, the depth of the excavation was measured, that is, the position of the ascocarp in relation to the current surface of the land, and then the raised soil of the land was carefully returned to its original position. The extracted ascocarps, after preliminary observation, were sent to the lab for a standard determination procedure. In the following days, the collection of data on natural truffle-grounds was started. The sample element is circular in shape, which does not have the "corner effect" of square or rectangular test surfaces (Ljubojević et al., 2015). A circular area with a radius of 5.65 m and a surface area of 100 m^2 is staked out from each marker. The plant species present on the plot and the predominant plant formation were determined. Also, the most probable host-plant was determined and the distance between the marker or truffle and the host plant was measured. On each plot, the closure of the canopy was evaluated, the elevation and slope of the terrain were measured, exposure was determined, rockiness was evaluated, soil depth was measured, soil pH and humidity were measured. In order to

Table 3. Basic characteristics of tr+++uffle sites (BHD, breast-height diameter).

Ascoma #	Site #	Species	Closure of canopy	Host tree (H.t.)	Host tree BHD (cm)	Distance H.t. - ascoma (cm)	Potential host tree (P.h.t.)	Potential host tree BHD (cm)
1	1a	*Tuber aestivum*	0.5	beech	40	40	beech	30
2	1a		0.5	beech	28	250	beech	28
3	1a		0.5	beech	65	200	beech	33
4	1a	Tuber excavatum	0.6	beech	42	120	beech	38
5	1b	*Tuber aestivum*	0.5	beech	36	120	beech	40
6	1b		0.5	hornbeam	35	140	hornbeam	21
7	1b		0.5	oak	40	30	-	-
8	1b	Tuber excavatum	0.4	hornbeam	40	220	hornbeam	36
9	2	Tuber aestivum	0.1	beech	20	190	white pine	28
10	2		0.1	beech	40	80	-	-
11	3	Tuber aestivum	0.7	spruce	52	160	spruce	46
12	3		0.8	spruce	28	120	spruce	28

get a comprehensive picture of the vegetation conditions, in the middle of June next year, the flora was re-determined. On this occasion, the soil moisture was measured once again.

The determination of plant material was carried out using keys in basic and standard floras (Javorka and Csapody, 1975; Josifović, 1970–1986; Sarić and Josifović, 1986). The set of all circular surfaces from one "natural truffiéra" constitutes a sample by means of which the relative representation of plant formations is evaluated. Other ecological-vegetation characteristics are evaluated in the same way. Host plants are trees or shrubs that are reliably known to form ectomycorrhizal (EcM) associations with fungi from the genus *Tuber*. The EcM associations are sometimes called ectotrophic associations or sheathing mycorrhizas (Brundrett and Cairney, 2002). Host plants are normally identified by observations of the natural position of the ascocarp in relation to the rhizosphere of the plant, which is assumed to have established mycorrhiza with it. Other plants that have been reliably known to form EcM with truffles have been marked as potential host plants. Plants that form some other types of mycorrhizas and those that do not form any well recognized types of mycorrhizas are labeled as co-occurring, i.e., companion plants. The breast-height diameter of the host plant was measured using a caliper, while the marker distance from the base of the host plant was measured with a field tape and expressed in cm. The degree of canopy closure was determined based on the shading of the soil by tree tops on two orthogonal diameters of the sample surface in the directions N-S and E-W (Ljubojević et al., 2015).

The altitude, terrain slope and exposure were determined using a GPS device. The stoniness and rockiness of the terrain were determined by measuring the horizontal projection of stones and rocks in relation to the total observed area, expressed as a percentage. In this case, by stoniness, we mean individual, broken pieces of rock that can be moved on the surface of the ground, while rockiness is the appearance of hard stone masses or cliffs that cannot be moved (Löffler, 1984). Soil depth was measured

Table 4. Further basic characteristics of truffle sites.

Ascoma #	Site #	Distance P.h.t - ascoma (cm)	Depth ascoma (cm)	Slope of the terrain (%)	Orientation of the terrain	Stoniness and rockiness (%)	Soil depth (cm)	Soil pH	Soil moisture (%)	
									02.11. 2020	17.06. 2021
1	1/1	180	5	21	E	10	18	7.8	72	76
2	1/1	390	5	21	E	10	18	7.8	72	76
3	1/1	325	5	10	E	5	20	7.8	72	76
4	1/1	140	7	12	E	8	22	7,8	71	77
5	1/2	200	6	5	SE	0	25	7.7	72	75
6	1/2	220	6	5	SE	0	22	7.7	71	75
7	1/2	-	6	5	W	0	23	7.7	71	75
8	1/2	250	8	0	0	0	22	7.7	68	71
9	3	350	5	8	SE	2	17	7.3	38	44
10	3	-	5	8	SE	2	17	7.3	39	44
11	4	310	5	6	SW	1	18	6.8	74	78
12	4	120	5	6	SW	1	18	6.8	75	79

using the auger hole method. We used a 120 cm soil probe (steel slender rod with a t-handle), with an auger Ø 3 cm. The hole is bored into the soil until it is stopped by barriers or an impenetrable layer. The depth of the hole thus created was measured with a meter ruler and expressed in cm. Soil pH and moisture *in situ* were measured using the ZD-05 Soil PH and Moisture Tester, made in China). In order to get a more complete picture of the vegetation conditions, in the middle of June of the next year, the flora was re-determined. On that occasion, soil moisture was measured once again. We recorded the visible changes in the natural state of the environment with the naked eye, briefly described them and positioned them in relation to the nearest detected ascocarp. We did not deal with a more detailed analysis of the possible influence of recent anthropogenic disturbances on the natural potential of truffles in the observed habitats. Also, we only recorded the occurrences of ruderal vegetation on anthropogenically influenced sites, without analyzing the floristic composition. In doing so, we observed the ruderal vegetation as defined by Faber-Langendoen et al. (1998) and Faber-Langendoen et al. (2014).

5. Proposed Methods of Growing Truffles

5.1 Truffle Ground # 1 - Planting Mycorrhizal Seedlings

Truffle ground # 1 is represents: "Mountain beech forest on soils rich in bases (*Fagetum montanum*)". Within the community itself, as well as in bare locations outside it, and at altitudes above 885 m, the following host trees are recommended: European beech (*Fagus sylvatica*), common hornbeam (*Carpinus betulus*) and oriental hornbeam (*Carpinus orientalis*). At altitudes below 885 m, the following are recommended: *Carpinus betulus*, *C. orientalis*, *Fagus sylvatica* and *Quercus petraea* (sessile oak). In the future, the newly formed truffle production area should follow the existing plant formation with its thickness and height structure, as well as its floristic composition. In addition to truffle ascocarps, this truffle production area will eventually produce a significant amount of forest wood assortments as well as non-wood forest products in the form of wild medicinal and edible herbs and mushrooms. The establishment of this new formation would be carried out using seedlings of the mentioned species, aged 1–3 years, mycorrhized with the species *Tuber aestivum*. The planting density for beech and sessile oak should be around 280 trees per hectare, which corresponds to the average density of 6 × 6 m in a typical truffle orchard. Hornbeams would be planted in a slightly denser arrangement of about 400 trees per hectare, which corresponds to a 5x5 m layout in a typical truffle orchard. The preferred solution is a combination of species in order to achieve a stable mixed stand.

5.2 Truffle ground # 2 - Planting Mycorrhizal Seedlings

Truffle ground # 2 consists of grassland with individual representatives of trees and shrubs, with a completely open structure (canopy openness 0.1). The following species are recommended as host trees: *Carpinus betulus* and *Quercus petraea*. In the future, the formation of the truffle production area should, with its thickness and height structure, correspond to a truffle orchard with a schematic arrangement of trees, with a density of 5 × 5 m. The establishment of this new formation would be carried out using seedlings of the mentioned species, aged 1–3 years, mycorrhized with *Tuber aestivum* species and *T. melanosporum*. Although *T. melanosporum* has not yet been found in this locality, the field conditions in everything agree with the environmental conditions in which this type of truffle normally occurs or is grown. In the case of the impossibility of providing the mentioned reproductive material, common hazel (*Corylus avellana*) seedlings, mycorrhizal with *T. aestivum* and *T. melanosporum*, are considered. In this case, the arrangement of trees (shrubs) would be 4 × 4 m.

5.3 Truffle Ground # 3 – Applaing Truffle Trapping

Truffle ground # 3 constitutes " a forest of spruce and white pine (*Piceo - Pinetum* sens. lat.)". Given that this is a highly economical forest, with dense growth and structure (canopy openness 0.7–1.0 cm), it would not make sense to introduce mycorrhized seedlings. Instead, the trees would be underseeded with spores of *T. aestivum*, with the aim of enhancing the number of mycorrhized spruce trees, and thus increasing the yield of the stand, both per host tree and per unit area. The spore inoculum would be introduced into holes with a diameter of 20 cm. 5–10 cm deep. The inoculum would contain at least 1 g of spores per inoculated tree. One hole would be dug for the selected tree, at a distance of 50 cm from the trunk in the SW direction. The density of inoculated trees should approximately follow a 5 × 5 m layout scheme.

Acknowledgments

We would like to thank academician Dr. Vladimir Beus, University professor in the Faculty of Forestry in Sarajevo, for providing useful information about the ecological-vegetational characteristics of the researched area, which he obtained by working on this issue for many years.

6. Conclusion

Within the West Bosnian limestone-dolomite area in the Interior Dinarides region, three truffle grounds and plant communities, were detected: A) Mountain beech forest on soils rich in bases (*Fagetum montanum*); B) Forest-grassland mosaic; C) Forest of spruce and white pine (*Piceo - Pinetum sens. lat.*). Three vegetational communities have been addressed. 1) Within the first community and in adjacent open positions, truffières would be arranged with host trees: European beech (*Fagus sylvatica*), common hornbeam (*Carpinus betulus*), oriental hornbeam (*Carpinus orientalis*) and sessile oak (*Quercus petraea*) mycorrhized with the species *Tuber aestivum*. The planting density for beech and sessile oak should be around 6 × 6 m in a typical truffle orchard. Hornbeams are planted in a slightly denser arrangement of about 5 × 5 m in a typical truffle orchard. The preferred solution is a combination of species in order to achieve a stable mixed stand. 2) The second community consists of grassland with individual representatives of trees and shrubs, with a completely open structure. The following species are recommended as host trees: common hornbeam (*Carpinus betulus*), sessile oak (*Quercus petraea*) and common hazel (*Corylus avellana*), mycorrhizal with the species *Tuber aestivum* and *T. melanosporum*. In the future, the formation of the truffle production area should correspond to a truffle orchard with a schematic arrangement of trees, with a density of 5 × 5 m, when it comes to common hornbeam and sessile oak, while in the case of common hazel, the arrangement of trees (shrubs) would be 4 × 4 m. 3) The third community is a highly economical forest, with dense growth and structure. Here, trees are underseeded with spores of *Tuber aestivum*, with the aim of increasing the number of mycorrhized spruce trees and thus increasing the yield of the stand, both per host tree and per unit area. The density of inoculated trees should approximately follow a 5×5m layout scheme.

References

Anonymous (1989). Basic geological map of SFRJ [Osnovna geološka karta SFRJ: 1958–1989 in Serbo-Croatian] Savezni geološki zavod. Beograd, p. 22.

Anonymous (1998). Forestry master plan for Forest economy region "Šipovo" for the period 01.01.1999 – 31.12.2008. [ŠPO za ŠPP "Šipovsko" za period 01.01.1999. – 31.12.2008 in *Serbian*] JPŠ Šume RS. Istraživačko-razvojni i projektni centar, Banja Luka, Bosnia and Herzegovina.

Anonymous (2010–2020). Geographycal and meteorogical data. Statistical Yearbook of Republika Srpska. Institute of Statistics of Republika Srpska. Issues for calendar years, 2009.

Anonymous (2023/1). Department of Climatology. The Republic Hydrometeorological Service of Serbia, Republic of Serbia.

Anonymous (2023). Truffles Market Outlook 2023 to 2033. Future Market Insights, Inc. Christiana Corporate. Delaware, Newark, USA, p. 250.

Bencivenga, M. and Granetti, B. (1990). Analisi comparativa delle caratteristiche ecologiche di *Tuber magnatum* Pico e *Tuber melanosporum* Vitt. nell' ltaiia centrale. pp. 433–434. In: Bencivenga, M. and Granetti, B. (eds.). Atti del II Congr Int sui Tartufo. Comunita Montana dei Monti Martani e del Serano, Spoleto, Italy.

Benucci, G.M.N., Bonito, G, Falini, L.B., Bencivenga, M. and Donnini D. et al. (2012). Truffles, timber, food, and fuel: Sustainable approaches for multi-cropping truffles and economically important plants. pp. 265–280. In: Zambonelli, A. and Bonito, G. (eds.). Edible Ectomycorrhizal Mushrooms, Springer, Berlin and Heidelerg, Germany.

Bratek, Z., Merenyi, Z., Illyes, Z., Völcz, G., Tamasko, G. et al. (2010). Studies on the ecophysiology of *Tuber aestivum* populations in the Carpatho-Pannonian region. Österreichische Zeitschrift für Pilzkunde, 19: 221–226.

Brundrett, C.M. and Cairney, W.G.J. (2002). Ectomycorrhizas in plant communities. pp. 105–150. In: Sivasithamparm, K., Dixon, K. and Barrett, R.L. (eds.). Microorganisms in Plant Conservation and Biodiversity. Kluwer Academic Publishers, Netherlands.

Calonge, F.D., García, F, Santos, J.C. and Juste, P. (1995). Contribución al estudio de los hongos de Valladolid y provincias limítrofes III. Algunas especies hipogeas interesantes. Boletin Sociedad Micológica de Madrid, 20: 291–299.

Čičić, S. (2002). Geological composition and tectonics of Bosnia and Herzegovina. Earth Science Institute Sarajevo, Bosnia and Herzegovina, p. 350.

Faber-Langendoen, D., Keeler-Wolf, T., Meidinger, D., Tart, D., Hoagland, B. et al. (2014). EcoVeg: A new approach to vegetation description and classification. Ecol. Monogr., 84: 533–561.

Faber-Langendoen, D., Weakley, A.S., Anderson, M., Bouirgeoron, P., Crawford, R. et al. (1998). International Classification of Ecological Communities: Terrestrial Vegetation of the United States, Volume I. The National Vegetation Classification System. The Nature Conservancy, Arlington, VA, USA, p. 126.

Gračanin, M. and Ilijanić Lj. (1977). Introduction to Plant Ecology [Uvod u ekologiju bilja *in Croatian-Serbian*] Školska knjiga, Zagreb, Croatia, p. 305.

Gryndler, M., Černá, L., Bukovská, P., Hršelová, H. and Jansa, J. (2014). *Tuber aestivum* association with non-host roots. Mycorrhiza, 24: 603–610.

Gryndler, M., Šmilauer, P., Štovíček, V., Nováková, K., Hršelová, H. et al. (2017). Truffle biogeography —A case study revealing ecological niche separation of different *Tuber* species. Ecol. Evol., 7: 4275–4288.

Harley, J.L. and Harley, E.L. (1987). A check-list of micorrhiza in British flora. New Phytol., 105: 1–102.

Harris, I.C. and Jones, P.D. (2017). Climatic Research Unit (CRU) Time-Series (TS) Version 3.24.01 of High Resolution Gridded Data of Month-by-month Variation in Climate (January 1901–December 2015). Centre for Environmental Data Analysis, University of East Anglia Climatic Research Unit, Republic of Srpska, Bosnia and Herzegovina.

Jamali, S. and Banihashemi, Z. (2012). Hosts and distribution of desert truffles in Iran, based on morphological and molecular criteria. J. Agric. Sci. Technol., 14: 1379–1396.

Javorka, S. and Csapody, V. (1975). Iconographia florae partis austro-orientalis Europae centralis. Akademiai Kiado. Budapest, Hungary, p. 649.

Josifović, M. (1970–1986). Flora SR Srbije (I–X). Srpska akademija nauka i umetnosti. Beograd. Lawrynowicz, M. (1990). Chorology of the European hypogeous *Ascomycetes*. II. *Tuberales Acta* Mycologica, 26(1): 7–76.

Lawrynowicz, M. (1990). Chorology of the European hypogeous Ascomycetes. II. Tuberales Acta Mycologica, 26(1): 7–76.

Ljubojević, S., Šumatić, N., Marčeta, D., Hrkić, Z., Petković, D. et al. (2015). Ecological production potential of medicinal and edible plants in the forests and forest lands of the Republic of Srpska [Ekološko-proizvodni potencijali ljekovitog i jestivog bilja u šumama i na šumskim zemljištima Republike Srpske *in Serbian*] Nezavisni Univerzitet Banja Luka, Bosnia and Herzegovina, p. 292.

Löffler, H.J. (1984). Terrain classification for forestry. Report TIM/EFC/WP.1/R.51, EU Timber Committee and FAO–ILO, p. 55.

Mello, A., Murat, C. and Bonfante, P. (2006). Truffles: much more than a prized and local fungal delicacy. FEMS Microbiol. Lett., 260: 1–8.

Milosavljević, M. (1980). Climatology [Klimatologija *in Serbian-Croatian*]. Naučna knjiga, Beograd, Hungary, p. 261.

Molina, R., Massicotte, H. and Trappe, J.M. (1992). Specificity phenomena in mycorrhizal symbiosis: community-ecological consequences and practical implications. pp. 357–423. In: Allen, M.F. (ed.). Mycorrhizal Functioning: An Integrated Plant-Fungal Process, Chapman & Hall, London.

Moreno, G., Díez, J. and Manjón, J.L. (2000). *Picoa lefebvrei* and *Tirmania nivea*, two rare hypogeous fungi from Spain. Mycol. Res., 104: 378–381.

Riousset, L., Riousset, G., Jalade, M. and Chevalier, G. (1996). Prima raccolta in Francia di *Picoa lefebvrei* (Patouillard) Maire 1906. Il Fungo Supplement # 3: 15–17.

Riousset, L., Riousset, G., Jalade, M., Chevalier, G. (1989). Prima raccolta in Francia di *Phaeangium lefebvrei* Patouillard 1894. Il Fungo, 9–11.

Riousset, L.G., Chevalier, G. and Bardet, M.C. (2001). Truffles d'Europe at de Chine. Paris: INRA. Sarić, M. and Josifović, M. (1986). Flora SR Srbije. Srpska akademija nauka i umetnosti. Beograd, Serbia. p. 400.

Sarić, M. and Josifović, M. (1986). Flora SR Srbije. Srpska akademija nauka i umetnosti. Beograd, Serbia, p. 400.

Šilić, Č. (1977). Forest Herbaceous Plants. Sarajevo, Zagreb, Beograd. Brightness, Schoolbook, Vuk Kardžić. p. 272.

Škorić, A., Filipovski, G. and Ćirić, M. (1973). Classification of soils of Yugoslavia [Klasifikacija zemljišta Jugoslavije *in Serbian-Croatian*] Akademija nauka i umjetnosti Bosne i Hercegovine, Posebna izdanja, Knjiga LXXVIII, Odjeljenje prirodnih i matematičkih nauka. Knjiga 13. Sarajevo, Bosnia and Herzegovina.

Spaargaren, O. (1994). World Reference Base for Soil Resources. Draft, ISSS-ISRIC-FAO, Wageningen/ Rome, p. 161.

Stefanović, V., Beus, V., Burlica, Č., Dizdarević, H. and Vukorep, I. et al. (1983). Ecological-vegetational reonization of Bosnia and Herzegovina [Ekološko-vegetacijska rejonizacija Bosne i Hercegovine *in Serbian-Croatian*]. Radovi Šumarskog fakulteta Univerziteta u Sarajevu, 17: 1–83.

Stobbe, U., Bűntgen, U., Sproll, L., Tegel, W., Egli, S. et al. (2012). Spatial distribution and ecological variation of re-discovered German truffle habitats. Fungal Ecol., 5: 591−599.

Stobbe, U., Stobbe, A., Sproll, L., Tegel, W., Peter, M. et al. (2013). New evidence for the symbiosis between *Tuber aestivum* and *Picea abies*. Mycorrhiza, 23: 669−673.

Škorić, A., Filipovski, G. and Ćirić, M. (1985). Classification of soils of Yugoslavia [Klasifikacija zemljišta Jugoslavije in Serbian-Croatian] Akademija nauka i umjetnosti Bosne i Hercegovine, Posebna izdanja, Knjiga LXXVIII, Odjeljenje prirodnih i matematičkih nauka. Knjiga 13. Sarajevo, Bosnia and Herzegovina

Vujnović, L. (1981). Basic Geological Map of SFRJ [Osnovna geološka karta SFRJ 1: 100 000, Tumač za list Bugojno (L 33−143, *in Serbian-Croatian*]. Savezni geološki zavod, Beograd, p. 81.

Bioactive Potential

Chapter 10

The Medicinal Properties of Truffles

Mehmet Akyüz[1,*] and *Sevda Kırbağ*[2]

1. Introduction

Macrofungi have been a source of fascination for humans since ancient times due to their ecological, environmental, health, and nutritional benefits. They are particularly important to many indigenous peoples, providing daily sustenance and commercial income (Li et al., 2021; Pérez-Moreno et al., 2021; Niego et al., 2023). For centuries, people have collected and consumed wild edible mushrooms for food and medicinal purposes. Certain macrofungal species also hold cultural significance in religious rituals, handicrafts, myths, and beliefs. This tradition dates back to many ancient civilizations, including Greek, Egyptian, Roman, Chinese, and Mexican. These civilizations were aware of the therapeutic benefits of mushrooms and used them in various ways (Dugan, 2011; Martins, 2016). Mushrooms were considered a delicacy by early civilisations such as Egyptian(Pharaohs), and Roman. While the ancient Greeks believed that mushrooms gave warriors strength in battle, the Romans saw them as the 'food of the gods' created by the lightning struck by Jupiter. The Egyptians thought they were a gift from the 'god Osiris'. The Chinese considered mushrooms to be a health food and referred to them as an 'elixir of life'. Similarly, Mexicans used mushrooms as a 'gift of love' in religious ceremonies (Manzi et al., 1999; Miles and Chang, 2004; Valverde et al., 2015). Mushrooms have attracted much attention due to their taste, nutritional value, and potential medicinal benefits. They are commonly referred to as 'superfoods' and are considered an important part of a healthy diet (Kour et al., 2022). These objective and factual characteristics make them a popular choice for consumption.

[1] Department of Biology, Faculty of Arts and Science, Bitlis Eren University, Bitlis, Türkiye.
[2] Department of Biology, Faculty of Science, Fırat University, Elazığ, Türkiye.
* Corresponding author: makyuz@beu.edu.tr

The edibility of mushrooms is determined by specific criteria, including the absence of toxic effects on humans as well as desirable taste, aroma, nutritional value, and medicinal characteristics. Ascocarp and basidiocarp refer to the fleshy fruit bodies of various macrofungi. These fruiting bodies can be found either below the soil (hypogeous) or above the soil (epigeous) and can be harvested by hand (Chang, 2001; Beever et al., 2014; Lee et al., 2020; Yadav and Negi, 2021; Chugh et al., 2022). Mushrooms have been important for five main reasons: as a food source, for income generation, for ecosystem maintenance, for their medicinal properties, and for industry (Miles and Chang, 2004; Raghavendra et al., 2017; Pérez-Moreno et al., 2021).

Mushrooms are a valuable food source due to their high protein content compared to most vegetables, delicate flavor, value for their typical meat-like texture, mineral richness, low calories, cholesterol, and fat content (mostly composed of unsaturated fatty acids), abundance of vitamins (B, D, K, and sometimes A and C), and their content of useful nutraceuticals. They also provide dietary fiber while contributing few calories to the diet (Manzi et al., 1999, 2001; Mattila et al., 2001; Sanmee et al., 2003; Srikram and Supapvanich, 2016; Dimopoulou et al., 2022; Ferraro et al., 2022). Mushrooms are considered functional foods due to their high nutritional, culinary, and pharmacological values. Their consumption has increased through both dietary intake and dietary supplements. Edible wild mushrooms have been valued for centuries due to their nutritional properties, flavor, aroma, and other characteristics. They are highly regarded for their nutritional value, attractive aroma, taste, and texture, making them an ideal ingredient for cooking (Perez-Montes et al., 2021; Jiang et al., 2023). Due to its high nutritional value, desirable taste, and aroma, mushroom consumption has increased significantly and continues to be consumed as food and a cure for various ailments (Mwangi et al., 2022).

What's more, it has been reported that mushrooms have beneficial effects on health and can be utilized in the treatment of certain diseases. Fungi are sources of mycochemical compounds that benefit human health. These compounds include proteins, polysaccharides, steroids, glycosides, vitamins, minerals, nucleotides, enzymes, flavonoids, glycoproteins, phenols, tannins, terpenes, lectins, peptides, glucan, folates, ergothioneine, amino and organic acids, alkaloids, volatile oils, carotenoids, and oxidized polyphenols. The compounds have a range of biological functions, including antimicrobial, antiviral, antioxidant, anticancer, antidiabetic, antiallergic, antiobesity, antidepressant, antihyperlipidemic, detoxification, cytotoxic, hypotensive, neuroprotective, hepatoprotective effects, anti-inflammatory activity, prebiotic, digestive, and immunomodulating properties, as well as cardiovascular protection.

Truffles have the ability to protect the kidneys and bones, lower cholesterol and blood lipids, enhance the immune system, and prevent or treat conditions such as Alzheimer's, Parkinson's, hypertension, and a high risk of stroke. These agents possess anti-tumor properties that not only prevent tumor development and inflammation but also decrease the probability of cancer invasion and metastasis (Lindequist et al., 2005; Valverde et al., 2015; Venturella et al., 2021; Bhambri et al., 2022; Kour et al., 2022; Mwangi et al., 2022; Hamza et al., 2023; Łysakowska et al., 2023; Niego et al., 2023: Yu et al., 2023; Devi et al., 2024). These properties make mushrooms important not only for nutrition but also for their therapeutic benefits. They are a significant natural food source for nutrition and disease treatment due to their diverse characteristics.

Edible fungi are categorized into two groups: epigeous fungi, often called 'mushrooms,' which grow above ground, and subterranean or hypogeous species, known as 'truffles,' which grow underground (Chang, 2001; Beever et al., 2014; Chugh et al., 2022). The previous paragraph discusses the extensive studies conducted by researchers on various species of epigeous fungi that contain bioactive substances and highlight their therapeutic properties. They are a natural food source that is rich in nutrients and bioactive components, making them a valuable source of healthy food. The concept of what a truffle is may change, just as, as Bonito and Smith (2016) put it, "we discover and learn about these exquisite fungi!". This will provide our readers with a different perspective. This chapter will focus primarily on hypogeous species and provide information on their medicinal and therapeutic properties. Thus, hypogeous species, such as forest truffles (*Tuber*) and desert truffles (*Terfezia*, *Tirmania*, and *Picoa*), are particularly noteworthy due to their distinctive characteristics. Recently, they have received more attention than other epigeous mushrooms. Food culture, or gastronomy, reflects the customs and traditions of various societies. Truffles are a type of food that has been consumed for centuries due to their distinct earthy flavor, delicate aroma, and unique texture. They are subterranean fungi that are popular in the culinary world for these reasons (Kagan-Zur et al., 2014; Khalifa et al., 2019; Lee et al., 2020; Chauhan et al., 2021; Ferreira et al., 2023; Tejedor-Calvo et al., 2023a).

Truffles are fruiting bodies that grow underground and belong to the phylum Ascomycota. They are highly valued for their unique flavour and aroma. Truffles are categorized into different taxa within Pezizales J. Schrot, which include *Terfezia* and *Tirmania* (Pezizaceae Dumort.), *Picoa* (Pyronemataceae Corda), *Tuber* (Tuberaceae Dumort.), and other genera, as documented by Moreno et al. (2014). There are two main types of truffles: forest truffles (*Tuber* spp.) and desert truffles (*Terfezia*, *Tirmania*, and *Picoa* species). Among these, *Terfezia*, *Tuber*, and *Tirmania* species are the most commonly consumed. *Tubers* are prevalent in the Mediterranean, Europe, South-East Asia, Australia, and North America (Payen et al., 2014; Nakamura et al., 2020; Choo et al., 2021; Wu et al., 2021). They are a type of ectomycorrhizal fungus in symbiosis with higher trees such as Pinaceae and Fagaceae (Wang et al., 2014; Bonito et al., 2013, 2016; Gryndler, 2016). Truffles are highly valued for their rarity, organoleptic properties, and flavour content (Romanazzi et al., 2015). They are commonly used as a flavour enhancer rather than a standalone food product due to their strong aroma and high price (Bonito and Smith, 2016; Vahdatzadeh et al., 2018; Mustafa et al., 2020; Piatti et al., 2024). Desert truffles, from the genera *Terfezia*, *Tirmania*, and *Picoa*, are predominantly found in the Mediterranean, the Middle East, Europe, North Africa, and the Arabian Peninsula. They form mycorrhizal associations, primarily growing on the roots of plants in the Cistaceae, which encompasses various species of the genus *Helianthemum* (L.) (Loizides et al., 2012; Owaid, 2016; Patel et al., 2017; Behzadi et al., 2021; Elfalla et al., 2021; Morte et al., 2021; Veeraraghavan et al., 2022; Zitouni-Haouar et al., 2023).

Desert truffles boast a longstanding history of utilization in traditional Mediterranean medicine. They were first mentioned in ancient Egyptian poetry. Historically, they were considered both food and medicine for royalty, and were served to the Pharaohs. Throughout history, only the wealthy and famous have been able to enjoy these precious delicacies (El Enshasy et al., 2013). The desert truffles are seasonal edible fungi that hold significant ecological and socio-economic importance (Mandeel and Al-Laith, 2007; Al-Laith, 2010; Louro et al., 2019; Tagnamas et al., 2020). Depending on the climate, altitude, vegetation, and soil structure, truffles can grow in different exclusive habitats (Castellano and Türkoğlu, 2012; Akyüz et al., 2012, 2015, 2017; Türkoğlu and Castellano, 2013, 2014; Gezer et al., 2014; Bradai et al., 2015b; Türkoğlu et al., 2015; Bermaki et al., 2017; Abdelsalam et al., 2023). Literature related to the subject provides detailed information on the taxonomy, distribution, species, vegetation, ecological characteristics, mycorrhizal relationships, soil properties, nutritional content, medicinal, and ethnomycological characteristics of forest and desert truffle species (Kagan-Zur et al., 2014; Zambonelli et al., 2016; Thomas et al., 2019). Our knowledge of truffles has increased due to the comprehensive information gathered from these sources.

Tuber species are highly prized in gastronomy and culinary cultures for their distinctive flavor (Mello et al., 2006; Bonito and Smith, 2016; Romanazzi et al., 2015; Yan et al., 2017; Vahdatzadeh et al., 2018; Phong et al., 2022; Korkmaz and Duru, 2023; Korkmaz et al., 2023; Piatti et al., 2023). They are rare and expensive, making them a sought-after delicacy among gourmets and the wealthy (Pacioni et al., 2014; Payen et al., 2014; Choo et al., 2021). Certain segments of culinary culture, particularly in Europe, have a great interest in *Tuber* species (Zambonelli et al., 2015; Patel et al., 2017; Bontempo et al., 2020; Pacioni et al., 2014). Truffles are a type of edible mushroom that grows underground symbiotically with the roots of higher plants. Due to their rarity, unique aroma, high nutritional value, and content of various medicinal and therapeutic bioactive metabolites, they are considered one of the world's most expensive foods (Cullere et al., 2013a,b; Hilszczańska et al., 2016; Li et al., 2019; Farag et al., 2021; Segneanu et al., 2021).

Forest truffles have many health benefits, including antioxidant, antiinflammatory, antiviral, hepatoprotective, antimutagenic, antituberculosis, immunomodulating, antitumour, antimicrobial, and aphrodisiac properties. The flavor and aroma of truffles are among the main reasons for their worldwide popularity as a food (Khalifa et al., 2019; Lee et al., 2020; Mustafa et al., 2020; Leonardi et al., 2021; Veeraraghavan et al., 2022). However, desert truffles are also highly valued in culinary culture for their abundance, delicious taste, aroma, and use as a meat substitute for centuries (Sawaya et al., 1985; Murcia et al., 2003; Akyüz and Kırbağ, 2018, 2019; Behzadi et al., 2021; Ferreira et al., 2023). They are affordable and appreciated by all members of society, making them attractive to everyone. Desert truffles are related to European truffles, but their aroma and flavor are less intense than those of European truffles (Mustafa et al., 2020). The quality of desert truffles, which reflect the food cultures of the Mediterranean region, Mesopotamia, and the Middle East, may be comparable to that of other types of tubers. This chapter will discuss the medicinal and therapeutic properties of different true truffle species in the context of both scientific studies and traditional knowledge from rural and indigenous communities. This chapter covers the medicinal and therapeutic properties of commercially and economically valuable true truffles, including those found in forests and deserts.

2. Medicinal Properties of Truffles

Truffles are a natural and ancient food source with significant nutritional value and economic potential. Very little is known about the medicinal and therapeutic properties of truffles. This section of the chapter focuses on the therapeutic and medicinal uses of truffles in both indigenous and scientific contexts. This text provides an overview of the medicinal and therapeutic properties of various truffle species, including *Terfezia*, *Picoa*, *Tirmani*, and *Tuber*.

2.2 Antimicrobial Activities

Microbial pathogens cause diseases globally and pose a significant threat to human health. Currently, the most prevalent diseases in society are infectious diseases, including bloodstream infections, respiratory infections, urinary tract infections, bacterial food poisoning, acute gastroenteritis, joint infections, gonorrhea, bacterial cellulitis, syphilis, colitis, meningitis, chlamydia, vaginosis, and streptococcal infections (Pandey et al., 2021). There is a growing demand for natural products to reduce the side effects of drugs used in disease treatment. These products can be used alone, as supplements, or in combination with synthetic drugs. Edible mushrooms can be recommended as a natural remedy for treating diseases in the human body. Therefore, it would be beneficial to investigate the medicinal and therapeutic properties of truffle species that are consumed as food (Table 1).

Desert truffle species are commonly used by indigenous people in traditional contexts as natural medicine and food. They are often used to treat eye infections. Diluted truffle extract is traditionally used by the Bedouin for the relief and treatment of eye pain, especially during thunderstorms (Al-Marzooky, 1981). The documented use of desert truffle juice extracted from *Terfezia boudieri*, *T. claveryi*, and *Tirmania nivea* in the Middle East for treating eye and skin conditions has been noted (Patel, 2012; Shavit and Shavit, 2014). According to Darwish and colleagues (2022), desert truffles are utilized in folk medicine to address inflammatory conditions, particularly those affecting the eyes and skin. Additionally, Khalifa et al. (2019) observed that historically, truffles were recognized as a therapeutic agent for eye diseases such as trachoma. Mandeel and Al-Laith (2007) demonstrated that boiled water truffle extract, which Bedouins claim to be an effective remedy for trachoma, inhibits the growth of *Chlamydia trachomatis*. Desert truffles are easily digestible, and their juice is used to treat eye diseases.

Other scientific studies have focused on whether truffle species support local people's knowledge of their use and whether they have different antimicrobial effects (Table 1). In this regard, Hammami et al. (2023) found that water screening and quencher extracts of *Terfezia boudieri* and *Tirmania nivea* inhibited the growth of *S. aureus*, *A. hydrophila*, *P. aeruginosa*, *S. typhimurium*, *E. coli*, *L. monocytogenes*, and *B. cereus* to different degrees. Both truffle extracts exhibited significant antimicrobial activity against all tested strains compared to water extracts. The

Table 1. Antimicrobial properties of forest and desert truffles.

Truffle	Sample	Tested microorganism	Medicinal effect	Reference
Terfezia boudieri	Acetone, chloroform and methanol extracts	Various pathogenic microorganisms	Antibacterial and Antifungal	Doğan and Aydın (2013)
Terfezia boudieri	Acetone, aqueous, chloroform, and methanol extracts	HSV-2 and VSV virus	Antiviral	Mekawey (2015)
Terfezia boudieri and *Tirmania nivea*	Water screening and quencher extracts	*S. aureus*, *A. hydrophila*, *P. aeruginosa*, *S. typhimurium*, *E. coli*, *L. monocytogenes* and *B. cereus*	Antimicrobial	Hammami et al. (2023)
Terfezia claveryi	Truffle extract	*S. aureus*, *E. coli*, and *S. epidermidis*	Treating eye infections	Janakat et al. (2004, 2005)
Terfezia claveryi	Aqueous extract	corneal ulcers pathogens	Antibacterial	Aldebasi et al. (2013)
Terfezia claveryi	chloroform,dichloromethane, ethyl acetate and methanol extracts	Gr (+) and Gr (-) bacteria, and yeast	Antimicrobial	Neggaz et al. (2015)
Terfezia claveryi	Truffle juice	*E. coli*, *P. aeruginosa*, *B. subtilis*, and *S. aureus*	Trachoma	Alhussaini et al. (2016)
Terfezia claveryi	Aqueous extract	*E. coli*, *S. epidermidis*, and *S. aureus*	Treating ocular infections	Casarica et al. (2016)
Terfezia claveryi	Methanol and ethyl acetate extract	same microorganisms and *A. oryzae*	Antibacterial, antifungal and antibiofilm	Khojah et al. (2022)
Terfezia claveryi and *Terfezia boudieri*	Truffle extract	Clinical bacteria isolates associated with ocular infections	For eye infections	Badger-Emeka et al. (2021)
Terfezia claveryi and *Tirmania nivea*	Aqueous extracts	*S. aureus* and *P. aeruginosa*	Treating eye infections	Gouzi et al. (2011)
Terfezia claveryi and *Tirmania nivea*	Aqueous extracts	*S. aureus* and other pathogens	Antibacterial	Malik et al. (2018)
Terfezia claveryi and *Tirmania nivea*	Truffle extract	*B. subtilis*, *E. coli*	Antimicrobial	Elsayed et al. (2019)

Terfezia claveryi and *Tirmania pinoyi*	With acid-soluble protein extracts	*P. aeruginosa* and *S. aureus*	Antibacterial	Schillaci et al. (2017)
Terfezia claveryi and *Tirmania pinoyi*	Truffle extract	*Staphylococcus, Streptococcus, Bacillus, Enterococcus, Escherichia, Salmonella, Aspergillus, Fusarium, Gliocladium* and *Candida*	Antimicrobial	Dib-Bellahouel and Fortas (2019)
Terfezia and *Picoa* spp.	Methanol, acetone extracts	Bacterial and ophthalmic infections	Antimicrobial	Fidan et al. (2022)
Terfezia magnusii, Tuber aestivum, Tuber gennadii and *Tuber melanosporum*	Truffle extract	Various bacteria	Antimicrobial	Tejedor-Calvo et al. (2021)
Terfezia (3 spp.) and *Picoa* (2 spp.)	Methyl alcohol extract	Various pathogenic microorganisms	Antimicrobial	Akyüz et al. (2015)
Terfezia and *Tirmania* species	Truffles extract	Gr (+) and Gr (-) bacteria, filamentous fungi, and yeasts	Antimicrobial	Sallam et al. (2022)
Tirmania nivea	Methanolic, chloroform aqueous and petroleum ether extracts	Gr (+) and Gr (–) bacteria	Antimicrobial	Hamza et al. (2016b)
Tirmania nivea	Truffle extract	*S. aureus*, *S. pyogenes*, *K. pneumoniae*, *E. coli* and *S. typhi*	Antibacterial	Hachim et al. (2020)
Tirmania nivea	Extract with synthesize gold nanoparticles	*S. aureus*, *E. coli*, *P. aeruginosa*, and *K. pneumoniae*	Antibacterial	Obaid et al. (2023)
Terfezia pinoyi	Ethyl acetate extract	*B. subtilis* and *S. aureus*		Dib-Bellahouel and Fortas (2011)
Terfezia arenaria	Maceration with methanol and dichloromethane extracts	*A. niger, Penicillium sp., and C. albicans, S. aureus, E. faecalis, E. coli,* and *P. aeruginosa*	Antibacterial and antifungal	Harir et al. (2019)
Desert truffle	Diluted truffle extract	Eye and skin microorganisms	Treat eye and skin infections	Al-Marzooky (1981); Patel (2012); Shavit and Shavit (2014); Darwish et al. (2022)
Desert truffle	Boiled water truffle extract	*Chlamydia trachomatis*	Trachoma	Mandeel and Al-Laith (2007); Khalifa et al. (2019)

research also revealed that *T. boudieri* and *T. nivea* contain elevated levels of phenolic compounds, flavonoids, flavanols, tannins, and proanthocyanidins, which may confer antimicrobial properties. Obaid et al. (2023) conducted a cost-effective bio approach using an extract derived from *T. nivea* to synthesize gold nanoparticles (AuNPs). Their findings indicated that the AuNPs displayed superior antibacterial efficacy against *S. aureus* when compared to *E. coli*, *P. aeruginosa*, and *K. pneumoniae*. Khojah et al. (2022) highlighted the antibacterial, antifungal, and antibiofilm properties of extracts from *T. claveryi*.

Notably, the methanolic extract exhibited the highest efficacy, while the ethyl acetate extract demonstrated promising activity against the same microorganisms and *A. oryzae*. Additionally, the methanolic extract of *T. claveryi* showed the most potent activity against Gr (–) *B. subtilis*, as evidenced by its lower MIC and MBC values compared to other extracts. Fidan et al. (2022) determined that the methanol and acetone extracts of *T. boudieri*, *T. claveryi*, and *Terfezia olbiensis* exhibited higher antimicrobial activity compared to *Picoa juniperi* and *Picoa lefebvrei*. These findings suggest that *Terfezia* species might offer potential therapeutic benefits in treating bacterial diseases, including ophthalmic infections. Sallam et al. (2022) conducted an in vitro assessment to evaluate the antimicrobial efficacy of truffle extracts against a range of pathogens, encompassing Gr (+) and Gr (–) bacteria, filamentous fungi, and yeasts, using agar-well diffusion. Notably, the findings indicated that both *Terfezia* and *Tirmania* extracts demonstrated higher antibacterial activity compared to their antifungal activity. Tejedor-Calvo et al. (2021) observed that extracts from *Terfezia gennadii*, *T. magnusii*, *Tuber aestivum*, and *Tuber melanosporum* exhibited a mild inhibitory effect when tested against various bacteria. Badger-Emeka et al. (2021) observed that extracts from *T. claveryi* and *T. boudieri* inhibited the growth of clinical bacteria isolates associated with ocular infections, including strains resistant to conventional antibiotics. Their antimicrobial activity was confirmed through SEM and molecular docking analysis, justifying the potential use of these extracts as a herbal remedy for eye infections.

Hachim et al. (2020) showed in their research that phenol, carboxylic acid, and alkanes, functional groups found in *T. nivea*, may be responsible for the antimicrobial and cytotoxic activity. They found that *T. nivea* extract had an antimicrobial effect against *S. aureus*, *S. pyogenes*, *K. pneumonia*, *E. coli*, and *S. typhi*. Elsayed et al. (2019) found that *T. nivea* and *Terfezia claveryi* extracts exhibited significant antimicrobial effects against *B. subtilis*, but the activity was less pronounced against *E. coli*. Dib-Bellahouel and Fortas (2019) observed that *Tirmania pinoyi* and *T. claveryi* exhibited different levels of antimicrobial activity against different microorganisms such as *Staphylococcus*, *Streptococcus*, *Bacillus*, *Enterococcus*, *Escherichia*, *Salmonella*, *Aspergillus*, *Fusarium*, *Gliocladium*, and *Candida*. These effects were attributed to cyclic compounds such as pyrazines and quinones present in fungal structures. Harir et al. (2019) found that the dichloromethane extract of *Terfezia arenaria* exhibited significant antibacterial and antifungal effects. The maceration with methanol and dichloromethane extracts of *T. arenaria* showed antifungal activity against all tested strains, including *A. niger*, *Penicillium* sp., and *C. albicans*. The antibacterial activities of the dichloromethane extracts of *T. arenaria* against all tested bacteria, namely *S. aureus*, *E. faecalis*, *E. coli*, and *P. aeruginosa*, were significantly higher than those of the methanol extracts.

Malik et al. (2018) found that aqueous extracts from both *T. nivea* and *T. claveryi* demonstrated antibacterial activity against methicillin-resistant *S. aureus* (MRSA). Moreover, these truffle extracts exhibited a higher antibacterial effect against MRSA compared to ciprofloxacin, azithromycin, ampicillin, and oxacillin. Notably, the extract derived from *T. claveryi* displayed superior and more potent antibacterial effects compared to the extract from *T. nivea*. Schillaci et al. (2017) demonstrated that *T. pinoyi* exhibits potent antibacterial activity against both *P. aeruginosa* and *S. aureus*, with acid-soluble protein extracts from both *T. pinoyi* and *Terfezia claveryi* showing a minimum inhibitory concentration of 50 μg/mL against the tested pathogens. Alhussaini et al. (2016) demonstrated the efficacy of *T. claveryi* against trachoma in its third phase. However, in vitro analysis revealed limited or no effect of truffle juice against *E. coli, P. aeruginosa*, *B. subtilis*, and *S. aureus*. However, patients treated with cortizone compounds in the form of drops along with truffle juice drops experienced a clear reduction in symptoms. This research suggests that incorporating truffle juice into traditional trachoma treatment at all stages could be beneficial. Casarica et al. (2016) found that an aqueous extract of *T. claveryi* possesses antibacterial properties against *E. coli*, *S. epidermidis*, and *S. aureus*. Therefore, it is suggested that this extract could be utilized for treating ocular infections caused by these bacterial species.

Hamza et al. (2016b) reported that extracts of *T. nivea* displayed varying degrees of antimicrobial activity against a broad spectrum of Gr (+) and Gr (-) bacteria. The methanolic extract emerged as the most effective among those tested, while the chloroform extract exhibited greater potential antimicrobial activity compared to the aqueous and petroleum ether extracts. This study suggests that extracts from this desert truffle could be utilized to preserve the microbiological quality of processed food products. Akyüz et al. (2015) found that methyl alcohol extracted from three *Terfezia* and two *Picoa* species inhibited the growth of various pathogenic microorganisms. The study also revealed that *Terfezia* species were more effective than *Picoa* species against the tested microorganisms. Neggaz et al. (2015) reported that chloroform, dichloromethane, ethyl acetate, and methanol extracts of *T. claveryi* exhibited superior antimicrobial activity against Gr (+) and Gr (–) bacteria, and yeast compared to the maceration extract. Particularly noteworthy was the chloroform extract of *T. claveryi*, which demonstrated the highest antimicrobial activity, with minimum inhibitory concentration values ranging from 12.5 to 100 mg/ml. Mekawey (2015) reported that acetone, aqueous, chloroform, and methanol extracts of *T. boudieri* exhibited moderate inhibitory activity against HSV-2 and VSV viruses when compared to the acyclovir standard.

Doğan and Aydın (2013) investigated the antimicrobial potential of acetone, chloroform, and methanol extracts of *T. boudieri* against various pathogenic microorganisms. Their findings revealed that the acetone extract demonstrated the most potent inhibitory effect against all tested microorganisms, especially *C. albicans*. Additionally, both the acetone and chloroform extracts of *T. boudieri* exhibited greater antibacterial and antifungal activity compared to the methanol extracts. Aldebasi et al. (2013) reported that *T. claveryi* exhibited a potent antibacterial effect against all clinical isolates of corneal ulcers tested, showing the highest efficacy against *P. aeruginosa*. Moreover, the study found that the aqueous extract of *T. claveryi* demonstrated bacteriostatic effects at lower concentrations and bactericidal activity at higher concentrations, with the MIC value being lower than the MBC value. Dib-Bellahouel and Fortas (2011) noted that the ethyl acetate extract derived from *T. pinoyi* demonstrated inhibitory effects on the growth of *B. subtilis* and *S. aureus*. However, it did not exhibit inhibitory effects on *P. aeruginosa*, *Enterococcus* sp., or *E. coli*. The observed antibacterial properties of the extract were attributed to the presence of various compounds, notably pyrazines. Gouzi et al. (2011) detected that aqueous extracts from *Terfezia claveryi* and *Tirmania nivea* displayed robust antibacterial effects against *S. aureus* and *P. aeruginosa*, as demonstrated through agar well diffusion assays. Conversely, no antibacterial activity was observed in the aqueous extract of *Terfezia leonis*. These findings suggest that *Terfezia claveryi* and *Tirmania nivea* could potentially serve as natural therapeutic agents for treating eye infections caused by resistant bacteria, such as *S. aureus* and *P. aeruginosa*. Janakat et al. (2004, 2005) observed that the extract of *T. claveryi* exhibited inhibitory action against *S. aureus*, *E. coli*, and *S. epidermidis*. The latter is a common cause of eye infections. As with other medicinal effects, the bioactive compounds in different truffle species appear to inhibit the growth of different pathogens to varying degrees. The aforementioned research has shown that these natural truffles products, when used as food, can inhibit the growth of various pathogenic microorganisms.

2.3 Antioxidant, Antiinflammatory, Antitumor and Cytotoxic Effects

Oxidative stress, which has been linked to severe neurological problems, aging, cirrhosis, diabetes, atherosclerosis, and cancer, can be caused by free radicals (Askarova et al., 2020; Mwangi et al., 2022). They disrupt homeostasis and metabolic processes in cells and tissues (Kehrer and Klotz, 2015). The substances can be produced by both internal and external sources, such as ultraviolet light, ionizing radiation, environmental toxins, haemotherapeutic agents and proinflammatory cytokines (Widel et al., 2014). In view of this data, it is important to highlight the medicinal effects of seasonal truffle species, such as their antioxidant, inflammatory, and cytotoxic properties (Tables 2, 3).

2.3.1 Antioxidant Activities

Piatti and colleagues (2023) detected that *T. magnatum* contains varying amounts of total phenolics and flavonoids, as well as volatile and non-volatile profiles. Their studies revealed that *T. magnatum* extract is rich in total phenolic and flavonoid content, which are effective in DPPH radical scavenging ability (513–574 mg TE/kg dw). Additionally, this truffle exhibited a good antioxidant effect, making it a desirable choice for consumers due to its positive impact on human health. Hammami et al. (2023) discovered that both the aqueous extract and ground material of *T. boudieri* exhibited the highest capacity for DPPH scavenging (IC50 = 0.18 mg/ml) and chelating iron (IC50 = 0.22 mg/ml). Furthermore, *T. nivea* raw material exhibited the highest iron reduction capacity. The studies demonstrated that *T. boudieri* and *T. nivea* are rich in phenolic compounds and possess antioxidant activity, which can

Table 2. Antioxidant and antiinflammatory effects of truffles (–, not defined).

Truffle	Bioactive component	Analytical method	Medicinal effect	Reference
Terfezia boudieri, *Tirmania nivea*	Phenolic compounds	DPPH scavenging and chelating iron capacity	Antioxidant	Hammami et al. (2023)
Terfezia boudieri and*Terfezia claveryi*	Aqueous extracts	DPPH radical scavenging activity	Antioxidant	Sonji et al. (2022)
Picoa juniperii and *Terfezia claveryi*	Bioactive compounds	Lipid peroxidation (LOOx), deoxyribose (OHx), and peroxidase (H_2O_2) based assays	Antioxidant	Murcia et al. (2002)
Tuber magnatum	Phenolics, flavonoids volatile and non-volatile profiles	DPPH radical scavenging ability	Antioxidant	Piatti et al. (2023)
Tuber indicum	Phenolic and flavonoid content	ABTS, DPPH scavenging activity	Antioxidant	Li et al. (2022)
three *Terfezia* and two *Picoa*	Bioactive compounds	Free radical-scavenging, reducing capacities (FRAP, DPPH) and ion chelating activity	Antioxidant	Fidan et al. (2022)
Tirmania sp., *Terfezia* sp.	Truffle extract	Free radical (DPPH) scavenging activity	Antioxidant	Sallam et al. (2022)
Tuber indicum	Pectin, β-glucan, vitamin D2, ergosterol, phenolics, and flavonoids	DPPH, TPC, TFC, ABTS, and FRAP assays	Antioxidant and anti-inflammatory	Wu et al. (2021)
Tuber magnatum	Total phenolic compounds	Reducing power, DPPH, FRAP, and ORAC	Antioxidant	Tejedor-Calvo et al. (2021)
Tuber aestivum	Total phenolics, flavonoid, tannin and aroma profile	DPPH, BHA and ascorbic acid equivalents of ferric reducing power	Antioxidant	Shah et al. (2020)
Terfezia claveryi, *Picoa lefebvrei*, *Terfezia* sp.	Unsaturated fatty acids and phenolics	Free radical-scavenging activity (DPPH)	Antioxidant	Şahin et al. (2020)
Terfezia boudieri	Vitamins, sitosterol, sugars, flavonoids, phenol, and unsaturated fatty acids	ABTS, DPPH, MDA	Antioxidant	Akyüz et al. (2019)

Table 2. contd. ...

Table 2. contd. ...

Truffle	Bioactive component	Analytical method	Medicinal effect	Reference
Tuber indicum	Fatty acids, organic acids, free sugars, phenolic acids, flavonoids, polysaccharides	Radical-scavenging activity (DPPH, ABTS)	Antioxidant	Li et al. (2019)
Terfezia claveryi and *Terfezia boudieri*	Bioactive compounds		Antioxidant and cytotoxic	Khalifa et al. (2019)
Tuber pseudohimalayense, Tuber latisporum, Tuber subglobosum	Lycopene, flavonoids, β-carotene, phenols, and ascorbic acid	DPPH scavenging, ferricyanide/prussian blue, hydroxyl radical-scavenging, and ferrous ion-chelating	Antioxidant	Yan et al. (2017)
Terfezia claveryi	Phytosterols, triterpenes and vitamin	DPPH, FRAP, and ABTS assays	Antioxidant, anti-angiogenic and anticancer	Dahham et al. (2018)
Terfezia boudieri	Phenolics, flavonoids, ascorbic acid, carotenoids, and total anthocyanins	DPPH radical scavenging and lipid peroxidation inhibition	Antioxidant	Hamza et al. (2016a)
Tuber huidongense	Polysaccharides purified from	DPPH radical scavenging capacity	Antioxidant	Chen et al. (2016)
Terfezia	Ethyl acetate and hexane extracts	Radical-scavenging effect	Lipid peroxidation, free radicals, and moderate anticholinesterase	Kıvrak (2015)
Terfezia claveryi	Methanol, chloroform and ethyl acetate extracts	DPPH radicals	Antioxidant	Neggaz et al. (2015)
Terfezia, *Picoa* species	Methanol extract	DPPH radical	Antioxidant	Akyüz et al. (2015)
Tuber aestivum, *Tuber magnatum*	Phenols, tannins, and flavonoids	DPPH, superoxide anion, nitric oxide, lipid peroxides, and by reducing ferric ions	Antioxidant	Beara et al. (2014)
Terfezia boudieri	Myricetin, kaempferol, naringin, naringenin, resveratrol, and unsaturated fatty acids	DPPH	Antioxidant	Akyüz (2013)

Terfezia boudieri	-	Total antioxidant, reducing power, superoxide anion radical scavenging, DPPH scavenging, and metal chelating	Antioxidant, ferrous ion-chelating activity	Dündar et al. (2012)
Tuber aestivum, *Tuber melanosporum*, *Tuber indicum*	Ergosterol and phenolic contents	Oxidation inhibition	Antioxidant	Villares et al. (2012)
Tuber indicum	Phenolics and flavonoids	DPPH free radical, superoxide radical-scavenging, ferrous ion-chelating, hydroxyl radical scavenging activity and reducing power	Antioxidant	Guo et al. (2011)
Tuber indicum	Polysaccharide fractions TIP1-1 and TIP2-1	Free radical scavenging activity, H_2O_2-injured	Antioxidant	Luo et al. (2011)
Tirmania nivea	Polyphenols and ergosterol	DPPH, ferric reducing ability (FRAP), deoxyribose, and nitric oxide (NO)	Antioxidant and antiradical	Al-Laith (2010)
Terfezia claveryi	Polyphenols, flavonoids, and condensed tannins	Mycochemical test	Anti-inflammatory	Souna et al. (2023)
Terfezia boudieri	Flavonoids, steroids, saponins, tannins, β-sitosterol, gallic acid, and carbohydrates	In vitro	Anti-inflammatory	Alshawush et al. (2022)
Terfezia claveryi, *Tirmania nivea*	Phenolic acids, saturated and unsaturated fatty acids, alkanes, and carboxylic acids	In vitro	Anti-inflammatory	Darwish et al. (2021)
Tuber indicum	Pectin, β-glucan, vitamin D2, ergosterol, phenols, and flavonoid	Protease inhibitory rate	Anti-inflammatory	Wu et al. (2021)
Tuber aestivum, *Tuber magnatum*	Bioactive compounds	Ex vivo conditions by evaluating COX-1 and 12-LOX inhibition	Anti-inflammatory	Beara et al. (2014)

Table 3. Anticancer, antitumor and cytotoxic properties of truffles (–, not defined).

Truffle	Sample	Bioactive compound	Tested cancer cell line	Medicinal effect	Reference
Picoa and *Terfezia* species	Methanol and water extract	-	H1299 and HUVEC cells	Cytotoxic and anticancer	Akyüz et al. (2021)
Terfezia arenaria and *Terfezia gennadii*	Truffle extract	Polysaccharide fractions	NCI-H460, HepG2, HeLa, and MCF-7 cells	Antitumor, anti/proliferative, hepatotoxicity and human tumor cells	Tejedor-Calvo et al. (2021)
Terfezia boudieri	Various extracts (chloroform, acetone, and methanol)	-	HCT116, HEPG2, HEP2, MCF7, CACO2 and HELA cell lines	antitumor	Mekawey (2015)
Terfezia boudieri	N-hexane, ethyl acetate, and aqueous/methanol extracts	-	Lymphocyte proliferation	Apoptosis, angiogenesis and anticancer	Al-Obaydi et al. (2020)
Terfezia boudieri	Aqueous and ethanolic extracts	-	Colon cancer cells, HCT-116 and Caco-2	Cytotoxic and anticancer	Sawaya et al. (2023)
Terfezia boudieri *Picoa juniperi* and *Picoa lefebvrei*	-	-	HepG2 and MCF-7 cell	Cytotoxic and anticancer	Fidan et al. (2022)
Terfezia boudieri *Terfezia claveryi* and *Tirmania nivea*	-	Phenolic compounds, tocopherols, ascorbic acid, and oxidative enzymes	-	Aging process, diabetes, cardiovascular disease, and anticancer	Veeraraghavan et al. (2022)
Terfezia claveryi	Aqueous extract with silver nanoparticles	Amino acids and flavonoids	MCF-7	Cytotoxic	Khadri et al. (2017)
Terfezia claveryi	-	Polysaccharides	Ehrlich ascites carcinoma	Apoptotic, anti-carcinogenic and anticancer	Atila et al. (2018)

Terfezia claveryi	Solvent extracts	Stigmasterol, β-sitosterol, squalene, and lupeol	PC3, HT29, U-87 MG, and MCF-7	Anticancer	Dahham et al. (2018)
Terfezia claveryi and *Tirmania nivea*	Methanolic, water, hexane and ethyl acetate extracts	-	MCF-7 cancer cells	Anti-breast cancer activity	El-Sayed et al. (2019)
Tirmania nivea		Phenolic, carboxylic acid and alkane functional groups	Hep-2 cell line	Anticancer	Hachim et al. (2020)
Tirmania nivea	Synthesize gold nanoparticles (AuNPs	-	Prostate cancer cell line	Anticancer	Obaid et al. (2023)
Tuber aestivum, *Tuber indicum*, *Tuber melanosporum* and *Tuber sinense*	-	Several polysaccharides	A549, HepG2, SK-BR-3, HCT-116 and HL-60 cells	Antitumor effect	Zhao et al. (2014)
Tuber aestivum and *Tuber magnatum*	Aqueous extracts	-	MCF7, HeLa, and HT-29	Cytotoxic, and anticancer	Beara et al. (2014)
Tuber melanosporum	-	Metabolic enzymes	-	Anticancer	Pacioni et al. (2015)
Tuber rufum	Water-soluble heteroglycan	Polysaccharide	Human blood lymphocytes	Cytotoxic, and ameliorative effects	Pattanayak et al. (2017)

help protect the organism against oxidation. Li et al. (2022) observed that *T. indicum* collected in July exhibited the highest flavonoid content and ABTS scavenging activity, whereas samples collected in August showed the highest phenolic content, DPPH radical scavenging activity, and ferric reducing power. The study highlighted that *T. indicum* harvested at an early stage of maturity displayed significantly elevated phenolic and flavonoid contents, along with stronger antioxidant activity, compared to samples collected during other months.

Fidan et al. (2022) discovered that the methanol and acetone extracts of three *Terfezia* and two *Picoa* truffles contained bioactive compounds exhibiting significant antioxidant activity, as demonstrated through free radical scavenging and reducing capacities. Among the tested truffles, *T. boudieri* exhibited the highest FRAP and DPPH scavenging activities, as well as ion chelating activities. Sonji et al. (2022) reported that the aqueous extracts of *T. claveryi* and *T. boudieri* demonstrated nearly equal DPPH radical scavenging activity, with values of 0.69 mg/mL and 0.68 mg/mL, respectively. Sallam et al. (2022) discovered significant free radical (DPPH) scavenging activity in both *Tirmania* sp. and *Terfezia* sp. extracts. However, the extracts from *Terfezia* sp. exhibited greater potency compared to those from *Tirmania* sp.

Murcia et al. (2002) conducted a study comparing the antioxidant activity of *T. claveryi* and *Picoa juniperi* with common food antioxidants like BHA, α-tocopherol, BHT, and propyl gallate. Their analysis, utilizing lipid peroxidation (LOOx), deoxyribose (OHx), and peroxidase (H_2O_2)-based assays, revealed that truffles exhibited higher inhibition of oxidation compared to the food antioxidants tested. The findings suggested that *T. claveryi* demonstrates superior antioxidant activity compared to *P. juniperi*. Wu et al. (2021) reported the differential nutritional value, antioxidant, and anti-inflammatory effects of *T. indicum* from three regions in China. The study assessed antioxidant activity through DPPH, TPC, TFC, ABTS, and FRAP assays. The observed anti-inflammatory effects were attributed to functional groups present in *T. indicum*, such as pectin, β-glucan, vitamin D2, ergosterol, phenolics, and flavonoids. *T. indicum* samples from Yunnan Province exhibited the highest values for most parameters tested, including fat, pectin, energy, total phenolic and flavonoid content, ABTS and DPPH free radical scavenging capacity, and potent anti-inflammatory effects.

Tejedor-Calvo et al. (2021) determined that *Tuber magnatum* exhibited the highest levels of total phenolic compounds and surpassed other species in all four methods utilized to assess antioxidant activity (reducing power, DPPH, FRAP, and ORAC). Shah et al. (2020) discovered that water and methanol extracts of freeze-dried mature truffles contained higher levels of total phenolics and exhibited greater in vitro antioxidant activity compared to immature truffle extracts, while also having a lower tannin content. The study suggests that during *Tuber aestivum* maturation, the bioactive content, including lipids (fatty acids), phenolics, flavonoids, and tannins, as well as the aroma profile, increase. Methanol extracts of mature *T. aestivum* were found to effectively scavenge DPPH radicals and demonstrated higher BHA and ascorbic acid equivalents of ferric reducing power. Additionally, the iron-reducing capacity of immature *T. aestivum* extracts was superior, likely due to their higher flavonoid content. Şahin et al. (2020) observed that the species

T. claveryi, *Picoa lefebvrei*, and *Terfezia* sp. (probably *T. olbiensis*) exhibit different levels of unsaturated fatty acids and phenolics. Furthermore, the free radical scavenging activity (DPPH) of these species ranged from 73–79 μg/ml.

Akyüz et al. (2019) reported that *T. boudieri* is a highly nutritious food, rich in vitamins such as K1, K2, D1, D2, α-tocopherol, retinol, and retinolast, as well as sitosterol (ergosterol, stigmasterol, and β-sitosterol), sugars (glucose, sucrose, and arabinose), total flavonoids, total phenol, protein, and unsaturated fatty acids. The antioxidant activity of *T. boudieri* was found to be between 69–97% for ABTS, 42–86% for DPPH, and 7 nmol/mL for MDA. It has been demonstrated that *T. boudieri* can be a valuable source of antioxidants, helping to enhance an organism's overall antioxidant capacity and providing protection against lipid peroxidation. Li et al. (2019) demonstrated that the functional groups present in *T. indicum*, such as fatty acids, organic acids, free sugars, and bioactive compounds including phenolic acids, flavonoids, and polysaccharides, may contribute to its antioxidant effect. They also identified a significant correlation between the phenolic content and the radical scavenging activity (DPPH, ABTS) of *T. indicum* methanolic extract. Khalifa and colleagues (2019) confirmed the antioxidant activity of *T. claveryi* and *T. boudieri*. Additionally, *T. claveryi* exhibited strong cytotoxic activity.

Yan et al. (2017) documented that *Tuber pseudohimalayense*, *T. latisporum*, and *T. subglobosum* are rich in nutritional components such as unsaturated fatty acids, carbohydrates, sugars, and proteins. Moreover, these truffles contain various bioactive compounds, including lycopene, flavonoids, β-carotene, phenols, and ascorbic acid. They also demonstrated antioxidant effects through assays such as DPPH scavenging, ferricyanide/prussian blue, hydroxyl radical-scavenging, and ferrous ion-chelating. The methanol extract of *Tuber pseudohimalayense* displayed notable radical scavenging capacity and a high concentration of total phenols, while *T. subglobosum* exhibited the highest flavonoid content. These findings suggest the potential of these extracts as antioxidants for preventing diseases associated with oxidative damage. Dahham et al. (2018) noted that solvent extracts of *T. claveryi* showcased antioxidant properties, as evidenced by DPPH, FRAP, and ABTS assays. The analysis of *T. claveryi* unveiled the presence of phytosterols, triterpenes, and vitamins, which exhibit bioactive properties such as antioxidant, anti-angiogenic, and anticancer effects.

Hamza et al. (2016a) reported that *T. boudieri* contains varying amounts of phenolics, flavonoids, ascorbic acid, carotenoids, and total anthocyanins. The methanolic extract of *T. boudieri*, which contained the highest phenolic and flavonoid content, exhibited the highest DPPH radical scavenging and lipid peroxidation inhibition. Chen and colleagues (2016) assessed the DPPH radical scavenging capacity of water-soluble polysaccharides purified from *Tuber huidongense*. Kıvrak (2015) found that the ethyl acetate extract of *Terfezia* demonstrated the highest radical scavenging effect, while the hexane extracts exhibited the greatest lipid peroxidation inhibitory activity. The extracts from *Terfezia* species have the potential to serve as sources of protection against lipid peroxidation, free radicals, and moderate anticholinesterase agents in pharmaceuticals. Furthermore, they could be used as food additives to extend the shelf life of food products and in the dietary supplement industry. Neggaz et al. (2015) found that the methanolic extract of

T. claveryi displayed the most potent radical-scavenging activity against DPPH radicals (IC50 = 8.56 mg/mL). However, the chloroform and ethyl acetate extracts showed no antioxidant effect within the concentration range of 5–40 mg/ml.

Akyüz et al. (2015) found that the efficiency of *Terfezia* and *Picoa* species in removing the DPPH radical increased with the addition of 25–100 μl samples. The group that received a 100-μl sample exhibited the highest level of free radical scavenging effect, with *T. boudieri* showing 8.00% of the overall effect. Beara et al. (2014) found that extracts derived from both *Tuber aestivum* and *T. magnatum* exhibited in vitro antioxidant activity by scavenging various radicals, including DPPH, superoxide anion, nitric oxide, lipid peroxides, and reducing ferric ions. This antioxidative capacity was attributed to the presence of phenols, tannins, and flavonoids in the extracts. Akyüz (2013) suggested that *T. boudieri* contains high levels of myricetin, kaempferol, naringin, naringenin, resveratrol, and unsaturated fatty acids. The free radical scavenging activity (DPPH) of *T. boudieri* was found to be 22.24%. Dündar et al. (2012) observed that ethanolic extracts from *T. boudieri* exhibited notable antioxidant activity across multiple antioxidant systems, including total antioxidant, reducing power, superoxide anion radical scavenging, DPPH scavenging, and metal chelating, when compared to the standards. *T. boudieri* displayed the highest activity among the standards in terms of both total antioxidant activity and ferrous ion chelating activity.

Villares et al. (2012) detected various bioactive organic compounds, such as ergosterol and phenolic contents, in *Tuber aestivum*, *T. melanosporum*, and *T. indicum*. All truffles displayed lipid oxidation inhibition, with *T. aestivum* demonstrating the highest effectiveness (24%), followed by *T. indicum* (23%), while *T. melanosporum* exhibited the lowest antioxidant activity (18%). Guo et al. (2011) demonstrated the antioxidant properties of the extract from *T. indicum* and its four different solvent sub-fractions through various assays, including DPPH free radical, superoxide radical-scavenging, ferrous ion-chelating, hydroxyl radical-scavenging activity, and reducing power. The n-butyl alcohol fraction and the ethyl acetate fraction exhibited the most potent antioxidant properties among the four fractions. The study also established a correlation between the levels of phenolics and flavonoids and their efficacy in scavenging radicals and chelating ferrous ions. Luo et al. (2011) discovered that polysaccharide fractions TIP1-1 and TIP2-1 possess notable antioxidant activity, as determined by antioxidant activity analyses. TIP2-1, characterized by its inclusion of uronic acid and lower molecular weight compared to TIP1-1, exhibited superior free radical scavenging activity and demonstrated a protective effect on H_2O_2-injured PC12 cells. The study underscores the importance of lower molecular weight and the presence of uronic acid in TIP2-1, contributing to its enhanced antioxidant activity. Consequently, *T. indicum*, the Chinese truffle, emerges as a valuable source of natural antioxidants and a promising candidate for antioxidant-rich health food products.

Al-Laith (2010) employed four methods—DPPH, ferric reducing ability (FRAP), deoxyribose, and nitric oxide (NO)—to assess the antioxidant and antiradical activities of *T. nivea* from Bahrainian, Iranian, Moroccan, and Saudi Arabian origins. The study revealed that desert truffles are a promising source of natural antioxidants attributed to their high levels of polyphenols and ergosterol, which confer potent antioxidant and antiradical properties. Among the four samples tested, Iranian truffles displayed the highest values across several variables, whereas Moroccan truffles exhibited the lowest values. Significant correlations were observed between total and free phenolics and FRAP levels, as well as between flavonoids and the percentage of radical inhibition in DPPH, NO, and deoxyribose assays. Truffles have been extensively researched for their antioxidant properties and the presence of bioactive compounds that may offer health benefits to humans (Table 2). These fungi reportedly demonstrate in vitro antioxidant activity owing to the abundance of various compounds, including phenols, flavonoids, tannins, polysaccharides, unsaturated fatty acids, phytosterols, triterpenes, sugars, and vitamins such as ascorbic acid, β-carotene, lycopene, K1, K2, D1, D2, α-tocopherol, retinol, and retinolast. Truffles have the capability to scavenge radicals like DPPH, superoxide anion, lipid peroxides, and nitric oxide. Additionally, they can chelate ions and reduce iron. These diverse properties underscore the significance of truffles for both nutritional and therapeutic purposes. Tejedor-Calvo et al. (2021) propose that establishing a direct correlation between various bioactivities, including antioxidant, antimicrobial, and anti-proliferative activities, and specific bioactive molecules is a challenging task. Nevertheless, numerous bioactive compounds present in mushroom structures are believed to possess medicinal and therapeutic effects.

2.3.2 Anti-Inflammatory Activities

Although the exact effects of many bioactive components found in truffle species are not yet known, numerous studies have suggested that they possess significant therapeutic and anti-inflammatory properties (Table 2). Souna et al. (2023) investigated the mycochemical tests of *T. claveryi* and found a rich presence of secondary metabolites. The hydro-methanol extract exhibited high levels of polyphenols, flavonoids, and condensed tannins. At a concentration of 1.5 mg/mL, the extract inhibited denaturation by 83.53%, compared to diclofenac sodium (98.43%). This research confirms the anti-inflammatory properties of *T. claveryi*, supporting its traditional use. Alshawush et al. (2022) reported the presence of flavonoids, steroids, saponins, tannins, β-sitosterol, gallic acid, and carbohydrates in varying amounts in *T. boudieri*. The study confirms the traditional use of *T. boudieri* extract as a natural anti-inflammatory agent, supported by its significant antioxidant effect and the promising anti-lipoxygenase properties of both TME and pure isolated gallic acid. Darwish et al. (2021) found that *T. claveryi* and *T. nivea* demonstrated anti-inflammatory activity in vitro, which was attributed to their composition rich in a combination of phenolic acids, saturated and unsaturated fatty acids, alkanes, and carboxylic acids. Wu et al. (2021) conducted a protease inhibitory analysis to investigate the anti-inflammatory properties of *T. indicum*. Their results unveiled a direct correlation between concentration and protease inhibitory rate (IC50), signifying a dose-dependent protease inhibitory effect in *T. indicum* samples, with the most pronounced effects observed in those sourced from Yunnan. Further analysis revealed elevated levels of pectin, β-glucan, vitamin D2, ergosterol, phenols, and flavonoids in *T. indicum*, suggesting that these constituents may substantially contribute to its potent anti-inflammatory effects. Additionally, Beara et al. (2014) investigated the anti-inflammatory response of *Tuber aestivum* and *T. magnatum* in ex vivo conditions by evaluating COX-1 and 12-LOX inhibition. They observed that the methanol extract of *T. magnatum* demonstrated a significant anti-inflammatory effect (IC50 = 2.18 mg/ml) in comparison to standard compounds such as acetylsalicylic acid and quercetin. The consumption of these species, which contain bioactive components identified by the aforementioned researchers in truffles, provides both nutritional and therapeutic benefits as a natural food source.

2.3.3 Anti-Tumor and Cytotoxic Activities

Cancer is a highly prevalent disease in modern society. To reduce the risk of developing tumors, it is important to consume natural food sources that contain significant bioactive components. Forest and desert truffles are excellent examples of such sources. Sawaya et al. (2023) discovered that both the aqueous and ethanolic extracts of *T. boudieri* exhibited similar cytotoxic effects on colon cancer cells, HCT-116 and Caco-2. Treatment with the aqueous extract of *T. boudieri* resulted in a dose-dependent reduction in the cells' ability for wound healing and colony formation, suggesting its cytotoxicity and potential to inhibit proliferation, consequently hindering colony formation and wound healing. Obaid et al. (2023) conducted a cost-effective bio approach using an extract derived from *T. nivea* to synthesize gold nanoparticles (AuNPs). The cytotoxicity test showed toxicity towards the prostate cancer cell line (74.6%). The study suggests that gold nanoparticles have potential for prostate cancer therapy as a novel treatment. Veeraraghavan et al. (2022) conducted a review of the antioxidant activity of *Terfezia boudieri, T. claveryi*, and *Tirmania nivea*, emphasizing the role of phenolic compounds, tocopherols, ascorbic acid, and oxidative enzymes in preventing damage caused by free radicals and inhibiting the initiation, progression, and spread of cancer cells, thus suggesting that the consumption of naturally occurring antioxidants may offer protection against the aging process, diabetes, cardiovascular disease, and cancer.

Fidan et al. (2022) discovered that *T. boudieri* exhibits significant protective effects against DNA damage and cytotoxicity in both HepG2 and MCF-7 cell lines, indicating the potential of *Terfezia* species as protective agents. Additionally, extracts from *P. lefebvrei* and *P. juniperi* were found to have the least cytotoxic effects. Akyüz et al. (2021) noted that H1299 cells displayed heightened sensitivity to the water extract of *T. olbiensis* (4.66%), *P. juniperi* (12.04%), *P. lefebvrei* (21.93%), and the methanol extract of *T. boudieri* (20.93%). Conversely, extracts from *Terfezia* and *Picoa* species had the least cytotoxic effects on HUVEC cells, with the exception of the water extract of *P. juniperi* (38.46%). Tejedor-Calvo et al. (2021) found that extracts from *Terfezia* and *T. gennadii* inhibited growth in NCI-H460, HepG2, HeLa, and MCF-7 cells, displaying anti-proliferative activity, while *T. arenaria* inhibited PLP2 cell growth, suggesting potential hepatotoxicity. Overall, *Terfezia* species exhibited superior anti/proliferative effects compared to *Tuber* species across human tumor cells, leading to the conclusion that polysaccharide fractions could enhance antitumor activity. Hachim et al. (2020) noted that *T. nivea* contains phenolic, carboxylic acid, and alkane functional groups. The authors found that the crude organic compound of *T. nivea* was toxic to the Hep-2 cell line. Further research is needed to purify and identify *T. nivea*'s bioactive compounds and explore their potential pharmaceutical effects.

Al-Obaydi et al. (2020) investigated the effects of different solvent extracts of *T. boudieri* on cancer cell viability. They discovered that the N-hexane extract showed the highest potency against most cell lines, while the aqueous/methanol extract triggered apoptosis and inhibited angiogenesis. Additionally, the ethyl acetate extract promoted lymphocyte proliferation. These extracts, including N-hexane, ethyl acetate, and aqueous/methanol, demonstrated anticancer properties and boosted both innate and acquired immunity. Studies propose their potential for inhibiting angiogenesis and inducing apoptosis. El-Sayed et al. (2019) found that the hexane and ethyl acetate extracts of *Terfezia claveryi* and *Tirmania nivea* exhibited a dose-dependent effect against MCF-7 cancer cells. These extracts showed the highest efficacy in terms of anti breast cancer activity, while the methanolic, 70% methanolic, and water extracts demonstrated less effectiveness against the MCF-7 cancer cell line. Dahham et al. (2018) found that solvent extracts of *T. claveryi* demonstrated cytotoxic effects on various cancer cell lines, including PC3, HT29, U-87 MG, and MCF-7. The truffle's anticancer properties are attributed to bioactive compounds such as stigmasterol, β-sitosterol, squalene, and lupeol. The study suggests that the anticancer effect of *T. claveryi* may arise from the presence of these primary chemical constituents and the synergistic effects among them.

According to Attia et al. (2018), polysaccharides extracted from *T. claveryi* demonstrate potential anticancer properties against the Ehrlich ascites carcinoma (EAC) cell line, exhibiting an apoptotic effect in a dose- and time-dependent manner. Pattanayak et al. (2017) investigated the cytotoxic effects of a water-soluble heteroglycan from *T. rufum* on human blood lymphocytes. Their findings suggested that this polysaccharide might serve as an immunostimulant, as it exhibited ameliorative effects on maintaining redox balance. Khadri et al. (2017) synthesized silver nanoparticles using the aqueous extract of *T. claveryi*, renowned for its abundance of amino acids and flavonoids. These nanoparticles were assessed for their efficacy against MCF-7, revealing substantial in vitro cytotoxic effects as assessed by the colorimetric sulforhodamine B assay. Mekawey (2015) conducted a comprehensive study to evaluate the efficacy of various *T. boudieri* extracts against HCT116, HEPG2, HEP2, MCF7, CACO2, and HELA cell lines. The findings revealed that the aqueous extract demonstrated significant antitumor activity against CACO2, MCF7, HEP2, and HCT116 cells. Conversely, the chloroform extract exhibited a moderate effect, while the acetone and methanol extracts showed the least effectiveness.

Pacioni et al. (2015) identified various metabolic enzymes in *T. melanosporum*, such as N-acylphosphatidylethanolamine-specific phospholipase D, fatty acid amide hydrolase, monoacylglycerol lipase, and diacylglycerol lipase. Additionally, they discovered anandamide, a fatty acid neurotransmitter derived from arachidonic acid, which exhibits anticancer properties. Beara et al. (2014) explored the in vitro cytotoxic effects of *Tuber magnatum* and *T. aestivum* on various tumor cell lines, including MCF7, HeLa, and HT-29. Their findings highlighted notable effects of the aqueous extracts, particularly on MCF-7, with IC50 values ranging from 2.3 to 34. Zhao et al. (2014) isolated several polysaccharides from the fermentation systems of *Tuber indicum*, *T. sinense*, *T. melanosporum*, and *T. aestivum*. They detected that 12 of these polysaccharides displayed a significant in vitro antitumor effect against A549, HepG2, SK-BR-3, HCT-116, and HL-60 cells compared to those from *Tuber*. The aforementioned researchers have identified valuable information regarding the significance of several bioactive substances present in the structures of truffle species. The identification of mycochemical compounds in truffles provides important information for preventing and treating diseases, including cancer (Table 3).

2.4 DNA Protective Effect, Genotoxicity and Biofilm Inhibitors

Limited research has been conducted in recent years to determine the DNA-protective, genotoxic properties, and biofilm-inhibitory activity of truffles (Table 4). It is crucial to determine these medicinal effects in different truffle species. Akyüz et al. (2021) detected that increased concentrations of *P. lefebvrei* and *T. olbiensis* extracts provided DNA protection. Nonetheless, the protective effect of the water extract of *P. juniperi* was observed only at a concentration of 40 mg/mL, while the *T. boudieri* extract did not demonstrate DNA-protective effects at the tested concentrations. Fidan and colleagues (2022) found that acetonic and methanolic extracts from *Terfezia boudieri* and *T. claveryi* exhibited the most significant protective effect against H_2O_2-induced genotoxicity. In contrast, *T. olbiensis* extracts showed the least DNA damage effects on *S. cerevisiae* compared to the negative control. Furthermore, both the acetone extracts of *T. olbiensis* and the methanol extracts of *P. lefebvrei* and *P. juniperi* demonstrated the weakest protective effects. Khojah et al. (2022) explored the anti-biofilm potential of *T. claveryi* and determined it to be more effective than the ethyl acetate extract against the standard strain of *S. aureus*. This suggests that *T. claveryi* may serve as a valuable source of microbiologically active compounds for treating biofilm-resistant bacteria.

2.5 Hepatotoxic and Antidiabetic Properties

Truffle species have varying levels of medicinal and therapeutic properties in addition to their use as food. These properties demonstrate the significance of these species and the need to appreciate them. Janakat and Nassar (2010) reported that extracts of *T. claveryi* demonstrated potent hepatotoxic activity against the known hepatotoxin carbon tetrachloride. The study evaluated several factors, including aspartate aminotransferase, alkaline phosphatase, alanine aminotransferase, and serum bilirubin concentration. The results indicated that all three extracts significantly reduced liver function test parameters. Farag et al. (2021) identified *Terfezia boudieri* and *T. claveryi* as promising sources of functional foods for diabetes management due to their rich content of myoinositol and trehalose. Furthermore, the presence of oxygenated monoterpenes suggests that desert truffles could serve as effective antimicrobials against trachoma, supporting their traditional use for this purpose.

Zhong et al. (2019) extracted a polysaccharide named TIP from *T. indicum*. Their research suggests that TIP has the potential to alleviate exercise-induced fatigue by regulating the gut environment. Al-Ahmed and Khalil (2019) found that the methanol extract of *T. claveryi* had stronger inhibitory activity in the α-amylase inhibition compared to the positive control, and almost the same IC_{50} in the inhibition

Table 4. Other medicinal effect of truffles (–, not defined).

Truffle species	Sample	Medicinal effect	References
Terfezia boudieri and *Terfezia claveryi*	Truffle extract	Antidiabetic	Farag et al. (2021)
Terfezia claveryi	Ethyl acetate extract	Anti-biofilm	Khojah et al. (2022)
Terfezia claveryi	Truffle extract	Hepatotoxic activity	Janakat and Nassar (2010)
Terfezia claveryi	Methanol extract	Anti-diabetic activities	Al-Ahmed and Khalil (2019)
Terfezia (3 spp.) and two *Picoa* (2 spp.)	Truffle	Treating anemia and eye infections, promoting satiety, providing energy, and as an aphrodisiac	Akyüz et al. (2017)
Terfezia and *Picoa* spp.	Aqueous extract	DNA protective effect	Akyüz et al. (2021)
Terfezia and *Picoa* spp.	Acetonic and methanolic extracts	H_2O_2-induced genotoxicity and DNA damage effects	Fidan et al. (2022)
Tuber aestivum and *Tuber brumale*	Methanol extracts	Anti-diabetic	Tejedor-Calvo et al. (2023)
Tuber indicum	-	Fatigue	Zhong et al. (2019)
Desert truffles	Truffle	Aphrodisiac properties	Patel, 2012; El Enshasy et al. (2013); Shavit and Shavit (2014); Khojasteh et al. (2013)
Desert truffles	Truffle	Possess beneficial health properties	Mandeel and Al-Laith (2007)
Desert truffles	Truffle	Fertility enhancement, treatment of eye and skin diseases, fatigue, wound healing, weakness, vomiting, respiratory ailments, colds, rheumatism and arthritis	Bradai et al. (2015a)

of haemoglobin glycosylation compared to the positive control α-tocopherol. The study showed that the extract had significant anti-diabetic activities *in vitro* and *in vivo*. Tejedor-Calvo et al. (2023) discovered that specific truffle extracts possess the ability to inhibit enzymes linked to type II diabetes. Pressurized water extracts primarily target α-amylase, whereas ethanolic extracts demonstrate greater effectiveness against α-glucosidase. Notably, extracts derived from *Tuber brumale* and *T. aestivum* exhibit an IC_{50} of 29.22 mg/mL against α-amylase and 7.93 mg/mL against α-glucosidase. Consequently, the PLE method facilitates the extraction of bioactive enriched fractions from truffles, demonstrating potent anti-diabetic properties. The researchers mentioned above have identified truffle species with fatigue-reducing, antidiabetic, and hepatoprotective properties, highlighting their significance in nutrition (Table 4).

2.6 Other Medicinal Properties (Ethnomycological and Aphrodisiac)

This section of the study presents information on the local population's usage of these species and summarizes the use of local treatments (Table 4). Desert truffles are traditionally collected for their nutritional value, as well as their medicinal and aphrodisiac properties (Patel, 2012; El Enshasy et al., 2013; Shavit and Shavit, 2014). Akyüz et al. (2017) reported that indigenous people use three *Terfezia* and two *Picoa* species for various purposes, including hobbies, sports, traditions, income, and nutrition. They also use them for health benefits, such as treating anemia and eye infections, promoting satiety, providing energy, and as aphrodisiacs. Mandeel and Al-Laith (2007) proposed that desert truffles are highly nutritious and possess beneficial health properties. They are also known for their medicinal properties. Bradai and colleagues (2015a) reported that the indigenous people of the Sahara utilize truffles for a multitude of purposes, including consumption, fertility enhancement, and treatment of eye diseases and fatigue. They observed that truffles traditionally occupy a significant role in folk medicine for addressing diverse ailments such as eye and skin diseases, wound healing, weakness, vomiting, respiratory ailments, colds, rheumatism, and arthritis. Khojasteh and colleagues (2013) revealed that the extract from *T. boudieri* might exhibit aphrodisiac properties owing to its capacity to elevate androgen levels. This locally obtained insight presents valuable preliminary findings for academic exploration.

3. Conclusion

This chapter provides detailed information on the bioactive components of various truffle species and their medicinal properties, which vary depending on the species. True truffles, including forest truffles (*Tuber* spp.) and desert truffles (*Terfezia*, *Tirmania*, and *Picoa* species), exhibit structural and reproductive variation and inhabit diverse ecogeographic zones worldwide. Global ethnomycological surveys provide evidence for the nutritional and therapeutic importance of wild species in both dietary practices and traditional medicine. Concurrently, scientific investigations validate the nutritional and pharmacological attributes inherent in these truffles. Nutraceuticals, derived from natural sources and historically used therapeutically, have shown promise in effectively treating certain diseases. This involves consuming functional foods or isolated biologically active compounds, either as nutraceuticals or daily dietary supplements (Nandi et al., 2019). Certain epigeous fungi, as emphasized in this chapter's introduction, demonstrate significant potential for treating various diseases. Mushrooms and their active ingredients also provide crucial medicinal benefits. This section presents data illustrating the effectiveness of bioactive compounds isolated from commercially and economically significant forest and desert truffles in treating various diseases. Overall, truffles contain numerous biologically active compounds that have been identified as responsible for their therapeutic properties. These compounds include antimicrobial, anticancer, antioxidant, and antidiabetic effects, among others. However, the underlying mechanisms involved in the treatment process of these bioactive components in clinical applications are not yet fully understood. Identifying bioactive compounds in truffle species used as food could significantly contribute to the treatment of various diseases in clinical applications.

References

Abdelsalam, S.M., El-Mokadem, M.T., Mekawey, A.A., Sallam, F.E. and Sakr, E.A. et al. (2023). Characterization of bacterial-associated with Egyptian *Tirmania nivea* and *Terfezia canariensis*. Biocatal. Agric. Biotechnol., 52: 102804. 10.1016/j.bcab.2023.102804

Akyüz, M. (2013). Nutritive value, flavonoid content and radical scavenging activity of the truffle (*Terfezia boudieri* Chatin). J. Soil. Sci. Plant. Nutr. 13: 143–151. 10.4067/S0718-95162013005000013

Akyüz, M. and Kirbag, S. (2018). Nutritive value of desert truffles species of genera *Terfezia* and *Picoa* (Ascomycetes) from arid and semiarid regions of eastern Turkey. Inter. J. Med. Mushrooms., 20: 1097–1106. 10.1615/IntJMedMushrooms.2018028796

Akyüz, M., Akyüz, H.E. and Kırbağ, S. (2017). Ethnomycological aspects of traditional usage and indigenous knowledge about the arid-semi arid truffles consumed by the residents of the eastern anatolia region of Turkey. Gazi Univ. J. Sci. 30: 57–70.

Akyüz, M., Karagöz, I.D., Inci, Ş., Kılıç, İ.H., Kırbağ, S. et al. (2021). Cytotoxicity and DNA protective effects of the *Terfezia* and *Picoa* species from the eastern region of Turkey. Istanbul. J. Pharm., 51: 198–203. 10.26650/IstanbulJPharm.2020.0064

Akyüz, M., Kırbağ, S. and Bircan, B. (2015). Medical characteristics of arid-semi arid Truffle (*Terfezia* and *Picoa*) in the Elazığ-Malatya region of Turkey. Hacettepe. J. Biol. Chem., 43: 301–308. 10.15671/HJBC.20154315989

Akyüz, M., Kirbağ, S. and Kurşat, M. (2012). Ecological aspects of the arid and semi-arid truffle in Turkey: evaluation of soil characteristics, morphology, distribution, and mycorrhizal relationships. Turk. J. Bot., 36: 386–391. 10.3906/bot-1103-4

Akyüz, M., Kirbag, S., Bircan, B. and Gurhan, Y. (2015). Diversity and distribution of arid-semi arid truffle (*Terfezia* and *Picoa*) in Elazig-Malatya region of Turkey. Mycosphere, 6: 766–783. 10.5943/mycosphere/6/6/1

Akyüz, M., Kırbağ, S., Gürhan, Y. and Bircan, B. (2017). Evaluation of soil characteristics and environmental parameters of arid-semi arid desert truffles from eastern Turkey. Hacettepe. J. Biol. Chem., 45: 395–402. 10.15671/HJBC.2018.180

Akyüz, M., Kirecci, A.D.O., Gokce, Z., Kirbag, S., Yilmaz, O. et al. (2019). Biochemical constituents and antioxidant activities of some mushrooms from Turkey: *Agaricus* spp., *Pleurotus* spp., *Morchella esculenta* and *Terfezia boudieri*. Istanbul. J. Pharm., 49: 1–6. 10.26650/IstanbulJPharm.2019.25896

Al Obaydi, M.F., Hamed, W.M., Al Kury, L.T. and Talib, W.H. (2020). *Terfezia boudieri*: a desert truffle with anticancer and immunomodulatory activities. Front. Nutr., 7: 38. 10.3389/fnut.2020.00038

AlAhmed, A. and Khalil, H.E. (2019). Antidiabetic activity of *Terfezia claveryi*; an in vitro and in vivo study. Biomed. Pharm. J., 12: 603–608.

Aldebasi, Y.H., Aly, S.M., Qureshi, M.A. and Khadri, H. (2013). Novel antibacterial activity of *Terfezia claveryi* aqueous extract against clinical isolates of corneal ulcer. Afr. J. Biotechnol., 12: 6340–6346. 10.5897/AJB2013.12811

Alhussaini, M.S., Saadabi, A.M., Hashim, K. and Al-Ghanayem, A.A. (2016). Efficacy of the desert truffle *Terfezia claveryi* to cure trachoma disease with special emphasis on its antibacterial bioactivity. Trends. Med. Res., 11: 28-34.

Al–Laith, A.A.A. (2010). Antioxidant components and antioxidant/antiradical activities of desert truffle (*Tirmania nivea*) from various Middle Eastern origins. J. Food. Compost. Anal., 23: 15–22. 10.1016/j.jfca.2009.07.005

Al–Marzooky, M.A. (1981). Truffles in eye disease. pp 353–357. The International Conference on Islamic Medicine.

Alshawush, M.M., Burshed, H.A., Alasoom, A.J., Altaweel, A.A. and Khalil, H.E. et al. (2022). Chemical profiling, antioxidant and lipoxygenase enzyme inhibition activities of wild edible truffle (*Terfezia boudieri*) from Northern Borders of Saudi Arabia. Pharmacogn. J., 14: 319–326. 10.5530/pj.2022.14.109.

Askarova, S., Umbayev, B., Masoud, A.R., Kaiyrlykyzy, A., Safarova, Y. et al. (2020). The links between the gut microbiome, aging, modern lifestyle and Alzheimer's disease. Front. Cell. Infect. Microbiol., 10: 1–12. 10.3389/fcimb.2020.00104

Attia, W.Y., El-Naggar, R.E., Bawadekji, A. and Al Ali, M. (2018). Evaluation of some *in vitro* anti–carcinogenic activities of polysaccharides extracted from ascomata of the desert truffle *Terfezia claveryi* Chatin. J. Appl. Environ. Biol. Sci., 8: 152–159.

Badger-Emeka, L.I., Emeka, P.M., Aldossari, S. and Khalil, H.E. (2021). *Terfezia claveryi* and *Terfezia boudieri* extracts: An antimicrobial and molecular assay on clinical isolates associated with eye infections. Pharmacogn. Mag., 16: 780–788. 10.4103/pm.pm_199_20

Beara, I.N., Lesjak, M.M., Četojević-Simin, D.D., Marjanović, Ž.S., Ristić, J.D. et al. (2014). Phenolic profile, antioxidant, anti–inflammatory and cytotoxic activities of black (*Tuber aestivum* Vittad.) and white (*Tuber magnatum* Pico) truffles. Food. Chem., 165: 460–466. 10.1016/j.foodchem.2014.05.116

Beever, R.E. and Lebel, T. (2014). Truffles of New Zealand: a discussion of bird dispersal characteristics of fruit bodies. J. Auckland. Bot. Soc., 69: 170–178.

Behzadi, A.A., Zareie, M., Abbasi, A., Masoumi, B., Ashrafi-Dehkordi, E. et al. (2021). Physicochemical properties, nutritional composition, and phylogenetic analysis of black truffles grown in Fars Province, Iran. Int. J. Nutr. Sci., 6: 45–51. 10.30476/IJNS.2021.90183.1128

Bermaki, F.Z., Khabar, L. and Ezzanega, A. (2017). Bioecology of desert truffles in the province of Figuig in Eastern Morocco. Rev. Mex. Mic., 46: 29–36.

Bhambri, A., Srivastava, M., Mahale, V.G., Mahale, S. and Karn, S.K. et al. (2022). Mushrooms as potential sources of active metabolites and medicines. Front. Microbiol., 13: 837266. 10.3389/fmicb.2022.837266

Bonito, G., Smith, M.E., Nowak, M., Healy, R.A., Guevara, G. et al. (2013). Historical biogeography and diversification of truffles in the Tuberaceae and their newly identified southern hemisphere sister lineage. PloS One, 8: e52765. 10.1371/journal.pone.0052765

Bonito, G.M. and Smith, M.E. (2016). General systematic position of the truffles: Evolutionary theories. pp. 3–18. *In*: Zambonelli, A., Iotti, M., Murat, C. (eds.). True truffle (*Tuber* spp.) in the World. Soil ecology, systematics and biochemistry. Soil Biology, Springer, Cham. 10.1007/978-3-319-31436-5_1

Bontempo, L., Camin, F., Perini, M., Ziller, L. and Larcher, R. et al. (2020). Isotopic and elemental characterisation of Italian white truffle: A first exploratory study. Food. Chem. Toxicol., 145: 111627. 10.1016/j.fct.2020.111627

Bradai, L., Bissati, S., Chenchouni, H. and Amrani, K. (2015b). Effects of climate on the productivity of desert truffles beneath hyper–arid conditions. Int. J. Biometeorol., 59: 907–915. 10.1007/s00484-014-0891-8

Bradai, L., Neffar, S., Amrani, K., Bissati, S. and Chenchouni, H. et al. (2015a). Ethnomycological survey of traditional usage and indigenous knowledge on desert truffles among the native Sahara Desert people of Algeria. J. Ethnopharmacol., 162: 31–38. 10.1016/j.jep.2014.12.031

Casarica, A., Moscovici, M., Daas, M., Nicu, I., Panteli, M. et al. (2016). A purified extract from brown truffles of the species *Terfezia claveryi* Chatin and its antimicrobial activity. Farmacia, 64: 298–301.

Castellano, M.A. and Türkoğlu, A. (2012). New records of truffle taxa in *Tuber* and *Terfezia* from Turkey. Turk. J. Bot., 36: 295–298. 10.3906/bot-1106-10

Chang, S.T. (2001). Mushrooms and mushroom cultivation. Encyclopedia of life sciences. John Wiley & Sons, Ltd. 10.1038/npg.els.0000370

Chauhan, O.P., Vijay, V., Pandey, A.K. and Semwal, A.D. (2021). Biochemical and health properties of truffles. Def. Life. Sci. J., 6: 251–258. 10.14429/dlsj.6.15659

Chen, G., Zhang, S., Ran, C., Wang, L. and Kan, J. et al. (2016). Extraction, characterization and antioxidant activity of water–soluble polysaccharides from *Tuber huidongense*. Int. J. Biol. Macromol., 91: 431–442. 10.1016/j.ijbiomac.2016.05.108

Choo, K.S., Bollen, M., Dykes, G.A. and Coorey, R. (2021). Aroma-volatile profile and its changes in Australian grown black Périgord truffle (*Tuber melanosporum*) during storage. Int. J. Food. Sci. Technol., 56: 5762–5776. 10.1111/ijfs.15171

Chugh, R.M., Mittal, P., Mp, N., Arora, T., Bhattacharya, T. et al. (2022). Fungal mushrooms: a natural compound with therapeutic applications. Front. Pharmacol., 13: 925387. 10.3389/fphar.2022.925387

Culleré, L., Ferreira, V., Venturini, M.E., Marco, P. and Blanco, D. (2013a). Chemical and sensory effects of the freezing process on the aroma profile of black truffles (*Tuber melanosporum*). Food. Chem., 136: 518–525. 10.1016/j.foodchem.2012.08.030

Culleré, L., Ferreira, V., Venturini, M.E., Marco, P. and Blanco, D. et al. (2013b). Potential aromatic compounds as markers to differentiate between *Tuber melanosporum* and *Tuber indicum* truffles. Food. Chem., 141: 105–110.

Dahham, S.S., Al-Rawi, S.S., Ibrahim, A.H., Majid, A.S.A. and Majid, A.M.S.A. et al. (2018). Antioxidant, anticancer, apoptosis properties and chemical composition of black truffle *Terfezia claveryi*. Saudi. J. Biol. Sci., 25: 1524–1534. 10.1016/j.sjbs.2016.01.031

Darwish, R.S., Shawky, E., Nassar, K.M., ElSayed, R.M.R., Hussein, D.E. et al., Ghareeb, D.A. and El Sohafy, S.M. (2021). Differential anti–inflammatory biomarkers of the desert truffles *Terfezia claveryi* and *Tirmania nivea* revealed via UPLC–QqQ–MS–based metabolomics combined to chemometrics. LWT, 150: 111965. 10.1016/j.lwt.2021.111965

Devi, P. V., Islam, J., Narzary, P., Sharma, D. and Sultana, F. et al. (2024). Bioactive compounds, nutraceutical values and its application in food product development of oyster mushroom. J. Future. Foods., 4: 335–342. 10.1016/j.jfutfo.2023.11.005

Dib–Bellahouel, S. and Fortas, Z. (2011). Antibacterial activity of various fractions of ethyl acetate extract from the desert truffle, *Tirmania pinoyi*, preliminarily analyzed by gas chromatography mass spectrometry (GC–MS). Afr. J. Biotechnol., 10: 9694–9699. 10.5897/AJB10.2687

Dib–Bellahouel, S. and Fortas, Z. (2019). Antimicrobial activities of desert truffle extracts and their chemical identification. J. Pharm. Pharmacol., 7: 593–597. 10.17265/2328-2150/2019.11.005

Dimopoulou, M., Kolonas, A., Mourtakos, S., Androutsos, O. and Gortzi, O. et al. (2022). Nutritional composition and biological properties of sixteen edible mushroom species. Appl. Sci., 12: 8074. 10.3390/app12168074.

Dogan, H. and Aydin, S. (2013). Determination of antimicrobial effect, antioxidant activity and phenolic contents of desert truffle in Turkey. Afr. J. Tradit. Complement. Altern. Med., 10: 52–58. 10.4314/ajtcam.v10i4.9

Dugan, F.M. (2011). Conspectus of world ethnomycology: fungi in ceremonies, crafts, diets, medicines and myths. St. Paul, American Phytopathological Society, p. 151.

Dundar, A., Yesil, O.F., Acay, H., Okumus, V., Ozdemir, S. et al. (2012). Antioxidant properties, chemical composition and nutritional value of *Terfezia boudieri* (Chatin) from Turkey. Food. Sci. Technol. Int., 18: 317–328. 10.1177/1082013211427

El Enshasy, H., Elsayed, E.A., Aziz, R. and Wadaan, M.A. (2013). Mushrooms and truffles: Historical biofactors for complementary medicine in Africa and in the Middle East. J. Evid. Based. Complementary. Altern. Med., 2013: 1–10. 10.1155/2013/620451

Elfalla, Z., Mirwan, H.B., El Waer, H.N., Bajaj, S.R. and Shamekh, S. et al. (2021) Investigation on the growth parameters for Libyan wild truffles. Int. J. Eng. Appl. Sci., 8: 1–5.

Elsayed, E.A., Alsahli, F.D., Barakat, I.A., El Enshasy, H.A. and Wadaan, M.A et al. (2019). Assessment of in vitro antimicrobial and anti–breast cancer activities of extracts isolated from desert truffles in Saudi Arabia. J. Sci. Ind. Res., 78: 419–425.

Farag, M.A., Fathi, D., Shamma, S., Shawkat, M.S.A., Shalabi, S.M. et al. (2021). Comparative metabolome classification of desert truffles *Terfezia claveryi* and *Terfezia boudieri* via its aroma and nutrients profile. LWT, 142: 111046. 10.1016/j.lwt.2021.111046

Ferraro, V., Venturella, G., Pecoraro, L., Gao, W., & Gargano, M. L et al. (2022). Cultivated mushrooms: Importance of a multipurpose crop, with special focus on Italian fungiculture. Plant. Biosyst., 156: 130–142. 10.1080/11263504.2020.1837283

Ferreira, I., Dias, T., Mouazen, A.M. and Cruz, C. (2023). Using science and technology to unveil the hidden delicacy *Terfezia arenaria*, a desert truffle. Foods, 12: 3527. 10.3390/foods12193527

Fidan, M., Ali, M.M., Erez, M.E., Cigerci, I.H., Ozdemir, S. et al. (2022). Antioxidant, antimicrobial, cytotoxic and protective effects of truffles. Anal. Biochem., 641: 114566. 10.1016/j.ab.2022.114566

Gezer, K., Kaygusuz, O., Çelik, A. and Işıloğlu, M. (2014). Ecological characteristics of truffles growing in Denizli province, Turkey. J. Food. Agric. Environ., 12: 1105–1109. 10.1234/4.2014.5296

Gouzi, H., Belyagoubi, L., Abdelali, K.N. and Khelifi, A. (2011). In vitro antibacterial activities of aqueous extracts from Algerian desert truffles (*Terfezia* and *Tirmania*, Ascomycetes) against *Pseudomonas aeruginosa* and *Staphylococcus aureus*. Int. J. Med. Mushrooms., 13: 553-558. 10.1615/IntJMedMushr.v13.i6.70

Gryndler, M. (2016). True truffle host diversity. pp. 267–281. *In*: Zambonelli, A., Iotti, M. and Murat, C. (eds.). True Truffle (*Tuber* spp.) in the World. Soil Biology, Springer, Cham. 10.1007/978-3-319-31436-5_16

Guo, T., Wei, L., Sun, J., Hou, C.L. and Fan, L et al. (2011). Antioxidant activities of extract and fractions from *Tuber indicum* Cooke & Massee. Food. Chem., 127: 1634–1640. 10.1016/j.foodchem.2011.02.030

Hachim, A.K., Hateet, R.R. and Muhsin, T.M. (2020). Study cytotoxicity, antıbacterial activity, and composition by FT–IR and GC–MS for organic extract of white Iraqı desert truffle *Tirmania nivea*. Periódico. Tchê. Química., 17: 18–31.

Hammami, R., Alsaloom, A. N., Aouadhi, C., Alrokban, A., Mendili, M. et al. (2023). HPLC analysis, mycochemical contents and biological activities of two edible hypogeous ascomycetes: *Tirmania nivea* and *Terfezia boudieri*. Heliyon, 9: e14331. 10.1016/j.heliyon.2023.e14331

Hamza, A., Ghanekar, S. and Kumar, D.S. (2023). Current trends in health–promoting potential and biomaterial applications of edible mushrooms for human wellness. Food. Biosci., 51: 102290. 10.1016/j.fbio.2022.102290

Hamza, A., Zouari, N., Zouari, S., Jdir, H., Zaidi, S. et al. (2016a). Nutraceutical potential, antioxidant and antibacterial activities of *Terfezia boudieri* Chatin, a wild edible desert truffle from Tunisia arid zone. Arab. J. Chem., 9: 383–389. 10.1016/j.arabjc.2013.06.015.

Hamza, A., Jdir, H. and Zouari, N. (2016b). Nutritional, antioxidant and antibacterial properties of *Tirmania nivera*, A wild edible desert truffle from Tunisia arid zone. J. Appl. Res. Med. Aromatic. Plants., 5: 1–4. 10.4172/2167-0412.1000258

Harir, M., Bendif, H., Yahiaoui, M., Bellahcene, M., Zohra, F et al. (2019). Evaluation of antimicrobial activity of *Terfezia arenaria* extracts collected from Saharan desert against bacteria and filamentous fungi. 3 Biotech, 9: 281. 10.1007/s13205-019-1816-3

Hilszczańska, D., Siebyła, M., Horak, J., Król, M., Podsadni, P. et al. (2016). Comparison of chemical composition in *Tuber aestivum* Vittad. of different geographical origin. Chem. Biodivers., 13: 1617–1629. 10.1002/cbdv.201600041

Janakat, S., Al-Fakhiri, S. and Sallal, A.K. (2004). A promising peptide antibiotic from *Terfezia claveryi* aqueous extract against *Staphylococcus aureus in vitro*. Phytother, Res., 18: 810–813. https://doi.org/10.1002/ptr.1563

Janakat, S. and Nassar, M. (2010). Hepatoprotective activity of desert truffle (*Terfezia claveryi*) in comparison with the effect of *Nigella sativa* in the Rat. Pak. J. Nutr., 10: 52–56. 10.3923/pjn.2010.52.56

Janakat, S.M., Al-Fakhiri, S.M. and Sallal, A.K. (2005). Evaluation of antibacterial activity of aqueous and methanolic extracts of the truffle *Terfezia claveryi* against *Pseudomonas aeruginosa*. Saudi. Med. J., 26: 952–955.

Jiang, C., Duan, X., Lin, L., Wu, W., Li, X. et al. (2023). A review on the edible mushroom as a source of special flavor: Flavor categories, influencing factors, and challenges. Food. Front., 4: 1561–1577. 10.1002/fft2.263

Kagan–Zur, V., Roth–Bejerano, N., Sitrit, Y. and Morte, A. (2014). Desert truffles: Phylogeny, physiology, distribution and domestication. Soil Biology, Springer. Berlin. Heidelberg., 10.1007/978-3-642-40096-4

Kehrer, J.P. and Klotz, L.O. (2015). Free radicals and related reactive species as mediators of tissue injury and disease: implications for Health. Crit. Rev. Toxicol., 45: 765–798. 10.3109/10408444.2015.1074159

Khadri, H., Aldebasi, Y.H. and Riazunnisa, K. (2017). Trufe mediated (*Terfezia claveryi*) synthesis of silver nanoparticles and its potential cytotoxicity in human breast cancer cells (MCF–7). Afr. J. Biotechnol., 16: 1278–1284. 10.5897/AJB2017.16031

Khalifa, S.A., Farag, M.A., Yosri, N., Sabir, J.S., Saeed, A. et al. (2019). Truffles: from islamic culture to chemistry, pharmacology, and food trends in recent times. Trends. Food. Sci. Technol., 91: 193–218. 10.1016/j.tifs.2019.07.008

Khojah, H.M., Abdelhalim, O.B., Mostafa, M.A. and Habib, E.S.E. (2022). Antimicrobial efficacy of extracts of Saudi Arabian desert *Terfezia claveryi* truffles. Saudi. J. Biol. Sci., 29: 103462. 10.1016/j.sjbs.2022.103462

Khojasteh, S.M.B., Amiri, L. and Sheikhzadeh, F. (2013). Effect of the alcoholic extract of *Terfezia boudieri* on reproductive hormones in male rats. Int. J. Pharm. Bio. Sci., 3: 517–522.

Kıvrak, İ. (2015). Analytical methods applied to assess chemical composition, nutritional value and in vitro bioactivities of *Terfezia olbiensis* and *Terfezia claveryi* from Turkey. Food. Anal. Methods., 8: 1279–1293. 10.1007/s12161-014-0009-2

Korkmaz, C. and Duru, M.E. (2023). Investigation of aroma components of *Tuber aestivum* and *Tuber borchii* collected from different locations in Türkiye. Mantar Dergisi, 14: 82–91. 10.30708. mantar.1299554

Korkmaz, C., Hellal, K., Taş Küçükaydın, M., Çayan, F., Küçükaydın, S. et al. (2023). Volatile compound profiling of seven *Tuber* species using HS–SPME–GC–MS and classification by a chemometric approach. ACS Omega., 8: 34111–34119. 10.1021/acsomega.3c05292

Kour, H., Kour, D., Kour, S., Singh, S., Hashmi, S.A.J. et al. (2022). Bioactive compounds from mushrooms: an emerging bioresources of food and nutraceuticals. Food. Biosci., 50: 102124. 10.1016/j.fbio.2022.102124

Lee, H., Nam, K., Zahra, Z. and Farooqi, M.Q.U. (2020). Potentials of truffles in nutritional and medicinal applications: A review. Fungal. Biol. Biotechnol., 7: 1–17. 10.1186/s40694-020-00097-x

Leonardi, M., Iotti, M., Pacioni, G., Hall, I.R. and Zambonelli, A et al. (2021). Truffles: biodiversity, ecological significances, and biotechnological applications. Pp. 107–146. *In*: Abdel–Azeem, A.M., Yadav, A.N., Yadav, N. and Usmani, Z. (eds.). Industrially important fungi for sustainable development. Fungal Biology, Springer, Cham. 10.1007/978-3-030-67561-5_4

Li, H., Tian, Y., Jri N.M., Ye, L., Karunarathna, S.C. et al. (2021). Reviewing the world's edible mushroom species: A new evidence-based classification system. Compr. Rev. Food. Sci. Food. Saf., 20: 1982–2014. 10.1111/1541-4337.12708

Li, J.M., Liang, H.., Qiao, P., Su, K.M., Liu, P.G. et al. (2019). Chemical composition and antioxidant activity of *Tuber indicum* from different geographical regions of China. Chem. Biodivers., 16: e1800609. 10.1002/cbdv.201800609

Li, Y., Liang, H., Zhou, D., Xing, Y. and Chen, J. et al. (2022). Phenolics, flavonoids content and antioxidant activities of *Tuber indicum* at different maturity stages. Chem. Biodivers., 19: e202100830. 10.1002/cbdv.202100830

Lindequist, U., Niedermeyer, T.H. and Jülich, W.D. (2005). The pharmacological potential of mushrooms. J. Evid. Based. Complementary. Altern. Med., 2: 285–299. 10.1093/ecam/neh107

Loizides, M., Hobart, C., Konstandinides, G. and & Yiangou, Y. (2012). Desert truffles: The mysterious jewels of antiquity. Field. Mycol. 13: 17–21. 10.1016/j.fldmyc.2011.12.004

Louro, R., Santos-Silva, C. and Nobre, T. (2019). What is in a name? *Terfezia* classification revisited. Fungal. Biol., 123: 267–273. 10.1016/j.funbio.2019.01.003

Luo, Q., Zhang, J., Yan, L., Tang, Y., Ding, X. et al. (2011). Composition and antioxidant activity of water–soluble polysaccharides from *Tuber indicum*. J. Med. Food., 14: 1609–1616. 10.1089/jmf.2011.1659

Łysakowska, P., Sobota, A. and Wirkijowska, A. (2023). Medicinal mushrooms: their bioactive components, nutritional value and application in functional food production-A review. Molecules, 28: 5393. 10.3390/molecules28145393

Malik, H.M., Munazza Gull, M.G., Ulfat Omar, U.O., Kumosani, T.A., Al-Hejin, A.M. et al. (2018). Evaluation of the antibacterial potential of desert truffles (*Terfezia* spp) extracts against methicillin resistant *Staphylococcus aureus* (MRSA). J. Exp. Biol. Agric. Sci., 6: 652–660. 10.18006/2018.6(4).652.660

Mandeel, Q.A. and Al-Laith, A.A.A. (2007). Ethnomycological aspects of the desert truffle among native Bahraini and non–Bahraini peoples of the Kingdom of Bahrain. J. Ethnopharmacol., 110: 118–129. 10.1016/j.jep.2006.09.014

Manzi, P., Aguzzi, A. and Pizzoferrato, L. (2001). Nutritional value of mushrooms widely consumed in Italy. Food. Chem., 73: 321–325. 10.1016/S0308-8146(00)00304-6

Manzi, P., Gambelli, L., Marconi, S., Vivanti, V. and Pizzoferrato, L. et al. (1999). Nutrients in edible mushrooms: an interspecies comparative study. Food. Chem., 65: 477–482. 10.1016/S0308-8146(98)00212-X.

Martins, A. (2016). The numbers behind mushroom biodiversity. pp. 15–63. *In*: Ferreira, I.C.F.R., Morales, P. and Barros, L. (eds.). Wild plants, mushrooms and nuts: functional food properties and applications. John Wiley & Sons, Ltd. 10.1002/9781118944653.ch2

Mattila, P., Könkö, K., Eurola, M., Pihlava, J.M., Astola, J. et al. (2001). Contents of vitamins, mineral elements, and some phenolic compounds in cultivated mushrooms. J. Agric. Food. Chem., 49: 2343–2348. 10.1021/jf001525d

Mekawey, A.A.I. (2015). *Terfezia boudieri* as sources of antitumor and antiviral agent. World. J. Pharm. Pharm. Sci., 4: 294–315.

Mello, A., Murat, C. and Bonfante, P. (2006). Truffles: much more than a prized and local fungal delicacy. FEMS Microbiol. Lett., 260: 1–8. 10.1111/j.1574-6968.2006.00252.x

Miles, P.G. and Chang, S.T. (2004). Mushrooms: cultivation, nutritional value, medicinal effect, and environmental impact. Boca Raton: CRC Press. 10.1201/9780203492086

Moreno, G., Alvarado, P. and Manjón, J.L. (2014). Hypogeous desert fungi. pp. 3–20. *In*: Kagan–Zur, V., Roth–Bejerano, N., Sitrit, Y. and Morte, A. (eds.). Desert Truffles. Soil Biology. Springer, Berlin, Heidelberg. 10.1007/978-3-642-40096-4_1

Morte, A., Kagan–Zur, V., Navarro–Ródenas, A. and Sitrit, Y. (2021). Cultivation of desert truffles-A crop suitable for arid and semi–arid zones. Agronomy, 11: 1462. 10.3390/agronomy11081462

Murcia, M.A., Martínez-Tomé, M., Jiménez, A.M., Vera, A.M., Honrubia, M. et al. (2002). Antioxidant activity of edible fungi (truffles and mushrooms): losses during industrial processing. J. Food. Prot., 65: 1614–1622. 10.4315/0362-028X-65.10.1614

Murcia, M.A., Martínez-Tomé, M., Vera, A., Morte, A., Gutierrez, A. et al. (2003). Effect of industrial processing on desert truffles *Terfezia claveryi* Chatin and *Picoa juniperi* Vittadini): proximate composition and fatty acids. J. Sci. Food. Agric., 83: 535–541. 10.1002/jsfa.1397

Mustafa, A.M., Angeloni, S., Nzekoue, F.K., Abouelenein, D., Sagratini, G. et al. (2020). An overview on truffle aroma and main volatile compounds. Molecules, 25: 5948. 10.3390/molecules25245948

Mwangi, R.W., Macharia, J.M., Wagara, I.N. and Bence, R.L. (2022). The antioxidant potential of different edible and medicinal mushrooms. Biomed. Pharmacother., 147: 112621. 10.1016/j.biopha.2022.112621

Nakamura, N., Abe, J.P., Shibata, H., Kinoshita, A., Obase, K. et al. (2020). Genotypic diversity of the Asiatic black truffle, *Tuber himalayense*, collected in spontaneous and highly productive truffle grounds. Mycol. Prog., 19: 1511–1523. 10.1007/s11557-020-01642-z

Nandi, S., Sikder, R. And Acharya, K. (2019). Secondary metabolites of mushrooms: a potential source for anticancer therapeutics with translational opportunities. pp. 563–598. *In*: Satyanarayana, T., Deshmukh, S.K. and Deshpande, M.V. (eds.). Advancing Frontiers in Mycology & Mycotechnology: Basic and Applied Aspects of Fungi. Springer, Singapore. 10.1007/978-981-13-9349-5_23

Neggaz, S., Fortas, Z., Chenni, M., El Abed, D., Ramli, B. et al. (2015). In vitro evaluation of antioxidant, antibacterial and antifungal activities of *Terfezia claveryi* Chatin. Phytotherapie, 16: 20–26. 10.1007/s10298-015-0993-4

Niego, A.G.T., Lambert, C., Mortimer, P., Thongklang, N., Rapior, S. et al. (2023). The contribution of fungi to the global economy. Fungal. Divers., 121: 95–137. 10.1007/s13225-023-00520-9

Obaid, A.S., Hassan, K.T., Hassan, O.M., Ali, H.H., Ibraheem, I.J. et al. (2023). In-vitro antibacterial, cytotoxicity, and anti-prostate cancer effects of gold nanoparticles synthesized using extract of desert truffles (*Tirmania nivea*). Mater. Chem. Phys., 301: 127673. 10.1016/j.matchemphys.2023.127673

Owaid, M.N. (2016). Biodiversity and bioecology of Iraqi desert truffles (Pezizaceae) during season 2014. J. Aridland. Agric., 2: 22–25. 10.19071/jaa.2016.v2.3046

Pacioni, G., Cerretani, L., Procida, G. and Cichelli, A. (2014). Composition of commercial truffle flavored oils with GC–MS analysis and discrimination with an electronic nose. Food. Chem., 146: 30–35. 10.1016/j.foodchem.2013.09.016

Pacioni, G., Rapino, C., Zarivi, O., Falconi, A., Leonardi, M. et al. (2015). Truffles contain endocannabinoid metabolic enzymes and anandamide. Phytochem., 110: 104–110. 10.1016/j.phytochem.2014.11.012

Pandey, A.T., Pandey, I., Kerkar, P. and Singh, M.P. (2021). Antimicrobial activity and mycochemical profile of methanol extract from *Pleurotus flabellatus*. Vegetos, 34: 619–629. 10.1007/s42535-021-00242-w.

Patel, S. (2012). Food, health and agricultural importance of truffles: A review of current scientific literature. Curr. Trends. Biotechnol. Pharm., 6: 15–27.

Patel, S., Rauf, A., Khan, H., Khalid, S. and Mubarak, M.S. et al. (2017). Potential health benefits of natural products derived from truffles: A review. Trends. Food. Sci. Technol. 70: 1–8.10.1016/j.tifs.2017.09.009

Pattanayak, M., Samanta, S., Maity, P., Manna, D. K., Sen, I.K. et al. (2017). Polysaccharide of an edible truffle *Tuber rufum*: structural studies and effects on human lymphocytes. Int. J. Biol.. Macromol., 95: 1037–1048. 10.1016/j.ijbiomac.2016.10.092

Payen, T., Murat, C. and Bonito, G. (2014). Truffle phyloge–nomics: New insights into truffle evolution and truffle life cycle. pp. 211–234. *In*: Martin, F. (ed.). Genomics of plant–related fungi. Advances in Botanical Research, Elsevier Academic Press. 10.1016/B978-0-12-397940-7.00007-0

Perez-Montes, A., Rangel-Vargas, E., Lorenzo, J.M., Romero, L. and Santos, E.M. et al. (2021). Edible mushrooms as a novel trend in the development of healthier meat products. Curr. Opin. Food. Sci., 37: 118–124. 10.1016/j.cofs.2020.10.004

Pérez-Moreno, J., Guerin-Laguette, A., Rinaldi, A.C., Yu, F., Verbeken, A. et al. (2021). Edible mycorrhizal fungi of the world: What is their role in forest sustainability, food security, biocultural conservation and climate change?. Plants. People. Planet., 3: 471–490. 10.1002/ppp3.10199

Phong, W.N., Gibberd, M.R., Payne, A.D., Dykes, G.A. and Coorey, R. et al. (2022). Methods used for extraction of plant volatiles have potential to preserve truffle aroma: A review. Compr. Rev. Food. Sci. Food. Saf., 21: 1677–1701. 10.1111/1541-4337.12927

Piatti, D., Marconi, R., Caprioli, G., Zannotti, M., Giovannetti, R. et al. (2024). White acqualagna truffle (*Tuber magnatum* Pico): Evaluation of volatile and non–volatile profiles by GC–MS, sensory analyses and elemental composition by ICP–MS. Food. Chem., 439: 138089. 10.1016/j.foodchem.2023.138089

Raghavendra, V.B., Venkitasamy, C., Pan, Z. and Nayak, C. (2017). Functional foods from mushrooms. pp. 65–91. *In*: Gupta, V.K., Treichel, H., Shapaval, V.O.S., Oliveira, L.A. and Tuohy, M.G. (eds.). Microbial functional foods and nutraceuticals. John Wiley & Sons, Ltd., Chichester, UK. 10.1002/9781119048961.ch4

Romanazzi, G., Servili, A. and Murolo, S. (2015). Postharvest diseases of *Tuber melanosporum*. Acta Hortic., 1144, 129–132. 10.17660/ActaHortic.2016.1144.18

Şahin, A., Kaşık, G., Alkan, S. and Özcan, M.M. (2020). Total phenol, antioxidant activity, fatty acid composition and mineral contents of the species of fungi known as domalan. J. Agroalimentary. Proces. Technol., 26: 149–154.

Sallam, F.E., El-Mokadem, M.T., Mekawey, A.A.I. and Saker, E.A.E. (2022). Egyptian truffles as a source of antimicrobial and antioxidant agents. J. Sci. Res. Sci. 39: 49–61. 10.21608/jsrs.2022.275787

Sanmee, R., Dell, B., Lumyong, P., Izumori, K. and Lumyong, S. (2003). Nutritive value of popular wild edible mushrooms from northern Thailand. Food. Chem., 82: 527–532. 10.1016/S0308-8146(02)00595-2

Sawaya, K., Abou Najem, S., Khawaja, G. and Khalil, M. (2023). Proapoptotic and antiproliferative effects of the desert truffle *Terfezia boudieri* on colon cancer cell lines. Evid. Based. Complement. Alternat. Med. 1693332 : 1–13. 10.1155/2023/1693332

Sawaya, W.N., Al-Shalhat, A., Al-Sogair, A. and Al-Mohammad, M. (1985). Chemical composition and nutritive value of truffles of Saudi Arabia. J. Food. Sci., 50: 450–453. 10.1111/j.1365-2621.1985.tb13425.x

Schillaci, D., Cusimano, M.G., Cascioferro, S.M., Di Stefano, V., Arizza, V. et al. (2017). Antibacterial activity of desert truffles from saudi arabia against *Staphylococcus aureus* and *Pseudomonas aeruginosa*. Int. J. Med. Mushrooms., 19: 121–125. 10.1615/IntJMedMushrooms.v19.i2.30

Segneanu, A.E., Cepan, M., Bobica, A., Stanusoiu, I., Dragomir, I.C. et al. (2021). Chemical Screening of Metabolites Profile from Romanian *Tuber* spp. Plants, 10: 540. 10.3390/plants10030540

Shah, N., Usvalampi, A., Chaudhary, S., Seppänen-Laakso, T., Marathe, S. et al. (2020). An investigation on changes in composition and antioxidant potential of mature and immature summer truffles (*Tuber aestivum*). Eur. Food. Res. Technol. 246: 723-731. 10.1007/s00217-020-03438-7

Shavit, E. and Shavit, E. (2014). The medicinal value of desert truffles. pp. 323–340. *In*: Kagan–Zur, V., Roth–Bejerano, N., Sitrit, Y. and Morte, A. (eds.). Desert Truffles. Soil Biology, Springer, Berlin, Heidelberg. 10.1007/978-3-642-40096-4_20

Sonji, G.M., Assi, M., Sonji, N.M., Boukhary, R., Rahal, M. et al. (2022). Antioxidant activity, metal content, and essential oil composition of two desert truffles species: *Terfezia boudieri* and *Terfezia claveryi*. J. Pharmacogn. Phytochem., 11: 10–19. 10.22271/phyto.2022.v11.i4a.14438

Souna, M., Tefiani, C., Azzi, R., Chaouche, T.M., Habi, S. et al. (2023). Mycochemical constituents and anti–inflammatory activity of *Terfezia claveryi* Chatin from Algeria. Biol. Life. Sci. Forum., 27: 19. 10.3390/IECAG2023-14980

Srikram, A. and Supapvanich, S. (2016). Proximate compositions and bioactive compounds of edible wild and cultivated mushrooms from Northeast Thailand. Agric. Nat. Res., 50: 432–436. 10.1016/j.anres.2016.08.001

Tagnamas, Z., Bahammou, Y., Kouhila, M., Hilali, S., Idlimam, A. et al. (2020). Conservation of Moroccan truffle (*Terfezia boudieri*) using solar drying method. Renew. Energy., 146: 16–24. 10.1016/j.renene.2019.06.107

Tejedor-Calvo, E., Amara, K., Reis, F.S., Barros, L., Martins, A. et al. (2021). Chemical composition and evaluation of antioxidant, antimicrobial and antiproliferative activities of *Tuber* and *Terfezia* truffles. Food. Res. Int., 140: 110071. 10.1016/j.foodres.2020.110071

Tejedor-Calvo, E., Garcia-Barreda, S., Dambolena, J.S., Pelissero, D., Sánchez, S. et al. (2023). Aromatic profile of black truffle grown in Argentina: Characterization of commercial categories and alterations associated to maturation, harvesting date and orchard management practices. Food. Res. Int., 173: 113300. 10.1016/j.foodres.2023.113300

Tejedor-Calvo, E., Morales, D., Morillo, L., Vega, L., Caro, M. et al. (2023). Pressurized liquid (PLE) truffle extracts have inhibitory activity on key enzymes related to type 2 diabetes (α–Glucosidase and α–Amylase). Foods, 12: 2724. 10.3390/foods12142724

Thomas, P.W., Elkhateeb, W.A. and Daba, G. (2019). Truffle and truffle–like fungi from continental Africa. Acta. Mycol., 54: 1132. 10.5586/am.1132

Türkoğlu, A. and Castellano, M.A. (2013). New records of truffle fungi (Basidiomycetes) from Turkey. Turk. J. Bot., 37: 970–976. 10.3906/bot-1212-54

Türkoğlu, A. and Castellano, M.A. (2014). New records of some Ascomycete truffle fungi from Turkey. Turk. J. Bot., 38: 406–416. 10.3906/bot-1303-24

Türkoğlu, A., Castellano, M.A., Trappe, J. and Güngör, M.Y. (2015). Turkish truffles I: 18 new records for Turkey. Turk. J. Bot., 39: 359–376. 10.3906/bot-1406-42

Vahdatzadeh, M. and Splivallo, R. (2018). Improving truffle mycelium flavor through strain selection targeting volatiles of the Ehrlich pathway. Sci. Rep., 8: 9304. 10.1038/s41598-018-27620-w

Valverde, M.E., Hernández-Pérez, T. and Paredes-López, O. (2015). Edible mushrooms: improving human health and promoting quality life. Int. J. Microbiol., 2015: 376–387. 10.1155/2015/376387

Veeraraghavan, V.P., Hussain, S., Papayya Balakrishna, J., Dhawale, L., Kullappan, M. et al. (2022). A comprehensive and critical review on ethnopharmacological importance of desert truffles: *Terfezia claveryi*, *Terfezia boudieri*, and *Tirmania nivea*. Food. Rev. Int., 38: 846–865. 10.1080/87559129.2021.1889581.

Venturella, G., Ferraro, V., Cirlincione, F. and Gargano, M.L. (2021). Medicinal mushrooms: bioactive compounds, use, and clinical trials. Int. J. Mol. Sci., 22: 634. 10.3390/ijms22020634

Villares, A., García-Lafuente, A., Guillamon, E. and Ramos, Á. (2012). Identification and quantification of ergosterol and phenolic compounds occurring in *Tuber* spp. truffles. J. Food. Compost. Anal., 26: 177–182. 10.1016/j.jfca.2011.12.003

Wang, H.F., Zhao, Y.L. and Fan, Y.J. (2014). Ecological characters of truffles. pp. 2903–2908. *In*: Li, S., Jin, Q., Jiang, X. and Park, J. (eds.). Frontier and future development of information technology in medicine and education. Lecture Notes in Electrical Engineering, Springer, Dordrecht. 10.1007/978-94-007-7618-0_368

Widel, M., Krzywon, A., Gajda, K., Skonieczna, M. and Rzeszowska-Wolny, J. (2014). Induction of bystander effects by UVA, UVB, and UVC radiation in human fibroblasts and the implication of reactive oxygen species. Free. Radic. Biol. Med., 68: 278–287. 10.1016/j.freeradbiomed.2013.12.021

Wu, Z., Meenu, M. and Xu, B. (2021). Nutritional value and antioxidant activity of Chinese black truffle (*Tuber indicum*) grown in different geographical regions in China. LWT, 135: 110226. 10.1016/j.lwt.2020.110226

Yadav, D. and Negi, P.S. (2021). Bioactive components of mushrooms: Processing effects and health benefits. Food. Res. Int., 148: 110599. 10.1016/j.foodres.2021.110599

Yan, X., Wang, Y., Sang, X. and Fan, L. (2017). Nutritional value, chemical composition and antioxidant activity of three *Tuber* species from China. AMB Expr., 7: 1–8. 10.1186/s13568-017-0431-0

Yu, Y., Liu, Z., Song, K., Li, L., Chen, M. et al. (2023). Medicinal value of edible mushroom polysaccharides: A review. J. Future. Foods., 3: 16–23. 10.1016/j.jfutfo.2022.09.003

Zambonelli, A., Iotti, M. and Hall, I. (2015). Current status of truffle cultivation: Recent results and future perspectives. Ital. J. Mycol., 44: 31–40. 10.6092/issn.2465-311X/5593

Zambonelli, A., Iotti, M. and Murat, C. (2016). True Truffle (*Tuber* spp.) in the World: Soil ecology, systematics and biochemistry. Springer International Publishing, Springer, Cham. 10.1007/978-3-319-31436-5

Zhao, W., Wang, X.H., Li, H.M., Wang, S.H., Chen, T. et al. (2014). Isolation and characterization of polysaccharides with the antitumor activity from *Tuber* fruiting bodies and fermentation system. Appl. Microbiol. Biotechnol., 98: 1991–2002. 10.1007/s00253-013-5379-7

Zhong, L., Ma, N., Zheng, H., Ma, G., Zhao, L. et al. (2019). *Tuber indicum* polysaccharide relieves fatigue by regulating gut microbiota in mice. J. Funct. Foods., 63: 103580. 10.1016/j.jff.2019.103580

Zitouni-Haouar, F.E.H., Bidartondo, M.I., Moreno, G., Carlavilla, J.R., Manjón, J.L. et al. (2023). Bioclimatic origin shapes phylogenetic structure of *Tirmania* (Pezizaceae): New Species and New Record from North Africa. J. Fungi., 9: 532. 10.3390/jof9050532

Chapter 11

Biomedical Significance of Truffles

Insights into Human Health

Sandesh J. Marathe[1,2]

1. Introduction

Truffles are hypogeous ascomycetous macrofungi that have been traditionally associated with gourmet cuisine owing to their gustatory value. Various species of truffles have garnered attention for their diverse properties and potential applications through ongoing research. Their low productivity makes them valuable thus earning them epithets like 'underground gold' (Patel et al., 2017). The huge difference in the supply-demand chain leads to a surge in costs reaching as high as US $5,000 per kilogram for white truffles (Patel et al., 2017). Although various species of truffles can be found all over the world major occurrences of truffles belonging to the genus *Tuber* have been recorded in Europe particularly in Italy, Spain, France, Croatia, England, North America, Africa, Australia, New Zealand, and Asia in the Middle Eastern countries and China (Patel et al., 2017). On the other hand, desert truffles have been reported to grow in countries such as Israel, Tunisia, Iraq, Bahrain, Kuwait, Botswana, Saudi Arabia, Jordan, Syria, Libya, Egypt, Algeria, South Africa and Morocco (El Enshasy et al., 2013).

The use of truffles by humans can be traced to prehistoric times. For instance the traditional use of desert truffles by the native people of Kalahari desert in southern Africa has been seen for over a millennium. More specifically the native

[1] Division of Hematology and Oncology, Department of Medicine, College of Medicine, University of Tennessee Health Science Center (UTHSC), Memphis , USA.
[2] UTHSC Center for Cancer Research, College of Medicine, Memphis , USA.
* Corresponding author: sandeshmarathe92@gmail.com

people of North Africa and the Middle East have been consuming *Terfezia* and *Tirmania* species since prehistoric times (El Enshasy et al., 2013). Species such as *Mattirolomyces, Kalaharituber* and *Eremiomyces* have been consumed by the Khoisan people indigenous to Southern Africa (El Enshasy et al., 2013). Despite the age-old use of truffles in Africa relatively less attention has been given to the indigenous information on their medicinal use in comparison to mushrooms (El Enshasy et al., 2013). Nevertheless some reports have shed light on the ethnomedical use of truffles. For instance desert truffles of the species *Tirmania* and *Terfezia* have been reported to be used to treat ophthalmic diseases such as trachoma and as an aphrodisiac in folk medicine (El Enshasy et al., 2013).

Truffles are valuable not only due to their unique aroma but also because they exhibit a biochemical composition that encompasses a multitude of nutritional and medicinal benefits. They are abundant in nutritional components such as minerals, carbohydrates, fats, lipids, proteins, amino acids, phytosterols, phenolics and terpenoids (Lee et al., 2020; Li et al., 2022). The rich nutritional composition of truffles contributes to their various biological properties such as immunomodulatory, anti-inflammatory, antitumor, antioxidant, antibacterial, and hepatoprotective properties (Lee et al., 2020). Owing to their rich bioactive composition truffles are now finding applications in numerous cosmetic products and perfumes resulting in the filing of several patents for health supplements derived from truffles (Thomas et al., 2021). Considering the use of truffles in folk medicine and the vast arena of bioactive compounds found in truffles, this chapter aims at highlighting the latest developments on the potential of truffles in treating human ailments.

2. Truffles in Human Health

The swift changes in environmental conditions and the rapid modifications in modern lifestyle have led to an increase in body-mass index (BMI) as well as rates of obesity, which can carry significant ramifications for both public and global health (Fontes et al., 2022). Dietary factors are one of the most important causes of lifestyle diseases that exacerbate metabolic disorders and lead to increased morbidity and mortality in diseases like obesity, type II diabetes mellitus (T2DM), cardiovascular diseases (CVD), cancer and neurodegenerative disorders to name a few. Such non-communicable diseases that are associated with dietary risk factors have been reported to cause the highest mortality rates among all causes (Fontes et al., 2022).

A plethora of bioactive compounds have been identified in various truffle species. Truffles contain high amounts of compounds like terpenoids, flavonoids, phenolics,polysaccharides, phytosterols and ergosterol. Moreover, truffles are also a good source of fatty acids, amino acids, fibers and minerals (Wang and Marcone, 2011). The presence of these compounds confers upon truffles a multitude of biological activities.

2.1 Anti-Diabetic Potential

Some of the factors contributing to T2DM include hyperglycemia (insulin resistance), hypertension and dyslipidemia (Li et al., 2023). Various truffle species have been explored for their bioactive compounds that target these factors. For instance, the enzyme dipeptidyl peptidase-IV (DPP-IV) degrades the incretin hormone glucagon-like peptide-1 (GLP-1) which is involved in glucose homeostasis. Thus, the inhibition of DPP-IV is one of the possible modes of modulating glucose levels. Two bioactive peptides from the edible truffle *T. sinense* have been identified by peptidomics analysis that effectively inhibit DPP-IV (Li et al., 2023). These peptides were chosen among a total of 4,038 endogenous peptides isolated from fresh ascocarps of *T. sinense*, screened primarily using an *in silico* approach. These peptides showed a strong affinity towards DPP-IV, evidenced by the formation of hydrogen bonds and hydrophobic interaction with amino acid residues of DPP-IV with affinity values ranging between 8.8 and 8.4 kcal/mol as well as a low IC50 values of 643.14 ± 1.79 µM and 668.29±1.31 µM (Li et al., 2023). Furthermore, oral glucose tolerance tests in mice revealed substantial hypoglycemic effects of the peptides thus hinting at the anti-diabetic potential of *T. sinense* (Li et al., 2023). Similarly, key enzymes such as α-amylase and α-glucosidase involved in the breakdown of carbohydrates into simpler sugars have been targeted using truffle bioactives as a means to control hyperglycemia. For instance, extracts of *T. aestivum* have been shown to inhibit α-amylase and α-glucosidase using *in vitro* assays (Beara et al., 2021). Likewise, α-glucosidase inhibitory activity of extracts of *T. claveryi* a truffle traditionally used in the Middle East has also been shown *in vitro* (Abu-Odeh et al., 2022). The same study also reported the antihyperglycemic activity of *T. claveryi* extracts in a high-fat diet alloxan-induced diabetic mouse model proving the *in vivo* anti-diabetic potential of *T. claveryi* (Abu-Odeh et al., 2022).

Bioactives from truffles have also been studied for their anti-diabetic potential at the genetic level. For instance an aqueous extract of *T. melanosporum* has been shown to reduce glucose levels in a streptozotocin-induced hyperglycemic rat model via the regulation of Nrf2 and NF-κB gene expression (Zhang et al., 2018). This has been supported independently by another study where administration of *T. melanosporum* increased the expression of Nrf2 amongst other genes such as catalase manganese superoxide dismutase 2, heme oxygenase (HO)-1 and HO-2 (Jiang et al., 2018). The authors also reported a reduction in plasma glucose serum levels of total cholesterol, triglycerides and glycated hemoglobin as well as the body weight of the mice. Conversely an increase in the serum levels of high-density lipoprotein cholesterol was noted indicating potential hypoglycemic and hypolipidemic effects (Jiang et al., 2018). Other studies focusing on evaluating the anti-diabetic potential of truffles have been compiled in Table 1.

2.2 Anti-Inflammatory and Immunomodulatory Activity

Among the various biological activities reported bioactive compounds in truffles have also been reported to show immunomodulatory activity. Various *in vitro* and *in vivo* methods have been used to assess the anti-inflammatory potential of truffle bioactives. For instance the *in vitro* immunomodulatory potential of the desert truffles *T. claveryi* and *T. nivea* has been studied using lipopolysaccharides (LPS)-stimulated human WBCs (Darwish et al., 2021). The authors reported a significant reduction of 68.9%,78.8% and 59.4% respectively in the inflammatory

Table 1. Reports on various biological activities of truffles.

Truffle	Biological activity	Model system	Reference
Mattirolomyces terfezoides, Tuber brumale var. *moschatum* and *T. aestivum* var. uncinatum	Anti-diabetic	*In vitro* α-amylase and α-glucosidase inhibition assay	Tejedor-Calvo et al. (2023)
Terfezia and *Tuber* genera	Antioxidant	*In vitro* DPPH , FRAP and ORAC assays	Tejedor-Calvo et al. (2021)
Terfezia and *Tirmania* genera	Antioxidant	*In vitro* DPPH , FRAP, ABTS superoxide anion radical scavenging and ferricyanide/Prussian blue assays	Gouzi et al. (2013)
Tuber melanosporum	Anti-diabetic	STZ-induced diabetic rats	Zhang et al. (2020)
T. aestivum	Anti-diabetic	*In vitro* α-amylase and α-glucosidase inhibition assay	Mudliyar et al. (2019)
T. aestivum and *T. magnatum*	Anti-inflammatory	*Ex vivo* COX-1 and 12-LOX assay	Beara et al. (2014)
T. aestivum	Anti-inflammatory	*In vitro* LOX inhibition assay	Marathe et al. (2020)
T. aestivum	Antioxidant	*In vitro* DPPH superoxide anion radical scavenging, nitric oxide radical scavenging potential to inhibit lipid peroxidation (LP) and FRAP assay	Beara et al. (2014)
T. boudieri	Cytotoxicity	MTT assay to study cytotoxicity against cancer cell lines (HCT-116 and Caco-2)	Sawaya et al. (2023)
T. boudieri	Cytotoxicity	MTT assay to study cytotoxicity against cancer cell lines T47D, MCF-7, MDA-MB231 , HCT-116 and Hela	Al Obaydi et al. (2020)
T. boudieri	Antioxidant	*In vitro* DPPH and lipid peroxidation inhibition assay	Hamza et al. (2016)
T. boudieri	Anti-microbial	Minimum inhibitory concentration assessment against gram-positive *E. faecalis S. aureus*, *S. epidermis* and *B. subtilis* and gram-negative *S. typhimurium, E. coli* and *P. aeruginosa* bacteria.	Hamza et al. (2016)
T. claveryi	Anti-inflammatory	*In vitro* BSA denaturation assay	Mimoune et al. (2023)
T. claveryi	Cytotoxicity	MTT assay to study cytotoxicity against cancer cell lines HT29, U-87 MG, PC3 and MCF-7	Dahham et al. (2018)
T. claveryi	Anti-angiogenic	*Ex vivo* rats aortic ring assay	Dahham et al. (2018)

Table 1. contd. ...

Table 1. contd. ...

Truffle	Biological activity	Model system	Reference
T. claveryi	Cytotoxicity	MTT assay to study cytotoxicity against prostate cancer cell line VCaP.	Obaid et al. (2023)
T. claveryi	Antioxidant	*In vitro* DPPH, FRAP and ABTS antioxidant assays	Dahham et al. (2018)
T. claveryi	Antioxidant	*In vitro* DPPH , FRAP and ABTS assays	Dahham et al. (2018)
T. huidongense	Antioxidant	*In vitro* DPPH and ABTS antioxidant assays	Chen et al. (2016)
T. indicum	Antioxidant	*In vitro* DPPH FRAP, hydroxyl radical scavenging and superoxide anion radical scavenging assays	Guo et al. (2011)
T. latisporum, *T. subglobosum* and *T. pseudohimalayense*	Antioxidant	*In vitro* DPPH , FRAP , hydroxyl radical scavenging and ferricyanide/Prussian blue assays.	Yan et al. (2017)
T. magnatum	Cytotoxicity	Tumor cell lines HeLa, MCF7, HT-29	Beara et al. (2014)
T. magnatum	Antioxidant	*In vitro* DPPH superoxide anion radical scavenging nitric oxide radical scavenging potential to inhibit lipid peroxidation and FRAP assay	Beara et al. (2014)
T. magnatum	Antioxidant	*In vitro* DPPH and FRAP assays	Hsu et al. (2022)
T. magnusii	Anti-microbial	*In vitro* well diffusion assay against *S. aureus*	Tejedor-Calvo et al. (2021)
T. melanosporum	Antioxidant	*In vitro* DPPH, FRAP and hydroxyl radical scavenging assays	Xu et al. (2021)
T. melanosporum	Antioxidant	*In vitro* DPPH , ABTS and FRAP assays	Savini et al. (2017)
T. panzhihuanense and *T. pseudoexcavatum*	Antioxidant	Alleviation of H_2O_2-induced oxidative stress against porcine jejunal epithelial (IPEC-J2) cells and effects of polysaccharides on the activities of malondialdehyde , superoxide dismutase, catalase, and glutathione peroxidase-GSH-Px in IPEC-J2 cells.	Li et al. (2020)
T. pinoyi and *T. claveryi*	Anti-microbial	Minimum inhibitory concentration assessment against *S. aureus* and *P. aeruginosa*	Schillaci et al. (2017)

biomarkers TNF-α, IL-1β and IFN-γ with the use of *T. claveryi* extracts. On the other hand *T. nivea* extracts reduced the inflammatory biomarkers TNF-α, IL-1β and IFN-γ by 46.6%,67.3%, and 45.5% respectively Darwish et al. (2021). In an attempt to delineate the role of bioactive components in the extracts in the observed immunomodulatory activity the authors employed orthogonal projections to latent structures discriminant analysis (OPLS-DA) biplot. The analysis revealed that the downregulation of TNF-α was associated with naringenin oleic acid, margaric acid and resveratrol whereas the downregulation of IL-1β and IFN-γ was attributed to docosatetraenoic acid, lauric acid, behenic acid, and caffeic acid (Darwish et al., 2021). Similarly (the immunomodulatory potential of bioactive polysaccharides from the truffle *T. melanosporum* has shown anti-inflammatory activity as evidenced by reduced secretions of IL-1β and IL-6 by 40% and 60% respectively in THP-1 cells (Tejedor-Calvo et al., 2020). The authors also reported a significant reduction in proinflammatory cytokines IL-6 and TNF-α using the extracts from the truffles *T. aestivum* and *T. claveryi* (Tejedor-Calvo et al., 2022).

A meticulous study on the bioactive composition of extracts of the desert truffle *T. boudieri* and its anti-inflammatory potential has been attempted using *in vitro* assays (Al Obaydi et al., 2020). The authors reported the presence of various compounds belonging to the class of phytosterols, flavonoids, terpenoids, tannins, alkaloids and carbohydrates in the truffle extracts. An increase in the levels of the TH1 cytokines IFN-γ and IL-2 and a decrease in the level of the TH2 cytokine IL-4 were reported upon lymphocyte stimulation (Al Obaydi et al., 2020). The desert truffles *T. boudieri* and *T. claveryi* have also been studied mechanistically to understand the underlying anti-inflammatory mechanisms (Dawood et al., 2023). The authors reported a reduced production of nitric oxide (NO) as well as the production of mRNA and protein level expression of various pro-inflammatory markers such as NO synthase COX-2, TNF-α and IL-6 with the use of *T. boudieri* (Dawood et al., 2023). A comparable inhibitory effect of *T. claveryi* extracts was also seen. *T. boudieri* and *T. claveryi* extracts displayed anti-inflammatory effects at epigenetic levels by significant downregulation of microRNA expression levels of miR-146a, miR-155 and miR-21 in LPS/IFN-γ-stimulated RAW 264.7 cells (Dawood et al., 2023). Other studies focusing on the immunomodulatory potential of truffles are summarized in Table 1.

2.3 Anticancer Potential

The anticancer potential of various truffle species has been studied using *in vitro* and *in vivo* methods. Many researchers have used *in vitro* cell viability and proliferation assays using truffle bioactives to study the anticancer effect of truffles against cancer cell lines. More reliable *in vivo* or *ex ovo* methods such as the chorioallantoic membrane assay have also been utilized. Furthermore a mechanistic approach has also been utilized to understand the anticancer effect of truffles at the molecular level by studying the protein expression in cancer cell lines. For instance the levels of the Bax/Bcl-2 ratio which play a crucial role in the cell death switch leading to decreased cellular resistance to apoptosis and hence a lower incidence of tumor formation, have been seen with the use of *T. boudieri* extracts against HCT-116 and Caco-2 cell lines (Sawaya et al., 2023). The authors also reported elevated expression of p53 in both cell lines with the use of HCT-116 and Caco-2. On the contrary these cell lines showed a decrease in the levels of cyclin D1 and c-Myc with the use of *T. boudieri* extracts (Sawaya et al., 2023) thus hinting at a potential anticancer effect of *T. boudieri* extracts. Similarly (Saleh et al., 2022) showed the anticancer effect of various Terfezia species, *T. claveryi, T. boudieri* and *T. olbiensis*) on pancreatic cancer using the PANC-1 cell line. The progression of PANC-1 cell growth was reduced by each extract through the induction of apoptosis which was evident from the upregulation of the proapoptotic genes BAX, CDKN1A and TP53 and the downregulation of the antiapoptotic gene BCL2 (Saleh et al., 2022).

Some studies have also examined the anticancer activity of truffles in a more focused manner by isolating bioactives using chromatographic techniques to assess the anticancer potential. For instance (Çayan et al., 2022a) isolated various compounds from *Reddellomyces westraliensis* such as brassicasterol ergosta-7,9,22-trien-3β-ol, ergosterol peroxide, adenosine, D-ribitol, fumaric acid, and mannitol amongst which ergosta-7,9,22-trien-3β-ol exhibited the best cytotoxic activity against cancer cell lines H1299 and MCF-7 with an IC50 of 38.62 ± 0.46 μg/ml and 24.93 ± 0.80 μg/ml (Çayan et al., 2022a). In another study, several polysaccharides were isolated from mycelial cultures of various truffle species such as *T. melanosporum, T. sinense, T. indicum* and *T. aestivum* as well as the fruiting bodies of some truffles such as *T. indicum, T. sinense* and *T. himalayense* and were reported for their dose-dependent anticancer potential against cancer cell lines, e.g., SK-BR-3, HepG2. HL-60, A549 and HCT-116 (Zhao et al., 2014).

Although the majority of studies have focused on studying the cytotoxic activities of truffles as a means to evaluate anticancer activity, a few studies have attempted to employ models that are closer to an *in vivo* milieu. For instance the use of an angiogenesis model using chicken embryos to study the anticancer effect of the summer truffle *T. aestivum* has been attempted. Both polar and nonpolar extracts of *T. aestivum* were reported to reduce angiogenesis in chick chorioallantoic membrane assays pointing to the potential anticancer activity of the truffle (Marathe et al., 2020). The authors also studied the bioactive composition of the extracts using gas chromatography-mass spectrometry in an attempt to correlate the observed antiangiogenic effect with the *T. aestivum* bioactives (Marathe et al., 2020). Apart from the studies discussed above a few more researchers have attempted to evaluate the anticancer potential of truffles. These studies have been summarized in Table 1.

2.4 Antioxidant Potential

Among the various biological activities that have been explored for truffle bioactives; the antioxidant potential of truffles has been the most widely explored. The most common methods for assessing the antioxidant potential of truffles are performed using *in vitro* assays such as ABTS, DPPH and FRAP. Because food processing causes a significant loss in the antioxidant potential of bioactives certain studies have also assessed the impact of processing conditions on the antioxidant activity of truffles. For instance (Murcia et al., 2002) studied the effect of industrial processing conditions such as freezing and canning on the antioxidant activity of two truffles, viz. *T. claveryi* Chatin and *Picoa juniperi* Vittadini and compared them with commercial food antioxidants such as α-tocopherol (E-307), propyl gallate E-310, BHT E-321 and BHA E-320 (Murcia et al., 2002). The truffles presented a higher antioxidant potential than the commercial antioxidants as evident from lipid peroxidase (H_2O_2) peroxidation (LOO·) and deoxyribose (OH·) assays. Although freezing was inconsequential canning reduced the antioxidant potential of the truffles Murcia et al. (2002) thus making the careful assessment of the effect of processing conditions on truffle bioactivity an important factor to be considered for storage.

The difference in the geographical origin of truffles can also have a significant impact on the biological potential of truffles belonging to the same species. This can be attributed mainly to the differences in the environmental conditions of the truffle-bearing regions. For instance, a difference in the antioxidant capacity of the desert truffle *T. nivea* originating from various Middle Eastern regions has been seen by Al-Laith (2010). In that compared to *T. nivea* originating in Bahrain,Morocco and Saudi Arabia *T. nivea* from Iran showed higher antioxidant potential evident from a higher percent inhibition and lower IC50 values assessed using various assays such as FRAP, DPPH, nitric oxide (NO) radical scavenging activity and hydroxyl radical scavenging activity by a deoxyribose assay Al-Laith (2010). Likewise, truffles can show substantial differences in their antioxidant potential due to differences in geographical origin within a country. A study reported differences in the nutritional content and antioxidant properties of Chinese black truffles (*T. indicum*) collected from various geographical regions in China Wu et al. (2021). In that regard *T. indicum* samples from Yunnan showed significantly higher values for DPPH, ABTS and FRAP assays compared to *T. indicum* samples from Jilin and Sichuan Wu et al. (2021). This has been supported by another independent study on *T. indicum* collected from five different regions of four Chinese provinces viz. Yunnan, Tibet, Hebei and Liaoning where *T. indicum* from Qujing in Yunnan province showed superior antioxidant potential assessed with DPPH and ABTS radical-scavenging assays (Li et al., 2019).

The stage of maturity of truffles also plays a role in their nutritional profile and thus their antioxidant potential. This has been demonstrated in a few studies. For instance for Chinese black truffle *T. indicum* the month of truffle collection was shown to cause an effect on its antioxidant potential where the ascocarps collected in the month of July in immature stage showed the highest ABTS scavenging activity with an EC50 of 3.73 mg/ml; however the ascocarps collected in August (mature stage) displayed a more potent antioxidant activity in DPPH and FRAP assays with an EC50 of 3.73 mg/ml and a ferric reducing activity power of 243.63 μmoL FeSO4/g respectively Li et al. (2022). Similarly, Shah et al. (2020) showed that mature *T. aestivum* ascocarps possess a superior antioxidant potential than immature *T. aestivum* ascocarps as evident from their lower IC50 in ABTS and DPPH assays Shah et al. (2020). Although researchers prefer evaluating the antioxidant potential of truffles using their crude extracts or specific molecular entities such as polysaccharides, importance importance has also been given to isolating specific bioactives from truffles and evaluating their antioxidant potential. For instance, various bioactive compounds such as brassicasterol ergosta-7,9,22-trien-3β-ol ergosterol peroxide adenosine D-ribitol fumaric acid and mannitol from *R. westraliensis* have been isolated Çayan et al. (2022a). The authors studied the antioxidant potential of these compounds using DPPH and ABTS assays where D-ribitol displayed the most potent activity with an IC50 of 8.38 ± 0.40 µg/ml and 6.74 ± 0.22 µg/ml respectively for these assays Çayan et al. (2022a). In a similar study compounds such as fumaric acid, brassicasterol, mannitol and ergosterol peroxide were isolated from *R. parvulosporus* and reported for their comparable antioxidant potential Çayan et al. (2022b).

Fewer studies have given importance to assessing the antioxidant potential of truffles in the cellular system or *in vivo*. Guan et al. (2022) reported the antioxidant potential of purified polysaccharides from *T. aestivum* and *T. melanospourm* by evaluating their capabilities to alleviate H_2O_2-induced oxidative stress against porcine jejunal epithelial (IPEC-J2) cells Guan et al. (2022). Similarly the antioxidant potential of *T. indicum* polysaccharides in H_2O_2-induced oxidative stress against PC12 cells as well as their free radical scavenging activity have also been reported Luo et al. (2011). Several other studies have reported the potent antioxidant potential of truffles belonging to various species and obtained from various regions of the world. A thorough account of such reports has been compiled and presented in Table 1.

2.5 Antimicrobial Activity

Research focusing on antimicrobials is crucial in order to safeguard public health on a global level especially in recent times considering the rising concerns about antimicrobial resistance. Bioactive compounds from natural sources have the potential to be used as alternatives to conventional antibiotics. With the broad spectrum of secondary metabolites that fungi produce they are a potential source of bioactives that can exhibit bacteriostatic or fungistatic effects Sułkowska-Ziaja et al. (2023). Despite the abundant content of potent secondary metabolites in truffles, very few researchers have attempted to explore their antimicrobial properties which can perhaps be attributed to the rare occurrence of truffles and their costliness. The *in vitro* antimicrobial potential of three Algerian desert truffles viz. *T. claveryi, T. leonis,* and *T. nivea* against two opportunistic pathogens *Pseudomonas aeruginosa* and *Staphylococcus aureus* has been studied by Gouzi et al. (2011). The authors reported 86.48% and 99.09% inhibition of *S. aureus* by *T. claveryi* and *T. nivea* respectively. On the other hand *T. nivea* showed complete inhibition of *P. aeruginosa* growth whereas *T. claveryi* inhibited the growth of *P. aeruginosa* by 71% (Gouzi et al., 2011). The antimicrobial efficacy of *T. claveryi* against Gram-positive bacteria: *Bacillus subtilis* and *S. aureus*; Gram-negative bacteria: *P. aeruginosa* and *Escherichia coli* and also against fungi *Aspergillus oryzae* and *Candida albicans* has been independently reported in another study Khojah et al. (2022). In another interesting study the antimicrobial activity of silver nanoparticles synthesized using Hafr Al-Batin desert truffles of *Terminia* species against three Gram-positive bacteria viz. *B. subtilis, B. cereus* and *S. aureus* as well as Gram-negative bacteria such as *Salmonella enterica* and *E. coli* was reported (Alshammari and Abd El Aty, 2022). The authors also reported antifungal activity against the yeast *C. albicans* and the mold *A. niger* (Alshammari and Abd El Aty, 2022). A few other studies that report the antimicrobial potential of truffles have been presented in Table 1.

3. Conclusion

Besides their biological properties truffles are a promising source of bioactive compounds with several potential implications for human health. Truffles have been widely known for their aphrodisiac properties in many cultures. However, scientific research and evidence on the aphrodisiac potential of truffles are scarce. Similarly truffles have been explored for their anti-cholinesterase activity by a few researchers. This necessitates and paves the way for further exploring the biological activities of truffles. Apart from fresh truffles utilized in culinary applications a variety of truffle products including truffle oil,salt,butter, sauces, pastes and cheeses are readily available for consumption. Despite the rich potential of fresh truffles as a source of bioactive compounds with promising biological activities there is a need and scope to assess the biological potential of truffle food products. This is primarily because processing conditions significantly impact the stability of these bioactives. Exploring truffle food research for potential health applications or for the development of functional foods entails thorough investigations into the preservation of bioactive compounds during industrial processing as well as exploring the potential health benefits of consuming these products. This research could facilitate the development of possible truffle-based functional foods with enhanced nutritional and health-promoting properties.

References

Abu-Odeh, A., Shehadeh, M., Suaifan, G.A., Karameh, N., Abdel Rahman, D. et al. (2022). *In vitro* and *in vivo* antidiabetic activity (phenolic content and microscopical characterization of *Terfezia claveryi*. Molecules, 27: 4843. https://doi.org/10.3390/molecules27154843

Al Obaydi, M.F., Hamed, W.M., Al Kury, L.T. and Talib, W.H. (2020). *Terfezia boudieri*: A desert truffle with anticancer and immunomodulatory activities. Front. Nutr., 7: 38. https://doi.org/10.3389/fnut.2020.00038

Al-Laith, A/A/A. (2010). Antioxidant components and antioxidant/antiradical activities of desert truffle (*Tirmania nivea*) from various Middle Eastern origins. J. Food Compos. Anal., 23: 15–22. https://doi.org/10.1016/j.jfca.2009.07.005

Alshammari, S.O. and Abd El Aty, A.A. (2022). Statically controlled mycogenic-synthesis of novel biologically active silver-nanoparticles using Hafr Al-Batin desert truffles and its antimicrobial efficacy against pathogens. Saudi J. Biol. Sci., 29: 103334. https://doi.org/10.1016/j.sjbs.2022.103334

Beara I., Majkić, T. and Torović, L. (2021). Bioguided design of new black truffle (*Tuber aestivum* Vittad.) product enriched with herbs and spices. LWT, 138: 110637. https://doi.org/10.1016/j.lwt.2020.110637

Beara I.N., Lesjak, M.M., Četojević-Simin, D.D., Marjanović, Ž.S., Ristić, J.D. et al. (2014). Phenolic profile (antioxidant (anti-inflammatory and cytotoxic activities of black (*Tuber aestivum* Vittad.) and white (*Tuber magnatum* Pico) truffles. Food Chem., 165: 460–466. https://doi.org/10.1016/j.foodchem.2014.05.116

Çayan, F., Tel-Çayan, G., Deveci, E., Duru, M.E., Öztürk, M. et al. (2022a). Isolation and identification of compounds from truffle *Reddellomyces westraliensis* and their antioxidant (cytotoxic and enzyme inhibitory activities. Process Biochem., 121: 553–562. https://doi.org/10.1016/j.procbio.2022.08.001

Çayan, F., Tel-Çayan, G., Deveci, E., Duru, M.E., Türk, M. et al. (2022b). A detailed study on multifaceted bioactivities of the extracts and isolated compounds from truffle *Reddellomyces parvulosporus*. Int. J. Food Sci. Technol. (57: 1411–1419. https://doi.org/10.1111/ijfs.15190

Chen, G., Zhang, S., Ran, C., Wang, L. and Kan, J. (2016). Extraction (characterization and antioxidant activity of water-soluble polysaccharides from *Tuber huidongense*. Int. J. Biol. Macromol., 91: 431–442. https://doi.org/10.1016/j.ijbiomac.2016.05.108

Dahham, S.S., Al-Rawi, S.S., Ibrahim, A.H., Majid A.S.A., Majid A.M.S.A. et al. (2018). Antioxidant anticancer apoptosis properties and chemical composition of black truffle *Terfezia claveryi*. Saudi J. Biol. Sci., 25: 1524–1534. https://doi.org/10.1016/j.sjbs.2016.01.031

Darwish R.S., Shawky E., Nassar K.M., ElSayed R.M.R., Hussein D.E. et al. (2021). Differential anti-inflammatory biomarkers of the desert truffles *Terfezia claveryi* and *Tirmania nivea* revealed via UPLC-QqQ-MS-based metabolomics combined to chemometrics. LWT (150: 111965. https://doi.org/10.1016/j.lwt.2021.111965

Dawood, A.S., Sedeek, M.S., Farag, M.A. and Abdelnaser, A. (2023). *Terfezia boudieri* and *Terfezia claveryi* inhibit the LPS/IFN-γ-mediated inflammation in RAW 264.7 macrophages through an Nrf2-independent mechanism. Sci. Rep., 13: 10106. https://doi.org/10.1038/s41598-023-35612-8

El Enshasy, H., Elsayed, E.A., Aziz, R. and Wadaan, M.A. (2013). Mushrooms and truffles: Historical biofactories for complementary medicine in Africa and in the middle East. Evid. Based Compl. Alt. Med. 620451. https://doi.org/10.1155/2013/620451

Fontes A., Ramalho-Santos J., Zischka H. and Azul A.M. (2022). Mushrooms on the plate: Trends towards NAFLD treatment (health improvement and sustainable diets. Eur. J. Clin. Invest., 52. https://doi.org/10.1111/eci.13667

Gouzi, H., Belyagoubi, L., Abdelali, K.N. and Khelifi, A. (2011). *In vitro* antibacterial activities of aqueous extracts from Algerian desert truffles (*Terfezia* and *Tirmania* (ascomycetes) against *Pseudomonas aeruginosa* and *Staphylococcus aureus*. Int. J. Med. Mushrooms, 13: 553–558. https://doi.org/10.1615/IntJMedMushr.v13.i6.70

Gouzi, H., Leboukh, M. and Bouchouka, E. (2013). Antioxidant and antiradical properties of methanolic extracts from Algerian wild edible desert truffles (*Terfezia* and *Tirmania* (*Ascomycetes*). Int. J. Med. Mushrooms, 15(5): 471–486. https://doi.org/10.1615/IntJMedMushr.v15.i5.50

Guan, T., Wei, X., Xu, P., Chen, K., Zou, Y. et al. (2022). Comparison of structural and antioxidant activity of polysaccharide extracted from truffles. J. Food Sci., 87: 2999–3012. https://doi.org/10.1111/1750-3841.16207

Guo, T., Wei, L., Sun, J., Hou, C.L. and Fan, L. (2011). Antioxidant activities of extract and fractions from *Tuber indicum* Cooke & Massee. Food Chem., 127: 1634–1640. https://doi.org/10.1016/j.foodchem.2011.02.030

Hamza, A., Zouari, N., Zouari, S., Jdir, H., Zaidi, S. et al. (2016). Nutraceutical potential (antioxidant and antibacterial activities of *Terfezia boudieri* Chatin a wild edible desert truffle from Tunisia arid zone. Arab J. Chem. (9: 383–389. https://doi.org/10.1016/j.arabjc.2013.06.015

Hsu, J.Y., Chen, M.H., Lai, Y.S. and Chen, S.D. (2022). Antioxidant profile and biosafety of white truffle mycelial products obtained by solid-state fermentation. Molecules, 27: 109. https://doi.org/10.3390/molecules27010109

Jiang, X., Teng, S., Wang, X., Li, S., Zhang, Y. et al. (2018). The antidiabetic and antinephritic activities of *Tuber melanosporum* via modulation of Nrf2-mediated oxidative stress in the db/db mouse. Oxid. Med. Cell Longev., 2018: 7453865. https://doi.org/10.1155/2018/7453865

Khojah, H.M., Abdelhalim, O.B., Mostafa, M.A. and Habib, E.S.E. (2022). Antimicrobial efficacy of extracts of Saudi Arabian desert *Terfezia claveryi* truffles. Saudi J. Biol. Sci., 29: 103462. https://doi.org/10.1016/j.sjbs.2022.103462

Lee, H., Nam, K., Zahra, Z. and Farooqi, M.Q.U. (2020). Potentials of truffles in nutritional and medicinal applications: A review. Fungal Biol. Biotechnol., 7: 1–17. https://doi.org/10.1186/s40694-020-00097-x

Li, J.M., Liang, H.Q., Qiao, P., Su, K.M., Liu, P.G. et al. (2019). Chemical composition and antioxidant activity of *Tuber indicum* from different geographical regions of China. Chem. Biodivers., 16: e1800609. https://doi.org/10.1002/cbdv.201800609

Li, X., Zhu, Z., Ye, L., Kang, Z., Zhang, X. et al. (2020). Comparison of the partial structure and antioxidant activity of polysaccharides from two species of Chinese truffles. Molecules, 25: 4345. https://doi.org/10.3390/molecules25184345

Li, Y., Liang, H., Zhou, D., Xing, Y. and Chen, J. (2022). Phenolics (flavonoids content and antioxidant activities of *Tuber indicum* at different maturity stages. Chem. Biodivers., 19: e202100830. https://doi.org/10.1002/cbdv.202100830

Li, Y., Zhang, F., Gong, J. and Peng, C. (2023). Two novel dipeptidyl peptidase-IV (DPP-IV) inhibitory peptides identified from truffle (*Tuber sinense*) by peptidomics *in silico* and molecular docking analysis. J. Food Compos. Anal., 121: 105384. https://doi.org/10.1016/j.jfca.2023.105384

Luo, Q., Zhang, J., Yan, L.,Tang, Y., Ding, X. et al. (2011). Composition and antioxidant activity of water-soluble polysaccharides from *Tuber indicum*. J. Med. Food, 14: 1609–1616. https://doi.org/10.1089/jmf.2011.1659

Marathe, S.J., Hamzi, W., Bashein, A.M., Deska, J., Seppänen-Laakso, T. et al. (2020). Anti-angiogenic and anti-inflammatory activity of the summer truffle (*Tuber aestivum* Vittad.) extracts and a correlation with the chemical constituents identified therein. Food Res. Int., 137: 109699. https://doi.org/10.1016/j.foodres.2020.109699

Mudliyar, D.S., Wallenius, J.H., Bedade, D.K., Singhal, R.S., Madi, N. et al. (2019). Ultrasound assisted extraction of the polysaccharide from *Tuber aestivum* and its in vitro anti-hyperglycemic activity. Bioact. Carbohydrates Diet Fibre., 20: 100198. https://doi.org/10.1016/j.bcdf.2019.100198

Murcia, M.A., Martínez-Tomé, M., Jiménez, A.M., Vera, A.M., Honrubia, M. et al. (2002). Antioxidant activity of edible fungi (truffles and mushrooms): Losses during industrial processing. J. Food Prot., 65: 1614–1622. https://doi.org/10.4315/0362-028X-65.10.1614

Obaid, A.S., Hassan, K.T., Hassan, O.M., Ali, H.H., Ibraheem, I.J. et al. (2023). *In-vitro* antibacterial (cytotoxicity (and anti-prostate cancer effects of gold nanoparticles synthesized using extract of desert truffles (*Tirmania nivea*). Mater. Chem. Phys., 301: 127673. https://doi.org/10.1016/j.matchemphys.2023.127673

Patel, S., Rauf, A., Khan, H., Khalid, S. and Mubarak, M.S. (2017). Potential health benefits of natural products derived from truffles: A review. 70: 1–8. Trends Food Sci. Technol. https://doi.org/10.1016/j.tifs.2017.09.009

Saleh, K.K., Kirbag, S. and Dalkılıç, S. (2022). Inhibitory effects of *Terfezia* (Ascomycota) desert truffles on PANC-1 cell growth via upregulation of proapoptotic genes TP53 (CDKN1A (and BAX and downregulation of antiapoptotic Gene BCL2. Int. J. Med. Mushrooms, 24: 61–70. https://doi.org/10.1615/IntJMedMushrooms.2022044383

Savini, S., Loizzo, M.R., Tundis, R., Mozzon, M., Foligni, R. et al. (2017). Fresh refrigerated *Tuber melanosporum* truffle: effect of the storage conditions on the antioxidant profile antioxidant activity and volatile profile. Eur. Food Res., 243: 2255–2263. Technol. https://doi.org/10.1007/s00217-017-2927-x

Sawaya, K., Abou Najem, S., Khawaja, G. and Khalil, M. (2023). Proapoptotic and antiproliferative effects of the desert truffle *Terfezia boudieri* on colon cancer cell lines. Evidence-based Complement Altern. Med., 2023: 1693332. https://doi.org/10.1155/2023/1693332

Schillaci, D., Cusimano, M.G., Cascioferro, S.M., Di Stefano, V., Arizza, V. et al. (2017). Antibacterial activity of desert truffles from Saudi Arabia against *Staphylococcus aureus* and *Pseudomonas aeruginosa*. Int. J. Med. Mushrooms, 19: 121–125. https://doi.org/10.1615/IntJMedMushrooms.v19.i2.30

Shah, N., Usvalampi, A., Chaudhary, S., Seppänen-Laakso, T., Marathe, S. et al. (2020). An investigation on changes in composition and antioxidant potential of mature and immature summer truffle (*Tuber aestivum*). Eur. Food Res. Technol., 246: 723–731. https://doi.org/10.1007/s00217-020-03438-7

Souna, M., Tefiani, C., Azzi, R., Chaouche, T.M., Habi, S. et al. (2023). Mycochemical constituents and anti-inflammatory activity of *Terfezia claveryi* Chatin from Algeria. Biol. Life Sci. Forum, 27: 19. https://doi.org/https://doi.org/10.3390/IECAG2023–14980

Sułkowska-Ziaja, K., Trepa, M., Olechowska-Jarząb, A., Nowak, P., Ziaja, M. et al. (2023). Natural compounds of fungal origin with antimicrobial activity—potential cosmetics applications. Pharmaceuticals, 16: 1200. https://doi.org/10.3390/ph16091200

Tejedor-Calvo, E., Amara, K., Reis, F.S., Barros, L., Martins, A. et al. (2021). Chemical composition and evaluation of antioxidant, antimicrobial and antiproliferative activities of *Tuber* and *Terfezia* truffles. Food Res. Int., 140: 110071. https://doi.org/10.1016/j.foodres.2020.110071

Tejedor-Calvo, E., García-Barreda, S., Sánchez, S., Morte, A., Siles-Sánchez, M.D.L.N. et al. (2022). Application of pressurized liquid extractions to obtain bioactive compounds from *Tuber aestivum* and *Terfezia claveryi*. Foods, 11: 298. https://doi.org/10.3390/foods11030298

Tejedor-Calvo, E., Morales, D., Marco, P., Sánchez, S., Garcia-Barreda, S. et al. (2020). Screening of bioactive compounds in truffles and evaluation of pressurized liquid extractions (PLE) to obtain fractions with biological activities. Food Res. Int., 132: 109054. https://doi.org/10.1016/j.foodres.2020.109054

Tejedor-Calvo, E., Morales, D., Morillo, L., Vega, L., Caro, M. et al. (2023). Pressurized liquid (PLE) truffle extracts have inhibitory activity on key enzymes related to type 2 diabetes α-Glucosidase and α-Amylase). Foods, 12: 2724. https://doi.org/10.3390/foods12142724

Thomas, P., Elkhateeb, W. and Daba, G. (2021). Industrial applications of truffles and truffle-like fungi. pp. 82–88. *In*: Sridhar, K.R. and Deshmukh, S.K. (eds.). Advances of Macrofungi: Diversity (Ecology and Biotechnology. CRC Press, Boca Raton.

Wang, S. and Marcone, M.F. (2011). The biochemistry and biological properties of the world's most expensive underground edible mushroom: Truffles. Food Res. Int., 44: 2567–2581. https://doi.org/10.1016/j.foodres.2011.06.008

Wu, Z., Meenu, M. and Xu, B. (2021). Nutritional value and antioxidant activity of Chinese black truffle (*Tuber indicum*) grown in different geographical regions in China. LWT, 135: 110226. https://doi.org/10.1016/j.lwt.2020.110226

Xu, L., Su, Q., Zhang, Z., Guan, G., Huang, D. et al. (2021). Increased extracellular saponin production after the addition of rutin in truffle liquid fermentation and its antioxidant activities. Fermentation, 7: 103. https://doi.org/10.3390/fermentation7030103

Yan, X., Wang, Y., Sang, X. and Fan, L. (2017). Nutritional value, chemical composition and antioxidant activity of three *Tuber* species from China. AMB Express, 7: 1–8. https://doi.org/10.1186/s13568-017-0431-0

Zhang, T., Jayachandran, M., Ganesan, K. and Xu, B. (2018). Black truffle aqueous extract attenuates oxidative stress and inflammation in STZ-induced Hyperglycemic rats via Nrf2 and NF-κB pathways. Front. Pharmacol., 9: 342491. https://doi.org/10.3389/fphar.2018.01257

Zhang, T., Jayachandran, M., Ganesan, K. and Xu, B. (2020). The black truffle *Tuber melanosporum* (Ascomycetes) ameliorates hyperglycemia and regulates insulin signaling pathway in STZ-induced diabetic rats. Int. J. Med. Mushrooms, 22: 1057–1066. https://doi.org/10.1615/IntJMedMushrooms.2020036355

Zhao, W., Wang, X.H., Li, H.M., Wang, S.H., Chen, T. et al. (2014). Isolation and characterization of polysaccharides with the antitumor activity from *Tuber* fruiting bodies and fermentation system. Appl. Microbiol. Biotechnol., 98: 1991–2002. https://doi.org/10.1007/s00253-013-5379-7

Chapter 12

Desert Truffles

A Potential Bioresource for Bioactive Compounds of Medicinal Value

Siti Zulaiha Hanapi,[1] *Dalia Sukmawati,*[2] *Muktiningsih Nurjayadi,*[3] *Faiq H.S. Hussain*[4] and *Hesham Ali El Enshasy*[5,6,7,*]

1. Introduction

Desert truffles belong to the phylum Ascomycota and are classified as mycorrhizal fungi within the order Pezizales (formerly known as Tuberales) (Zniber et al., 2022), associating with *Helianthemum* spp. as host plants (Marqués-Gálvez et al., 2021). They reside within the Pezizaceae family, encompassing various genera

[1] Department of Deputy of Vice Chancellor (Research and Innovation), Universiti Teknologi Malaysia.
[2] Biology Department, Faculty of Mathematics and Natural Sciences, Universitas Negeri Jakarta, Indonesia.
[3] Chemistry Department, Faculty of Mathematics and Natural Sciences, Universitas Negeri Jakarta, Indonesia.
[4] Medical Analysis Department, Faculty of Applied Science, Tishk International University, Erbil, Iraq.
[5] Innovation Centre of Agritechnology in Advanced Bioprocessing (ICA), Universiti Teknologi Malayisa (UTM), Pagoh Campus, Johor, Malaysia.
[6] Faculty of Chemical Engineering, Universiti Teknologi Malaysia, 81310, UTM Skudai, Johor, Malaysia.
[7] City of Scientific Research and Technology Applications, New Burg Al Arab, Alexandria, Egypt.
* Corresponding author: hesham@utm.my

such as *Terfezia, Tirmania, Delastria, Picoa, Balsamia*, and *Melanogaster*. Within Arab countries, they are usually called "Kamé", "Kholassi", and "Zoubaïdi" but "truffles of sand" are also found in sandy desert soils; while in Spain, they are called "Criadilla de Tierra" and "Criadilla Vaquera". In the Middle East it is called "Kamaah" and "Fagaa" (Farag et al., 2021). In arid regions of North Africa, Western Asia and Turkey desert truffles or Mantariin Tukish, are known as "Terfas", which means earth mushrooms (Ghellam et al., 2020). While the term "truffle" traditionally refers to the large fruiting bodies of fungi belonging to the genus Tuber, its usage has broadened in practice. It now commonly encompasses the subterranean fruiting bodies (hypogeals) of many other fungal species. Distributions of desert truffles in Africa and the Middle East region and their historical biofactories for complementary medicine have been reviewed previously by El Enshasy et al. (2013). Desert truffles are hypogeous ascomycetes commonly presented by their two most common genera: *Trimania* (white terfez) and *Terfezia* (red terfez) (Dib-Bellahouel et al., 2019). These edible macrofungi are distributed naturally from North African countries such as Morocco, Tunisia, Algeria and Egypt to other Asian middle-eastern countries, including Iran, Iraq, Jordan, Kuwait, Lebanon, Saudi Arabia and Syria (Al-Laith, 2010; Dib-Bellahouel et al., 2019; Owaid, 2022).

Generally, truffles don't have any specific shape (Chauhan et al., 2021). However, they are normally found in brown, gray or white with spherical shapes, a spongy texture and different sizes ranging from 1 to 7 cm. Lee et al. (2020) reported that truffles are usually sturdy, dense, and woody in comparison to mushrooms, which are softer and more fragile. In most truffles studied, the flesh is initially white and becomes darker as it matures, with a marbling of lighter tissue (Ghorai et al., 2011). In Morocco, ascocarps collected showed different shapes of cordiform, round and ovoid measuring from 3 to 5 cm, with fresh weights ranging between 17 and 50 g (Zniber et al., 2022). Algerian Saharan desert truffles, *T. arenaria*, presented an outer shell whose thickness varies (0.5–1.0 mm) and was pale-yellow to brown (Harir et al., 2019). Truffles are known for their unusually large genome size, such as in *T. melanosporum*, which has 125 Mb (Chen et al., 2014). This is partly due to the presence of high transposable elements and repetitive DNA content, which account for more than 58% of the truffle genome. A schematic view of the three major clades (Pezizales, Truffles and their genus) and families is mainly introduced by Lee et al. (2020) (Fig. 1).

Desert truffles are considered expensive, with the costliest being European species. However, people in the Middle East region are considered the largest consumers of truffles despite being costly (Al-Laith, 2010). The most expensive truffles include: *Tuber melanosporum* (black truffle), *T. magnatum* (white truffle), *T. aestivum* (Burgundy truffle), *Tirmania nivea* and *Terfezia chlaveryi* (desert truffles) (Chauhan et al., 2021). The intense taste and aroma of commercially collected truffles, such as a species known as black truffles, are so potent that they are used as a flavoring instead of a cuisine (Mello et al., 2006). An analysis of *Trimania nivea*'s volatile profile revealed eight-carbon volatile compounds, with 1-octen-3-ol being the most abundant (64%); it also identified 29 novel compounds (Ferreira et al., 2023). An initial study on ethnomycological facets and traditional folklore of the desert truffle in Bahrain found that *Tirmania nivea* was the most preferred

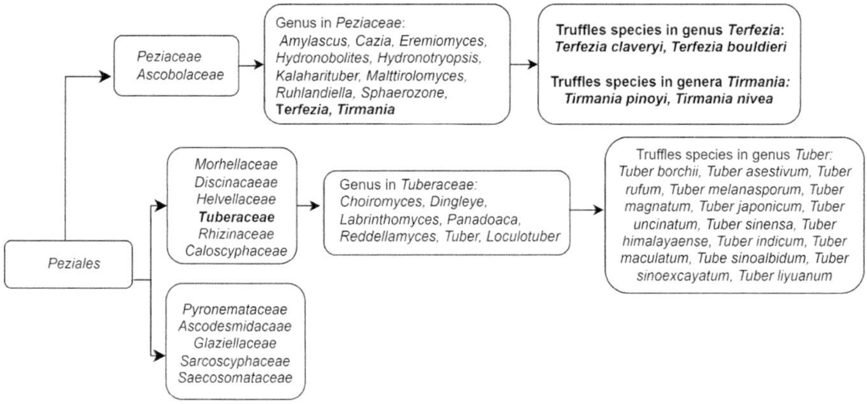

Fig. 1. A schematic of Pezizales. Truffles, their genus, and family (adopted from Lee et al., 2020).

and expensive truffle owing to its pleasant light aroma, delicacy, and white soft tissues (Mandeel et al., 2007). On the other hand in the native Sahara area of Algeria, *T. claveryi is* the most consumed and expensive species (Bradai et al., 2015a). This study documents the nutritional, medicinal and bioactive potential of desert truffles, along with their production and bioprospect potentials.

2. Nutritional and Medicinal Value of Truffles

Research going back into time has found that desert truffles are rich in proteins, minerals, dietary fibers, vitamins, unsaturated fatty acids and many other compounds with health benefits (Patel, 2012; Gadallah et al., 2016; Bouatia et al., 2018; Owaid, 2022). Desert truffles harbor rich phytochemicals such as phenolic compounds that possess several medical effects such as antimicrobial (Khadhri et al., 2017), anticancer (Dahham et al., 2018; Al Obaydi et al., 2020), anti-inflammatory and antioxidant effects (Chauhan et al., 2021). The desert truffle has also been reported to have hepatoprotective efficacy (Janakat et al., 2010). In China, the unique desert truffles, *Mattirolomyces terfezioides* were reported to contain sulfur volatiles (Lu et al., 2022). The abundance of myoinositol and trehalose in *T. claveryi* and *T. boudieri* showed potential as anti-diabetic functional foods (Farag et al., 2021). Bioactive compounds in different types of desert truffles are given in Table 1.

In the Sahara desert, desert truffles are used against eye infections and to promote male fertility (Bradai et al., 2015a). As reported by Malik et al. (2018), an aqueous extract from the brown truffle species *Terfezia claveryi* contains a potent antimicrobial with a low molecular weight less than 10kDa (MW < 10 kDa). The antimicrobial properties of this extract has potential applications in treating eye infections caused by *Escherichia coli, Staphylococcus aureus* and *S. epidermidis*. This finding suggests the extract's potential as a natural alternative to chemically synthesized antibiotics, thereby aiding in the fight against the development of drug-resistant bacteria. Some research has documented the importance of nutritional composition, particularly in the protein and amino acids of desert truffles, as shown in Table 2.

Table 1. Bioactive compounds in different types of desert truffles.

Type of truffles	Extraction process	Bioactive compounds	References
Terfezia claveryi	Extraction involved using solvents (hexane, ethyl acetate, ethanol, methanol, and water) with a hot maceration method at 40°C.	Stigmasterol, beta-Sitosterol, squalene, lupeol, octadecadienoic acid, and oleic acid.	Dahham et al. (2018)
Terfezia claveryi	70/30 solvent mix of ethanol and water.	Gallic acid, Sinapic acids and Chlorogenic acid	Khlaif et al. (2021)
Terfezia boudieri and *Terfezia claveryi*	Water and ethanol	Alcohols, ketones, aldehydes, esters, ethers and furans.	Farag et al. (2021)
Terfezia claveryi (black) and *Tirmania pinoyi* (White)	Maceration with water, methanol, and ethanol 70%.	p-Hydroxy benzoic acid, Syringic acid and trans-Cinnamic acid	Al-Atassi et al. (2022)
Tirmania pinoyi (Maire) Malencon	Dried ascoms added with Ethyl acetate in soxhlet/3 hr	Pyrazines Quinones	Dib-Bellahouel et al. (2019)
Tuber melanosporum, *T. aestivum*, *T. aestivum* var. *uncinatum*	Water and ethanol	9.1% β-glucan and 4.5% ergosterol	Tejedor-Calvo et al. (2020)

Table 2. Nutritional composition of desert truffles (NA, not available).

Truffle Type	Chemical Compositions	Minerals	Reference
Terfezia arenaria	Protein (14.04 to 24.15%) and Crude fat (5.10%)	NA	Tejedor-Calvo et al. (2021)
Terfezia arenaria	Carbohydrates (67%), proteins (14%) and dietary fibre (10%)	NA	Ferreira et al. (2023)
Terfezia boudieri	Carbohydrates (62%), Proteins (26%), Ash (4.49%) and Fat (8%)	Potassium, phosphorus, iron, and calcium	(Hamza et al. (2016)
Terfezia claveryi	Total Protein (16%), Total carbohydrate (28%), Total crude fibre (4%) and Total crude fat (2%)	NA	Bokhary et al. (1987)
Terfezia claveryi	Moisture (81.5%), Carbohydrates (64.7%), Crude proteins (26.2%), Lipids (3.0%), Fibers (16.3%) and Amino acids (18.5%)	NA	Khlaif et al. (2021)
Tirmani pinoyi	Moisture (81.5 g/100 g fresh mass), total carbohydrate (64.74 g/100 g, dry mass), Crude protein (26.96 g/100 g dry mass), Crude fat (3.01 g/100 g dry mass), Ash (5.28 g/100 g, dry mass), insoluble ash (0.0047 g/100 g, dry mass) and dietary fiber (13.3 g/100 g dry mass).	Copper (65.3 mg/kg). Cyanide, oxalate, mercury and arsenic were absent	Bouatia et al. (2018)

2.2 Antioxidant Activity

It was reported previously that dried truffles were found to vary in their antioxidant and antiradical activities (Al-Laith, 2010). *Terfezia boudieri, T. claveryi*, and *Tirmania nivea* have been reviewed to have strong antioxidant potential along with antibacterial and antiviral effects in vitro and in animal models (Veeraraghavan et al., 2022). The antioxidant activities of truffles are usually attributed to the high content of ascorbic acid, anthocyanins, bioactive phenolics, and many other compounds with known antioxidant properties. Table 3 summarizes some antioxidant properties of various types of desert truffles.

Table 3. Antioxidant activities of desert truffles.

Species	**Method**	**Bioactive compound of potential antioxidant property**	**Reference**
Terfezia arenaria	TPC and TFC	Crude extract	Harir et al. (2019)
Terfezia boudieri	DPPH	Catechin (20 mg/g), Ferulic acid (15 mg/g), P-coumaric acid (10 mg/g) and Cinnamic acid (6 mg/g).	Dogan et al. (2013)
Terfezia boudieri	DPPH TFC	Ascorbic acid (12.2%), Total carotenoids (1.43%) and Total anthocyanins (35.40 %)	Hamza et al. (2016)
Terfezia claveryi	DPPH, FRAP and ABTS methods	Crude extract using different solvent system: water, hexane, ethanol, methanol and ethyl acetate	Dahham et al. (2018)
Terfezia claveryi	TPC, TFC, DPPH, FRAP	Crude extract	Al-Atassi et al. (2022)
Tirmania nivea	Ferric reducing ability (FRAP), DPPH, deoxyribose, and nitric oxide (NO	Ascorbic acid: 9.6 ± 0.15 g/100 g; Anthocyanins: 12.0 ± 8.34 g/100 g; Total esterified phenolics: 1860 ± 361 mg/100 g; total free phenolics: 1328 ± 167 mg/100 g; Total flavonoids: 293 ± 2 mg/100 g; Total carotenoids: 681 ± 245 mg/ 100g	Al-Laith (2010)

2.3 Antimicrobial and Antifungal Activity

A growing body of literature has suggested that desert truffles may offer a promising alternative source of safe and effective commercial antibiotics. *Tirmania pinoyi* extract showed the highest virucidal impact and pre-treatment antiviral activity, while *Tuber indicum* extract has the highest post-treatment antiviral activity (El Khateeb et al., 2024). The extract of *T. claveryi* possesses antibacterial activity and demonstrates efficacy in healing corneal ulcers in rabbit eyes. Antibacterial potential was also reported in *T. claveryi*, which showed more effectiveness against methicillin-resistant *Staphylococcus aureus* (MRSA) than *T. nivea,* which is also in agreement with the study of Casarica et al. (2016). While these results are promising, further investigation is warranted to compare their effectiveness to established treatments

(Aldebasi et al., 2012). As shown in Table 4, different types of desert truffles showed antimicrobial activities. However, in most studies, they reported higher antibacterial properties than antifungal. More interestingly, some of the extracts of desert truffles, such as *Terfezia nivea* and *T. claveryi*, showed antibacterial activity and were able to inhibit the growth of MRSA.

Table 4. Antibacterial and antifungal activities of truffles (NA, not available).

Species of Truffle	Antibacterial activity	Antifungal activity	Reference
Terfezia arenaria	No activity against *Enterococcus faecalis, Escherichia coli, Staphylococcus aureus,* but low to Pseudomonas aeruginosa	*Aspergillus niger, Candida albicans* and *Penicillium* sp.	Harir et al. (2019)
Terfezia clavery	*Staphylococcus aureus*	NA	Dib-Bellahouel et al. (2019)
Terfezia claveryi	*Bacillus subtilis* (IFO3007), *Escherichia coli* and (IFO3301), *Pseudomonas aeruginosa* IFO3448) and *Staphylococcus aureus* (IFO3060). Methanolic extract showed potent anti-biofilm activity (IC50, 2.0 ± 0.18 mg/mL) against *Staphylococcus aureus.*	*Aspergillus oryzae* (IFO4177) and *Candida albicans* (IFO0583)	Khojah et al. (2022)
Terfezia nivea	*Escherichia coli, Staphylococcus aureus* and *S. epidermidis*	NA	Casarica et al. (2016)
Terfezia nivea *T. claveryi*	Active against Methicillin Resistant *Staphylococcus epidermidis aureus* (MRSA)	NA	Malik et al. (2018)
Terfezia nivea *T. claveryi*	*Bacillus subtilis* and *Escherichia coli*	NA	El Sayed et al. (2019a)
Tirmania pinoyi	*Bacillus subtilis, Enterococcus faecalis, Escherichia coli* (ATCC 25922), *Salmonella typhimurium* and *Streptococcus* sp. (ATCC 29212)	*Aspergillus niger* (ATCC 16404), Candida *albicans, Fusarium oxysporum* sp. *albedinis* and *Gliocladium roseum*	Dib-Bellahouel et al. (2019)
Tirmania sp.	*Pseudomonas aeruginosa* and *Staphylococcus aureus*	-	Owaid et al. (2018)

2.4 Anti-Inflammatory Activity

The anti-inflammatory property is one of the main functional characteristics of desert truffles, as reported by many authors. A recent study reported that the hydro-methanol extract of the desert truffle, *Terfazia claveryi*, exhibited in vitro anti-inflammatory activity using the heat denaturation protein inhibition method. This was attributed to the existence of phenols, flavonoids and tannins in the tested extract (Souna et al., 2023). The study extracts of *T. boudieri* and *T. claveryi* appear to reduce inflammation by suppressing the production of inflammatory molecules (Dawood et al., 2023). Furthermore, a recent study reported that *Tirmania nivea* and *Tirmania ponoyi* display the lowest inhibitory effect but higher wound closure compared to *Tuber indicum*, suggesting wound healing capabilities (Elkhateeb et al., 2024).

2.5 Anticancer Activity

The cytotoxicity test of *T. claveryi* showed moderate anticancer activities in the brain cell line (U-87 MG) (IC_{50} of 50 μg/ml) and significant apoptosis through the mitochondrial pathway and DNA fragmentation (Dahham et al., 2018). In another study, the anticancer properties of *T. claveryi* were also reported by Bawadekji (2018), who determined that the apoptotic effect of *T. claveryi* polysaccharide extract on Ehrlich's ascites carcinoma (EAC) cells was dose and time-dependent. The *in vitro* assessment study for the anticancer properties of hexane and ethyl acetate extracts of two types of desert truffles, *Trimania nivea* and *Terfezia claveryi*, collected from Saudi Arabia was carried out by Elsayed et al. (2019a, 2019b). The results showed clear dose-dependent anticancer properties for both extracts against MCF-7, HepG2 and L929 cells. A similar study by Saleh et al. (2022) discovered that *Terfezia* desert truffles inhibit pancreatic cancer cell growth via apoptosis induction, making them potential therapeutic foods.

Extracts derived from different types of truffles, such as *Terfezia claveryi* and *Terfezia gennadii*, demonstrated promising anti-proliferative activity (Tejedor-Calvo et al., 2021). This is evidenced by their ability to inhibit the cellular growth of different cancer cell lines, including HepG2 (liver cancer), HeLa (cervical cancer), MCF-7 (breast cancer) and NCI-H460 (lung cancer). In the same study, *Terfezia arenaria* extract was found to exhibit potential hepatotoxicity, as indicated by its inhibition of growth in PLP2 cells (a primary porcine liver cell culture) at a GI50 concentration of 220 μg/ml. *Terfezia boudieri, T. claveryi*, and *Tirmania nivea* were reported for their anti-cancer and antidiabetic properties (used as folk medicine) after being proven efficacious by various *in vivo* studies (Veeraraghavan et al., 2022).

3. Cultivation of Truffles and Bioprospects

Nowadays, the demand for innovative culinary experiences using truffles is contributing to the growing interest in desert truffle plantations. In the year 2000, mushroom and truffle production combined reached 8.78 million tons, and significantly increased to 44.20 million tons in the following year (Ferreira et al., 2023). The widespread collection of desert truffles in various regions without due consideration for conservation poses a threat to biodiversity, emphasizing the importance of conducting basic and applied research for the sustainable utilization of these valuable natural resources (Al-Qarawi et al., 2012). However, the cultivation of desert truffles has not been given importance in agricultural R & D for decades, but global warming demands an urgent need to develop new crops suitable to arid conditions, which are current steps in domestication efforts (Morte et al., 2020). Andrino et al. (2019) mentioned that this might be due to limited knowledge on the environmental factors concerned with desert truffle fructification. The land-use changes, along with shifts in rain patterns as a result of climate change, have led to a decline in the wild production of truffles (Morte et al., 2021). The most important constraint for their cultivation is the difficulty in the optimization and stabilization of fruit body production over time (Morte et al., 2012).

The morphology of desert truffles usually varies depending on moisture, temperature, the physicochemical properties of the soil and intraspecific differences (Zniber et al., 2022). Al-Qarawi et al. (2012) provided a comprehensive overview of truffle research, taxonomy, biotechnology, physiology, preservation, mycorrhizal relationships for sustainability, cultivation method associations with Helianthemum species for sustainable use, and cultivation techniques . The cultivation of the first desert truffle, *Terfezia claveryi* mutualistic with *Helianthemum almeriense* took place in 1999 in the southeast of Spain (Morte et al., 2008a). This has prompted many researchers to conduct research. This includes details of both biotechnological potential and practical plantation management; it has been compiled in numerous publications (Morte et al., 2008a; Morte et al., 2012; Morte et al., 2017; Slama et al., 2010). Recent studies demonstrated that *Terfezia claveryi* and *Tirmania nivea* desert truffles thrive in sandy, slightly alkaline (pH 7.00 ± 8.01), mildly saline (38.70 ± 75.70) soils with moderate levels of bicarbonates, sulfates, chlorides, and calcareous content (Salama et al., 2023). These soils also contain moderate organic matter (1.4 ± 10.3%) and low levels of magnesium, sodium, and potassium.

Favorable conditions for the desert truffle species have been elucidated through recent investigations. Earlier studies reported that factors that influence desert truffle production are the frame of the plantation, planting season, soil characteristics, water availability (irrigation) and weed management (Morte et al., 2012). In general, desert truffles thrive in calcareous, well-aerated, sandy soil, exhibiting a uniform texture. Occasionally, they are found in calcareous sandy loam soil but seldom in acidic soil. Desert truffles primarily require little water (Turgeman et al., 2011), both in quantity and timing, with a preference for winter and spring precipitation. They thrive in diverse bioclimatic stages, from arid to Saharan regions. Recent studies highlight the prevalence of desert truffles in regions like Beni Guil and Abou Lakhal, characterized by moderate, year-round rainfall (170 mm), averaging temperatures of 17.9–18.6°C. These areas, including the Great Sahara in southern Morocco, feature favorable conditions like evening rain, dewfalls, thunder, and lightning, crucial for truffle growth (Zniber et al., 2022).

Geomorphological zones are crucial factors that collect rainwater and facilitate the growth of both truffles and their host plants (Salama et al., 2023). In other studies, it was also reported that when more rain falls, the chances of having a high yield of truffles increase (Malik et al., 2018). The fruit bodies emerge during the rainy season, typically spanning from January/February to April/May, varying by location (Bradai et al., 2015b). Insufficient precipitation, the physicochemical properties of the soil, meteorological conditions, and ambient temperature rates can impact the physiology of the host plant, leading to a reduction in spore production and subsequent ascocarp production (Hashem et al., 2018). Studies by Andrino et al. (2019) on the cultivation of *Terfezia claveryi* indicated that their production has two key seasons: autumn (September through October) and spring (end of March). This study highlights that managing the aridity index and soil water potential is crucial when implementing irrigation strategies during these seasons seasons . Soil characteristics also influence the growth behaviors, chemical composition and distribution pattern of plants (Salama et al., 2023). In general, desert truffles mostly inhabit sandy soils. In terms of mineral requirements, desert truffles necessitate the presence of calcium, iron, nitrogen, phosphate and potassium, with reports indicating that a low phosphate and iron content enhances mycorrhizal potential (Bermaki et al., 2017). The presence of Alkaline Phosphatase (ALP) in desert truffles acts as a key indicator of metabolic activity as it is produced in drought conditions (Hashem et al., 2018). According to Hilszczanska et al. (2016), favorable environmental conditions for truffle growth-high average temperatures, abundant precipitation, and mild weather may stimulate protein biosynthesis in truffles, potentially leading to enhanced nutritional value.

Morte et al. (2008a) describe the importance of host plants in maximizing the production of desert truffles. They revealed that the nature of the fungal inoculum (mycelium or spores), the plant source (micropropagated plantlets or seedlings) and the culture conditions (*in vitro* or *in vivo*) used are the factors that should be considered to obtain continuous desert truffle production. *Helianthemum* species have been widely used as the mycorrhizal plant host for the production of desert truffles such as *H. sessiliflorum* Desf. Pers. (Slama et al., 2010), *Helianthemum almeriense*, *H. violaceum*, *H. canariense*, *H. hirtum* and *H. lippii* (Morte et al., 2012). According to Bermaki et al. (2017), *Helianthemum* genus species, such as *H. lipii*, *H. ledifolium*, *H. hirtum* and *H. asperum* are known to symbiotically coexist with desert truffles.

Truffle plantations aim to replicate the natural environment where it grows and sporocarps are produced spontaneously (Piñuela et al., 2024). However, the focus lies on orchard management to promote the production of fruiting bodies from the introduced truffle species. The establishment of seedling procedures with suitable management procedures such as irrigation, fertilization or mulching and targeted fungus growth are crucial for truffle orchard management (Čejka et al., 2020). Mycorrhizal technology is a profound technology used to cultivate desert truffles on a commercial scale as a source of antibiotics (Hashem et al., 2018).

A high water content required by truffles renders them highly perishable, limiting their long-term storage potential (Ghellam and Koca, 2020). This has necessitated the exploration of various preservation techniques and alternative methods for transforming them into sustainable products through bioprocessing. However, studies on desert truffle bioprocessing and production are scant. Arenas et al. (2018) reported a pioneering achievement in the cultivation of desert truffles, specifically *Terfezia claveryi*. They found that cultivation using mycelium instead of spores resulted in a higher efficiency of ectomycorrhizal colonization in the *Heliantyemum* roots of more than 50% of the produce. The best results were enhanced with glucose as a carbon source at pH 5 to produce 0.3 to 0.5 g/l biomass and 10.7 to 95.8 mg/l/d dry weight.

4. Conclusion

Truffles have been used for centuries in different cultures, not only as a rich food but also as part of folk medicine. However, future studies are needed to explore in depth the biotherapeutic value and the complementary action of bioactive compounds in truffles. In spite of the existence of much research on desert truffles and proof of their medicinal values, more evidence is needed for the safety and efficacy proven by animal models and clinical studies. On the other hand, most desert truffles are not available in high quantities and are mainly collected from wild life. Therefore, for the sustainability of desert truffle supply, more research is needed to study in depth factors affecting the growth of truffles in deserts. These studies will help in the design of new truffle plantation methods for a continuous supply. By conducting robust authentication studies and implementing effective plant management practices while considering biotic and abiotic factors, we can ensure the sustainable utilization of native truffles. This approach not only promotes their long-term survival but also contributes to food security. These phases are essential to balance the nutritional and health needs of consumers and resource conservation to promote responsible truffle utilization.

References

Al-Atassi, A.Z., Al-Khatib, R.A. and Othman, M.A. (2022). Comparison of phenolic contents and antioxidant activities for black and white desert truffles spread in Syria. Iraqi J. Pharmaceut. Sci., 31: 184–192.

Al-Laith, A.A. (2010). Antioxidant components and antioxidant/antiradical activities of desert truffle (*Tirmania nivea*) from various Middle Eastern origins. J. Food Comp. Analys., 23: 15–22.

Al Obaydi, M.F., Hamed, W.M., Al Kury, L.T. and Talib, W.H. (2020). *Terfezia boudieri*: A desert truffle with anticancer and immunomodulatory activities. Front. Nutr., 7: 38.

Al-Qarawi, A.A and Mridha, M.A.U. (2012). Status and need of research on desert truffles in Saudi Arabia. J. Pure Appl. Microbiol., 6: 1051–1062.

Aldebasi, Y.H., Nouh, W.G., Abdel Atti, N.M., Bekhit, M.S., Qureshi et al (2012). Comparative pathological studies on the healing effect of natural (*Terfezia claveryi*) and synthetic (Vigamox) antimicrobials on corneal ulcers in rabbits. J. Pharm. Biomed. Sci., 2: 66–77.

Andrino, A., Navarro-Ródenas, A., Marqués-Gálvez, J.E. and Morte, A. (2019). The crop of desert truffle depends on agroclimatic parameters during two key annual periods. Agron. Sust. Dev., 39: 1–11.

Arenas, F., Navarro-Ródenas, A., Chávez, D., Gutiérrez, A. and Pérez-Gilabert, M. (2018). Mycelium of *Terfezia claveryi* as inoculum source to produce desert truffle mycorrhizal plants. Mycorrhiza, 28: 691–701.

Bawadekji, A. (2018). Evaluation of some *in vitro* Anti-carcinogenic activities of polysaccharides extracted from ascomata of the desert truffle *Terfezia claveryi* Chatin. J. Appl. Environ. Biol., Sci., 8: 152–159.

Bermaki, F., Khabar, L. and Ezzanega, A. (2017). Bioecology of desert truffles in the province of Figuig in Eastern Morocco. Scientia Fungorum, 46: 29–36.

Bokhary, HA, Suleiman, A.A.A., Basalah, M.O. and Parvez, S. (1987). Chemical composition of desert truffles from Saudi Arabia. Canadian Inst. Food Sci. Technol. J., 20: 336–341.

Bouatia, M., Touré, H., Cheikh, A., Eljaoudi, R., Rahali, Y. et al. (2018). Analysis of nutrient and antinutrient content of the truffle (*Tirmania pinoyi*) from Morocco. Int. Food Res. J., 25: 174–178.

Bradai, L., Neffar, S., Amrani, K., Bissati, S. and Chenchouni, H. et al. (2015a). Ethnomycological survey of traditional usage and indigenous knowledge on desert truffles among the native sahara desert people of Algeria. J. Ethnopharmacol., 162: 31–38.

Bradai, L., Bissati, S., Chenchouni, H. and Amrani, K. (2015b). Effects of climate on the productivity of desert truffles beneath hyper-arid conditions. Int. J. Biometeorol., 59: 907–915.

Casarica, A., Moscovici, M., Daas, M., Nicu, A., Panteli, M. et al (2016). A purified extract from brown truffles of the species *Terfezia claveryi* chatin and its antimicrobial activity. Farmacia, 64: 298–301.

Čejka, T., Trnka, M., Krusic, P.J., Stobbe, U., Oliach, D. et al. (2020). Predicted climate change will increase the truffle cultivation potential in central Europe. Sci. Rep., 10: 21281.

Chauhan, O.P., Vijay, V., Pandey, A.K. and Semwal, A.D. (2021). Biochemical and health properties of truffles. Def. Life Sci. J., 6: 251–258.

Chen, P.Y., Montanini, B., Liao, W.W., Morselli, M., Jaroszewicz, A. et al. (2014). A comprehensive resource of genomic, epigenomic and transcriptomic sequencing data for the black truffle Tuber melanosporum. Gigascience, 3: 25.

Dahham, S.S., Al-Rawi, S.S., Ibrahim, A.H., Abdul Majid A.S. and Abdul Majid, A.M.S. et al. (2018). Antioxidant, anticancer, apoptosis properties and chemical composition of black truffle *Terfezia claveryi*. Saudi J. Biol. Sci., 25: 1524–1534.

Dawood, S.S., Sedeek, M.S., Farag, M.A. and Abdelnaser, A. (2023). *Terfezia boudieri* and *Terfezia claveryi* inhibit the LPS/IFN-γ-mediated inflammation in RAW 264.7 macrophages through an Nrf2-independent mechanism. Sci. Rep., 13: 10106.

Dib-Bellahouel, S. and Fortas, Z. (2019). Antimicrobial activities of desert truffle extracts and their chemical identification. J. Pahrm. Pharmacol., 7: 593–597.

Dogan, H. and Aydın, S. (2013). Determination of antimicrobial effect, antioxidant activity and phenolic contents of desert truffle in Turkey. Afr. J. Trad. Compl. Altern. Med., 10: 52–58.

El Enshasy, H., Elsayed, E.A., Aziz R. and Wadaan, M.A. (2013). Mushrooms and truffles: historical biofactories for complementary medicine in Africa and in the middle East. Evid. Based Compl. Altern. Med., 620451

Elkhateeb, W.A., Soliman, A.A.F., Shaheen, M.N., Elmahdy, E.M. and Daba, G.M. et al. (2024). Bioactive potentials of the truffle mushrooms *Tirmania nivea*, *Tirmania pinoyi* and *Tuber indicum*. Egyptian Pharm. J., 23: 94–102.

Elsayed, E.A., Alsahli, F.D., Barakat, I.A., El Enshasy, H.A. and Wadaan, M.A. et al. (2019a). Assessment of *in vitro* antimicrobial and anti-breast cancer activities of extracts isolated from desert truffles in Saudi Arabia. J. Sci. Ind. Res., 78: 419–425.

Elsayed, E.A., Alsahli, F.D., El Enshasy, H.A. and Wadaan, M.A. (2019b). Cytotoxic activities of different solvent extracts of *Trimania nivea* and *Terfezia claveryi* against HepG2 and L929 cells. J. Sci. Ind. Res., 78: 454–457.

Farag, M.A., Fathi, D., Shamma, S., Shawkat, M.S.A., Shalabi, S.M. et al. (2021). Comparative metabolome classification of desert truffles *Terfezia claveryi* and *Terfezia boudieri* via its aroma and nutrients profile. LWT, 142: 111046.

Ferreira, I., Dias, T., Mouazen, A.M. and Cruz, C. (2023). Using science and technology to unveil the hidden delicacy *Terfezia arenaria*, a desert truffle. Foods, 12: 3527.

Gadallah, M.G.E. and Ashoush, I.S. (2016). Value addition on nutritional and sensory properties of biscuit using desert truffle (*Terfezia claveryi*) powder. Food Nutr. Sci., 7: 1171–1181.

Ghellam, M. and Koca, I. (2020). Modelling of rehydration kinetics of desert truffles (*Terfezia* spp.) dried by microwave oven. Turkish J. Agric. Food Sci. Technol., 8: 407–415.

Ghorai, S., Banik, S.P., Verma, D., Chowdhury, S., Mukherjee, S. et al. (2011). Fungal biotechnology in food and feed processing. Food Res. Int., 42: 577–587.

Hamza, A., Zouari, N., Zouari, S., Jdir, H., Zaidi, S. et al. (2016). Nutraceutical potential, antioxidant and antibacterial activities of *Terfezia boudieri* Chatin, a wild edible desert truffle from Tunisia arid zone. Arabian J. Chem., 9: 383–389.

Harir, M., Bendif, H., Yahiaoui, M., Bellahcene, M., Zohra, F. and Rodríguez-Couto, S. (2019). Evaluation of antimicrobial activity of *Terfezia arenaria* extracts collected from saharan desert against bacteria and filamentous fungi. 3 Biotech, 9: 281.

Hashem, A., Alqarawi, A.A., Shah, M.A., Wirth, S., Egamberdieva, D. et al. (2018). Desert truffles in Saudi Arabia: diversity, ecology, and conservation. pp. 353–369. In: Sharma, S. and Varma, A. (eds.). Microbial Resource Conservation. Springer Cham.

Hilszczanska, D., Siebyła, M., Horák, J., Król, M., Podsadni, P. et al. (2016). Comparison of chemical composition in *Tuber aestivum* Vittad. of different geographical origin. Chem. Biodivers., 13: 1617–1629.

Janakat, S. and Nassar, M. (2010). Hepatoprotective activity of desert truffle (*Terfezia claveryi*) in comparison with the effect of *Nigella sativa* in the rat. Pakistan J. Nutr., 9: 52–56.

Khadhri, A., Aouadhi, C. and Aschi-Smiti, S. (2017). Screening of bioactive compounds of medicinal mushrooms collected on Tunisian territory. Int. J. Med. Mushr., 19: 127–135.

Khlaif, D.K., Kadum, H. and Abod, N.S. (2021). Nutritional and chemical compositions of the Desert truffle (*Terfezia claveryi*) in Samawa city of Iraq. IOP Conf. Ser. Earth Environ. Sci., 932(1): 012050.

Khojah, H.M.J., Abdelhalim, O.B., Mostafa, M.A.H. and Habib, E.E. (2022). Antimicrobial efficacy of extracts of Saudi Arabian desert *Terfezia claveryi* truffles. Saudi J. Biol. Sci., 29: 103462.

Lee, H., Nam, K., Zahra, Z. and Farooqi, M.Q.U. (2020). Potentials of truffles in nutritional and medicinal applications: a review. Fungal Biol. Biotechnol., 7: 9.

Lu, B., Zhang, F-M., Yu, F-Q. and Rinaldi, A.C. (2022). Ethnobiological notes and volatile profiles of two rare Chinese desert truffles. Mycology, 13: 177–184.

Malik, H.M, Gull, M., Omar, U., Kumosani, T.A., Al-Hejin, A.M. et al. (2018). Evaluation of the antibacterial potential of desert truffles (*Terfezia* spp.) extracts against methicillin resistant *Staphylococcus aureus* (MRSA). J. Exp. Biol. Agric. Sci., 6: 652–660.

Mandeel, Q. and Laith, A. (2007). Ethnomycological aspects of the desert truffle among native Bahraini and non-Bahraini peoples of the Kingdom of Bahrain. J. Ethnopharmacol., 110: 118–129.

Marqués-Gálvez, J.E., Miyauchi, S., Paolocci, F., Navarro-Ródenas, A., Arenas, F. et al. (2021). Desert truffle genomes reveal their reproductive modes and new insights into plant–fungal interaction and ectendomycorrhizal lifestyle. New Phytol., 229: 2917–2932.

Mello, A., Murat, C. and Bonfante, P. (2006). Truffles: much more than a prized and local fungal delicacy. FEMS Microbiol. Lett., 260: 1–8.

Morte, A., Honrubia, M. and Gutiérrez, A. (2008a). Biotechnology and cultivation of desert truffles. pp. 467–483. In: Varma, A. (ed.). Mycorrhiza: State of the Art, Genetics and Molecular Biology, Eco-Function, Biotechnology, Eco-Physiology, Structure and Systematics, Springer Berlin Heidelberg.

Morte, A., Andrino, A., Honrubia, M. and Navarro-Ródenas, A. (2012). *Terfezia* cultivation in arid and semiarid soils. pp. 241–263. In: Zambonelli, A. and Bonito, G. (eds.). Edible Ectomycorrhizal Mushrooms. Springer, Berlin, Heidelberg.

Morte, A., Arenas, F., Marqués-Gálvez, J.E., Andrino, A., Guarnizo, Á.L. et al. (2021). Desert Truffles (*Terfezia* spp.) Breeding. pp. 479–504. In: Al-Khayri, J.M., Jain, S.M. and Johnson, D.V. (eds.). Advances in Plant Breeding Strategies: Vegetable Crops. Springer Cham.

Morte, A., Gutiérrez, A. and Ródenas, A.N. (2020). Advances in Desert Truffle Mycorrhization and Cultivation. pp. 205–219. In: Pérez-Moreno, J., Guerin-Laguette, A., Flores Arzú, R., Yu, FQ. (eds.). Mushrooms, Humans and Nature in a Changing World. Springer Cham.

Morte, A., Pérez-Gilabert, M., Gutiérrez, A., Arenas, F., Marqués-Gálvez, J.E. et al. (2017). Basic and Applied Research for Desert Truffle Cultivation. pp. 23–42. In: Varma, A., Prasad, R. and Tuteja, N. (eds.). Mycorrhiza - Eco-Physiology, Secondary Metabolites, Nanomaterials. Springer Cham.

Owaid, M.N., Muslim, R.F. and Hamad, H.A. (2018). Mycosynthesis of silver nanoparticles using *Terminia* sp. desert truffle, pezizaceae, and their antibacterial activity. Jordan J. Biol. Sci., 11: 401–405.

Owaid, M.N. (2022). Biosynthesis of silver nanoparticles from truffles and mushrooms and their applications as nanodrugs. Curr. Appl. Sci. Technol., 22: 1–11.

Patel, S. (2012). Food, health and agricultural importance of truffles: A review of current scientific literature. Curr. Trends Biotechnol. Pharm., 6(1): 15–27.

Piñuela, Y., Alday, J.G., Oliach, D., Castaño, C., Büntgen, U. et al. (2024). Habitat is more important than climate for structuring soil fungal communities associated in truffle sites. Fungal Biol., 128: 1724–1734.

Salama, H.M, Abd El gawad, A.M., Zaghloul, A.A. and Yassin, M.A. (2023). Ecological studies of *Helianthemum* plants associated with desert truffles in the Mediterranean north coastal area, Egypt. Arab. Int. J. Environm. Sci., 2: 1–22.

Saleh, K.K., Kirbag, S. and Dalkılıç, S. (2022). Inhibitory effects of Terfezia (Ascomycota) desert truffles on PANC-1 cell growth via upregulation of proapoptotic genes TP53, CDKN1A, and BAX and downregulation of antiapoptotic gene BCL2. Int. J. Med. Mushr., 24: 61–70.

Slama, A, Fortas, Z., Boudabous, A. and Neffati, M. (2010). Cultivation of an edible desert truffle (*Terfezia boudieri* Chatin). Afr. J. Microbiol. Res., 4: 2350–2356.

Souna, M., Tefiani, C., Azzi, R., Chouche, T.M. and Habi, S. et al. (2023). Mycochemical constituensts and anti-inflammatory activity of *Terfezia claveryi* Chatin from Algeria. Biol. Life Sci. Forum, 27: 19.

Tejedor-Calvo, E., Amara, K., Reis, F.S., Barros, L., Martins, A. et al. (2021). Chemical composition and evaluation of antioxidant, antimicrobial and antiproliferative activities of Tuber and Terfezia truffles. Food Res. Int., 140: 110071.

Tejedor-Calvo, E., Morales, D., Marco, P., Sánchez, S., Garcia-Barreda, S. et al. (2020). Screening of bioactive compounds in truffles and evaluation of pressurized liquid extractions (PLE) to obtain fractions with biological activities. Food Res. Int., 132: 109054.

Turgeman, T., Asher, J., Roth-Bejerano, N., Kagan-Zur, V., Kapulnik, Y. et al. (2011). Mycorrhizal association between the desert truffle *Terfezia boudieri* and *Helianthemum sessiliflorum* alters plant physiology and fitness to arid conditions. Mycorrhiza, 21: 623–630.

Veeraraghavan, V.P., Hussain, S., Balakrishna, J.P., Dhawale, L., Kullappan, M. et al. (2022). A comprehensive and critical review on ethnopharmacological importance of desert truffles: *Terfezia claveryi*, *Terfezia boudieri*, and *Tirmania nivea*. Food Rev. Int., 38: 846–865.

Zniber, I., Boukcim, H., Khabar, L., Ducousso, M., Henkrar, F. et al. (2022). Characterization of desert truffles in the great Moroccan sahara: A review. Environ. Sci. Proc., 16: 55.

Nutraceutical and Economic Perspectives

Chapter 13

Traditional Knowledge on the Culinary and Therapeutic Value of Truffles in Morocco

Hakkou, S.,[1,*] *Sabir, M.*[2] and *Machouri, N.*[1]

1. Introduction

Desert truffles are hypogeal wild fungi belonging to the Ascomycetes, that develop in the typically arid and semi-arid climate zones of the Mediterranean region, some of which are endemic (Alsheikh, 1994; Morte et al., 2008; Rodriguez, 2008; Trappe et al., 2008). They present a source of nutrition and natural therapy through their gastronomic value, distinctive aroma, and therapeutic properties. Prepared meals and mixes are special practices and know-how inherited from generation to generation. They have been known and consumed since ancient times, and their mention appears in the writings of all civilizations. The most classic tales relate to species of *Terfezia* and *Tirmania* rather than species of *Tuber* (Loizides et al., 2012).

Truffles have been known in the Middle East for 6,000 years. Their virtues were appreciated by the oldest civilizations (Mesopotamian, Sumerian, and Aramaic). They were harvested long ago by the Amorites and Sumerians in the Syrian desert. Clay tablets dating back 4,000 years and bearing cuneiform writing have been found in the archaeological archives of the Mesopotamian region, attesting to their existence under the name "Kam'atu" (Shavit, 2008; Nowak, 2015). The truffle was mentioned in a poem recorded in Egyptian temples: "Without leaves, without buds, without flowers, yet they form fruit. The entire creation is precious as a food, tonic, and medicine (El Enshasy et al., 2013).

[1] Faculty of Letters and Human Sciences, Rabat, Morocco.
[2] National School of Forest Engineers, Sale, Morocco.
* Corresponding author: soukaina.hakkou@gmail.com

Desert truffles are rich in protein and antioxidants. Several studies have focused on evaluating their nutritional value. In fact, more than 75% of the mass of the truffle consists of water (it can reach up to 81%), while the rest is dry matter composed of carbohydrates (60%), protein (20–27%), fat (3–7.5%), fiber (7–13%), and ascorbic acid (2–5%). These average figures may vary according to species, age, region, soil type, and climatic factors (Rodriguez, 2008; Wang and Marcone, 2011; El Enshay et al., 2013; Benmouloud, 2017). It has been indicated by some authors that the protein and amino acid content of desert truffles approaches that of animal protein and that 250 g of truffles would represent 23–27% of the recommended daily human consumption of protein. This would explain their consumption as a meat substitute in certain regions of the world (Morte et al., 2008; Owaid, 2018). The following table shows the chemical composition of certain species of truffles from different regions of the world (Table 1).

Other studies have confirmed that truffles contain considerable amounts of essential minerals, such as phosphorus, potassium, sodium, calcium, magnesium, manganese, iron, copper, and zinc (Al-Delaimy, 1977; Ahmed-Ashour et al., 1981; El Enshay, 2013; Owaid, 2018). Desert truffles contain a significant number of amino acids (~18 types). These amino acids play an important role in growth activities through the control of ion communication channels, hormone secretion, cell growth, and replication, in addition to forming certain types of proteins that are only found in these types of fungi (Al-Delaimy, 1977; Ahmed-Ashour et al., 1981; Sawaya et al., 1985; Al-Rawi and Taha, 2010).

In Morocco, truffles exist in their natural state in four major geographical regions, namely the Northeast region, the Maâmora forest, the Doukkala-Abda Sahel, and the Moroccan Sahara. They are of the order of 12 species of the genera: *Terfezia, Tirmania, Tuber, Delastria*, and *Picoa* (Hakkou et al., 2021; Hakkou et al., 2022a). During their production season, they constitute an income-generating activity for the benefit of the populations of the productive areas who use them to meet their needs. In addition to their socioeconomic role, desert truffles also represent a source of food and natural therapy. Thus, this study focuses on desert truffles as a natural product

Table 1. Proximal composition of truffle species from different origins.

Truffle	Fresh weight (%)	Percent (dry mass basis)					Reference
	Moisture	Protein	Fat	Carbohydrate	Fiber	Ash	
Picoa juneperi (Spain)	63.78	22.53	19.94	36.66	13.04	8.21	Murcia et al. (2002)
Terfezia boudieri (Libya)	77.70	17.19	6.40	59.73	3.8	12.88	Ahmed-Ashour et al. (1981)
Terfezia claveryi (Saudi Arabia)	75.44	24.96	4.20	-	7.02	6.39	Sawaya et al. (1985)
Tirmania nivea (Tunisia)	77.63	28.81	6.78	57.83	-	5.06	Hamza et al. (2013)
Tirmania pinoyi (Morocco)	81.5	26.96	3.00	64.76	-	5.28	Bouatia et al. (2018)

with both nutritional values and therapeutic properties. Its objective is to contribute to the promotion of these products by emphasizing the know-how related to their culinary and medicinal uses.

2. Culinary Uses of Moroccan Truffles

The use of truffles in Moroccan cuisine differs from one region to another, according to cultural specificities and traditional dishes known in each area. Common dishes found in almost all regions are "tagines" and Moroccan "couscous". However, it should be mentioned that the marketing of truffles constitutes an important source of income for the collecting population. These dishes are considered delicacies that the population indulges in when production is abundant or when there are unsold *Terfez*, including broken pieces and smaller truffles.

2.1 Northeast Region

The Northeast region is renowned for producing the white truffles, including *Tirmania nivea* and *T. pinoyi* as well as the the red truffle, *Terfezia claveryi* (Fig. 1a). There are other species, such as *T. boudieri*, *T. leptoderma*, *T. olbiensis*, and *Picoa juneperi*. However, white truffles and red truffles are the most commonly collected and consumed, with a particular preference for white truffles. Truffles are prepared according to several recipes, but the most common is that of boiling truffles in salted water. Once cooked, the truffles are either cut or crushed and mixed with butter, olive

Fig. 1 Truffles from different forests: a, *Terfezia claveryi* from the northeast region; b, *Terfezia arenaria* from the Maamora forest; c, *Terfezia boudieri* from the Maamora forest; d, *Tirmania pinoyi* from the Maamora forest.

oil, or eggs. Since the region is known for livestock, the consumption of meat-based dishes is the mainstay of its inhabitants, thus, truffles are used in the preparation of traditional Moroccan tagines with and without meat. In the first case, the truffle is used to flavor the dish, and in the second, it replaces the meat. People say truffles taste like meat. They can be used raw or cooked in salads. The arid climate of the region results in varying truffle abundance each year. Enthusiasts of this delicacy dry the truffles, slicing them to extend their enjoyment. Dried truffles can be stored for a year or more.

2.2 Maâmora Forest

The Maâmora forest is known for the production of several species. Thus, we have highlighted some of them: *Terfezia arenaria* (Fig. 1b), *T. leptoderma*, *Tuber asa*, *T. Borchii* var. *sphaerosperma*, *T. gennadii*, T. *oligospermum*, and *Delastria rosea*. The most collected truffles are *T. arenaria*, the pink Maâmora truffle, and the white truffles *T. oligospermum* and *T. asa*. In terms of consumption, it is the pink truffle that is favored, as the other two are considered inedible by the local population. The region boasts several recipes including Terfez Tajines *Terfez* with meat, *Terfez* omelets, *Terfez* boiled in salted water, and grilled *Terfez*. Another regional speciality mentioned by respondents is "*Medhousa*", a traditional dish particularly popular in the Gharb region. It consists of crumbled semolina galette, drizzled with sauce, topped with chicken, onion, and pieces of *Terfez*.

2.3 Doukkala-Abda Sahel

The Sahel is known to produce three species of truffles, *T. boudieri*, *T. oligospermum,* and *T. pinoyi* (Fig 1c). The blackish truffle, *T. boudieri*, is the most dominant. In this region, truffles are used in the preparation of traditional Moroccan tajines, with or without meat, salads, soups, and Moroccan couscous. Grilled truffles are common in the region. But the most popular recipe is the grated truffle omelet. The inhabitants of the Sahel attribute aphrodisiac virtues to this dish.

2.4 Moroccan Sahara

The Moroccan Sahara is characterized by the presence of four species of truffles, *T. nivea, T. pinoyi, T. claveryi*, and *Picoa juneperi* (Fig. 1d). In the Northeast region, the most commonly consumed truffles are those with whitish and reddish hues. Truffles are incorporated into several dishes including meat tajines or *Tichtar* (sun dried strips of camel meat), omelets, and boiled rice or wheat flour. The most common preparation method, especially during the collection days, involves lighting a fire to heat the sand thoroughly., The truffles are then placed on the ground and covered with the hot sand for an hour until they are toasted. They are eaten with olive oil, butter, or "*Smen*" (goat butter). Desert truffles are also sliced and dried, as they can be preserved for a long time and used later for meals. According to the collectors and the "Ahl Baganna" cooperative, truffles are used for making good-quality wine, considered better than a 20-year-old wine.

2.5 Relationship of the Jewish Community with Truffles

Almost all the truffle literature invokes a relationship between Jews and truffles. Some literature has even mentioned a connection between the Jewish religion and this product. This suggests that the "*manna*" provided by Allah as sustenance to the *Bani Israil* (the Hebrew people) during their exodus from Egypt was actually truffles. Field surveys with all stakeholders in the marketing chain, from collectors to exporters, revealed that a large quantity is destined for Jews. A significant number of wholesalers are Jews. They manage the supply for their community living in Morocco, particularly in Bab Marrakech, Casablanca, and handle exports for the benefit of Jews worldwide.

The above information caught our attention. Thus, we went in search of Jews in Rabat, Essaouira, and Marrakech to try to clarify this matter. All the Jews confirmed that the truffle is a precious product consumed during the *Purim* celebrations in March and *Passover* in April to honor the guests. A Moroccan Muslim bookseller we met in Essaouira shared that his father rented a house to Jews in Bab Marrakech, Casablanca. During Jewish festivals and the New Year, the house was filled with the delightful aroma of truffles, which everyone enjoyed immensely. The market in Bab Marrakech at the time knew of the arrival of large quantities of truffles intended for the Jews.

A Moroccan Jew who runs a Jewish cuisine restaurant in the *Melah* in Marrakech shared that since her childhood, their family has been accustomed to receiving black truffles from Safi (*Terfezia boudieri*), which they consider to be real truffles and superior to any other truffle they have tasted. For culinary use, they take large quantities, wash, peel, and cut them. Then they are cooked in water with lemon; once the foam starts to form, the truffles are removed from the water and stored in the fridge until the holidays. The meal consists of truffles cooked with olive oil, lemon, chicken or meat juice, and spices. Finally, they are served with pieces of mutton.

3. The Truffle in Traditional Moroccan Medicine

Since truffle collectors live in rural areas, they are used to treating themselves with natural products. Thus, they opt for the consumption of truffle-based dishes and the preparation of certain mixtures to cure certain diseases.

3.1 Northeast Region

The people surveyed revealed that drinking the water in which the truffles were cooked every morning is beneficial for the stomach, intestines, and kidneys. They also believe that the truffle is good for diabetes and eye diseases and helps fight laziness. For cases of fever, they beleive that eating truffles helps the body sweat and therefore heal quickly.

3.2 Maamora Forest

The local people of the forest believe that truffles can cure eye diseases, allergies, and body colds, such as rheumatism. Some say they boost the immune system and fight cancerous tumors.

3.3 Doukkala-Abda Sahel

Locals in the region believe that truffles can treat various ailments, such as body colds, eye diseases (including cataracts (*Jlala*)), and asthma. A common regional preparation involves mixing truffles with milk, and honey to boost energy and combat the cold. Additionally,truffles are used in cosmetics, with the powder from dried and ground desert truffles being used in facial masks.

3.4 Moroccan Sahara

In the Moroccan Sahara, desert truffles are beneficial for humans as well as animals. Sahrawi say that they can cure body coldness, rheumatism, and tumors.Their juice cures skin and eye diseases. Eye diseases are treated by using their juice as eye drops. Their components are also good for people with diabetes and high blood pressure. Truffle porridge water is also drunk to cleanse the intestines. For livestock, truffle powder is used to cure udder tumors in camels, cows, sheep, and goats. This is done by applying it on the swollen part and covering it with a band or a piece of cloth. The powder can be put in the dry state (talc) or as an aqueous mixture that has a cream texture. The truffle cooperative in Laayoune, "Ahl Baganna", makes creams based on desert truffles and goat and camel fat and beeswax to treat osteoarthritis (Fig. 2).

Fig. 2. Creams prepared from truffles produced by the Ahl Baganna cooperative, Laayoune, Morocco.

4. Discussion

The culinary use of desert truffles dates back to ancient times. Those in power liked to serve truffles to their guests. Thus, they appeared on the most luxurious tables of pharaohs, kings, presidents, emperors, and tsars (Olivier et al., 2018). In Egypt, around 2,600 B.C., the pharaoh considered truffles to be an exclusive royal dish. They were served to visiting delegations and foreign ambassadors intending to honor the pharaoh as truffle-based dishes (Rayss, 1959; Pagnol, 1983; Alsheikh, 1994; Zitouni-Haouar, 2016). In North Africa and the Middle East, truffles have been consumed since prehistoric times by wealthy people. It is very likely that their consumption was prohibited for the normal population. In ancient times, they were valued and frequently imported from Libya, Lesbos, and Cartage (Feeney, 2002; Loizides et al., 2012).

Nowadays, truffles have become widely accessible and sold in local markets. They are used in various dishes, with the primary consumers being in the Maghreb countries, the Middle East (Saudi Arabia, Kuwait, and Iraq), and parts of southern Europe (Spain, and Italy) (Morte et al., 1994). In Morocco, when production is high prices are low, the local population enjoys consuming them. In the Rabat-Salé region, where the Maâmora forest is located, truffle tajines are a popular dish. Truffles are also boiled like potatoes, used in stews with sauce and condiments, and incorporated into the preparation of *Rfissa* (Belakhder, 1997; Hakkou et al., 2022b).

In the Moroccan Eastern region (Oriental), four truffle recipes were reported by Serrhini et al. (1995) and confirmed by our field observations. These include truffle puree with olive oil, butter, and salt, tajines, spicy truffle dumplings known as *Maatouka*; and truffles cooked directly on embers or skewered with butter and salt. These preparations are similar to those found in other Arab countries. For example, in Algeria, truffles are enjoyed both raw and cooked, depending on the region. In M›sila and Ghardaïa they are prepared in white or red sauce, sometimes with meat. The most common dish involves baking or boiling truffles in water and seasoning them with salt. In Naàma, Mécheria, Ain-Sefra, Al-Bayadh, and Béchar, truffles are consumed as *Reffssa à Dhan*, or "sautéed mushrooms in butter". They are also used as a garnish for couscous, tajines, or pizzas, and some people dry them in the sun to use as a condiment, especially for pasta-based dishes during Ramadan (Boufeldja, 2017). In Kuwait, truffles are boiled in camel or cow›s milk, or roasted in melted butter (Kermani, 2013).

Nomad and Saharawi have long relied on truffles as a food source due to their meat-like flavor. The most common preparations include cutting and cooking them in embers with goat or camel meat, or boiling and eating them with butter or camel fat. They also serve as a meat substitute (Omer, 1994; Volpato et al., 2013). Truffles are particularly valuable during food shortages or scarcity. For instance, during the 1970 war in Iraq, truffles met the population's food needs. Khalastchy (2003) notes that food was scarce and for four months, people in the desert used truffles instead of meat in traditional meat dishes. Gast (2000) reported that in the Sahrawi language, "When there was nothing left, the dried terfez was our bread and our meat". To preserve them, truffles were sliced and dried, allowing them to be stored for one or two years.

As for medicinal use, truffles have long been part of pharmacopeial products around the world. The mention of truffles as medicinal food dates back to Chinese, Greek, and Egyptian civilizations and has been called the miracle of nature in Mesopotamia (Dahham et al., 2018). They are attributed with antioxidant, antimutagenic, anticarcinogenic, antimicrobial, antiviral, immunomodulatory, anti-inflammatory, antidepressant, and sedative bioactive constituent properties. They have the ability to protect the human body from oxidative stress induced by the formation of free radicals (Hannan et al., 1989; El Enshasy et al., 2013; Zitouni-Haouar, 2016).

Our results concluded that the truffle is used to cure rheumatism, osteoarthritis, and asthma agree with Volpato et al. (2013), sharing that the truffle is used by the inhabitants of the Sahara to cure colds and respiratory ailments and to treat rheumatism and arthritis by applying a poultice made from boiled and mashed truffles and consuming hot truffle soup. These two practices are supposed to extract the cold from the body. Aqueous extracts of truffles are used by people in the Middle East (Iraq, Saudi Arabia, and Eastern Jordan) and North Africa to treat eye infections that are very common in desert areas due to sandstorms such as conjunctiva and trachoma by boiling the truffle and applying its juice as eye drops. Truffles also help in healing skin infections caused by bacteria, healing wounds, and treating fatigue, vomiting, and gout (Al-Rahmah, 2001; Mandeel and Al-Laith, 2007; Hall et al., 2008; Shavit, 2008; Volpato et al., 2013; Boufeldja, 2017). The French dermatologist Marie-Guillaume-Alphonse Devergie (1798–1879), prescribed truffle water in 1849 for nausea, cholera, diarrhea, consumption, and gastralgia (Pagnol, 1983).

These findings align with ours, indicating that truffles are beneficial for treating eye infections, allergies, tumors and skin diseases, combating laziness and fatigue, and alleviating stomach pains. Prophetic medicine highlights the antimicrobial properties of truffle extracts for eye diseases. Said Ibn Zayd (may Allah be pleased with him) reported that the Prophet *Mohamed* (may the prayer of Allah and His salvation be upon him) said: "*The truffle is part of the manna, and its juice is a cure for the eye*" (Reported by Bukhari in his Sahih n°4478 and Muslim in his Sahih n°2049). Around the 10th century, the Muslim physician and philosopher Ibn Sina (Avicenna) recommended the truffle for various ailments such as vomiting, wounds, and fatigue. He used to truffle juice to treat common eye inflammation and recommended it in his book Al Qanûn fi Al-Tibb ("book of medical laws") for eye care (Shavit and Shavit, 2014). Our research also shows that truffle powder is used to cure udder tumors in camels, cows, sheep, and goats in the Moroccan Sahara, and this practice is common among Sahrawi in neighboring countries. They use it to treat metritis (inflammation of the lining of the uterus) by applying a poultice of boiled and mashed truffles with onion, garlic, and oil inside the uterus after childbirth. It is also used to treat mastitis (inflammation of the udders) or other external inflammations by washing the affected area with truffle boiling water or by applying a poultice of boiled or raw truffles heated over embers and crushed with oil or animal fat (Volpato et al., 2013). The following table summarizes the medicinal properties of certain truffle species for certain diseases and disorders (Table 2).

Table 2. Biomedical applications of certain truffle species.

Truffle	Biomedical application	Reference
Terfezia boudieri	Antibacterial, antifungal, antioxidant, anticandidal and radical scavenging activities	Akyuz (2013); Dogan and Aydin (2013); Gouzi et al. (2013)
Terfezia claveryi	Anti-cancer and antioxidant activities	Dahham et al. (2018)
Terfezia sp.	Treatment of eye infections, antibacterial, antifungal, antiviral, immunomodulating, hepatoprotective and antidepressant activities	Al-Mahamdi (2011); Al-Rahmah (2014); Gajos et al. (2014)
Tirmania nivea	Antioxidant and anticandidal activities	Al-Laith (2010)
Tirmania pinoyi	Antioxidant and antibacterial activities	Stojkovik et al. (2013)
Tirmania sp.	Treatment of eye infections, antibacterial, antifungal, antiviral, immunomodulating, hepatoprotective and antidepressant activities	Al Mahamdi (2011); Al-Rahmah (2014); Gajos (2014)

6. Conclusion

This study highlighted the significance of desert truffles in culinary preparations and traditional pharmacopeia recipes in Morocco. The perceptions of the collecting population about the nutritional values and healing capacities of this mushroom, which they use for food and treatment, are in line with the results of scientific studies carried out in this direction. Awareness of the distinction enjoyed by this product on several levels can be a starting point for its promotion for the benefit of the user population.

References

Ahmed-Ashour, A., Mohamed, M.A. and Hamid, M.A. (1981). Libyan truffles *Terfezia boudieri Chatin*: Chemical composition and toxicity. J. Food. Sci., 46(3): 927–929.

Akyuz, M. (2013). Nutritive value, flavonoid content and radical scavenging activity of the truffle (*Terfezia boudieri* Chatin). J, Soil Sci. Pl. Nutr., 13(1): 143–151.

Al-Delaimy, K.S. (1977). Protein and amino acid composition of truffles. Can. Inst. Food Sci. Technol. J., 10: 221–222.

Al-Laith, A.A.A. (2010). Antioxidant components and antioxidant/antiradical activities of desert truffle (*Tirmania nivea*) from various Middle Eastern origins. J. Food Comp. Anal., 23(1): 15–22.

Al-Mahamdi, W.A.A. (2011). Study the Biological Activity of Some Iraq Truffle Extracts Tuber (Rutba Region) Against Microorganisms and Tomato Mosaic Virus. M.Sc. Dissertation, College of Science, University of Anbar, Iraq.

Al-Rahmah, A.B.N.M. (2014). Desert and Forest Truffles, (Fuga) Food and Drug (in Arabic). 2nd Edition, Saud Kingdom University, Saudi Arabia, p. 288.

Al-Rahmah, A.N. (2001). Truffle of Deserts and Jungles. King Saud University Publication Riyadh, Saudi Arabia, p. 272.

Al-Rawi, A-A. and Taha, A-M. (2010). Chemical study for three Iraqi truffles types. Anbar J. Agric. Sci., 1: 33–41 (in Arabic).

Alsheikh, A.M. (1994). Taxonomy and mycorrhizal ecology of the desert truffles in the Genus *Terfezia*. Ph.D. dissertation. Oregon State University, Corvallis, p. 239.

Belakhdar, J. (1997). La pharmacopée marocaine traditionnelle. Ed. Ibis Press, p. 746.

Benmouloud, A. (2017). Les truffes : apport nutritionnel et thérapeutique. FMP Rabat, p. 83.

Bouatia, M., Touré, H.A., Cheikh, A., Eljaoudi, R., Rahali, Y. et al. (2018). Analysis of nutrient and antinutrient content of the truffle (*Tirmania pinoyi*) from Morocco. Int. Food Res. J., 25(1): 174–178.

Boufeldja, W. (2017). Evaluation du potentiel nutritionnel et antioxydant de quelques variétés de Truffes du Sud-ouest algérien. Effets antimicrobien et antiinflammatoire. Thèse de doctorat. Université Djillali Liabes, p. 112.

Dahham, S.S., Al-Rawi, S., Ibrahim, A.H., Abdul Majid, A. Sh. and Abdul Majid, A.M. Sh et al. (2018). Antioxidant, anticancer, apoptosis properties and chemical composition of black truffle Terfezia claveryi. Saudi J. Biol. Sci., 25(8): 1524–1534. https://doi.org/10.1016/j.sjbs.2016.01.031

Doğan, H. and Aydın, S. (2013). Determination of Antimicrobial Effect, Antioxidant Activity and Phenolic Contents of Desert Truffle in Turkey. Afr. J. Trad. Compl. Alt. Med., 10(4): 52–58.

El Enshasy, H., Elsayed, E.A., Aziz, R. and Wadaan, M.A. (2013). Mushrooms and Truffles: Historical Biofactories for Complementary Medicine in Africa and in the Middle East. Evid. Based Compl. Alt. Med., 2013: 1–10.

Feeney, J. (2002). Desert Truffles Galore. Saudi Aramco World, pp. 22–27.

Gajos, M., Ryszka, F. and Geistlinger, J. (2014). The therapeutic potential of truffle fungi: A patent survey. Acta Mycol., 49: 305–318.

Gast, M. (2000). Moissons du désert. Utilisation des ressources naturelles au Sahara Central. Ibis Press, Paris, p. 160.

Gouzi, H., Leboukh, M. and Bouchouka, E. (2013). Antioxidant and antiradical properties of methanolic extracts from Algerian wild edible desert truffles (*Terfezia* and *Tirmania*, Ascomycetes). Journal International des champignons médicinaux, 15 (5): 471–486.

Hakkou, S., Sabir, M. and Machouri N. 2021. Truffles from the northeast region of the kingdom of Morocco: geographical distribution and production potential. Int. J. Curr., Res., 13(12): 19817–19822. https://doi.org/10.24941/ijcr.42701.12.2021

Hakkou, S., Sabir, M. and Machouri, N. (2022a). Le Maroc truffier: la répartition géographique des truffes et truffes du désert et leur productivité. Rev. Mar. Sci. Agron. Vét. 10(3): 395–407.

Hakkou, S., Sabir, M. and Machouri, N. (2022b). Valorisation écotouristique des ressources naturelles : plantes aromatiques et médicinales de la région Rabat-Salé-Kénitra. Livre. Éditions Universitaires Européennes (EUE), p. 146.

Hall, I. R., Brown, G. and Zambonelli, A. (2008). Taming the truffle: the history, Lore and Science of the Ultimate Mushroom. Ed. Timber Press, Oregon, USA, p. 304.

Hamza, A., Zouari, N., Zouari, S., Jdir, H., Zaidi, S. et al. (2013). Nutraceutical potential, antioxidant and antibacterial activities of *Terfezia boudieri* Chatin, a wild edible desert truffle from Tunisia arid zone. Arab. J. Chem., 9(3): 383–389.

Hannan, M.A., Al-Dakan, A.A., Aboul-Enein, H.Y. and Al-Othaimeen, A.A. (1989). Mutagenic and antimutagenic factor(s) extracted from a desert mushroom using different solvents. Mutagenesis, 4(2): 111–114.

Kermani, I. (2013). Mycorhization contrôlée d'une Cistacée pérenne par les terfez en conditions génotoxiques et essai de transplantation sur le terrain. Faculté des sciences de la nature et de la vie. Université d'Oran, Algerria, p. 115.

Khalastchy, E. (2003). Desert truffles- Eileen Khalastchy writes. The Scribe - J Babylonian Jewry, p. 76.

Loizides, M., Hobart, C., Konstandinides, G. and Yiangou, Y. (2012). Desert truffles: the mysterious jewels of antiquity. Field Mycol., 13(1): 17–21.

Mandeel, Q.A. and Al-Laith, A.A.A. 2007. Ethnomycological aspects of the desert truffle among native Bahraini and non-Bahraini peoples of the Kingdom of Bahrain. J. Ethnopharmacol., 110(1):118–129.

Morte, A., Honrubia, M. and Gutierrez, A. (2008). Biotechnology and cultivation of desert truffles. pp. 467–483. In: Varma A (ed.), Mycorrhiza: State of The Art Genetics and Molecular Biology, Eco-Function, Biotechnology, Eco-Physiology, Structure and Systematics. Heidelberg: Springer Verlag, Berlin, Germany.

Morte, M.A., Cano, A., Honrubia, M. and Torres, P. (1994). In vitro mycorrhization of micropropagated *Helianthemum almeriense* plantlets with *Terfezia claveryi* (desert truffle). Agricultural Science in Finland, 3: 309–314. https://doi.org/10.23986/afsci.72700

Murcia, M.A., Martinez-Tomé, M., Jiménez, A.M., Vera, A.M., Honrubia, M et al. (2002). Antioxidant activity of edible fungi (truffles and mushrooms): losses during industrial processing. J. Food Protect., 65(10): 1614–1624.

Nowak, Z. (2015). Truffle - A Global History. The edible series, Reaktion Books, UK, p. 128.

Olivier, J.-M., Savignac, J.-C. and Sourzat, P. (2018). Truffe et Trufficulture. Livre Fanlac, Périgueux, France, p. 352.

Omer, E.A., Smith, D.L., Wood, K.V. and El-Menshawi, B.S. (1994). The volatiles of desert truffle: *Tirmania nivea*. Pl. Foods Hum, Nutr., 45(3): 247–249. https://doi.org/10.1007/BF01094093

Owaid, M.N. (2018). Bioecology and uses of desert truffles (Pezizales) in Middle Eastern. Walailak J Sci. Tech., 15(3): 179–188. https://doi.org/10.48048/wjst.2018.3058

Pagnol, J. (1983). La truffe. Ed Avignon. Aubanel, France, p. 184.

Rayss, T. 1959. Champignons hypogés dans les régions désertiques d›Israël. Bulletin de la Société Botanique de France, 112: 7–8, 480–487.

Rodriguez, A. (2008). Desert truffles. http://www.trufamania.com/desert-truffles.htm

Sawaya, W.N., Al-Shalhat, A., Al-Sogair, A. and Al-Mohammad, M. (1985). Chemical composition and nutritive value of truffles of Saudi Arabia. J. Food Sci., 50(2): 450–453. https://doi.org/10.1111/j.1365-2621.1985.tb13425.x

Serrhini, M., El Youssoufi, K. and Allali, K. (1995). Une ressource naturelle exceptionnelle des milieux pastoraux de l'Oriental et du centre Sud Marocain: le "terfèze", situation actuelle et perspectives d'amélioration. Ecole nationale d'agriculture, Meknès, Morocco, p. 33.

Shavit, E. (2008). Medicinal mushrooms: Truffles roasting in the evening fires, Pages from the History of desert truffles. Truffles, 1(3): 18–23.

Shavit, E. and Shavit, E. (2014). The Medicinal Value of Desert Truffles. pp. 323–340. In: Kagan-Zur, V., Roth-Bejerano, N., Sitrit, Y. and Morte A. (eds.). Desert Truffles Phylogeny, Physiology, Distribution and Domestication. Soil Biology, Springer, Berlin, Germany.

Stojković, D., Reis, F.S., Ferreira, I.C.F.R., Barros, L., Glamočlija, J. et al. (2013). *Tirmania pinoyi*: Chemical composition, in vitro antioxidant and antibacterial activities and *in situ* control of *Staphylococcus aureus* in chicken soup. Food Res. Int., 53(1): 56–62.

Trappe, J.M., Claridge, A.D., Arora, D. and Smit, W.A. (2008). Desert truffles of the African Kalahari: Ecology, ethnomycology, and taxonomy. Econ. Bot., 62: 521–529.

Volpato, G., Rossi, D. and Dentoni, D. (2013). A reward for patience and suffering: ethnomycology and commodification of desert truffles among Sahrawi refugees and nomads of Western Sahara. Econ. Bot., 67(2): 147–160. https://doi.org/10.1007/s12231-013-9234-7

Wang, S. and Marcone, M.F. (2011). The biochemistry and biological properties of the world's most expensive underground edible mushroom: Truffles. Food Res. Int., 44(9): 2567–2581. https://doi.org/10.1016/j.foodres.2011.06.008

Zitouni-Haouar, F.E-H. (2016). Etude de la diversité des truffes du désert et de leurs associations mycorhiziennes. Thèse de doctorat. Université d'Oran 1 Ahmed Ben Bella, p. 95.

Chapter 14

The Truffle Value Chain in Morocco and its Socioeconomic Significance

Hakkou, S.[1,*] *Sabir, M.*[2] and *Machouri N.*[1]

1. Introduction

Truffles are a valuable resource due to their economic potential and high demand in both local and international markets, presenting opportunities for economic development particularly in rural areas (Davies et al., 2012; Boufeldja, 2017). Historically, they were extensively traded wth the Greeks and the Romans importing them from Africa in jars filled with sand (Feeney, 2002). Today, truffles are a significant part of the trade industry involving both large and small traders. They are actively traded locally and internationally with prices influenced by supply and demand, variety, and origin (Moynier, 1835; Alsheikh and Trappe, 1983; Mandeel and Al-Laith, 2007). Truffles also provide supplementary incomes for families, especially Bedouins who sell them at high prices in towns (Shavit, 2008; Zitouni, 2010; Hakkou et al., 2022a). However, their production is heavily dependent on climatic conditions, particularly rainfall, which poses a challenge to the industry's development and income sustainability. Consequently, truffle cultivation has gained popularity in various countries, aiming to expand truffle plantations, promote socioeconomic development, and enhance the cultural identity associated with truffle production (Alsheikh, 1994; Olivier et al., 2018).

[1] Faculty of Letters and Human Sciences, Rabat, Morocco.
[2] National School of Forest Engineers, Sale, Morocco.
* Corresponding author: soukaina.hakkou@gmail.com

Moroccan truffles are a rich and diverse resource reflected in the extensive production that span almost the entire country and the variety of species. There are twelve species known as "desert truffles" and five species known as "true truffles". Desert truffles are found in the North-East, the Maâmora forest, the Doukkala-Abda Sahel, and the Moroccan Sahara. True truffles have been reported in the forests of the Middle Atlas (Malençon, 1973; Abourouh, 2011, 2020; Khabar, 2016; Hakkou et al., 2022b). Morocco was also the first North African country to introduce black *Tuber melanosporum* truffle cultivation in the Debdou and Imouzzer du Kandar zones (Laqbaqbi, 2020; Hakkou, 2022; Hakkou, 2023). However, this natural resource is exploited without regulation, allowing intermediaries to profit excessively. This chapter aims to study the national truffle chain and its various actors, explain the marketing chain from collection to export, and emphasize its socioeconomic importance.

2. Truffle-Growing Regions in Morocco

There are four known truffle-producing areas in Morocco: The North-East region, the Maâmora, the Doukkala-Abda Sahel, and the Moroccan Sahara. Their geographical locations are as follows (Fig. 1).

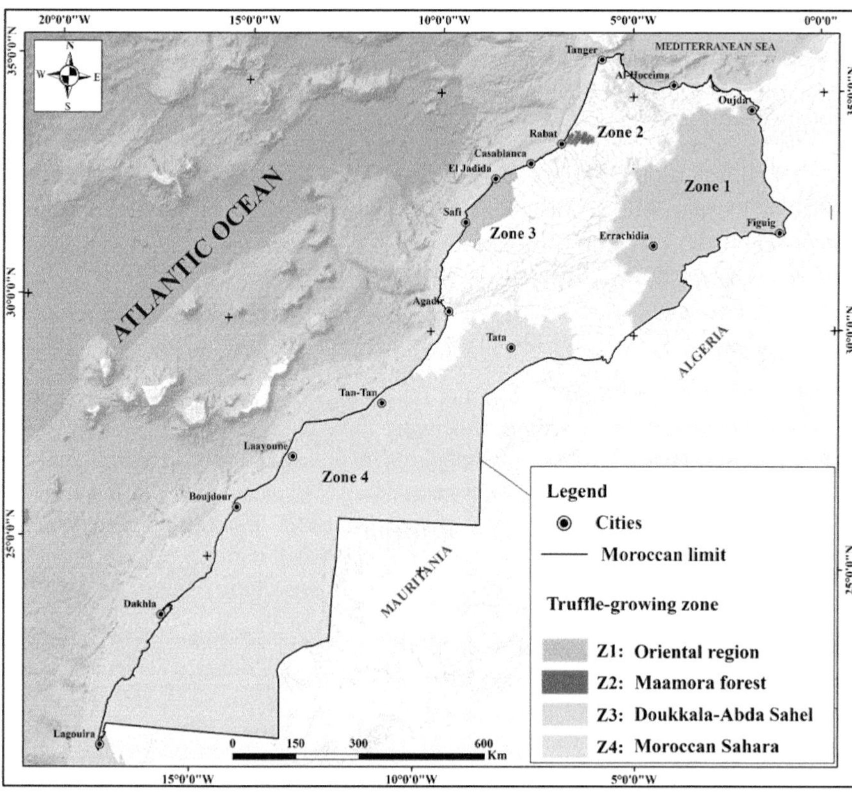

Fig. 1. Geographical locations of truffle-growing areas in Morocco.

2.1 North-East Region

It includes the high plateaus and pastoral plains of the Oriental, Upper Moulouya, and part of northern Tafilalet (DGCL, 2015). It is characterized by a diversity of bioclimates, with a predominance of arid and semi-arid bioclimates, steppe vegetation, and a relief of plains and plateaus (DRAO, 2019).

2.2 Maamora Forest

It is located in the north-west of Morocco in the Atlantic zone, between the Atlantic Ocean to the west, the Gharb plain to the north, the Zemmour and Zaër plateaus to the east, and the Chaouia to the south. It extends over the provinces of Kénitra, Salé, and Khémisset. It is characterized by a coastal sub-humid bioclimate in the western part and a semi-arid bioclimate in the central and eastern Maâmora, forest vegetation based on cork oak, and a flat relief (Aafi, 2007; Motib and Al Karkouri, 2020).

2.3 Doukkala-Abda Sahel

It is located in western Morocco and is part of the Atlantic domain (Tellal, 1989). It lies between El Jadida and Safi cities. The Sahel is part of the semi-arid bioclimatic zone, with degraded pastoral vegetation (Retama, dwarf palm) and a more or less undulating coastal relief (Bammi et al., 2014; Mouhiddine, 2016).

2.4 Moroccan Sahara

It is located in the south of the Kingdom of Morocco, bordered to the north by the southern Anti-Atlas Mountains and to the south by Lagouira. It is characterized by a Saharan climate, more or less sparse steppe vegetation, a flat relief with hamadas, and wide, shallow wadis (Obda, 2020; Sabir, 2020).

3. Truffle Industry in Morocco

The truffle chain involves several actors from collection to final destination (Fig. 2). On a regional scale, it is organized between the collectors of truffles in the wild, the intermediaries who buy them and supply them to semi-wholesalers or directly to wholesalers. The latter determines prices according to the availability of the product, the size of the order, and the urgency of delivery to the last buyers abroad. On a national scale, the truffle export business is common to all regions. Indeed, exporters obtain their supplies from all the truffle-growing regions of the Kingdom.

3.1 Local Stakeholders

3.1.1 Collectors

During the truffle-growing season, everyone participates in collecting truffles. Unlike other non-wood forest products like rosemary, there are no regulations specifying collection locations or who have the right to collect truffles (Fig. 3a, b). People of all backgrounds, genders, and ages are involved, including shepherds, nomads,

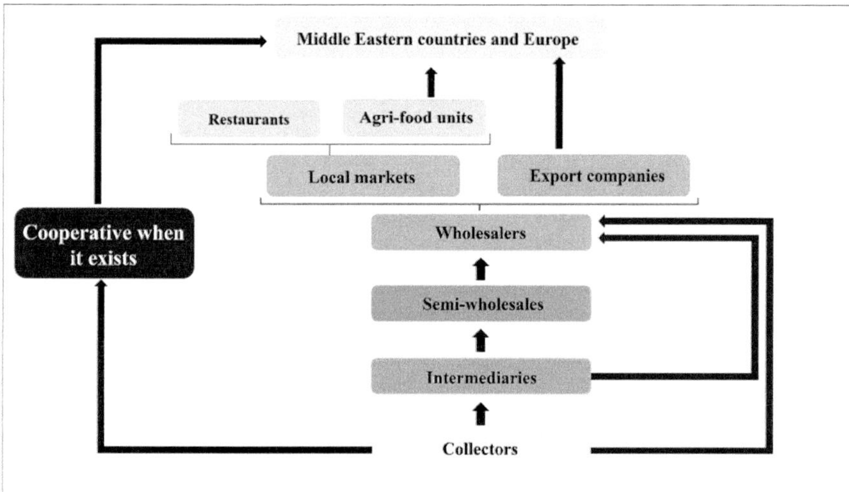

Fig. 2. Different players in the Moroccan truffle chain.

farmers, small-scale stockbreeders, shopkeepers, retired people (military), young fishermen, and the unemployed. Collecting is not limited to the inhabitants of the truffle-growing region; people from other areas also come to take advantage of this seasonal opportunity.

The working mode differs from region to region, depending on the nature of the field and the distances involved. In the Maâmora forest and the Doukkala-Abda Sahel, harvesting is mainly carried out by women, generally from morning until evening. The day starts at 7 or 9 am and ends around 6 pm. Those who do not have anyone to look after the children and do the household chores, work from 7 am to midday and from 3 pm to 7 pm. They reserve time to prepare lunch for their families. Collectors go in all directions to look for truffles, and the distances covered can exceed 20 km. Most of the women set off in groups and take advantage of the opportunity to spend some quality time together, especially at lunchtime, when they share the meals they have brought along. Working in a group also gives them a sense of security. Some women prefer to go it alone or with a family member. This category doesn't like to collect in areas where everyone goes. In the North-East region and the Moroccan Sahara, collection is a male mission. The type of work is determined by the conditions of the collectors and the distance between their homes and the collection sites. Those who live close to productive areas spend the day collecting and return home in the evening. They carry water, tea-making equipment, and food for lunch. Some of them spend the night in the collecting areas to start collecting early in the morning.

Collectors typically travel in groups of 4 to 10 people often from the same locality (Douar). Before setting off, they contribute 300 to 1000 DH per person to cover fuel and supplies. Authorities note that 60 to 70 tents are set up in each truffle-growing zone for a period of 2 to 3 months. Collectors move from one area to another based on truffle availability. Each group usually has a car and a motorcycle or scooter. The car is parked in a specific spot, and the three-wheeled vehicle circles the area, collecting the truffles collected by the collectors. Searching for truffles is a delicate and tiring task that requires expertise and local knowledge. Collectors spend the entire day bent over, looking for clues such as the host plant and cracked earth. They use a stake to dig up the truffles and a bag to carry them.

3.1.2 Intermediaries

Intermediaries form the second link in the truffle marketing chain after collectors. Typically, they are locals who engage in a variety of professions with most being

Fig. 3. Truffle collectors (men and women) (a and b); Truffles stored in a basket at the souk of Sebt El Oualidia, Doukkala-Abda Sahel c); Weighing a bucket of truffles by an intermediary at the Sidi Boukhalkhal souk, Maâmora forest in the north-west of Morocco (d); Truffle sellers at the souk in Sebt El Oualidia, Doukkala Sahel (e and f) (© Hakkou).

shopkeepers, poultry dealers, herbalists, and grocers (Fig. 3c–f). These intermediaries usually have networks of collectors from whom they gather truffles in the field, often at the end of the day. They operate on behalf of semi-wholesalers or wholesalers who supply transportation and purchasing capital. Intermediaries earn a commission of 2–20 DH/kg, depending on the truffle species and availability.

3.1.3 Semi-Wholesalers

Semi-wholesalers are often vegetable and fruit traders or merchants. They use pick-up trucks to transport the quantities collected. Most of them travel throughout the production season to all truffle-growing regions. They buy from intermediaries or directly from collectors and supply wholesalers. Some of them sell directly to exporters without going through wholesalers. They may be independent or hired by a wholesaler. In the first case, they have the money and the means of transport and earn a percentage of the purchase price. In the second case, the wholesaler provides the means of transport and the purchase money, and they receive a commission of between 15 DH to 50 DH/kg, depending on the truffle species and availability.

3.1.4 Wholesalers

Wholesalers are the final local actors in the truffle market. They include large-scale breeders, farmers, traders, and butchers who maybe natives of the region or come from other areas. Most wholesalers hail from Casablanca, Rabat, El Kamouni, Tétouan, Béni Mellal, Agadir, Tadla, Khénifra, Meknès, and the Oriental region. They purchase truffles from semi-wholesalers, intermediaries, and even directly from collectors. In the North-East region and the Moroccan Sahara, wholesalers rent garages in urban centers to store truffles, sort them by color and size, pack them in crates, and transport them quickly to the place of order. Often, this means shipping to Casablanca for export or to Kénitra, Fès, and Meknès for processing plants. Wholesalers control the purchase price based on external demand.

3.2 Stakeholders in the Truffle Export

Exporters are most often mushroom or fruit and vegetable company owners who work with approved packers to prepare truffles for export. Packaging plants and agri-food factories also export on their own accounts. Most packaging units are based in Deroua, very close to Mohammed V International Airport, to avoid delays in shipment and consequent rotting of the truffles. Others are in Casablanca, Settat, and Berrechid. Truffles are generally exported legally. When they are not, they are exported under other categories, such as vegetables, often for reasons of timing.

3.2.1 Export Companies

Export companies can be found in most Moroccan towns, especially in agricultural areas. They do not specialize in truffle exports alone, as truffles are not available year-round. Most of them export fruit and vegetables. They collaborate with wholesalers in the various truffle-growing regions of Morocco. As early as December, the owners start contacting wholesalers to make sure that production has begun and to get an idea of prices at the weekly souks (market). They then contact their customers, most of whom are in the Gulf countries. The aim of this operation is to specify the quantities to be ordered from wholesalers and the purchase price. Companies do not buy quantities of less than 100 kg.

A proforma invoice indicating the quantities to be delivered, the purchase, packaging, transit, and transport costs is sent by the company to the foreign buyer (Fig. 4). The latter transfers an advance payment, agreed upon between the two parties; this is often 50% of the amount of the proforma invoice. Once the specified quantities have been purchased, they are sent to one of the packaging units the company works with in Deroua (near Casablanca city) for processing and export, often on the same day. The necessary export documents are prepared by the packing plant and the forwarding agent appointed by the company. Once the truffles are on board the plane, the shipper (company) sends the invoice to the consignee, indicating quantities and export status. The consignee makes a bank transfer at the agreed price to the export company. Packaging involves cleaning the truffles, sorting them, and packing them in crates according to species and size. The crates are labeled with the species, weight, export company name, and consignee references (name, address, telephone number, country). The crates are then wrapped in a net to prevent the truffles from falling out.

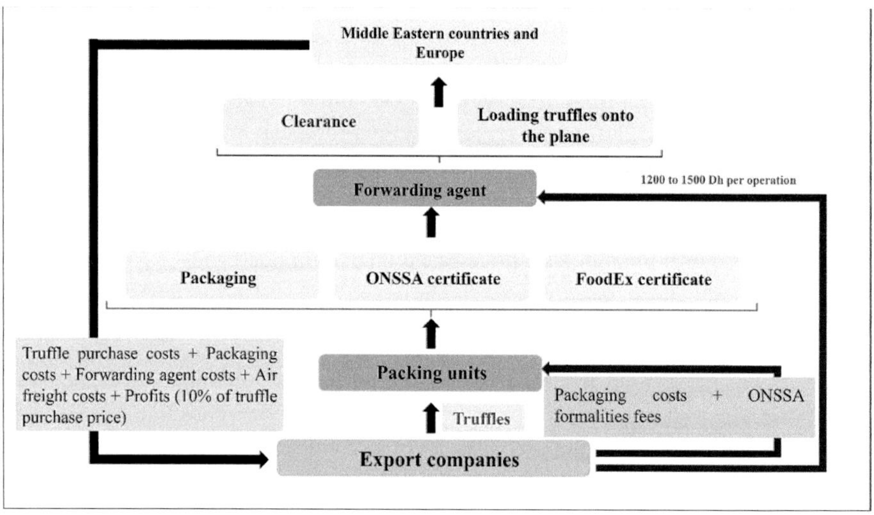

Fig. 4. Different steps in the export of Moroccan truffles.

3.2.2 Packaging Units

Packaging units are establishments approved by the National Office of Food Safety (NOSSA) to package food products. In most cases, they do not carry out any processing but simply clean and package products for export. The packaging operation consists of making a package according to the exporter's request adhering to ONSSA standards. In most cases, it is the exporter who provides the necessary equipment typically crates and nets. Some even provide manpower and utilize the premises only for work and storage. The exporter cleans the truffles of sand, sorts them by species and size, and weighs them. Any damaged or rotten truffles are trimmed with the healthy parts preserved, either in salt water in 1 kg tins or sliced for drying.

Each species requires a different treatment. For example, for white truffles, the most susceptible to rot, a single layer of truffles is packed in crates containing no more than 2–3 kg. Other species are packed in 6–8 kg crates. Export quantities vary according to production. At present, with the drop in production due to drought, these quantities are of the order of 1 to 1.5 tons per day. They require between 10 to 20 people to pack the product overnight. The conditioning unit is responsible for preparing the documents certifying that the truffles meet the standards required for export. The documents in question are a phytosanitary certificate from ONSSA and a certificate from Morocco FoodEx-EACCE (Autonomous Export Control and Coordination Agency).

3.2.3 Forwarding Agents

This is the final step before the truffles are loaded onto the plane. After the packing unit completes its work, the exporter collects the packed goods and the ONSSA certificate and heads for the airport, where the forwarding agent is waiting. The forwarder books the airline in advance, with collects an Air Waybill (AWB) confirming the receipt of the goods, and weighs the parcels to add the weight to the AWB. They then proceed with customs clearance. Truffles must be received three hours before the plane's departure. Moroccan products destined for the Middle East are exempt from customs duties.

3.3 Destination of Moroccan Truffles

Moroccan truffles are highly sought-after in the Middle East and Europe. In the Middle East, white truffles of the *Tirmania* genus, found in the Moroccan Sahara and the Oriental region, are the most exported. Key importers include Saudi Arabia, Kuwait, Bahrain, the United Arab Emirates, Qatar, Syria, and Jewish communities living in Palestine. These countries once producers of these truffle species, now face reduced production due to drought and war, leading them to import Moroccan truffles. Truffles are considered a prestige food, increasing their demand among wealthy families. Their nutritional value and aphrodisiac qualities also contribute to their popularity. In Europe, species of the *Terfezia* genus are in demand, particularly in France, Spain, Italy, and Germany, where they are used as garnishes. The Maâmora *Tuber oligospermum* a species of the *Tuber* genus is often exported to Spain and Italy due to its resemblance to the true white truffle, *Tuber magnatum.*

4. Socioeconomic Significance of Moroccan Truffles

The truffle season is a highly anticipated event, especially in recent years with irregular production. It provides an opportunity for the unemployed to earn money and stay occupied in a society where unemployment is stigmatized. Truffles are a significant source of income for people living in truffle-growing regions. Earnings at the end of the season vary based on the year's productivity and one's position in the marketing chain. Production abundance depends on rainfall, and daily collection quantities rely on the number of collectors and their expertise.

4.1 Collected Quantity

4.1.1 Collectors

Quantities collected are influenced by abundance during the production season, by the time spent collecting, by the number of collectors per family, and, above all, by the detection capacity of productive areas. In the Maâmora region, we note a dominance of those collecting 2–3 kg/day (36.6%) (Fig. 5a). Those collecting more than 4 kg/day represent only 5.2% of all collectors. Respondents told us that during the 1970s and 80s, when the season was good, people could collect up to 15 kg/day. In Sahel Doukkala-Abda, the majority collect from 0.25 to 1 kg/week, with the lowest percentage in the 2–3 kg/week range (Fig. 5b). In the Nord-East region, those collecting less than 5 kg/day dominate (42.85%) (Fig. 5c).

Those who collect large quantities either work in groups (families) or monopolize productive areas exclusively for themselves (nomads). According to the respondents, in 1995, 3 people collected 75 kg during a 2-hour search northeast of Talsint. As for the Moroccan Sahara, during a year of good production, the quantities collected by one person are in the order of 4 to 10 kg per day. A group can collect up to 7–30 kg, since the quantities collected by each person are not the same. In highly productive areas, people can collect up to 20–30 kg/day. In years when production is scarce, quantities range from 700 g to 4 kg/person/day.

4.1.2 Intermediaries

The quantities collected by intermediaries depend on the number of collectors present at the weekly souk and those from whom they obtain supplies in the douars. Quantities vary from 50–100 kg/intermediary/week in the Maâmora region, from 30 to 50 kg/intermediary/week in Sahel Doukkala-Abda, and from 50 kg to 4 quintals/intermediary/day in the North-East region and the Moroccan Sahara.

4.1.3 Semi-Wholesalers

The quantities collected depend on the number and efficiency of the intermediaries and collectors involved. The semi-wholesaler's commercial history in the area is also a determining factor, as it determines the trust that intermediaries can place in them to work. Quantities vary from 100 to 200 kg/semi-wholesaler/week in the Maâmora region and from 200 to 600 kg/semi-wholesaler/day in the North-East region. One notes that in Sahel Doukkala-Abda and the Moroccan Sahara, the marketing chain does not often include semi-wholesalers.

4.1.4 Wholesalers

The quantities purchased by each wholesaler depend above all on external demand, the productivity of the season, and the number of weekly souks they visit, as well as the number of intermediaries and semi-wholesalers they deal with. The quantity is thus of the order of 600 kg to 1 tonne per week in the Maâmora region, 50 kg to 3 quintals per week in the Sahel Doukkala-Abda region, and 2 to 10 tons per day in the

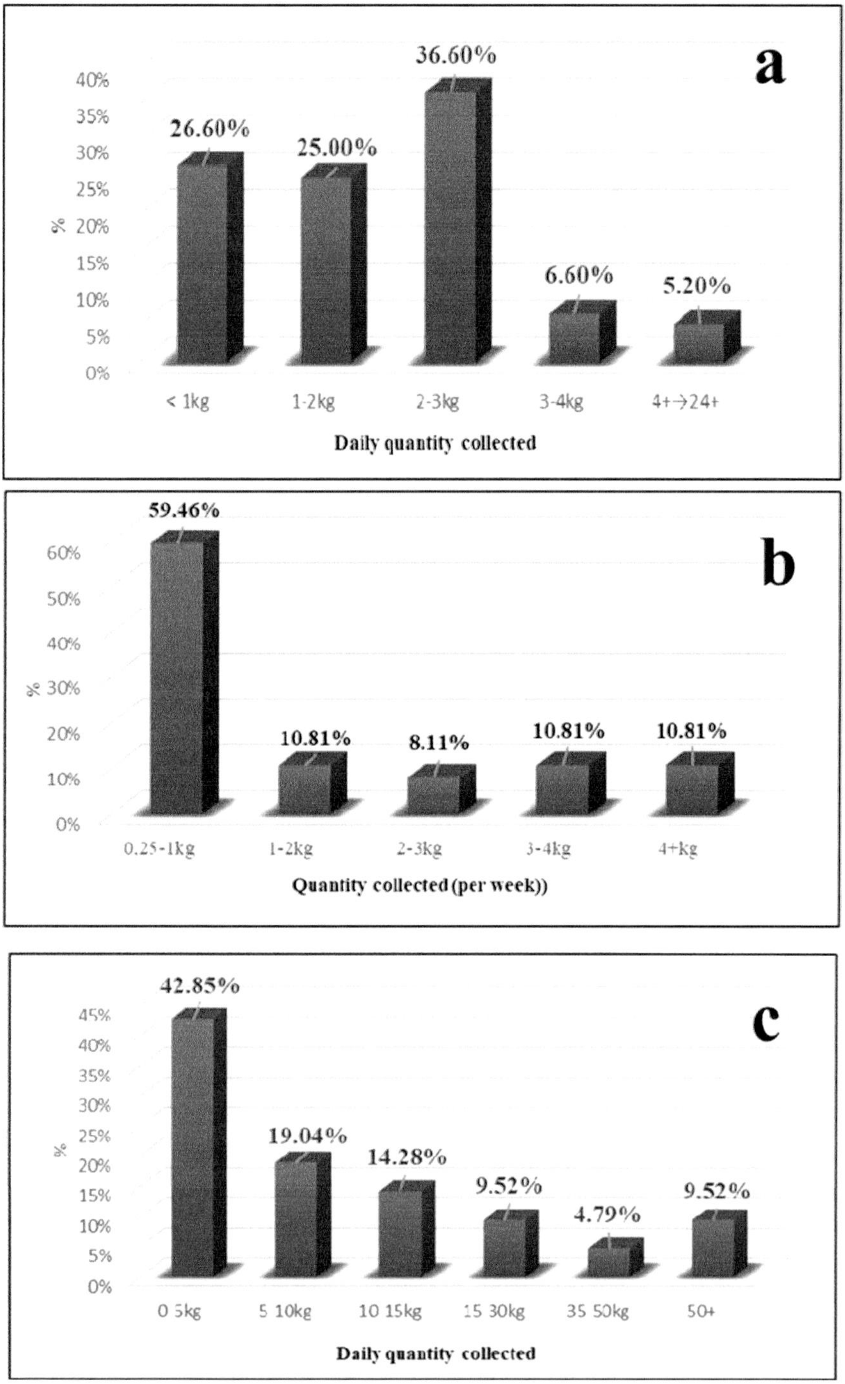

Fig. 5. Quantity of truffles collected in different regions: Maâmora Forest (a), Doukkala-Abda Sahel (b), and North-East Region (c).

North-East region and the Moroccan Sahara. The total daily quantity collected by all wholesalers can reach 50 to 60 tonnes.

4.2 Truffle Sales

Truffles are sold either at weekly souks, douars (villages), or collection sites. The latter option is more common in the northeastern region and the Moroccan Sahara, due to the vastness of the area. Collectors often go to the souk closest to where they live. The collectors often sit together in a place at the entrance to the souks and display their truffles contained in small containers to the various buyers. Intermediaries, semi-wholesalers, and wholesalers buy truffles in all truffle souks to meet their needs. The souks where truffles are sold cover almost the whole week, making the truffle trade active every day. Sales start very early in the morning, often just at dawn, and generally end no later than 11 am.

4.2.1 The Selling Price

The selling price of truffles depends on demand, particularly external to the country, and production. Thus, scarcity generates a remarkable increase in price. The number of intermediaries and wholesalers also plays a part in price oscillation. When they are numerous, competition is fierce, and prices rise. When they are few, which is almost always the case, they come to an agreement among themselves and set the price, which is always low, to the detriment of the collectors. Prices vary within the same region, depending on the species and size of the truffles. The largest and intact ones are the most prized.

In the Maâmora region, the pink Maâmora truffle (*Terfezia arenaria*), considered to be of better quality, is often more expensive than the white truffles (*Tuber oligospermum* and *Tuber asa*). The price difference can range from 20 to 50 DH/kg. The selling price of pink truffles from collectors varies from 30 to 300 DH/kg. Prices are highest at the beginning and end of the production season, when truffles are scarce. Towards the middle of the season, the truffle is abundant, and the price drops. The average price is 100 DH/kg. The average price for white truffles is 50–75 DH/kg. In Sahel Doukkala-Abda, the price varies between 50 and 150 DH/kg and rarely exceeds 200 DH/kg when production is scarce and demand is high. The average price is 80 DH/kg. In the North-East region, selling prices range from 20 to 70 DH/kg for red truffles (*Terfezia claveryi*) and 100 DH to 300 DH/kg for white truffles (*Tirmania nivea* and *pinoyi*). Prices can reach up to 500 DH/kg when production is low, as in 2022. In the Moroccan Sahara, the price varies from 25 to 500 Dh/kg. White truffles (*Tirmania nivea and T. pinoyi*) are the most expensive and in great demand. Prices range from 200 to 500 DH/kg. They rarely sell for less than 200 DH, unless production is abundant. In that case, the selling price is 90–100 DH/kg. As for the red truffle (*Terfezia claveryi*), it fetches between 25 DH and 200 DH/kg and can reach as much as 350 DH/kg in years of scarcity.

The sales margin varies from one player to another. At the time of purchase, each of the parties involved tries to buy at less than 10 DH or 20 DH difference from the selling price. For example, if the price set by the wholesaler is 100 DH/kg, the semi-wholesaler will buy from the intermediaries at 80–90 DH/kg, who in turn will buy from the collectors at 60–70 DH/kg. The wholesaler can then sell at 200–300 DH/kg or even more. Some wholesalers sell at a price between 400 and 1000 DH/kg.

5. Impact of Truffle Marketing on the Local Population

Marketing truffles result in variable monetary returns, depending on the quantities collected, the length of the production season, the price, and the number of collectors per family. Earnings are higher in the northeast and Sahara regions than in the Maâmora and Sahel of Doukkala-Abda. In the Maâmora, seasonal incomes (3 months) range from less than 2,400 DH to more than 12,000 DH (Fig. 6a). The highest income bracket (30%) is accounted for by respondents earning between 4,800 and 7,000 DH per season. In Sahel Doukkala-Abda, the majority (59.46%) of collectors earn less than 650 DH for the entire collection season (2 months) (Fig. 6b). A further 10.81% earn more than 2500 DH per season. These are people who collect more than 4 kg of truffles per week. In the north-east region, seasonal earnings range from a few hundred to thousands of dirhams (Fig. 6c).

A third of those surveyed earn between 20,000 DH and 40,000 DH over the entire harvesting season (two and a half months), followed by those earning between 5,000 DH and 15,000 DH (26.6%) and those earning between 40,000 DH and 80,000 DH (20%). As for the Moroccan Sahara, the average is 15,000–30,000 DH per person during the production season (1–1.5 months). Experience and expertise in the search for truffles determine the quantities harvested and, therefore, the profits from truffle sales. A good connoisseur of the land and of collecting techniques can earn 70,000 to 80,000 DH/month in a year of good production, and 3,000 to 4,500 DH/15 days in a year of poor production.

Collectors use the income from truffle sales to improve their living conditions and meet needs that have been put off for long due to a lack of means. In the Maâmora and Sahel Doukkala-Abda regions, where incomes are not high enough, earnings are spent the same day at the souk (market) to meet family needs (food, clothing, and kitchen utensils). Women who are a little better off use the money for social visits (weddings, births, and visits to the sick). There are also those who save money for difficult times.

In the north-east region and the Moroccan Sahara, the money earned is spent according to the needs of each family. Some use it to build new houses, repair existing ones, or buy new furniture. Others opt for profitable activities, such as trading, buying stores, cars, or livestock, or digging wells to irrigate crops. In addition, this income enables the local population to cover household expenses for food, water and electricity, clothing, children's schooling, and health care. The earnings of other actors in the marketing chain far exceed the income of the collectors, depending on the quantity and selling price. These earnings can range from 15,000–50,000 DH per season for intermediaries and semi-wholesalers in the Maâmora and Sahel Doukkala-Abda, and from 40,000–80,000 DH per season for intermediaries and semi-wholesalers in the North-East region and the Moroccan Sahara. Wholesalers are the biggest winners, and it's difficult to estimate their income as selling prices cannot be known precisely due to their discretion. Most of them invest in major projects in urban centers (coffee shops and restaurants).

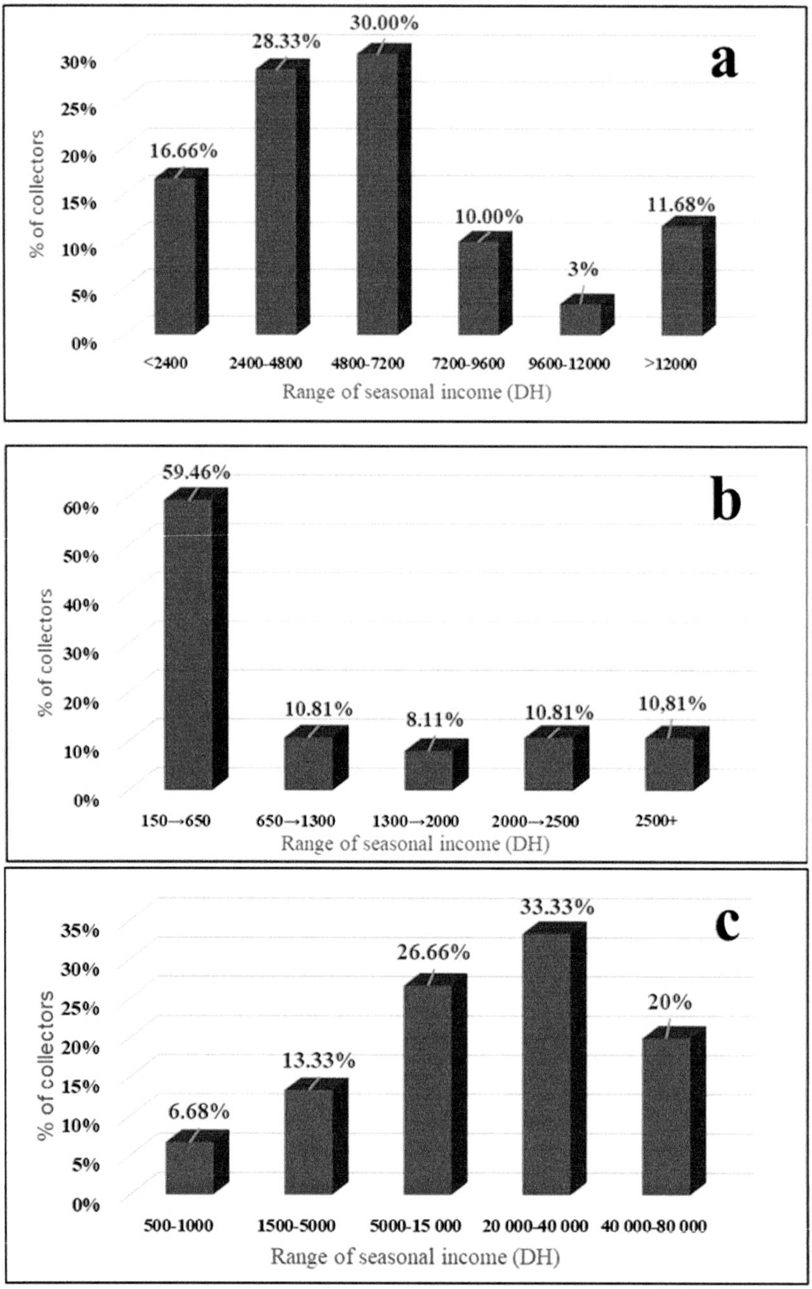

Fig. 6. Seasonal income of truffle collectors by region: Maâmora Forest (a), Doukkala Abda Sahel (b), and North-East region (c).

6. Conclusion

The Moroccan truffle chain plays a key role in improving the socioeconomic conditions of people living in truffle-growing regions. It involves a chain of actors from collection to export: collectors, intermediaries, semi-wholesalers, and wholesalers. Truffles are destined for local markets (restaurants and food processing units) and mainly for export through export companies. The main destinations for Moroccan truffles are the Middle East and certain European countries. Collection methods and collector characteristics vary from region to region. In the Maâmora forest and the Doukkala-Abda Sahel, most collectors are women. In the northeastern region and the Moroccan Sahara, on the other hand, it's the men who do the collecting. The truffle chain is an important source of income for the populations of these four regions. Selling prices vary from region to region, depending on species, size, and demand. They range from 20 to 70 DH/kg for red truffles and 100 to 300 DH/kg for white truffles in the Nord-East region. In the Maâmora region, prices range from 50 to 300 DH/kg, with an average of 100 DH/kg for pink truffles and 30 to 75 DH/kg for white truffles. In the Sahel of Doukkala-Abda, prices range from 30 to 200 DH/kg, with an average of 80 DH/kg. In the Moroccan Sahara, prices range from 25 to 500 DH/kg, with remarkable differences between species.

White truffles are the most expensive, with prices ranging from 200 to 600 DH/kg; they rarely sell for less than 200 DH. Red truffles, on the other hand, fetch between 25 and 200 DH/kg, and up to 350 DH/kg. The income derived from this activity and its share in the income of the actors vary according to their position in the truffle collection and marketing chain. Collectors earn between 5,000 and 80,000 DH per season in the northeast region; from 650 to 2,500 DH per season in the Doukkala-Abda Sahel; from 2,400 to 12,000 DH per season in the Maâmora forest; and from 15,000 to 80,000 DH per season in the Moroccan Sahara. In Maâmora and Doukkala-Abda Sahel, where incomes are not high, they are mainly used to meet daily needs, whereas in the Moroccan Sahara and the North-East region, people have been able to buy or build houses, livestock, and cars. Wholesalers make the most profit, given the size of the quantities sold and the selling prices. The collectors, who put in the most effort, benefit the least from the chain and are the weakest link in the marketing chain. The truffle chain lacks organization and regulation, particularly for collection, marketing, and exports. Particular attention needs to be paid to the organization of this sector on a national scale.

References

Aafi, A. (2007). Etude de la diversité floristique de l'écosystème de chêne liège de la forêt de la Maâmora. Thèse de Doctorat d'Etat Es-Sciences Agronomiques, Institut Agronomique et Vétérinaire Hassan II, Rabat, Morocco, p. 203.

Abourouh, M. (2011). "Truffes du désert" du Maroc: diversité et modes d'exploitation. 6ème Rencontre de MICOSYLVA, Mértola (ADPM), Portugal, 15–18 Mars 2011, p. 23.

Abourouh, M. (2020). Terfess et truffes de la Maâmora: importances biologique, écologiques et socio-économique. Consultation nationale "Restauration de la forêt de chêne-liège pour le développement et la valorisation des truffes dans la forêt de la Maâmora, FAO, p. 78.

Alsheikh, A.M. and Trappe J.M. (1983). Desert truffles: The genus *Tirmania*. Trans. Br. Mycol. Soc., 8(1): 83–90.

Alsheikh, A.M. (1994). Taxonomy and mycorrhizal ecology of the desert truffles in the Genus Terfezia. Ph.D. dissertation. Oregon State University, Corvallis, USA, p. 239.

Bammi, J., Mouhiddine, M., Fassi, D. and Douira, A. (2014). Contribution à la connaissance de la végétation des Doukkala-Abda (Maroc Atlantique): Approche écogéomorphologique. J. Anim. Pl. Sci., 20(3): 3202–3211.

Boufeldja, W. (2017). Evaluation du potentiel nutritionnel et antioxydant de quelques variétés de Truffes du Sud-ouest algérien. Effets antimicrobien et antiinflammatoire. Thèse de doctorat. Université Djillali Liabes, Algeria, p. 112.

Davies, J., Poulsen, L., Schulte-Herbrüggen, B., Mackinnon, K., Crawhall, N. et al. (2012). Conserving dryland biodiversity. International Union for the Conservation of Nature (IUCN), p. 100.

DGCL (2015). La région de l'Oriental: monographie générale. Direction générale des collectivités locales. Rabat, Morocco, p. 66.

DRAO (2019). Elaboration du schéma directeur d'aménagement des parcours et du développement de l'élevage dans la région de l'Oriental. Etude de base. Marché N° 24/2019. Direction régionale de l'agriculture de l'Oriental, Oujda, Morocco, p. 180.

Feeney, J. (2002). Desert truffles galore. Saudi Aramco World, 53(5): 22–27.

Hakkou, S. (2022). La trufficulture au Maroc, une success story à reproduire. La feuille et l'aiguille, 127.

Hakkou, S. (2023). Les truffes marocaines: étude socio-géographique et environnementale. Thèse de Doctorat. Faculté des Lettres et des Sciences Humaines. Université Mohamed V., Rabat, Morocco, p. 478.

Hakkou, S., Sabir, M. and Machouri, N. (2022b). Le Maroc truffier: la répartition géographique des truffes et truffes du désert et leur productivité. – Rev. Mar. Sci. Agron. Vét., 10(3): 395–407.

Hakkou, S., Sabir, M. and Machouri, N. (2022a). La filière truffe dans le Sahara marocain: une opportunité pour le développement socio-économique de la région. Revue espace géographique et société marocaine, 63: 57–89.

Khabar, L. (2016). Les Terfess et les truffes du Maroc, biodiversité et valorisation. Éd. Univ. Européenne, p. 276.

Laqbaqbi, A. (2020). La truffe entre France et Maroc ou d'un coté à l'autre de la Méditerranée. Imprimerie GRAPHO 12 Aveyron, p. 80.

Malençon, G. (1973). Champignons hypogés du Nord de l'Afrique. I. Ascomycètes. Persoonia, 7: 261–288.

Mandeel, Q.A. and Al-Laith, A.A.A. (2007). Ethnomycological aspects of the desert truffle among native Bahraini and non-Bahraini peoples of the Kingdom of Bahrain. J. Ethnopharmacol., 110(1): 118–129.

Motib, I and Al Karkouri, J. (2020). La forêt de la Maâmora occidentale dans son cadre géographique et juridique. Revue Espace Géographie et Société Marocaine, 33–34: 417–438.

Mouhiddine, M., Bammi, J. and Fassi, D. (2016). L'écosystème littoral et la végétation de l'Atlantique central marocain Approche éco-géomorphologique. Revue Marocaine de Géomorphologie, 2016–2017: 113–128.

Moynier, M.M. (1835). De la truffe, traité complet de ce tubercule. Gallica/Bibliothèque nationale de France, France, p. 476.

Obda, K., Obda, I. and Chanyour, Y. (2020). Hydrologie des milieux présahariens et sahariens marocains: dégradation des hydrosystèmes fragiles. pp. 105–128. In: Sahara Atlantique Marocain: Environnement, espaces et défis de développement territorial durable. UM5. FLSH Rabat, Morocco.

Olivier, J.-M., Savignac, J.-C. and Sourzat, P. (2018). Truffe et trufficulture. Périgueux: Editions Fanlac, France, p. 352.

Sabir, M. (2020). Les ressources en eau dans les zones Sud du Maroc. pp. 75–103. In: Watfeh, A. and Faaras, A. (eds.). Sahara Atlantique marocain: Environnement, espaces et défis de développement territorial durable. UM5. FLSH Rabat, Morocco.

Shavit, E. (2008). Medicinal mushrooms: Truffles roasting in the evening fires, Pages from the History of desert truffles. Truffles, 1(3): 18–23.

Tellal, R. (1989). Etude des potentialités pastorales du Sahel Nord d'Abda (Province de Safi). Thèse. Université Mohamed V, faculté des sciences de Rabat, Morocco, p. 139.

Zitouni, F.H. (2010). Etude des associations mycorhiziennes entre quatre espèces de terfez et diverses plantes Cistacées et ligneuse en conditions contrôlées. Mém. Magister, Univ. Oran, Algeria, p. 262.

Industrial Applications

Chapter 15

Small-Scale Truffle Cultivation

Srdjan D. Ljubojević

1. Introduction

Truffles are a group of fungi known for forming fruiting bodies below the surface of the earth. They play an important role in forest ecosystems as partners in a symbiotic relationship with trees, known as mycorrhiza. Truffles are valued and expensive fungi. They are used in human nutrition and have economic significance. They are also an important source of food for some animal species. Over 200 species of truffles have been described in the world (Wie et al., 2021; Mrak et al., 2024). Truffles are generally divided according to the color of the fruiting body, white or black. The most famous types of white truffles are: *Tuber magnatum* (white truffle, Italian white truffle, Piedmont truffle and Alba white truffle) and *Tuber borchii* (whitish truffle, white spring truffle and Bianchetto truffle). The most famous types of black truffles are *Tuber melanosporum* (French black truffle, Périgord truffle and precious black truffle), *Tuber aestivum* (Burgundy truffle and summer truffle) and *Tuber brumale* (black winter truffle, Muscat truffle and winter truffle) (Hall et al., 2007; Bencivenga et al., 2009; Bissanti, 2022).

The reduction of natural habitats and global climate change have significantly contributed to the decline in their production in recent decades, which at the same time led to an increase in demand and prices on the market. It is the value of truffles that has made their cultivation more and more attractive in many parts of the world day by day. Growing truffles is significantly different from growing other crops. To begin with, they are grown as a long-term economic crop that achieves its first full harvest only after 7–8 or more years. Before and after the establishment

Faculty for Ecology and Institute for Scientific Research, The Independent University of Banja Luka, Bosnia and Herzegovina.
* Corresponding author: ljuboje@blic.net

of the plantation, it is necessary to carry out numerous cultivation operations and interventions, which imply detailed planning and timely execution. In other words, all things need to be considered. When someone decides to establish a truffle orchard (truffle plantation), they may not fully grasp the intricate relationships between truffles, the host plant and the surrounding environment. However, by adhering to the principle of learning by doing, they will gain new insights daily, which will help them enhance their production. So far, a large number of truffle cultivation methods and organizational solutions for their implementation have been developed around the world. Bearing in mind the previously mentioned points, the cultivation of truffles on a small property is imposed as an objectively acceptable initiative. Under small-scale truffle farming, it means the cultivation of truffles on a property no larger than 2 ha, which is the smallest land size class according to generally accepted FAO standardization (Wiggins et al., 2010). Typically,the host trees in an artificial truffière are arranged at equal distances from each other. Depending on the type of truffle and local ecological conditions, the density of trees usually varies between 3×3 and 6×6 m, that is, between 277 trees/ha in a sparse arrangement and 1,111 trees/ha in a dense one. Small-scale truffle farming can also be organized as hobby gardens with a small number of trees, the production of which is solely for the satisfaction of one's own needs.

Small-scale truffle farming has certain advantages but also some disadvantages, which should be considered over time. Advantages: the expected income per household is twice as high as the income from traditional agriculture; truffle certification is possible, based on the planting of certified seedlings, which opens up the possibility of accessing subsidies and other forms of financial support; it is easier to find suitable land and choose the one that best suits the local conditions among several types of truffles. Disadvantages: small-scale truffle growers are faced with high crop input costs and significant plantation maintenance costs until the first year of full harvest; one of the solutions for these drawbacks can be a joint venture with a large supplier, which would provide planting material and guarantee purchase at agreed prices. Bearing in mind global climate change, the lack of irrigation water during critical periods of time can have a direct impact on yield and quality. The solution to this problem is not simple and requires additional financial resources. Up to 70% of small farms in Europe do not have water for irrigation (Guarína et al., 2020). This chapter deals with various aspects of small-scale truffle cultivation, maintenance and harvesting, with an emphasis on *Tuber* spp.

2. Growing Truffles

Based on the cultivation areas and the yields achieved on them, *T. aestivum* and *T. melanosporum*, followed by *T. borchii*, are predominantly used for artificial truffle cultivation (Bencivenga and Falini, 2012). The planting materials include spores (gamic inoculum), pure cultures of mycelium (vegetative mycelial inoculum) and colonized roots (symbiotic inoculum) (Iotti et al., 2012). Currently, truffles are grown in monoculture, although there is significant advocacy for the introduction of multicropping systems. These systems could allow for the production of food, fuel and fibers from the same area, in addition to truffles (Porter et al., 2009).

2.1 Growing Truffles with Non-Mycorrhizal Planting Material

Truffles are subterranean fungi, and because of this, their spores cannot be dispersed by the wind. Instead, wild animals which can locate, unearth, and consume them, spread the spores. These spores, typically found in the gleba (truffle's flesh), pass through the animal's digestive tract of the animal and are expelled with the excrement. This natural process allows truffles to reproduce without human intervention. Although the exact timeline is unknown, humans realized the value of truffles long ago. Since then, efforts have been made not only to harvest them in the wild but also to cultivate them in more controlled conditions, a task that has proven to be quite challenging.

2.1.1 Growing Truffles by Sowing Tree Seeds from Active Truffières

The first successful attempt to grow truffles dates back to 1810 and was linked to a French farmer named Joseph Talon. Talon's method, also known as the traditional method, is a passive approach to growing truffles through indirect mycorrhization (Sourzat, 2009). The essence of the method is to find a natural truffle spot, detect the host tree and the symbiotic truffle species, and then carefully collect the seeds of the host tree. Naturally, the seeds that are ripe and have already fallen to the ground are collected, increasing the probability that they have come into contact with truffle mycelium.

Collecting seeds on the tree makes no sense, because such seeds could not come into contact with the fungus mycelium. The seeds are then sown on bare land, whose main pedological and climatic characteristics correspond to both symbiotic partners, both the truffle and the host tree. In this way, Talon formed plantations of Périgord truffle (*T. melonsporum*) using oak acorns. There are no concrete written traces for other species, but it is quite reasonable to think about the use of hazelnut as a host tree, especially since it is known that Périgord truffle and winter truffle (*T. brumale*) naturalize and fruit within hazelnut orchards in southern Europe (Reyna, 2007). However, there are sources that state that Talon actually used wild oak seedlings, taken from under a verified host tree, and then transplanted them to a new location for the needs of the future plantation (Zambonelli et al., 2017). In this way, colonized roots appear in the role of symbiotic inoculum.

Acorn germination is an individual feature, that also depends on a number of external factors. Thus, Özderin et al. (2018) found the following germination percentages on a sample of over 1,000 seeds: holly oak (*Quercus ilex*), 90%; English oak (*Q. robur*), 62%; kermes oak (*Q. coccifera*), 59%. In order to increase the germination rates, it is recommended that the newly collected acorns be kept in plastic containers in perlite at 4°C for 2 months. After that, 10 days before planting, the seeds are soaked in lukewarm distilled water to swell. Before planting, the seeds are sterilized in 5% hydrochloric acid (Özderin et al., 2018). When the young plants sprout, permanent tillage begins, adapted to the age of the plants. Treatment is mainly done with hand tools. Irrigation is not practiced, but sometimes, before high heat, mulching is done to limit evaporation. Pruning is rarely used, and cut branches are used as mulch. Full production begins somewhat later than with new methods, somewhere around the age of 15 (Sourzat, 2009). By the end of the 19th century, there were 190,000 ha of truffle plantations in France, based on the traditional method, with yields up to 2,000 tons/year. Meanwhile, the traditional method gave way to other cultivation methods, with annual yields falling below 50 tons (Diette and Lauriac, 2005). So far, no rational explanations have been given for such a drastic drop in yields.

2.1.2 Growing Truffles by Inoculation of Spores

Spores are used in the production of truffles in two ways: 1) as an inoculum for mycorrhization of host tree seedlings, with which the truffle plantation will be founded; 2) as a corrector-improver in an already established truffle plantation, that has already germinated. In both cases, fresh, ripe, healthy, normally developed fruiting bodies are used. The species identification must be correct. It must not be mixed with the ascocarps of other types of truffles. If the material will not be used immediately, it must be stored in an adequate manner. Ascocarps can be stored in sterile moist sand at 4ºC for up to 2 years. (Zambonelli et al., 2010).

The production of mycorrhizal seedlings using the controlled mycorrhization process is entrusted to specialized, licensed nurseries. The essence of production consists of: germinating the seeds of the host plant on a special substrate, in containers in the greenhouse; transplanting germinated seeds into larger containers filled with substrate mixed with fungal spores; leaving the seedlings in the greenhouse for 1−2 years, in order to develop a quality mycorrhiza; mycorrhizal control before planting material in the field (Özderin et al., 2018; Pruett, 2008). The placement of previously chopped carpophores, mixed with a specific substrate (carrier), with the aim of intensifying and accelerating natural production, has been known in the practice of truffle farming since the mid-18th century (Murat et al., 2016). To date, various variants of this procedure have been described under equally different names, such as: Catalan holes, truffle nests, truffle traps and spore traps. Thus, truffle nests imply the introduction of truffle spores into an existing truffle area. This can be done with purchased (ready-made) material (Fig. 1a, b) and with material prepared according to one`s own recipe. Readymade materials differ from manufacturer to manufacturer, both in terms of the components used and the amount of spores.

Fig. 1. Different forms of reproductive material for establishing a truffle orchard: a, Substrate with spores, 50 liters (Truf-Up, Spain); b, Gleba with spores, frozen, 100 g (Truffefrance, France); c, Mycorized hazelnut seedling, 1−3 years (Robin Nurseries, France).

There are two methods of preparing truffle inoculum: 1) 250 g of crushed ascocarps (gleba with spores) + 350 g of honey + 50 liters of horticultural vermiculite + 50 liters of organic compost. Mix the components well. The resulting mixture is poured into packages of different volumes. At 50 cm from the tree, four holes are opened on the four sides of the globe. The diameter of the holes is 20 cm and the depth is 5–8 cm. They are located about 50 cm from the tree. 250 ml of the mixture is poured into each hole, which is then covered with excavated soil (Murat et al., 2016). 2) A pack of 100 g contains: 1 g of spores (equivalent to 20,000–40,000 spores) + organic matter + vermiculite + calcium + sugars. One package is enough for a single treatment of one host tree. At the edge of the contour of the tree crown, one hole is opened, 25 cm in diameter and 25 cm deep. The contents of one package are poured into the hole and then covered with excavated soil. In newly established plantations, which have not yet entered the production cycle, a hole is not dug, but the package with spores is scattered around the young tree and then the soil is hoed to a depth of about 10 cm (Anonymous, 2024).

2.2 Growing Truffles by Planting Mycorrhizal Seedlings

Unlike the natural forest, where truffles are a typical non-wood forest product, spontaneously distributed here and there, depending on the affinity for certain types of trees and bushes, truffle plantations are a schematically organized environment, typical for agriculture, above all for fruit growing. The host trees are originally forest trees, but here their habits are significantly different, due to regular pruning. So, the truffle plantation corresponds to a perennial economic crop in all respects. Before the host seedlings are introduced into the plantation, their roots must first be inoculated with truffle mycelia or spores and mycorrhizal (Fig. 1c) (Benucci et al., 2012). The first industrially produced mycorrhizal seedlings came from France in the 1970s of the last century. It is considered that they significantly contributed to the (re)affirmation of the artificial cultivation of truffles in France and Italy, then in Spain and other countries (Reyna and Garcia-Barreda, 2008). According to data from the French National Institute for Agricultural Research, around 300,000 certified seedlings are planted annually in France on about 1,000 ha, which gives an average planting density of about 300 truffle orchard trees/ha (Sourzat, 2017).

2.2.1 Pallier's Method

Raymond Pallier, a French expert on truffles, was one of the supporters of growing truffles according to the concepts of arboriculture. The essence of the method is the cultivation of mycorrhizal seedlings or trees in orchard-type plantations, with active care measures during the entire life cycle of the plantation. The method was first applied to the cultivation of *Tuber melanosporum* in the province of Périgord, and today it is applied throughout France, Europe, and even beyond (Sourzat, 2009). The basic features of this method are (Hall et al., 2001): planting of certified mycorrhizal seedlings on suitable soil, primarily limestone; land cultivation, adapted to the age of the plant, during the entire production period; controlled irrigation of host trees; pruning of trees during the entire production period (most often, trees are pruned to an inverted cone shape); use of chemical means for destroying weeds (glyphosate and others); use of chemical means to repel and destroy snails (snail killer and others).

The Pallier method allows a higher planting density, from 400 to 800 host trees/ha (Sourzat, 2009), which makes it suitable for use in cases of limited cultivation space. On the other hand, higher planting density requires more intensive pruning and stricter control of the living space of each individual host tree. Expected plantation yields from the 15th year onward range from 15–30 kg/ha. One of the characteristics of the method is the early entry into the production cycle, already in the 4th year for *Tuber melanosporum* plantations on hazelnut.

2.2.2 Tanguy Method

This model of truffle cultivation is named after Marcel Tanguy, a French truffle grower. It is also called "the method of managing truffle space" (Sourzat, 1994), "the method of truffle cultivation on a limestone meadow" (Sourzat, 2002) or "the method of truffle cultivation in calcareous lawn" (Sourzat, 2017). The special feature of this model is the planting of mycorrhizal trees followed by the natural recolonization of the environment with herbaceous vegetation, thanks to the fact that two to three years after planting, soil cultivation is stopped without any intervention on the ground. The main characteristics of this method are: planting of certified mycorrhizal seedlings on previously treated (plowed) or untreated (grassed) soil; processing and/or cleaning the soil around young trees using herbicides in the first two years after planting; cessation of these activities in the third year and maintenance of grass vegetation by mowing; occasional tillage of the soil in order to encourage the growth of new roots on the host trees; watering plants as needed; possible pruning of trees in the initial period in order to achieve an upright form of the habitus; subsequent corrective pruning, as indicated.

One of the characteristics of this method is its late entry into the production cycle, somewhere between 10−12 years after establishment, which is explained by the competition of grass vegetation (Sourzat and Taburet, 2008). This delay in production is allegedly compensated by high yields, which go up to 100 kg/ha (Sourzat, 2017). The stated statement cannot be confirmed or denied because the owners of these highly productive plantations simply do not want to go public. It is interesting to note that their plantations, as a rule, are not larger than one hectare, most often, they are between 0.1−0.6 ha. According to the same author and another source, the average yield per tree is up to 1 kg, and sometimes more (Sourzat, 2009), which at an average planting density of 300 trees/ha gives a three-fold higher yield. From the perspective of the labor involved, the "Tanguy" method is less labor-intensive than the "Pallier" method, due to less intensive pruning and tillage.

2.2.3 Mycorrhizal Seedlings Introduced in Wild Habitat

The implied purpose of the mycorrhizal seedlings is the formation of an artificial "truffle orchard". When resources such as space, material and technical support are available, mycorrhizal seedlings can be planted in their natural habitat. In doing so, one must take care that the chosen locations best suit the ecological needs of the selected truffle type, and also the host plant. So, for example, in deciduous forests with a dense structure, such as oak, hornbeam or beech forests, it is possible to introduce seedlings mycorrhizal with summer truffle (*T. aestivum*), while brighter and more open positions have seedlings mycorrhizal with Périgord truffle (*T. melanosporum*). When applying such solutions, the introduced seedlings will be exposed to the pressure of the competing mycorrhizal population *in situ*, which will probably result in slightly lower yields compared to classic truffle orchards.

2.3 Truffle Growing by In Situ Inoculation

Procedures for *in situ* inoculation of adult trees are described in several sources (Lo Bue et al., 1990; Reyna, 1992; Frochot et al., 1999), but they have not been widely used in the modern practice of truffle growing. Reyna et al. (2001) developed a method that proved successful in Portuguese oak (*Quercus faginea*) thickets interspersed with *Q. ilex* and *Pinus nigra* individuals in the Castellón region of Spain. Using a tractor with a plow, a trench 30–40 cm wide and 20–30 cm deep was formed around the groups of Portuguese oaks, into which the inoculum was inserted. The inoculum consisted of crushed fruiting bodies of Périgord truffle (*T. melanosporum*), mixed with vermiculum and water, with the addition of hydrogel for better humidity maintenance. The dose of ascocarp varied from 10–50 g per running meter of furrow. After closing the trenches, a mulch film was spread to avoid excessive evaporation and weed growth. After a certain time, the roots of the trees reached the zone of inoculation, which created the prerequisites for the establishment of mycorrhiza and then the beginning of fructification (Reyna, 1999; Reyna et al., 1999; Reyna et al., 2001). The same source states that at the same location, the inoculation of Portuguese oak was successfully carried out using the truffle trap method.

3. Establishing a Truffle Plot

Creating a truffle plot involves a deliberate series of activities aimed at selecting the truffle type that best suits the local conditions. The chosen method should have the most favorable conditions for implementation, considering the availability of reproduction material and possibly materials for land repair.

3.1 Ecological Characteristics of the Planting Area

The initial basis for planning truffle fields is represented by the analysis of the ecological characteristics of the selected locality, namely: land and parent substrate, geographical location, climatic conditions, as well as the plant formations found. This work should be approached very seriously, and in the absence of one's own competences, it is advisable to ask for help from qualified authorities.

3.1.1 The Land

The geological background and the type of soil largely determine the choice of truffle species and host plant. The type of parent substrate and the type of soil can be found out from the relevant planning documents, and in the absence of such a source, a field observation by a qualified pedologist is required, which will include the opening of the profile. Analysis of the soil profile provides insight into the horizons where truffles and the roots of symbiotic plants will develop. The main analytical parameters useful for suitability assessment are: soil reaction (pH), soil structure, soil texture, content of organic matter, carbon/nitrogen ratio (C/N ratio), total and soluble calcium carbonate and stoniness. The C/N ratio reflects the degree of soil mineralization. In natural truffières, this coefficient takes values in a wide range between 0.1–26 (Bencivenga et Granetti, 1988). C/N ratio 10 is considered the optimal size for man-made truffières (Fischer et al., 2017). An important parameter of the truffle ground is the presence of a fine fraction of calcium carbonate with a diameter of <50 microns (so-called active $CaCO_3$). This fraction, together with total calcium carbonate, regulates the soil pH (Jaillard et al., 2014).

The soil reaction corresponding to the largest number of commercial truffles ranges between 6.8–8.0 (Hrka, 1988). According to the USDA (2017) classification, this range includes: slightly acidic, neutral, very slightly alkaline and moderately alkaline soils. A favorable structure for truffle development is one that allows soil aeration and efficient water drainage, thus facilitating easier penetration of host tree roots (Poitou, 1988). For example, the ideal structure for growing black Périgord truffle (*T. melanosporum*) is granular or crumbly with water-stable aggregates below 4 mm (Jaillard et al., 2014). As for the texture, combinations with a balanced content of sand, loam and clay are favorable. Soils with a higher sand content are not suitable due to their poor ability to retain water. Likewise, soils with a pronounced clay content are not suitable, due to poor aeration and difficult drainage (Raglione et al., 2001; Bencivenga and Falini, 2012).

With the exception of the white truffle (*T. magnatum*), both black and white commercially important truffles are found on rocky terrain in the wild. Rockiness is even considered a positive soil characteristic, as it is responsible for good drainage and airiness. The stone layer creates a mulch effect and helps regulate the temperature of the upper soil layer (Olivera et al., 2014). According to García-Montero et al. (2007), stony soil is positively correlated with truffle yields. However, lands that are too stony and have little precipitation or where the volume of stony exceeds the volume of fine soil fractions are not suitable for growing truffles (Sourzat, 1997).

3.1.2 Geographical Position

The geographical position of the truffle ground is mainly determined by the altitude, exposure and slope of the terrain. These parameters are important for evaluating the suitability of the habitat for growing truffles as well as for choosing possible cultivation methods. At the same time, the geographical location should be considered along with the climate. In principle, at lower altitudes, closer to the sea, cooler exposures (NW, N, NE) are more favorable, while at higher altitudes in continental areas, warmer exposures (SE, S, SW) are more favorable. When it comes to the slope of the terrain, completely flat terrains should be avoided due to drainage problems. The most favorable positions are with gentle slopes, below 15% (Delmas and Poitou, 1973; Reyna, 1992).

3.1.3 Climatic Conditions

Unlike soil, whose characteristics can vary from one microlocation to another, climate characteristics have a more constant character over a wider area. This enables analysis based on data on precipitation and minimum and maximum temperatures from the nearest meteorological station and comparison with the climatic requirements of the species to be cultivated, as well as its mycorrhizal partner. Although truffles can survive in extreme temperature conditions, from –25°C to + 43°C (Reyna, 1992; Sáez and De Miguel, 1995; Reyna, 2000), it is recommended to choose a truffle plot devoid of extreme temperature oscillations (Trappe et al., 2009). In terms of rainfall, natural truffle habitats receive from 485 mm to over 1,500 mm of water precipitation per year. In this regard, the summer months are considered critical, when the precipitation should not fall below 72 mm per month (Reyna, 2000; Ricard, 2003). If this happens, it is necessary to intervene with irrigation. Mulching can also affect the temperature and water status of the soil by mitigating temperature extremes to a depth of about 10–15 cm (Michels, 2003; Sourzat, 2003; Olivera et al., 2014).

3.1.4 Previous Vegetation

Establishing a truffle plantation on agricultural land where cereals, legumes or fodder plants were previously grown is considered a good choice (Reyna, 2000). Abandoned orchards and vineyards are also considered favorable locations (Sourzat, 1997). If the plantation is intended to be based on forest land that was previously cleared or burned, a phytopathological analysis for root pathogen residues is recommended, in order to eliminate any risks to future host trees.

3.2 Choice of Reproduction Material

3.2.1 Choice of Truffle Type and Host Plant

Each truffle type has a certain preference for one or more symbiont plants. Bearing this in mind, and before establishing a plantation, it is advisable to carry out a detailed reconnaissance of the surrounding terrain. On this occasion, all species that are suitable for the role of host tree in the future plantation should be recorded. After that, the truffle type that will best suit the specific local conditions should be chosen. It is equally important that quality reproductive material can be obtained for the chosen type of truffle, which is also perfectly suited to local conditions. The most common symbiotic species in natural habitats are shown in Table 1. They can serve as one of the footholds when making a final decision.

3.2.2 Acquisition of the Host Plant

Propagation material should be purchased exclusively from certified nurseries. If the seedlings are not certified, it is worthwhile to submit the plants for analysis in a qualified laboratory to ascertain the quality of the seedlings and their mycorrhizal status before planting. Truffle tree seedlings should have a well-established root system. There must be no mycorrhizae of any other type of fungus (Fischer and Colinas, 1996). As for age, seedlings up to three years old are recommended. Four and more-year-old seedlings should not be used to establish a plantation. They are considered overgrown plants and their resumption can be unsatisfactory. If there is an opportunity to choose, it is better to get planting material produced from seeds, and only as a last resort, planting material produced from cuttings should be bought (cloned material).

Table 1. The most common symbiotic species (*Tuber* spp.) in the natural truffières in the Mediterranean and Sub-Mediterranean aresa.

Types of truffles	Symbiotic species							
	Oaks	**Horn-beams**	**European beech**	**Common hazel**	**Linden trees**	**Poplars**	**Willows**	**Pine trees**
Tuber aestivum (Summer truffle)	+	+	+	+				+
Tuber borchii (Whitish truffle)	+	+		+				+
Tuber brumale (Winter truffle)	+	+		+	+			+
Tuber magnatum (White truffle)	+	+		+	+	+	+	
Tuber melanosporum (Périgord truffle)	+	+		+				+

3.3 Preparation of the Land

Land preparation for planting is aimed at creating optimal conditions for the development of truffles and their symbiotic hosts. In order to achieve this, various interventions are carried out, primarily of a mechanical nature and in cases of need, some chemical measures as well.

3.3.1 Mechanical Measures

Mechanical measures include the removal of woody plant species, rippering, surface plowing, harrowing and stone removal. In the event that there are certain types of trees in the area intended for the establishment of a truffle field, it is essential to uproot them, in order to remove or at least reduce the presence of mycorrhizal fungal species, which are certainly present in their root systems. In practice, in addition to the removal of trees and bushes, it is necessary to carry out several soil treatments during the summer, in order to remove the largest amount of remaining parts of the roots. In order to enable drainage and aeration, it may be required to rip the soil with a sub-soiler or chisel to break up a possible hardpan from previous land use. After rippering, surface plowing follows at a depth of 20–25 cm, so that poor soil layers do not accumulate on the surface, in order to plow the surface organic mass and promote microbial flora. In autumn, before planting the mycorrhizal plants, the land is harrowed to level the soil and remove the herbaceous species that emerged at the end of summer. Digging and tilling the soil can bring large stones to the surface that make it difficult or prevent further cultivation measures. In these cases, it is recommended to break it or completely remove it. However, the presence of stones of such dimensions that do not interfere with cultivation procedures is considered favorable, because it forms a certain reserve of calcium and, even more, acts as a mulching material on the surface (Bencivenga and Falini, 2012; Fischer et al., 2017).

3.3.2 Chemical Measures

In a situation where the previous crop was a tree formation and the presence of pathogenic fungi in the residual roots was detected during clearing, it is necessary to apply fungicides in order to prevent the spread of infection to the new crop. The type and dose of fungicide ought to be prescribed by a qualified phytopathologist. If in the selected soil the pH is lower than required by the selected type of truffle or there is an evident lack of active calcium carbonate ($CaCO_3$), calcification measures can be carried out with the professional help of a pedologist, while the use of some chemical agents is not excluded.

3.4 Planting a Host Tree

In accordance with local weather conditions, planting should be carried out from the initiation of the autumn rains until the following spring. It is better to plant at a time when precipitation is forecast, in any case before the beginning of the driest and hottest months (Fischer et al., 2017).

3.4.1 Planting Density

Planting density is determined based on the dimensions of the trees that affect the shading of the land where truffles are grown, as well as on the basis of the care measures that need to be implemented. Table 2 shows the planting density of truffle trees in plantations in Italy (Bencivenga and Falini, 2012). The density can vary depending on the selected cultivation system and the intensity of the pruning of the truffle trees. There are three most common planting arrangements: square arrangement, flattened arrangement and a hedge formation. The choice depends on the owner's desire: to make the best use of the available space, to have as much room as possible for maneuvering the vehicle around the plantation (Bencivenga and Falini, 2012; Fischer et al., 2017).

3.4.2 Planting Technique

Planting consists of the following activities: making holes in the ground that are slightly deeper than the length of the container or lining with which the root of the mycorrhizal seedling is protected; removing the container or lining, taking care not to damage the root; positioning the seedling in the hole so that the sod is lowered 1–2 cm below the soil level; filling the hole with fine soil to ensure good adhesion of the sod to the surrounding soil; light watering to make the sod and soil blend as well as possible; installation of shelter for seedlings if there is a risk of animal attacks (Bencivenga and Falini, 2012; Fischer et al., 2017). In areas where woody species were previously present, certainly mycorrhizal with other types of mushrooms, it is useful to additionally scatter truffle spores during planting as if they were on the soil with which the holes with the seedlings will be filled. This intervention should limit the competition of mycorrhizal fungi present in the soil.

The preparation of sporadic inoculum can be followed in two ways: 1) The fruiting bodies are cut into leaves, dried in the sun, ground to a fine powder and mixed with sand and earth, so that a well-dosed and even application is possible; 2) The

Table 2. Planting density of truffle trees used in truffle cultivation in Italy (Source: Bencivenga and Falini, 2012).

Distance between truffle trees	**Downy oak (*Quercus pubescens*)**	**Holly oak (*Quercus ilex*)**	**Turkey oak (*Quercus cerris*)**	**English oak (*Quercus robur*)**	**Common hazel (*Corylus avellana*)**	**Hop-hornbeam (*Ostrya carpinifolia*)**	**Oriental hornbeam (*Carpinus orientalis*)**	**Linden (*Tilia* sp.)**	**Stone pine (*Pinus pinea*)**	**Aleppo pine (*Pinus halepensis*)**	**Poplar (*Populus* sp.)**	**Willow (*Salix* sp.)**
Minimum (m × m)	4×3	4×3	4×3	4×4	3×3	4×3	4×3	4×4	4×4	4×4	4×3	3×3
Maximum (m × m)	9×9	8×8	9×9	10×10	7×6	7×7	8×8	9×9	8×8	7×7	7×7	7×6
Average (m × m)	6×5	5×5	6×5	6×6	5×4	5×5	5×5	6×6	5×5	5×5	5×5	5×4

fruiting bodies, fresh or frozen, are slightly soaked in water, crushed in an ordinary blender, mixed with water at the moment of use and then applied (Bencivenga and Falini, 2012).

4. Care and Maintenance Measures

Once a truffle plantation is established, it should be properly managed in order to achieve satisfactory yields. We distinguish between two stages: the development stage and the fruiting stage. The first stage includes the period from the planting of the host tree until the appearance of the first fruit bodies, while the second stage includes the entire period of fructification. When the first phase ends and the second phase begins, one cannot follow with the naked eye, given that truffles are hypogean fungi. However, with some types of truffles (*T. aestivum*, *T. melanosporum* and *T. borchii*), when entering the fruiting stage, external manifestations may appear in the form of "scorched" ground vegetation around the host tree. This phenomenon is called "brûlé" or "burnt". It is usually circular in shape because it reflects the development of the mycelium in the form of concentric circles. It is interpreted as an allelopathic role of inhibitory volatile organic compounds produced by truffle ascocarps (Pacioni, 1991). The largest number of care and maintenance measures are aimed at ensuring the best possible environmental conditions for the host plant, from which the truffle, as a mycorrhizal partner, will also receive significant benefits. On the other hand, sporadic inoculation is one of the measures that directly helps to strengthen the truffle population in the zone of the host plant.

4.1 Water Supply

Water supply has the task of providing soil moisture in the dry period, when there is not enough natural precipitation. In addition, it contributes to the reduction of excessive soil heating. This cultivation measure has especially gained importance in recent years due to frequent droughts caused by climate change (Tramblay et al., 2020). According to forecasts from the European Environment Agency, further reductions in water resources in Europe are expected (Anonymous, 2012, 2017). Newly planted host trees need to be provided with enough moisture to survive and to maintain the mycorrhiza in an active state. At this stage of development, the root system is not yet sufficiently developed to draw moisture accumulated in the lower layers of the soil. If the planting was done in early spring, and if there are periods without precipitation lasting seven or more days in the period April–September, it is necessary to intervene with localized irrigation in the root zone. This task can be carried out by applying two irrigation systems: drop-by-drop and microsprinkler. Experiences from truffle plantations in Spain show that there are no differences between the two systems in terms of truffle yield, truffle number, quality (based on ascocarp size) or truffle harvest date. However, when other parameters, such as economic and ecological, are taken into account, drip irrigation proves to be superior because it requires less water consumption (Magarzo et al., 2023). If local conditions allow, it is preferable to use spring and underground water because, as a rule, they do not contain substances harmful to the host plant, as well as to truffles. In the absence of these resources, one can resort to collecting and using rainwater.

It is not possible to give a clear answer to the question of when and how much to water without taking into account a number of influential factors in each specific case. Also, the required amounts of water (the so-called irrigation regimes) can be determined with greater or lesser precision using different methodological procedures. Determining the irrigation regime is more delicate in the fruiting stage than in the development stage. If adequate data sources are available, irrigation needs can be determined as a ratio of precipitation to potential evapotranspiration. For the successful cultivation of agricultural crops, a ratio of precipitation to potential evapotranspiration between 0.7 and 1.3 is considered ideal. If it is less than 0.7, irrigation is necessary, and if it is greater than 1.3, drying is necessary (Bačani, 2006). For truffle plantations, especially black truffles, lower and differentiated ratios are recommended, namely: 0.5 for the period April–July and 0.3–0.5 for the period August–October (Olivera, et al., 2014). In practice, watering norms are more often expressed by the amount of water per host plant. Thus, the Italian manual for truffle growers predicts 7–8 liters per host plant every 7–10 days (Bencivenga and Falini, 2012), while the Spanish manual recommends 3–4 liters per host plant every 14–21 days (Sourzat, 2002; Fischer et al., 2017).

Estimates of water needs can also be determined by direct measurement of soil moisture using a hygrometer, taking into account the tolerance needs of the host plant for moisture. Thus, oaks, pines, spruces and hazels, as host plants, tolerate dry to well-drained soil in an interval of 21–40%, while willows tolerate moist soil in an interval of 41–60% (Anonymous, 2022). If and when there is a reduction in summer precipitation, followed by heatwaves, and on those occasions the measured values are lower than those mentioned, the irrigation system is activated without hesitation. Finally, one must not lose sight of the fact that excessive irrigation can lead to negative consequences, above all for the fungus as a mycorrhizal partner. Studies have shown that due to excess water in the soil, there is a decrease in the mass of fine roots and, therefore, the mycorrhizal mass from which the ascocarp mass is nourished (Bonet et al., 2006; Olivera et al., 2011). Excess water or moisture can also favor the settlement of some other mycorrhizal fungi on the root system of the host plant (Merényi et al., 2016).

4.2 Pruning

Pruning has several goals: the formation of a regular crown, which achieves better shading and insolation control. Development of the root system, assuming that it follows the development of the crown. Easier access to the host-tree when it comes to the installation of the irrigation system, easier implementation of other care measures such as hoeing, mowing and mulching and most importantly, the unhindered movement of the truffle dog during the harvest. Mycorrhizal trees need a strong leaf cover with which they provide photosynthetic products important for truffle development. Pruning begins cautiously in the second year. The procedure consists of removing the first ground branches and one-year shoots (so-called water suckers). In the following years, continue to clean the branches up to a height of about 50–60 cm.

The method of pruning largely depends on the type of host-tree and the type of truffle. In species with a bushy habit, such as common hazel (*Corylus avellana*), the 3–5 strongest sprouting branches are selected and the rest are cut to the ground. If the mycorrhizal partner is *T. melanosporum*, then all selected branches are freed from the side branches up to 1/3 of their height, with the aim of exposing the roots to insolation as much as possible. When the host tree has a typical tree habit (Fig. 2a), with one pronounced trunk, the pruning strategy is mainly reduced to the formation of either a spherical or conical crown form. The spherical shape of the crown (Fig. 2b) achieves the early creation of shade around the center of the tree, which is more suitable for the ecological needs of the summer truffle (*T. aestivum*). The conical form of the crown (Fig. 2c) increases the inflow of light that reaches the ground, which is more suitable for the ecological needs of the Périgord truffle (*T. melanosporum*), (Bruhn and Hall, 2011; Carbajo, 1999; Grente and Delmas, 1974).

With regard to the execution time, it is possible to distinguish between winter and green pruning. Winter pruning is performed during vegetation dormancy, usually at the end of winter or the beginning of spring. It achieves faster healing of cut wounds and better preservation of carbohydrates in the roots (Bencivenga and Falini, 2012; Sáez and De Miguel, 1995). Green pruning is performed on active vegetation. Somewhere in the eighth year, when the host tree develops to the extent that normal ascocarps begin to form on its mycorrhizal roots, activities should be started to limit the further growth of the crown, i.e., the expansion of the root system, and avoidance of crown closing. During the production phase, pruning is carried out every 2–5 years. (Fischer, 2017; Ricard, 2003).

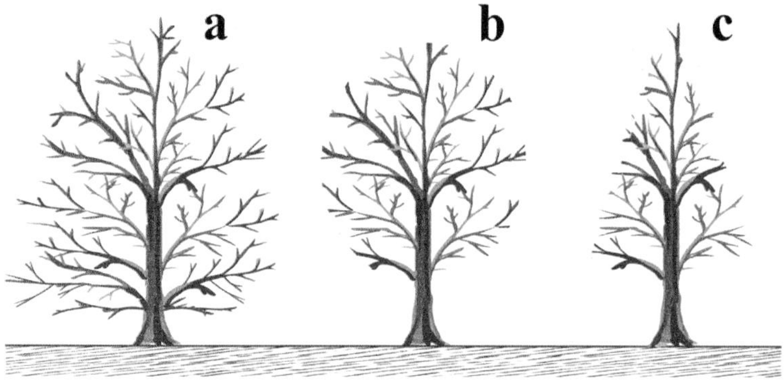

Fig. 2. a, young tree (not pruned); b, pruned (spherical crown); c, pruned (conical crown) (Bencivenga and Falini, 2012).

4.3 Weeding, Hoeing and Mulching

Weeding and hoeing aim to disrupt or slow down the development of competing herbaceous vegetation, reduce moisture loss and temperature fluctuations in the soil, and encourage root development of the host plant at a greater depth than that of neighboring non-mycorrhizal plants. The first hoeing is carried out in the period

February–March. The depth of penetration near the tree should not exceed 5 cm, and at about half a meter from the tree, a depth of 5–10 cm should be maintained. A second hoeing is preferable in July–August, and in cases of stronger and longer dry periods, hoeing can be repeated in the autumn. When hoeing, care should be taken about the depth of the intervention, because deeper digging can damage the mycorrhizal roots of the host plant and the mycelium of the mycorrhizal partner, thus putting the truffle crop itself into question. Hoeing can be omitted in the case of very permeable deep soils, capable of preserving their porosity throughout the season (Bencivenga and Falini, 2012). In large plantations, where intensive use of mechanization is practiced, the depth of hoeing goes up to 15–20 cm (Reyna, 2000). After the "brûlé" appears, careful processing of the site is suggested. Since "brûlé" is an indicator of active mycelium in the soil, at that moment it is important to spare the soil from stronger interventions (Sáez and de Miguel, 1995; Reyna, 2000). Excessive tillage at a frequency of 3–4 or more annually can even have a negative effect on soil structure and porosity due to aggregate destruction, increased compaction and reduced microbial activity (Ricard, 2003).

Mulching is a process by which the land is covered with different materials with the aim of limiting the moisture loss from the soil, preventing germination and preventing the development of competing vegetation (Olivera et al., 2014). The area covered roughly corresponds to the projection of the host tree's canopy, that is, the area penetrated by its root system. When it comes to young, newly planted seedlings, the mulch should cover an area of 50 × 50 cm around the plant (Bencivenga and Falini, 2012). Different materials are used for mulching, distict from those found on site to those of industrial origin. The mulching material should be permeable to moisture, easily removable and should not leave traces. The most common natural materials are wood greenery and thin branches, straw and limestone breccia with a thickness of 5–7 cm. In synthetic materials, various types of fabrics, plastic film and geotextiles are used. Among the materials, straw and plastic film are not considered ideal solutions (Zambonelli et al., 2005). Straw leaves too many marks behind which are difficult to remove. The plastic film is not permeable to rain and irrigation, for which reason it often has to be installed and then removed. On the other hand, jute sacks and reed mats are considered reliable materials (Bencivenga and Falini, 2012).

4.4 Refilling and Sporadic Inoculation

After a certain number of years, truffle plantations begin to show signs of exhaustion, which is primarily reflected in yield decline. According to Hrka (1988), in oak truffle plots, productivity begins to decline 50–60 years after establishment, and in poplar and willow truffle plots, it starts declining earlier after 30 years. The reasons for the decline in productivity, mainly the depletion of truffle fields, are numerous and varied. In order to renew the production potential of the truffle grove, it is refilled. Refilling of the truffle plot consists of the successive introduction of fresh planting material at the positions between the existing host trees, which are also successively removed from the plantation. Restoring an existing truffle plot is generally considered more advantageous than establishing a new one in the same location. When renewing, first of all, you gain time, because one need not wait several years for the production of new seedlings to begin. The continuity of truffle production is not interrupted. In addition, the volume of earthworks is smaller, resulting in lower costs. The same species are used for the restoration as when establishing the truffle plot. Mycorrhizal and non-mycorrhizal seedlings can be used, with the advantage of the former being unquestionable. For non-mycorrhizal seedlings, it is assumed that they will be inoculated with mycelium from existing neighboring trees. To be safe, they need to be disinfected before planting so that they do not introduce other competing fungi into the plantation.

4.5 Weed Control

In general, weed control can be carried out mechanically or by applying chemical agents. Mechanical weed control methods include mowing, which leaves the underground organs of the weeds in the soil, and weeding and hoeing, which remove the underground parts from the soil. It is not necessary to explain the advantages of weeding and hoeing in comparison to mowing aloneFor mowing competitive vegetation in truffle orchards, it is recommended to use self-propelled gasoline mowers with an adjustable cutting height and wheel drive. These devices do not need to be pushed or pulled using their own strength. Work with them is possible even in unfavorable conditions, on heavily overgrown and uneven terrain. Weeding is done manually, using garden hoes. In large truffle plantations, a gyroscopic crumbler can be used, a modern device that regulates the weeds mechanically by means of two horizontally rotating tools, one of which performs the function of mowing around the trunk and the other the function of hoeing. Herbicides should not be used in truffle orchards for several reasons, among others, because of the emphasized adsorption capacity of ascocarps for different compounds from the soil (Ljubojević et al., 2014). Contrary to the stated opinion, a number of relevant sources do not exclude the possibility of applying herbicides, primarily glyphosate, in the fight against weeds in truffle orchards (Bencivenga and Falini, 2012; Bonet et al., 2006; Fischer et al., 2017; Olivera et al., 2011; Ricard et al., 2003).

4.6 Protection Against Diseases and Pests

Protection measures in the truffle orchards are primarily directed towards the truffle ascocarps, but due attention is also paid to the health of the host trees. A large number of animals feed on the fruiting bodies of truffles, including: wild boars, foxes, wolves, deer, roe deer, bears, badgers, goats, squirrels, dormouse, mice and rabbits. It is accepted that wild boars, badgers and mice cause the greatest damage in truffières (Hrka, 1988). More recent findings indicate that moles and several species of birds, including the blackbird (*Turdus merula*), also consume truffles. On the one hand, animals that feed on truffles help their spread, because they throw out spores with their excrement, from which, under favorable conditions, the truffle mycelium germinate. On the other hand, if they multiply and become scarce in other food sources, these animals can cause serious damage to the truffle area. By erecting a fence around the truffle orchard, access to a large proportion of mentioned animals can be physically prevented. The use of traps is a laborious measure and is not an animal-friendly method, and in many countries, it is prohibited by law. With ultrasonic animal repellent devices, large areas can be secured easily without complications. These devices are equipped with a motion sensor, flashes of light and a short acoustic siren. These units all work on the same principle by producing an ultrasonic noise that an animal simply wants to move away from. It is not always an instant cure, but in most cases, results can be seen after only a few days.

The Roman snail (*Helix pomatia*) is one of the great gluttons of truffles. If traced under the canopy of the host tree, it is better to sprinkle snail and slug killers immediately along the entire edge of the projection of the canopy. Truffles are also eaten by common cockchafer larvae (*Melolontha melolontha*) and truffle fly larvae (*Helomyza tuberiperda*, *Leiodes cinnamomea*, *Suillia fuscicornis*, *S. gigantean* and *S. tuberivora*) (Hrka, 1988; Bencivenga and Falini, 2012). For example, the truffle fly *Helomyza tuberiperda* (syn. *Suillia tuberiperda*) has an extraordinary sense of smell. It lays eggs in the ground where the ascocarp lies, from which the larvae hatch, which then pierce the entire ascocarp and practically destroy it (Hrka, 1988). If a significant presence of any of the above-mentioned insects is observed, it is necessary to request expert help of a forestry entomologist.

The most common and most dangerous host-tree diseases often occur on the leaves. They can be of fungal origin or the result of insect activity. Mildews are a species of fungus widely distributed in the world. Nine types of powdery mildew have been recorded on different types of oak (Karadžić, 2010). The most conspicuous among them is *Microsphaera alphitoides*, i.e., its conidial form, *Oidium quercinum*. The first signs of infection appear in the spring in the form of cinnamon spots, which later turn gray-white and gradually spread to cover the entire surface of the leaf. With the development of the disease, the leaves are deformed and become stunted. The consequences of severe infections are reduced tree growth, drying of shoots, increased sensitivity to secondary parasites and decay of young plants. The disease is controlled with potassium permanganate and sulfur powder, with expert help from a forestry phytopathologist. In addition to mildews, anthracnose (*Apiognomonia* sp.) and bark cancer (*Nectria* sp.) often occur in oak plantations.

There are a large number of insects capable of causing leaf damage and defoliating trees in truffle orchards. Among the most important harmful insects in orchards where oaks are the host trees are *Thaumetopoea processionea* (oak processionary moth), *Lymantria dispar* (gypsy moth) and *Tortrix* sp. (oak moth). Caterpillars of the oak moth feed on leaves. Caterpillars are overgrown with hairs, which cause allergic reactions on human skin. In addition, irritation of the respiratory tract, dizziness, temperature and inflammation of the eye mucosa may occur. In more severe cases, symptoms of asthma and bronchitis may appear. Green oak rollers (*Tortrix viridana*) and yellow oak rollers (*T. loeflingiana* syn. *Aleimma loeflingiana*) also cause complete defoliation of plants. They are controlled by coating the trees with eggs, using petroleum, or using an insecticide recommended by forest protection experts. According to Bencivenga and Falini (2012), the fight can be conducted using specific bacteria, such as *Bacillus thuringiensis* Berliner, which attacks defoliator larvae. Regular monitoring and timely and appropriate interventions can significantly improve the health of the host trees and the quality of the produced truffles, and contribute to strengthening the sustainability of the plantation itself.

4.7 Fertilization

The ectomycorrhizal symbiosis between truffle fungi and truffle trees has a mutualistic character, considering that both parties benefit from this interaction (Bronstein, 2015). Truffle fungi extract nutrients from the soil and deliver them to the host plant via the fine root ends, from which, in the same way, it receives synthesized carbohydrates and sugars necessary for its own growth. The introduction of nitrogen, phosphorus and other ingredients into the truffle area reduces the affinity of the truffle tree for the nutrients normally donated by truffle fungi, which results in a reduction in the flow of carbohydrates and sugars, as well as mycorrhizal colonization as a whole (Treseder, 2004; Olivera et al., 2011). Therefore, it is not recommended to introduce NPK fertilizers into truffle orchards (Bencivenga and Falini, 2012) because this can potentially threaten the balance between truffle fungi and the host. Fertilizers specially designed for oak trees in truffle fields did not show clear improvements in truffle production (Olivier et al., 2002). They did not lead to a stronger growth of trees, nor did they affect the mycorrhizal status of the plantation (Bonet et al., 2006). Considering that most of the land has enough nutrients for the development of truffles, it is recommended to fertilize only on plantations with an extremely low content of a certain nutrient, in order to eliminate the specific deficiency (Olivier et al., 1996). Regular analysis of the state of nutrients in truffle plots may indicate the need for a certain intervention, especially if corrections to the pH of the soil were made before that (Fischer et al., 2017).

5. Harvesting

Truffles are hypogeous fungi, which makes finding and extracting them from the ground more difficult than harvesting mushrooms. In addition to the previously mentioned brûlé the uncertain visual indicators of truffles are the truffle fly, snails and cracks in the ground. The truffle fly (*Helmoyza tuberiperda*) is a *Hymenoptera*, about 1 cm long, bronze in color with a metallic sheen. The fly lays its eggs in ripe black truffles (*T. melanosporum*, *T. aestivum*) and then sits on the ground or hovers in the air right above the ascocarp. It is hard to notice except when it is floating in the air. An indicator of truffles can be the Roman snail (*Helix pomatia*), which feeds on truffles. An indicator of truffles can be cracks in the ground that appear above the ascocarp after rain (Hrka, 1988). Truffles can also be located using electronic equipment for their detection The equipment utilizes polymeric detectors, whose electrical resistance differs with the amount of volatile material in the air. This equipment was patented back in 1992 by Talou T., Krishna C.P. and Gaset A. at the National Institute of Industrial Property in France (Anonymous, 1992). To date, its application in practice has not been recorded.

The surest way to find truffles is with dogs. A good dog can find a ripe truffle at a distance of up to 50 m. It is considered that the best dogs for finding truffles are the Lagotto Romagnolo breed. Dogs of the following breeds are also considered to be excellent truffle hunters: Beagle, Golden Retriever, Setter and Springer Spaniel. Different types of truffles ripen at different times of the year (Table 3). No species ripens all at once, but in certain time intervals, so the harvest is adjusted to the ripening time. It is enough to pick truffles once or twice a week. Since the ripening season usually lasts 3–4 months, this means collections every 12–20 harvests.

The basic tool for picking truffles is a small square spatula about 7 cm wide and about 10 cm long, with a handle about 50 cm long. On stony terrain, where black truffles usually grow, in addition to the mentioned trowel, a smaller pickaxe and a narrow rake with 3–4 teeth are used. Proper digging and extraction of truffles from the ground is achieved only by digging as little as possible and returning the soil to the dug hole. Rotten truffles should not be picked, because their degradation releases spores that germinate at the right time and strengthen the mycorrhiza. Truffles should be cleared of soil immediately after extraction. Cleaning should not scrape or damage the skin (peridium), which is quite soft and sensitive with white truffles, less so with black truffles.

Table 3. Ripening time of the most important commercial types of truffles (*Tuber* spp.) (source: Hrka, 1988).

Types of truffles	I	II	III	IV	V	VI	VII	VIII	IX	X	XI	XII
Tuber aestivum (Summer truffle)				+	+	+	+	+				
Tuber borchii (Whitish truffle)			+	+	+							
Tuber brumale (Winter truffle)	+	+	+	+							+	+
Tuber magnatum (White truffle)	+							+	+	+	+	+
Tuber melanosporum (Périgord truffle)	+	+	+	+							+	+

6. Cultivation of the Most Important Commercial Truffles

The choice of location of the plantation in accordance with the needs of the truffle and the host plant is a basic prerequisite for success. This is achieved through the consistent implementation of appropriate management measures.

6.1 Cultivation of Tuber aestivum

Fruiting bodies are dark-brown to black, ornamented with pyramidal warts, generally intersected with angular fissures and occasionally with less pronounced transverse wrinkles. Their aroma is mild, similar to mushrooms, and resembles the smell of porcini mushrooms. They taste like nuts with a characteristic note of chives. The predominant geometric shape of fruit bodies is ellipsoid, followed by spherical (geoid) forms, although complex geometric bodies can also be found, such as heart-shaped and kidney-shaped ascoma (Ljubojević et al., 2022) (Fig. 3a).

Summer truffles belong to the very top of the truffles that dominate the world market, right behind the most exclusive Piedmont truffle (*Tuber magnatum*) and Périgord truffle (*Tuber melanosporum*) (Stobbe et al., 2013). The summer truffle

Fig. 3. Peridium of *Tuber* spp. with gleba: a, *Tuber aestivum*; b, *Tuber borchii*; c, *Tuber brumale*; d, *Tuber magnatum*; e, *Tuber melanosporum*.

is an autochthonous species in a large number of countries, which speaks to its wide ecological valence. The southern border of the area reaches an altitude of about 1600 m (Morocco), while the northern border descends to sea level (Sweden). The soil of most summer truffle habitats is calcareous with a pH > 7 (Chevalier and Frochot, 1989), while a pH of 7.4 is considered optimal (Stobbe et al., 2013). Summer truffle plantations can also be based on soils with a lower pH, but not below 6.5. In this case, possible corrections should be taken into account, especially if the concentration of active $CaCO_3$ is very low. A significant relationship was found between soil pH and the level of mycorrhization of the host plant, which increases with an increasing soil pH (Thomas, 2012). The preferred soil is permeable and well aerated in the surface layers. Considering texture, lands with a balanced content of sand, loam and clay should be located. Clayey lands should be ignored because they induce a surface arrangement of fruiting bodies, in which case animals damage them more often with ease. (Bencivenga and Falini, 2012).

In their natural habitats, summer truffles live in symbiosis with a large number of species, both deciduous and coniferous (Stobbe et al., 2013). This variability offers a wide choice of symbiotic partners. For the cultivation of summer truffles, the following are recommended: sessile oak, downy oak, Turkey oak and holly oak; hop-hornbeam and common hornbeam; common hazel, European beech, Austrian pine, Aleppo pine and stone pine (Hrka, 1988; Bencivenga and Falini, 2012). In the period until the beginning of fruiting, the recommended measures are: 1) irrigation—in order to encourage the host plants to take root as well as possible; 2) mulching—to maintain moisture in the soil; 3) weeding or hoeing near the plant, both manually to a depth of 5–6 cm between the rows and mechanically at a depth of about 10 cm; 4) pruning—in order to shape the crown; 5) dispersal of spores—in order to strengthen the production potential of the truffle orchard. With the beginning of fructification, summer truffles form "brûlé", which are more or less noticeable depending on the degree of shading of the soil, or the density of the crown canopy. After that, the following cultivation measures are carried out: irrigation, removal of weeds, mulching, mowing of ground vegetation and scattering of spores. If a drought occurs, irrigation should be carried out on the entire surface of the orchard, because severe drought causes the dehydration of ascocarps and premature ripening. When the land is sufficiently supplied with water, irrigation is considered an insignificant intervention for the formation of fruiting bodies.

The summer months (June to August) are considered a critical period. Weed removal is a measure that is carried out no later than one month before the summer harvest, to a depth of about 8–10 cm. Weeds are removed only from the "brûlé" surface. Mowing of ground vegetation is done according to the situation on the ground, leaving the mowed biomass on the spot as mulch. The scattering of spores contributes to greater plant fertility. Therefore, it is advisable to use that portion of the summer harvest that is not ideal for consumption at a dose of at least 5 g of gleba per host plant (Bencivenga and Falini, 2012). The harvesting of summer truffles in the Mediterranean area occurs in the summer months (June toAugust) (Bencivenga and Falini, 2012). In the interior of Europe, the fruiting season has two peaks in summer (July) and in autumn (November) (Diamandis and Perlerou, 2008; Weden et al., 2004).

6.2 Cultivation of Tuber borchii

Whitish truffles have a ball-shaped ascocarp, the size of a hazelnut to a hen's egg, often gibbous or irregular. The color of the skin is initially whitish and slightly velvety (covered with fine tiny hairs). With age, it becomes light brown and smooth (Fig. 3b). The ascocarp smells like garlic. In the beginning, this smell is mild and pleasant, but with age it becomes strong and unpleasant. The same is the case with the taste, which some associate with garlic and others with hazelnuts.

The whitish truffle is far less known and appreciated than the other white truffle (*Tuber magnatum*). However, due to the low supply of the Piedmont truffle, *T. borchii* found its place on the market. The earliest plantations of this truffle were established in 2000 in Italy (Zambonelli et al., 2000). According to the same source, it has high ecological adaptability and low host specificity. It is found all over Europe, from Britain in the west to Poland in the east, from Finland in the north to Sicily in the south (Salerni et al., 2014). In whitish truffle habitats, some other species of white truffles are not infrequently encountered, such as: *T. dryophilum*, *T. maculatum* and *T. puberulum* (Bencivenga and Falini, 2012). For this reason, it is mandatory to ask for a certificate on the origin of the planting material before bringing it to the truffle farm. Namely, determining the differences between white truffle and the mentioned species is possible only by means of microscopic analysis of spores or by molecular analysis. The implementation of this analysis exceeds the capacity of the owner or founder of the truffle orchard. White truffles can be found in soils with variable carbonate content and a broad pH range, starting from slightly acidic (pH 6.5) to basic (pH 7.8), (Gardin, 2005; Mello et al., 2006; Hall et al., 2007). The soil suitable for *T. borchii* is permeable, mostly sandy and sunny. In very clayey soils, the yield is poor and the fungal specimens are very small. It can be cultivated on marginal lands in areas where other economically valued truffle species grow with less or little success (Zambonelli et al., 2002; Bencivenga and Falini., 2012).

In its natural habitats, *Tuber borchii* lives in symbiosis with a large number of species, both deciduous and conifers (Mrak et al., 2024), off which downy oak, holly oak, Turkey oak, common hazel, hop-hornbeam, Aleppo pine, stone pine and Scots pine are recommended for cultivation. In the pre-fructification period, the recommended measures are: irrigation—only in the first two years, to help the host plants establish strong roots; weeding two to three times during spring and summer, to remove herbaceous vegetation, improve rainwater penetration, and reduce soil evaporation; pruning once per year, focusing on removing first order branches; spore dispersal to enhance the production potential of truffle fields. Whitish truffle, unlike *Tuber aestivum*, does not form "brûlé", which is why the beginning of fruition can only be determined with the utilization of a truffle dog. Since then, the following cultivation measures have been necessary: a) weed removal; b) pruning; c) trimming/mowing; d) antiparasitic treatments against the pine processionary (*Thaumetopoea pityocampa*). Weed removal is carried out with light intervention at a depth of 5–6 cm, usually in May and June. Pruning is carried out once a year in order to remove branches of the first order; cutting is stopped when the lower branches reach a height of 2–3 m. Trimming or mowing of herbaceous vegetation is undertaken in cases of excessive development. Antiparasitic treatments are implemented against the pine processionary, whose caterpillars cause damage and slow down plant growth. The whitish truffle harvest lasts from the beginning of February to the middle of April (Bencivenga and Falini, 2012).

6.3 Cultivation of Tuber brumale

The winter truffle has a round or tuberous fruiting body, often with a hollow at the bottom. The peridium is reddish-brown, almost black, and made of numerous polygonal warts, without transverse wrinkles (Fig. 3c). The ascocarp emits a strong, pleasant aroma, which, varies by variety, ranging from musky like fermented fruit to hazelnut or turnip. Its taste is pleasant and slightly bitter. Compared to other truffles the smaller size of *T. brumale* makes their collection and processing more difficult, which is reflected in their market demand (Merényi, 2016).

Tuber brumale is prevalent throughout most of Europe, excluding boreal and arctic regions (Merényi et al., 2014). Its natural populations outside Europe can be found in Iran (Molkabadi, 2023), while it was probably introduced to Australia and New Zealand with the inoculum of other truffle species (Guerin-Laguette et al., 2013; Linde and Selmes, 2012). *Tuber brumale* thrives in calcareous soils with a neutral or nearly neutral pH. It can tolerate slightly higher humidity and clay content. Unlike *T. melanosporum,* it requires less light and is best suited to the illumination at forest edges and gardens. Additionally, it has the highest tolerance for low winter temperatures among truffles (Bissanti, 2022), making it suitable for cultivation even in northern European countries. The most common symbiotic species in natural habitats are: common hazel, small-leaved lime, large-leaved lime, hop-hornbeam and oriental hornbeam, Turkey oak, sessile oak, downy oak and English oak, Austrian pine and stone pine (Merényi, 2016).

Recommendations are made to farmers to plant the winter truffle on soils whose physical and chemical characteristics are similar to those required by the whitish truffle, but are slightly more humid and poorer in calcium carbonate, primarily using hazel as a host, and applying the same cultivation techniques as for the whitish truffle (Bencivenga and Falini, 2012). When transitioning from the development stage to the fruiting stage, it is important to note that "brûlé" are much less noticeable compared to *T. aestivum* and *T. melanosporum* growth sites (Olivier et al., 2012).

The harvesting of winter truffles lasts from the beginning of November to the end of March (Hrka, 1988).

6.4 Cultivation of Tuber magnatum

The fruiting body of the Piedmont truffle can be of different sizes, ranging from the size of a hazelnut to the size of an apple; the average size is the closest to that of a hen's egg. Sometimes extremely large specimens can be found, which fetch a very high price on the market. The skin (peridium) is smooth, yellowish-white, yellowish-green or light-brown (Fig. 3d). Its Gleba is dotted with veins and varies from white to light brown, brown, pink and dull red (Hrka, 1988). White truffle is valued for its unique taste and aroma, which is described as similar to mushrooms, with nutty notes and a discreet hint of garlic.

White truffles are the highest quality and most expensive among all known truffle varieties. So far, the natural habitats of the white truffle have been registered in the area between Switzerland in the west and Bulgaria in the east, and between Sicily in the south and Hungary in the north (Čejka et al., 2023), excluding one suspected location in Thailand (Suwannarach et al., 2017). White truffle thrives on limestone clay-loamy, loamy-sandy and marly moist terrains, as well as on alluvial sediments of rivers and streams that have an alkaline or neutral reaction (pH 6.8–8.0). These are usually very young soils that are continuously enriched with new sediments due to floods or small landslide fronts. The fine structure of the land is balanced between sand, loam and clay; soil particles are not aggregated into stable structures. The moisture content is good in all seasons, without water stagnation (Hrka, 1988; Bencivenga and Falini, 2012). White truffles mainly grow in association with deciduous species.

Among the oaks, there are both Mediterranean (*Quercus cerris*, *Q. ilex* and *Q. pubescens*) and temperate species (*Q. robur* and *Q. petraea*).Poplars include *Populus alba*, *P. nigra*, *P. tremula*, *P. canadensis* and *P. deltoides*.Willows consist of *Salix alba*, *S. apenina*, *S. caprea* and *S. purpurea*.Lindens include *Tilia cordata*, *T. platyphyllos*, and *T. vulgaris*. Hornbeams are represented by *Carpinus betulus*, *Carpinus orientalis* and *Ostrya carpinifolia*. White truffles also form mycorrhizal partnerships with common hazel (*Corylus avellana*), Italian alder (*Alnus cordata*), European beech (*Fagus sylvatica*), wild pear (*Pirus pyraster*), field elm (*Ulmus minor*) and silver fir (*Abies alba*), the sole conifer representative (Čejka et al., 2023). When it comes to the Mediterranean, the literature specifies that poplars are typical of Italian white truffle sites, whereas sites on the other side of the Adriatic Sea (mainly the Istrian peninsula) are usually represented by oaks (Montacchini and Caramiello, 1968). Symbiotic species recommended for growing white truffles are poplars (*Populus alba, P. nigra* and *P. tremula*), oaks (*Quercus pubescens* and *Q. robur*), large-leaved linden (*Tilia platiphyllos*), common hazel (*Corylus avellana*), oriental hornbeam (*Carpinus orientalis*) and willows (*Salix alba* and *S. caprea*).

It is interesting to note that linden plantations were raised to create avenues, gardens and parks with non-mycorrhizal plants, often producing white truffles. During the period leading to fructification, the following measures are recommended: planting seedlings; irrigation in order to encourage strong root development in host plants; trimming or mowing herbaceous vegetation; pruning to quickly create wide canopies that provide land shading; mulching and/or very light weeding around symbiotic plants. When planting, special attention should be paid to the planting density,to ensure early shading of the entire soil. If the the land is covered with perennial herbaceous vegetation, it is advisable to avoid plowing. Instead, dig holes at the transplanting sites for the symbiotic plants. This method preserves the soil structure. In the second case, when plowing, the moisture needs of truffles should be considered. In this method high and wide beds are created. The mycorrhizal plants are planted at the top of these beds with deep depressions between them. With the beginning of fructification, the following cultivation measures are carried out: irrigation, pruning, mowing, stopping the development of bushes with adequate ecological measures and leaving spores in the soil. If a drought occurs, irrigation should be carried out on the entire surface of the truffle orchard, because severe drought causes the dehydration of ascocarps and their premature ripening. Summer irrigation, which maintains a constant and sufficient level of moisture in the soil, has proven to be effective in increasing yields. Pruning maintains the necessary shade and enables movement through the truffle field. Mowing of ground vegetation is done according to the situation on the ground, leaving the mowed biomass as mulch. The white truffle ripens from September to December. Truffles that ripen in September are not for harvesting because they have little gastronomic value.

6.5 Cultivation of Tuber melanosporum

The ascocarp of Périgord truffle is round, kidney-shaped or irregular in shape, the size of a walnut or an orange, but sometimes it can weigh up to 0.5 kg. Peridium (skin) is reddish-black or brown-black with polygonal warts 3–5 mm in diameter. Gleba (flesh) is black-purple to dark brown with fine white veins, which are irregularly branched around the edge (Hrka, 1988) (Fig. 3e).

It is considered the best-quality black truffle. In terms of quality and market value, it is not significantly behind the species *T. magnatum*. It is especially appreciated in France (Hrka, 1988). Like *T. magnatum*, it has a limited distributional range. Its populations are naturally found in Italy, France, Spain, and some Balkan areas (Riccioni et al., 2008). Today, black truffle plantations can be found in Australia, Chile, New Zealand and the USA (Belfiori et al., 2012). Périgord truffles grow on dry limestone terrain, with a pH 7.5–8.5 and a significant carbonate content. These are well-aerated and drained soils with a fine structure. A good clay content (20–40%) contributes to the preservation of humidity during dry periods. It has a relatively wide range of mycorrhizal hosts (Ceruti et al., 2003). The most common host trees in the wild are *Quercus ilex*, *Q. pubescens,* and *Corylus avellana*. Besides them, symbiotic partners appear: *Quercus robur*, *Q. coccifera*, *Q. faginea*, *Ostrya carpinifolia*, *Carpinus betulus* and *C. orientalis* (Hrka, 1988; Reyna and Garcia-Barreda, 2008). The recommended species for establishing truffle orchards are *Quercus ilex*, *Q. pubescens*, *Corylus avellana*, *Carpinus betulus, C. orientalis* and *Ostrya carpinifolia*. Planting density in Périgord truffle plantations carries significant weight. In principle, the planting density should be low because natural production is noticeably reduced when the surface is shaded, i.e., when the closure of the crown canopy exceeds 0.7–0.75 (Bencivenga and Falini, 2012).

Two choices are possible: a) low density that does not require interventions in the future; b) high density, which includes drastic pruning operations or gradual thinning of the plant (dynamic density). Low densities usually range from 200–300 trees/ha and foresee less pruning intervention. Higher densities ranging from 500–700 trees/ha require greater efforts and maintenance costs, but also give a higher yield. The decision for one or the other option should be considered on a case-by-case basis, primarily based on the availability of manpower for pruning. When choosing a location for Périgord truffles, you should also take these points into consideration: a) Altitude can vary from 300–1200 m. Due to recent changes in climate conditions that have led to rising temperatures and drought, summers at higher altitudes are more suitable for growing this truffle. With regard to exposure, it is preferable to have exposure to the Sun at high altitudes (900–1000 m) and shade at lower altitudes (< 400 m). b) Périgord truffle naturally inhabits many Mediterranean limestone mountains with less agricultural and forestry potential. This is why one should not run away from solutions that include marginal agricultural lands (Reyna and Gacrcia-Barreda, 2008). c) Most of the best natural truffières are located in areas with a low degree of canopy openness (about 0.3), which is a good practical indicator for estimating the density of plantations at a given location (Olivier et al., 1996; Reyna et al., 2004).

The development stage includes the first 4–7 years, during which the host plant adapts to the terrain conditions and the fungus mycelium spreads. In this period, in addition to performing routine interventions, it is important to perform pruning and dispersal of spores. The shape of the crown should resemble an inverted cup in colder and less sunny locations and be spherical or conical in warmer areas. The fruiting stage begins somewhere around the 10th year after the establishment of the plantation, when the first fruiting bodies and "brûlé" appear. The first few years of the fruiting stage are also called the consolidation period because the plantation still does not give maximum yields during that time. In the fruiting stage, the following activities are planned: irrigation, weeding or hoeing, pruning, mulching and spore inoculation. The number of summer days devoid of rain that Périgord truffle ascocarps can withstand depends on the characteristics of the soil and the maximum temperatures, but the knowledge of truffle growers and climate analysts says that they range around 30–35 days (Roux, 2001). In order to prevent the negative consequences of global warming and summer droughts, it is recommended to program irrigation as follows: a) carry out the first irrigation intervention at the beginning of summer, when the soil warms up and starts to dry. This provides moisture to the soil layer up to 20–25 cm deep; b) subsequent interventions must soak the first 5–10 cm of soil depth and have variable application dynamics, in accordance with specific meteorological conditions; c) irrigation should be stopped only when the autumn rains appear. It is considered appropriate to remove weeds during or immediately after the spring vegetation movement from April to the beginning of May.

On the basis of several years of observations, weeding can be limited only to the "brûlé", allowing the development of vegetation on the surrounding land, which is mowed once a year. It has been noticed that the "brûlé" expands by several centimeters every year, and larger truffles appear towards the peripheral parts of the clearing, which is actually the transition zone between dug-up lots and un-digged sods (grass formations). Hoeing the clearing requires a lot of manual work, which can be a problem if there is not enough manpower. In the case of common hazel with a bushy habit, which develops numerous annual shoots, pruning consists of removing excess shoots and leaving 5–6 per plant; in the following years, 2–3 old shoots are left and they produce numerous young shoots. In the case of oaks and hornbeams, winter pruning consists of removing shoots formed at the base of the trunk and topping the plant when it reaches a height of about 150 cm. In cases where the tree top land covers exceed 30% of the "brûlé", and crown pruning does not give the desired results, the trees are cleared (extracted). Particularly those with low or no ascocarp production are extracted (Reyna et al., 2004; Gregori et al., 2001). Mulching should start in mid-June and end in mid-October. About half of the "brûlé" should be covered with the appropriate material. In extremely cold conditions, in order to protect the truffles from freezing, the mulch can remain over the winter. Unfortunately, the issue of winter mulching has not been supported by concrete field research. It is recommended that the first and last wave of truffles not be extracted from the ground, but used as an inoculum from spores. Ascocarps of the first wave are mostly small, largely underripe, and as such they are not optimal either for the market or for personal consumption. The ascocarps of the last wave are mostly few and of poor quality. If it is estimated that the last wave cannot meet the set task, the additional inoculum is distributed in the ratio of about 10 g of gleba per "brûlé". The Périgord truffle ripens and is collected from the beginning of December to the middle of March.

7. Economic Aspects of Small-scale Truffle Farming

7.1 Expected Yield of Truffles

The yield possibilities of truffles are highly variable. Differences already occur between populations of one species, not to mention variability between those of different species. The reason for this lies in the fact that the yield size is influenced by a large number of factors, which change from place to place, and even more from climate to climate. This is why yield figures have only an indicative character. Truffle yield can be expressed in several ways: 1) per unit area of truffle ground (kg/ha or g/m^2) (Mannozzi-Torini, 1970); 2) per host tree (g (kg)/tree) (Chetan and Morar, 2011); 3) per "brûlé" (g (kg)/ "brûlé") (Garcia-Montero et al., 2014); 4) per unit of surface area of "brûlé" (g(kg)/m^2 "brûlé" (Garcia-Montero et al., 2014).

Most often, the yield is expressed per unit area of the truffle field. In addition to this information, it would be desirable to indicate the density, that is, the number of trees per unit area. Namely, the same host tree-truffle combination can be realized in a layout of 3 × 3 m and 6 × 6 m. In the first case, the yield is achieved by 1,111 trees and in the second case by 277 trees. This means that in the first case, 1,111 trees need to be dug up, pruned and watered, and in the end "turned" and in the second case only 277, which is not the same, neither from the aspect of the work involved nor from the aspect of costs. It has been observed that the truffle yield does not occur simultaneously for all planted trees. Instead the trees gradually enter the production cycle as the culture ages even though all the seedlings are of the same age and were mycorrhizalized in one place by a certified producer. By the tenth year of a *T. melanosporum* plantation on *Quercus ilex* in southern France, only 10% of the trees were productive. In terms of yield continuity on the same property, only 15% of the trees bore fruit every consecutive year, while 85% of the trees experienced one or more fallow years after their initial truffle yield. (Shaw et al., 1996).

The *Tuber aestivum* truffière in the Nouvelle-Aquitaine region of France, owned by an experimental network of truffle growers, is a good example of the yield potential of a small-scale plantation. In 2004, 117 trees of *Carpinus betulus*, *Corylus avellana*, *Quercus pubescens*, *Tilia cordata* and *Pinus nigra* were planted on this site in a random arrangement. Half of the trees are arranged 4x3 m (equivalent to 833 trees/ha) and half are arranged 4x6 m (equivalent to 417 trees/ha). The seasonal dynamics of yield after the beginning of fruiting, regardless of host species and planting density were as follows: in 2009, fructification began; in 2014, yield was 266 kg/ha; in 2017, yield was 320 kg/ha; in 2018, yield was 73 kg/ha (Todesco et al., 2019). There are two characteristics of truffle fields: seasonal growth of yields and occurrences of non-bearing years, i.e., poorly bearing years. In the same observed period (2009–2018), the average yield of truffles per tree of *Corylus avellana, Quercus pubescens, Tilia cordata* and *Pinus nigra* was approximately similar, and ranged between: 100–200 g/tree. However, the yield per tree of *Carpinus betulus* was significantly higher >400 g/tree (Todesco et al., 2019). From here, it is possible to draw the conclusion that an adequate choice of host tree can have a positive effect on the size of the yield, with the same factors.

When it comes to *Tuber borchii*, in stands of *Pinus pinea* and *P. pinaster* in Tuscany, Italy, a yield of 10.12 kg/ha was registered (Salerni et al., 2014). Apart from the fact that the yield was modest, the collected ascocarps had a small mass, mostly around 10 g/piece; specimens over 15 years old were rare. In the absence of relevant sources, it could be cautiously assumed that the average yield potential of *T. brumale* is roughly the same as that of *T. borchii*. According to experiences from Italy, *Tuber magnatum* begins to bear fruit only 15–20 years after the establishment of the plantation, and then in limited quantities of 2–4 kg/ha (Bencivenga et al., 2009). Also in Italy, plantations of *Tuber melanosporum* on oak (*Quercus* sp.), with a planting density of 400 trees/ha, gave an average yield of 100 kg/ha 15 years after planting (Mannozzi-Torini, 1970). According to the same source, the yield increases with the age of the plantation. In *Q. ilex* plantations in southern France, the ascocarp yield in the 12th year varied between 20 and 90 kg/ha (Shaw et al., 1996). In truffle fields of *T. melanosporum* in New Zealand, yields were achieved in a wide range between 20–350 kg/ha in the 10th year after establishment (Hall et al., 2017).

7.2 Expected Market Value of Truffles

Today, more than half of truffles obtained in the whole world are harvested from truffle orchards (Mello et al., 2006). Truffles collected in the wild and produced on plantations are sold on the markets of Europe and the world as a mixture. Because of this, it is not possible to accurately distinguish their origin, and therefore not even distinguish the prices of one and the other (Hall et al., 2007). Also, an undetermined amount of truffles is sold unofficially and is not quantified in official statistics (Gregori, 2007), which additionally affects price variability (Reyna and Garcia-Barreda, 2008).

Tuber aestivum is the most common type of truffle on the European market. In Spain, a kilogram of this truffle costs €25–70 (Oliach et al., 2020), while in Switzerland it costs €200–600 (Molinier et al., 2016). In Italy, *Tuber borchii* can reach a price of up to €200–250/kg in case the seasonal yield of other truffles has failed (Oliach et al., 2020). By the way, *T. borchi* owes its good price to the increase in its demand on the market (Lancellotti et al., 2016). The price of *Tuber brumale* in Spain ranges from 60–120 €/kg (Oliach et al., 2020). The lower price compared to *T. borchii* is influenced by the characteristic musky odor and the small size of the fruits (Hall et al., 2008).

Tuber magnatum rarely comes to the market from plantation cultivation as usual. The mass of pieces from plantations is significantly less than the mass of ascocarps found in the wild, which is why their price is incomparably lower, roughly in the range of average pieces of *Tuber melanosporum*. By the way, well-developed ascocarps of *T. magnatum* reach a price of €1,000–1,500/kg (Bach et al., 2021; Riccioni et al., 2016), while the largest ones are sold by free negotiation or auction. The price of *T. melanosporum* on the international market in the last few years ranged from €170 to 550/kg (Oliach et al., 2020).

7.3 Productive Period of the Plantation

The time that elapses from the moment of planting the mycorrhizal seedlings to the moment when the production of the plantation stops, can be viewed as one multi-year production cycle. Depending on the type of host and other influencing factors, the cycle can last from 30–35 years (Hall et al., 2007; Olivera et al., 2011; Fisher et al., 2017) to 50 years (Morcillo et al., 2015). However, some truffle fields, over a hundred years old, are still in full production (Morcillo et al., 2015). It is to be expected that the cycle will be somewhat shorter in the hazelnut plantations and somewhat longer in the oak plantations.

Different sources see different structures in the multi-year production cycle. One more widespread approach distinguishes: 1) the birth stage, which includes the time needed to prepare the ground before planting and the year in which mycorrhizal seedlings were planted; 2) development stage, which includes the passage of time from the moment the plantation is established to the appearance of the first fruit-bearing body; this period can last from 4–8 years; 3) progressive fruiting stage, which includes the period of time in which ascocarps appear as a result of a successfully established mycorrhizal symbiosis between the truffle tree and truffle fungi, whereby the seasonal yields generally increase from year to year, eventually reaching their maximum; this period can last 15–20 years; 4) the regressive fruiting stage is a period in which the plantation continues to bear fruit but yields gradually decrease from year to year and eventually stop completely; this period can last approximately 15–20 years, like the progressive one; 5) at the explanting or replanting stage, in the end, the owner or investor has to decide whether to clear the land and bring for another culture, or to replace the existing "worn-out" trees with fresh ones.

8. Conclusion

In the last twenty years, more than half of the truffles produced have come from artificial cultivation. Growing truffles in small quantities is an objectively acceptable initiative that provides rural landowners with the opportunity to make a significant profit. Small-scale truffle farming can also be organized as hobby gardens with a small number of trees, whose production is solely for the satisfaction of personal needs. In addition to being profitable, truffle plantations have a beneficial socioeconomic impact on neighboring areas, and compared to conventional agriculture, they have a significantly lower impact on the environment. The cultivation of truffles differs in many ways from the cultivation of other crops. They are grown as a long-term economic crop that achieves its first full crop after 10 or more years. Before and after establishing the plantation, it is necessary to carry out numerous cultivation operations, which imply detailed planning and timely execution.

References

Anonymous (1992). Electronic equipment for detecting truffles—uses polymeric detectors whose electrical resistance varies with amount of volatile material in air. Patent # 2696236 od 15.12.1992, Institut National de la Properiete Industrielle, France, p. 16.

Anonymus (2012). Water Resources in Europe in the Context of Vulnerability. State of Water Assessment. EEA Report # 11, Luxembourg.

Anonymus (2017). Climate Change, Impacts, and Vulnerability in Europe 2016. An Indicator- Based Report. EEA Report # 1, Luxembourg.

Anonymus (2022). Guide: Soil Moisture Recommendations for Flowers, Plants, and Vegetables. AcuRite. com. https://www.acurite.com/blog/soil-moisture-guide-for-plants-and-vegetables.html

Anonymus (2024). Truffle Spores. Tottaly Truffles. https://www.totallytruffles.co.uk

Bačani, A. (2006). Hydrogeology I. Faculty of Mining, Geology and Petroleum, University of Zagreb, Hidrogeologija I. Rudarsko-geološko-naftni fakultet, Sveučilište u Zagrebu, *in Croatian*, Zagreb, p. 198.

Bach, C., Beacco, P., Cammaletti, P., Babel-Chen, Z., Levesque, E. et al. (2021). First production of Italian white truffle (*Tuber magnatum* Pico) ascocarps in an orchard outside its natural range distribution in France. Mycorrhiza 2021, 31: 383–388.

Belfiori, B., Riccioni, C., Tempesta, S., Pasqualetti, M., Paolocci, F. et al. (2012). Comparison of ectomycorrhizal communities in natural and cultivated *Tuber melanosporum* truffle grounds. FEMS Microbiol. Ecol., 81: 547–561.

Bencivenga, M. and Granetti, B. (1988). Ricerca comparativa sulle esigenze ecologiche di *Tuber* magnatum Pico e *Tuber melanosporum* Vitt. dell'Italia centrale. Annali della Facolta di Agraria, Universita degli Studi di Perugia, 42: 861−872.

Bencivenga, M. and Falini, L.B. (2012). Manuale di tartuficoltura Esperienze di coltivarizione dei tartufi in Umbria. Universita degli studi di Perugia, p. 137.

Bencivenga, M., Di Massimo, G., Donnini, D. and Baciarelli Falini, L. (2009). The cultivation of truffles in Italy. Acta Botanica Yunnanica, Supl., 126: 21–28.

Benucci, G.M.N., Bonito, G, Falini, L.B. and Bencivenga, M. (2012). Mycorrhization of Pecan trees (*Carya illinoinensis*) with commercial truffle species: *Tuber aestivum* Vittad. and *Tuber borchii* Vittad. Mycorrhiza, 22: 383–392. https://doi.org/10.1007/s00572-011-0413-z

Bissanti, G. (2022). Species fungi *Tuber brumale*. Un Mondo Ecosistenibile. https://antropocene.it/en/2022/12/30/tuber-brumale-2/

Bonet, J.A., Fischer, C. and Colinas, C. (2006). Cultivation of black truffle to promote reforestation and land-use stability. Agronomy for Sustainable Development, 26: 69–76.

Bronstein, J. (2015). Mutualism. Oxford University Press, p. 312.

Bruhn, J. and Hall, M. (2011). Burgundy Black Truffle Cultivation in an Agroforestry Practice. Agroforestry in action. The Center for Agroforestry at the University of Missouri, p. 20.

Carbajo, P. (1999). Plantación de encinas micorrizadas para la producción de trufas (*Tuber melanosporum*) en la provincia de Soria. pp. 107–111. In: Micorrización en áreas mediterráneas de la Península Ibérica. Publ. Junta de Extremadura, Mérida.

Čejka, T., Trnka, M. and Büntgen, U. (2023). Sustainable cultivation of the white truffle (*Tuber magnatum*) requires ecological understanding. Mycorrhiza, 33: 291–302.

Ceruti, A., Fontana, A. and Nosenzo, C. (2003). Le Specie Europee del genere *Tuber* - Una Revisione Storica. Museo Regionale de Scienze Naturali, Torino.

Chetan, S.C. and Morar G. (2011). Economically and alimentary important truffle species, establishment and development of the truffle culture in Xiromero region. Agricultura, 3–4: 74–79.

Chevalier, G. and Frochot, H. (1989). Ecology and possibility of culture in Europe of the burgundy truffle (*Tuber uncinatum* Chatin). Agric. Ecosys. Environ., 28: 71–73.

Delmas, J. and Poitou, N. (1973). La truffe et ses exigences écologiques. Pépiniėristes Horticulteurs Maraichers, 144: 33–39.

Diamandis, S. and Perlerou, C. (2008). Recent records of hypogeous fungi in Greece. Acta Mycol., 43:139–142.

Diette, S. and Lauriac, A. (2005). La sylviculture truffière: aperçus historiques, apports techniques et enjeux pour la région méditerranéenne. Forêt méditerranéenne, 26: 157–168.

Fischer, C.R. and Colinas, C. (1996). Methodology for certification of *Quercus ilex* seedlings inoculated with *Tuber melanosporum* for commercial application. In: Proceedings of the 1st International Conference in Mycorrhizae. Berkeley, California, USA.

Fischer, C.R., Oliach, D., Bonet, J.A. and Colinas, C. (2017). Best Practices for Cultivation of Truffles. Forest Sciences Centre of Catalonia, Solsona, Spain; Yaşama Dair Vakıf, Antalaya, Turkey, p. 68.

Frochot, H., Chevalier, G., Barbotin, P. and Beaucamp F. (1999). Avancees sur la culture de la truffe de Bourgogne (*Tuber uncinatum* Chatin). Actes du V congres international Science et culture de la truffe. Aix en Provance, 387–392.

Garcia-Montero, G.L., Moreno, D., Monleon, J.V. and Arredondo-Ruiz, F. (2014). Natural production of *Tuber aestivum* in central Spain: *Pinus* spp. versus *Quercus* spp. brûles. Forest Systems, 23: 394–399.

García-Montero, L.G., Manjón, J.L., Pascual, C. and García-Abril, A. (2007). Ecological patterns of Tuber *melanosporum* and different *Quercus* Mediterranean forests: Quantitative production of truffles, burn sizes and soil studies. For. Ecol. Manage., 242: 288–296.

Gardin, L. (2005). I tartufi minori in Toscana. Gli ambienti di crescita dei tartufi marzuolo e scorzone. ARISA Regione Toscana, Italy. https://doi.org/10.13140/RG.2.1.2509.5844.

Gregori, G., Elisei, S., Pasquini, L., Sacchi, A., Spezi, D. et al. (2001). Rigenerazione di una vecchia tartufaia coltivata di *Tuber melanosporum* Vitt. In: International Science et Culture de la Truffe, Centre des Congrès, Aix-en-Provence, 4–6 March 1999.

Gregori, G. (2007). Principales aspectos de la trufa y la truficultura en Italia. In: Reyna, S. (ed.). Truficultura, Fundamentos y técnicas, Mundi-Prensa, Madrid, Spain.

Grente, J. and Delmas, J. (1974). Perspectives pour une trufficulture moderne. Ed. INRA. Clermont-Ferrand, 1–65.

Guarína, A., Riverab, M., Pinto-Correiab, T., Guiomarb, N., Šūmanec, S. et al. (2020): A new typology of small farms in Europe. Global Food Security, 26: 1–9

Guerin-Laguette, A., Cummings, N. Hesom-Williams, N. Butler, R., Wang, Y. et al. (2013). Mycorrhizas analyses in New Zealand truffières reveal frequent but variable persistence of Tuber *melanosporum* in co-existence with other truffle species. Mycorrhiza 23: 87–98.

Hall, I., Brown, G. and Byars, J. (2001). The black truffle: its history, uses and cultivation. 2nd Edition, New Zealand Institute for Food and Crop Research Limited, Christchurch.

Hall, I., Brown, G. and Zambonelli, A. (2007). Taming the truffle. The history, lore, and science of the ultimate mushroom. Timber Press, Portland, p. 304.

Hall, I., Fitzpatrick, N., Miros, P. and Zambonelli, A. (2017). Counter-season cultivation of truffles in the Southern Hemisphere: an update. Italian J. Mycol., 46: 21–36.

Hall, I., Zambonelli, A. and Nelson, P. (2008). Distinguishing the Perigord black truffle from the winter truffle and other look-alikes in the laboratory and field, 3rd Editkon, Truffles and Mushrooms, Dunedin.

Hrka, J. (1988). Truffles - knowing, picking, growing, preparing. Mladost, Zagreb, p. 75.

Iotti, M., Piattoni, F. and Zambonelli, A. (2012). Techniques for host plant inoculation with truffles and other edible ectomycorrhizal mushrooms. pp. 145–161. In: Zambonelli, A. and Bonito, G.M. (eds.). Springer Berlin Heidelberg. 10.1007/978-3-642-33823-6_9

Jaillard, B., Barry-Etienne, D., Colinas, C., de Miguel, A.-M., Genola, L. et al. (2014): Alkalinity and structure of soils determine the truffle production in the Pyrenean Regions. For. Sys., 23: 364–377.

Karadžić, D. (2010): Forest phytopathology. University of Belgrade. Faculty of Forestry, p. 774.

Lancellotti, E., Iotti, M., Zambonelli, A. and Franceschini, A. (2016). The puberulum group sensu lato (whitish truffles). pp. 105–124. In: Zambonelli, A., Iotti, M. and Murat, C. (eds.). True Truffle (*Tuber* spp.) in the World: Soil Ecology, Systematics and Biochemistry, Springer International Cham, Switzerland.

Linde, C.C. and Selmes, H. (2012). Genetic diversity and mating type distribution of *Tuber melanosporum* and their significance to truffle cultivation in artificially planted truffières in Australia. Appl. Environ. Microbiol., 78: 6534–6539.

Ljubojević, S., Marčeta, D. and Stupar, V. (2014). Habitat conditions and usability of the hog truffle (*Choiromyces meandriformis* Vitt.) found in Bosnia and Herzegovina. Proceedings. V International scientific agricultural symposium, Agrosym 2014, Jahorina, October 23–26: pp. 109–117.

Ljubojević, S., Vasilišin, L., Vučić, G. and Velemir, A. (2022). Morphological characteristics of summer truffle (*Tuber aestivum* Vittad.) from Bosnia and Herzegovina. EUREKA: Life Sciences, 2: 9–20.

Lo Bue, G., Gregori, G.L., Pasquini, L. and Maggiorotto, G. (1990). Sintesi micorrizica in campo fra piante adulte e tartufi pregiati mediante frammenti radicali. pp. 459–466. In: Bencivenga, M., Granetti, B. (eds.). Atti del Secondo Congresso Internazionale sul tartufo, Spoleto.

Magarzo, A., Alba, S., del Blanco, L.S., Manchón, I.F., Olaizola, J. et al. (2023). Effect of Irrigation Methods on Black Truffle Production. Agronomy, 13(10): 2505. https://doi.org/10.3390/agronomy13102505

Mannozzi-Torini, L. (1970). Manuale di tartuficoltura. Edagricole, Bologna, p. 119.

Mello, A., Murat, C. and Bonfante, P. (2006). Truffles: much more than a prized and local fungal delicacy. FEMS Microbiol. Lett., 260: 1–8.

Merényi, Z., Varga, T. and Bratek, Z. (2016). *Tuber brumale*: A Controversial Tuber Species. pp. 49–68. In: Zambonelli, A., Iotti, M., Murat, C. (eds.). True Truffle (*Tuber* spp.) in the World: Soil Ecology, Systematics and Biochemistry, Springer International Cham, Switzerland.

Merényi, Zs., Varga, T., Geml, J., Orczán, Á.K., Chevalier, G. et al. (2014). Phylogeny and phylogeography of the *Tuber brumale* aggr. Mycorrhiza, 24: 101–113.

Michels, C. (2003). Conditions climatiques et production truffière. In: Resumes des interventions, Journée Nationale de la trufficulture, 28 march 2003, Martel, France.

Molinier, V., Peter, M., Stobbe, U. and Egli, S. (2016). The Burgundy truffle (*Tuber aestivum* syn. *uncinatum*): A truffle species with a wide habitat range over Europe. pp. 33–47. In Zambonelli, A., Iotti, M., Murat, C. (eds.). True Truffle (*Tuber* spp.) in the World: Soil Ecology, Systematics and Biochemistry, Springer International Cham, Switzerland.

Molkabadi, R.S., Bonito, G., Ghasemi, K., Ghanbary, M.A.T. and Fard, F.R. et al. (2023). Optimising in vitro culture conditions for the truffle *Tuber brumale*. Botanica Serbica, 47: 259–269.

Montacchini, F. and Caramiello, R. (1968). Ecologia del *Tuber magnatum* Pico in Piemonte. Allionia, 14: 1–29.

Morcillo, M., Sánchez, M. and Vilanova, X. (2015). Truffle Farming Today a Comprehensive World Guide. Forestal Micologia & Aplicada, Spin, p. 352.
Mrak, T., Grebenc, T., Friedrich, S. and Münzenberger, B. (2024). Description, identification, and growth of *Tuber borchii* Vittad. mycorrhized *Pinus sylvestris* L. seedlings on different lime contents. Mycorrhiza, 34: 85–94.
Murat, C., Bonneau, L., Varga, H. De, Olivier, J.M., Sandrine, F. et al. (2016). Trapping truffle production in holes: a promising technique for improving production and unravelling truffle life cycle. Italian J. Mycol., 45: 47–53.
Oliach, D., Morte, A., Sánchez, S., Navarro-Ródenas, A., Marco, P. et al. (2020). Las trufas y las turmas. In Los Productos Forestales No Madereros en España: Del Monte a la Industria. INIA, Ministerio de Ciencia e Innovación; Gobierno de España: Madrid, Spain, pp. 283–324.
Olivera, A., Bonet, J.A., Palacio, L., Liu, B. and Colinas, C. et al. (2014). Weed control modifies *Tuber* melanosporum mycelial expansion in young oak plantations. Ann. For. Sci., 71: 495–504.
Olivera, A., Fischer, C.R., Bonet, J.A., Martınez de Aragon, J., Oliach, D. et al. (2011). Weed management and irrigation are key treatments in emerging black truffle (*Tuber* melanosporum) cultivation. New Forests, 42: 1–13.
Olivier, J.M., Savignac, J.C. and Sourzat, P. (1996). Truffe et trufficulture. Fanlac, Périgueux, France, p. 263.
Olivier, J.P.M., Savignac, J.C. and Sourzat, P. (2012). Truffe et trufficulture Fanlac, Périgueux, France, p. 400.
Özderin, S., Yılmaz, F. and Allı, H. (2018). Determining mycorrhiza rate in some oak species inoculated with *Tuber aestivum* Vittad. (summer truffle). Tur. J. For., 19: 226–232.
Pacioni, G. (1991). Effects of *Tuber* metabolites on the rhizospheric environment. Mycol. Res., 95: 1355–1358.
Poitou, N. (1988). Les sols truffiers. Choix du sol: prélèvement, analyse, correction, oligo- éléments. In: Journées Nationales de la Truffe, St. Paul Trois Châteaux, Drome, France.
Porter, J., Costanza, R., Sandhu, H., Sigsgaard, L., Wratten, S. et al. (2009). The value of producing food, energy, and ecosystem services within an agro-ecosystem. Ambio, 38: 186–193.
Pruett. G., Bruhn. J. and Mihail. J. (2008). Temporal dynamics of ectomycorrhizal community composition on root systems of oak seedlings infected with Burgundy truffle. Mycol. Res., 112: 1344–1354.
Raglione, M., Spadoni, M., Cavelli, S., Lorenzoni, P. and De Simone, C. et al. (2001). Les sols des truffi ères naturelles de *Tuber melanosporum* Vitt. dans l'Apennin central (Italie). pp. 276–280. In: Actes du Ve Congrès International "Science et culture de la truffe", Aix en Provence (France), 4–6 Mars 1999. IPSO, Marseille, France.
Reyna, D.S. (2007). Truficultura: Fundamentos y Técnicas. Mundi-Prensa, Madrid, Spain, p. 686.
Reyna, S. (1992). La trufa. Mundi-Prensa, P., Universidad Politécnica, Madrid, Spain, p. 115.
Reyna, S. (1999). Aproximación a una Selvicultura trufera. Tesis Doctoral. Valencia-Madrid, spin.
Reyna, S. (2000). La trufa, truficultura y selvicultura trufera. Mundi-Prensa, Madrid, Spain, p. 229.
Reyna, S. and Garcia-Barreda, S. (2008). European black truffle: its potential role in agroforestry development in the marginal lands of Mediterranean calcareous mountains. pp. 295–317. In: Rigueiro-Rodroguez, A., McAdam, J. and Mosquier-Losada, M.R. (eds.). Agroforestry in Europe: Current Status and Future Prospects. Springer Nature B.V.
Reyna, S., Barreal, J.A.R., Folch, L., Pérez-Badía, R., Domínguez, A. et al. (2001). Techniques for inoculating mature trees with *Tuber melanosporum* Vitt. Second International Conference on Edible Mycorrhizal Mushrooms, Christchurch, New Zealand.
Reyna, S., Garcia, S., Folch, L., Pérez-Badia, R., Galiana, F. et al. (2004). Selvicultura trufera en montes mediterráneos. pp. 523–246. In: Vallejo, R. and Alloza, J.A. (eds.). Avances en el estudio de la gestión del monte mediterránea, Fundación CEAM, València, Spain.
Reyna, S., Rodriguez, B., Saíz De Omecaña, J.A., Zazo, J., Pérez, R. et al. (1999). Producción de plantas micorrizadas de calidad: implantación, mantenimiento y mejora de rodales productores de trufa y otras setas. Programa de investigación y desarrollo en relación con la cubierta vegetal: Reunión de Coordinación del Centro de Estudios Ambientales del Mediterráneo. Castellón.
Ricard, J.M. (2003). La truffe. Guide technique de trufficulture. pp. 1–268. Centre Technique Interprofessionnel des Fruits et Légumes, París, France.

Riccioni, C., Belfiori, B., Rubini, A., Passeri, V., Arcioni, S. et al. (2008). *Tuber melanosporum* outcrosses: analysis of the genetic diversity within and among its natural populations under this new scenario. New Phytol., 180: 466–478.

Riccioni, C., Rubini, A., Belfiori, B., Gregori, G., Paolocci, F. et al. (2016). *Tuber magnatum*: The special one. What makes it so different from the other *Tuber* spp.? pp. 87–103. In: Zambonelli, A., Iotti, M., Murat, C. (eds.). True Truffle (*Tuber* spp.) in the World: Soil Ecology, Systematics and Biochemistry, Springer International Cham, Switzerland.

Roux, J.J. (2001). Les relations pluie-récolte sur 17 années dans le Var: *T. melanosporum* 1982–1998. In: International Science et Culture de la Truffe, Centre des Congrès, Aix- en-Provence, 4–6 March 1999.

Sáez, R. and De Miguel, A. (1995). Guía práctica de truficultura. Ed. I.T.G. Agricola S.A. and Universidad de Navarra. Pamplona, Spain, p. 94.

Salerni, E., Perini, C. and Gardin, L. (2014). Linking climate variables with *Tuber borchii* sporocarps production. Natural Resources, 5(8): 408–418.

Shaw, A.J.P., Lankey, K. and Jourdan, A. (1996). Factors affecting yield of *Tuber melanosporum* in a *Quercus ilex* plantation in southern France. Mycol. Res., 100: 1176–1178.

Sourzat, P. (1994). Guide pratique de trufficulture. Station d'experimentations sur la truffe, Lycée professionnel agricole Cahors-Le Montat, Le Montat, France, p. 119.

Sourzat, P. (1997). Guide pratique de trufficulture. Station d'expérimentations sur la truffe, Le Montat, France, p. 96.

Sourzat, P. (2002). Guide pratique de trufficulture. Station d'expérimentations sur la truffe, Le Montat, France, p. 119.

Sourzat, P. (2003). Les écosystèmes truffiers naturels: quels enseignements en tirer? Résumes des interventions, Journée Nationale de la trufficulture, 28 March 2003, Cuzance, France.

Sourzat, P. (2009). The truffle and its cultivation in France. Acta Botanica Yunnanica Suppl. 16: 72–80.

Sourzat, P. (2017). Truffle cultivation in the south of France: technical progress and prospects. Scientia Fungorum, 46: 63–72.

Sourzat, P. and Tabouret, P. (2008). La truffe *Tuber brumale*. Fédération Française des Trufficulteurs, Paris, France.

Stobbe, U., Egli, S., Tegel, W., Peter, M., Sproll, L. et al. (2013). Potential and limitations of Burgundy truffle cultivation. Appl. Microbiol. Biotechnol., 97: 5215–5224.

Suwannarach, N., Kumla, J., Meerak, J. and Lumyong, S. (2017). *Tuber magnatum* in Thailand, a first report from Asia. Mycotaxon 132: 635–642.

Thomas, W.P. (2012). The role of pH in *Tuber aestivum* syn. *uncinatum* mycorrhiza development within commercial orchards. Acta Mycologica, 47: 161–167.

Todesco, F., Belmondo, S., Guignet, Y., Laurent, L., Fizzala, S. et al. (2019). Soil temperature and hydric potential infuences the monthly variations of soil *Tuber aestivum* DNA in a highly productive orchard. Sci. Rep., 9: 12964: https://doi.org/10.1038/s41598-019-49602-2

Tramblay, Y., Llasat, M.C., Randin, C. and Coppola, E. (2020). Climate Change Impacts on Water Resources in the Mediterranean. Reg. Environ. Change, 20: 83. https://doi.org/10.1007/s10113-020-01665-y

Trappe, J., Molina, R., Luoma, D., Cazares, E., Pilz, D. et al. (2009). Diversity, Ecology and Conservation of Truffle Fungi in Forests of the Pacific Northwest. PNW-GTR-772, United States Department of Agriculture, Pacific Northwest Research Station, Portland, p. 194

Treseder, K.K. (2004). A meta-analysis of micorrhizal responses to nitrogem, phosphorous and atmospheric CO_2 in field studies. New Phytolol., 164: 347–355.

USDA (2017). Soil Survey Manual. Handbook No. 18, United States Department of Agriculture, p. 603.

Weden, C., Chevalier, G. and Danell, E. (2004). *Tuber aestivum* (syn. *T. uncinatum*) biotopes and their history on Gotland, Sweden. Mycol. Res., 108: 304–310.

Wie, J., Gao, W. and Huang, C.-Y. (2021). A checklist of edible ectomycorrhizal mushrooms in China. Mycosystema, 40:1938–1957.

Wiggins, S., Kirsten, J. and Lambi, L. (2010). The future of small farms. World Development, 38: 1341–1349.

Zambonelli, A., Iotti, M., Giomaro, G., Hall, I.R. and Stocchi, V. (2002). *T. borchii* cultivation: an interesting perspective. Edible mycorrhizal mushrooms and their cultivation. In: Hall, I., Wang,

Y., Danell, E. and Zambonelli, A. (eds.). Edible mycorrhizal mushrooms. Proceedings of 2nd international workshop on edible ectomycorrhizal mushrooms, New Zealand Institute for Crop and Food Research Limited Current status of truffle cultivation: Recent results and future perspectives, CD room. July 3–6, 2001, Christchurch, New Zealand.

Zambonelli, A., Iotti, M., Rossi, I. and Hall, I. (2000). Interaction between *Tuber borchii* and other ectomycorrhizal fungi in a field plantation. Mycol. Res., 104: 698–702.

Zambonelli, A., Iotti, M., Zinoni, F., Dallavalle, E., Hall, I.R. et al. (2005). Effect of mulching on Tuber *uncinatum* ectomycorrhizas in an experimental truffière. New Zealand J. Crop Hort. Sci., 33: 65–73.

Zambonelli, A., Leonardi, P., Iotti, M. and Hall, R.I. (2017). Ecological and genetic advances in the cultivation of *Tuber* spp. Revista Fitotecnia Mexicana, 40: 371–377.

Zambonelli, A., Plattoni, F., Iotti, M. and Hall, R.I. (2010). What makes a good truffle infected tree? Österreichische Zeitschrift für Pilzkunde, 19: 201–207.

Chapter 16

Industrial Prospects of Truffles

Kirsten Allen

1. Introduction

Truffle species and truffle-like fungi are central to the local gastronomy and cultural identity of cuisines throughout the Mediterranean regions and others outside Europe (Thomas et al., 2021). Lauded for their desirable flavor and aroma profiles, truffles are consistently ranked as the most expensive underground edible mushroom in the world (Wang and Marcone, 2011). Jacobs (2019) estimates that the European truffle market is valued at approximately €300 million. Truffles are currently harvested in orchards as cultivated varieties and collected from the wild in natural forests, contributing to the multi-million-euro industry in France, Italy, and Spain (Reyna and Garcia-Barreda, 2014). The truffle sites in the Aragón region of Spain, for example, stimulate the local economy by producing 20% of the global truffle market and hosting 11% of the world's black truffle plantations (ÁgAgueda et al., 2014). These mycorrhizal fungi, belonging to the Pezizales (Ascomycota), contribute significant economic value. *Tuber magnatum* (white truffle), *T. melanosporum* (black truffle), and *T. aestivum* (Burgundy truffle) are the most valued and expensive in the European markets (Hilszczańska et al., 2016). Desert truffles such as the *Terfezia* and *Tirmania* species have a long recorded culinary history in the Middle East, in addition to well-developed markets (Shavit, 2013). A pilot study in Zambia reported the untapped potential for the distribution and export of truffles from Western Zambia that may be competitive with their European truffle counterparts (Siachoono et al., 2015). The

New Use Agriculture and Natural Plant Products Program (NUANPP), Rutgers University, Department of Plant Biology-Foran Hall, New Jersey, USA.
Email: kirsten.allen719@gmail.com

aim of this chapter is to provide insight into several of the industrial uses for truffles. Specialized truffle metabolites, including exopolysaccharides, lectins, pigments, sterols, and volatiles, exhibiting promising applications in the food, pharmaceutical, and packaging industries, among others.

2. Bioactive Metabolites

Distinguished by their value in the culinary industry, truffle fungi also represent a rich repository of specialized compounds for human health (Fig. 1). Gajos et al. (2014) cite numerous bioactive metabolites produced by *Tuber* spp., including quinones, isoflavones, steroids, ceramides, and catechols. Several therapeutic benefits have been attributed to truffles, such as antioxidant, antimicrobial, immunomodulating, antitumor, antimutagenic, and antidepressant properties (Patel, 2012). The antioxidant and radical scavenging activities of Algerian truffles, *Terfezia claveryi* and *Tirmania nivea*, were evaluated, and both species demonstrated significant bioactivity along with high polyphenol, tannin, and flavonoid content (Wahiba et al., 2016). The *Terfezia boudieri* truffle from Turkey, contained high quantities of fatty acids such as linoleic, oleic, and palmitic acid, in addition to dietary flavonoids such as myricetin, kaempferol, naringin, naringenin, and resveratrol (Akyüz, 2013). Bioactive compounds, including phenolic and organic acids, isolated from extracts of the under investigated truffle, *Reddellomyces parvulosporus,* exhibited potent antioxidant, cytotoxic, and anticholinesterase activities (Çayan et al., 2022).

2.1 Mycosterols

Mycosterols, another class of bioactives found in truffle species, can promote significant health benefits by lowering low-density lipoprotein (LDL)-cholesterol levels (Corrêa et al., 2017). The sterols, ergosterol and brassicosterol, are commonly found in *Tuber* species in addition to β-glucan polysaccharides (Tejedor-Calvo et al., 2020). For example, the presence of ergosterol was detected in Chinese black truffle (*Tuber indicum*), along with glucan, vitamin D_2, and the dietary fiber, pectin. In the truffle samples collected from Sichuan, Wu et al. (2021) detected 0.76 mg/g ergosterol, 3.96 g/100 g pectin, 14.72 g/100 g β-glucan, and 46.37 μg/g vitamin D_2. Sommer and Vetter (2020) detected and characterized 27 sterols in two truffle species. The sterol content varied from 130 to 590 mg/100 g in *Tuber borchii* samples and 110 to 420 mg/100 g in *Tuber aestivum* samples. Tang et al. (2012) compared the sterol composition between the fruit bodies of several *Tuber* species and fermentation mycelia. In the study, the addition of soybean flour significantly increased the sterol content of the fermentation media. Also, the data revealed that the sterol content was much higher in the soybean mycelia than in the *Tuber* fruit bodies.

3. Nutritional Potential

Truffles possess significant nutritive value resulting from concentrated carbohydrates, proteins, fats, minerals, lipids, and amino acids (Wang and Marcone, 2011; Lee et al., 2020). Nutrient-rich truffles are appreciated across many cultures and geographic

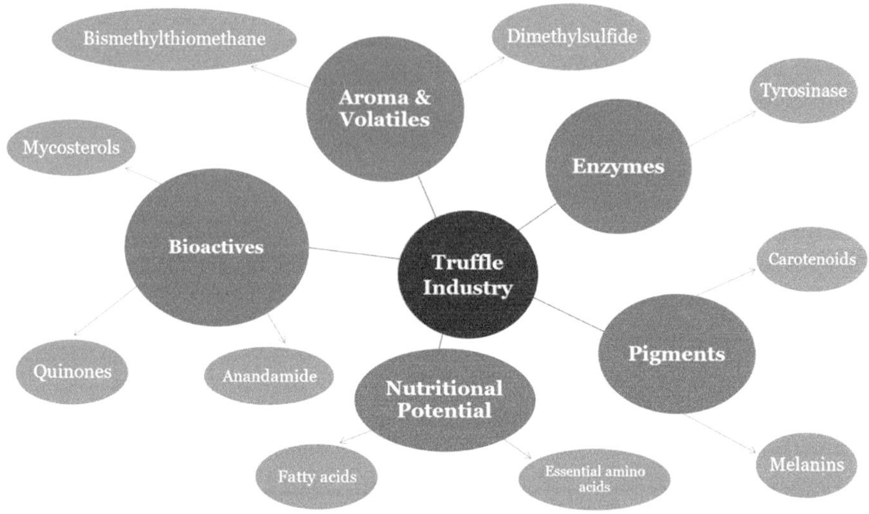

Fig. 1. Examples of industrial applications for truffles.

regions. Traditionally substituted for meat in local cuisines, desert truffles are prized throughout the Middle East and North Africa. Sawaya et al. (1985) reported that the *Terfezia claveryi* and *Tirmania nivea* truffles from Saudi Arabia contained all the essential amino acids and high amounts of potassium and phosphorus, in addition to other minerals. In comparison, Hamza et al. (2016) presented the nutritional composition of *T. nivea* truffles from the Tunisian arid zone. Carbohydrates were the most abundant macronutrient (57.83 g/100 g) followed by proteins (28.81 g/100 g). Widely consumed in Morocco, *Tirmania pinoyi* is an important food for the local population with high protein, fiber, and mineral content (Bouatia et al., 2018). Edible truffle species from China, *Tuber latisporum*, *T. subglobosum*, and *T. pseudohimalayense*, were evaluated for their nutritional value, and all three displayed low-fat contents and high contents of protein and unsaturated fatty acids (Yan et al., 2017). A study surveying the nutritional composition of the Turkish wild mushroom, *Tuber nitidum*, reported the presence of essential amino acids, valine, leucine, and phenylalanine, and several fatty acids, including linoleic, palmitic, oleic, stearic, and cis-11,14-eicosadienoic acids (Kivrak and Kivrak, 2018).

4. Exopolysaccharides

Truffles, like other fungi, also synthesize a diverse array of exopolysaccharides. Fungal exopolysaccharides (EPSs) are long-chain polysaccharides composed of branched and repeating units of sugar moieties in either flexible or rigid polymer forms (Kanmani and Lim, 2013; Osemwegie et al., 2020). EPSs are produced intracellularly and are subsequently secreted into their extracellular environment (Donot et al., 2012). These biopolymers can be easily extracted and purified from fungal sources and then implemented in industrial applications for cosmetic, food, pharmaceutical, and packaging products. Examples of EPSs include β-glucans, chitin, chitosan, sacchachitin, and pullalan (Hamidi et al., 2022). The variety of

EPSs produced in laboratory fermentation culture depends on the type of fungal strain, the medium composition, and the physical conditions maintained during fermentation (Mahapatra and Banerjee, 2013). Chen et al. (2023) conducted a study utilizing a submerged fermentation technique to cultivate high-molecular-weight exopolysaccharides from *Tuber borchii*. The optimized fermentation media included sucrose (80 g/l) and yeast extract (20 g/l) which maximized the mycelial growth and increased the production of β-(1–3)-glucan. Other studies involving the submerged fermentation of *Tuber melanosporum* have found that the addition of Dioscorea saponins can increase EPS production in liquid culture (Zhang et al., 2019; Liu et al., 2020).

5. Pigments

Mushrooms do not produce the same pigments that dominate in higher plants; however, many of the pigments that are widespread in higher fungi are quinones or related structures (Velíšek and Cejpek, 2011). De Angelis et al. (1996) reported quinonoid and polyphenolic biopolymers as significant constituents of truffe pigments. Quinones undergo subsequent non-enzymatic reactions that lead to darker melanin pigments (Weijn et al., 2013). Black pigments were isolated from *Tuber melanosporum* samples and were identified as allomelanins derived from the polyketide biosynthesis pathway. The peridium, or outer envelope of the sporophore, from the black truffle accumulates pigments throughout its development phases. In the beginning, the peridium is red, then evolves from brown to dark brown and eventually black (Pacioni et al., 1995). Melanin formation can cause brown pigmentation in fungi, and one of the melanin biosynthesis pathways requires the tyrosinase enzyme to produce melanin polymers (Eisenman and Casadevall, 2012; Weijn et al., 2013). Despite their albino appearance, white truffles (*Tuber magnatum* and *T. excavatum*) are tyrosinase positive like black truffles (*T. melanosporum*) (Miranda et al., 1996). Data from a melanogenesis study suggested the presence of yellow sulfur-containing melanins in white truffles (Miranda et al., 1997). The ascomata of *T. magnatum* Pico also frequently forms red patches. Amicucci et al. (2018) characterized the pigment profile responsible for this red coloration and found several carotenoids, including bixin, β-carotene, cis-1-glycosyl-apo-8′-lycopene, and fucoxanthin, in pigmented samples. The data identified possible bacterial species associated with the carotenoid production, including *Chryseobacterium* and *Microbacterium*.

6. Enzymes

Fungal enzymes from a variety of species possess a wide range of industrial applications, from food and cosmetic products to detergents and textiles (Gurung et al., 2013). Numerous studies have shown the enzymatic functions and capabilities of truffles. A commonly studied enzyme in truffles is the polyphenol oxidase known as tyrosinase (L-tyrosine, L-3,4-dihydroxyphenyl-alanine: oxygen oxidoreductase, EC 1.14.18.1). The enzyme is valued for its ability to oxidize phenolic compounds, producing melanin pigments (Swan, 1974). An early investigation indicated true tyrosinase activity in various *Tuber* spp. (*Tuber brumale*, *T. macrosporum*,

T. melanosporum, and *T. mesentericum*) (Miranda et al., 1992). Benaceur et al. (2020) purified and characterized polyphenol oxidase (PPO) from *Terfezia arenaria*. The PPO oxidized monophenolic, o-diphenolic, and triphenolic substrates, suggesting a promising pharmaceutical application.

To determine the bioactive potential of enzymes derived from *Tuber maculatum* and *T. aestivum*, Nadim et al. (2015) evaluated the extracellular enzymatic activities of the truffle mycelia. *Tuber maculatum* produced seven extracellular enzymes (amylase, catalase, cellulase, laccase, lipase, peroxidase, and xylanase), whereas *T. aestivum* produced only three enzymes (amylase, catalase, and peroxidase). Pacioni et al. (2015) investigated the presence of the endocannabinoid system in the winter black truffle (*Tuber melanosporum*). Transcriptional and translational expression results revealed the presence of the major endocannabinoid metabolic enzymes, including arachidonoylphosphatidylethanolamine-specific phospholipase D (NAPE-PLD) and fatty acid amide hydrolase (FAAH). However, the truffles did not express the relevant endocannabinoid-binding receptors. The fruiting bodies also contained the endocannabinoid, anandamide, which may be involved in the truffle maturation process and might have acted as an ancient attractant to truffle eaters.

7. Lectins

Another promising class of proteins with diverse industrial applications are truffle lectins. Fungal lectins recognize and bind carbohydrates reversibly and specifically to mediate cell recognition events in many biological and pathological developments. This may include cellular signaling, differentiation, host-pathogen interactions, and tissue metastasis (Varrot et al., 2013). Given their defensive properties and pathogenicity, lectins are useful markers in inflammations and cancer (Carrizo et al., 2005). The first fungal lectin was discovered in the fly agaric mushroom (*Amanita muscaria*), and its bioactivity was associated with the toxicity of the fungi (Ford, 1911; Kaliyaperumal et al., 2018). Few studies, however, have characterized lectins derived from truffle sources. Cerigini et al. (2008) reported TBF-1 (*Tuber borchii* fruit body-1) lectin as the main soluble protein in the aqueous extract of *T. borchii.* fruit bodies. The assays also revealed that TBF-1 can interact specifically with *Tuber*-associated *Rhizobia* species. As a defensive mechanism, truffle lectins recognize bacterial polysaccharides and expel them externally (Veeraraghavan et al., 2022). A novel lectin isolated from *Terfezia claveryi* exhibited significant antibacterial, antiviral, and anticarcinogenic effects in addition to hemagglutinating activity (Mahdi et al., 2020). Another study investigating the bioactive capabilities of lectins from *Terfezia* spp. demonstrated that the purified lectins can agglutinate with both A and O blood groups (Hamid and Al-Meani, 2021).

8. Aroma and Volatiles

An array of volatile organic compounds (VOCs) contributes to the complex aromas and flavors characteristic of truffles. More than 200 volatile and nonvolatile compounds have been extracted, isolated, and identified from truffle fruit bodies (Splivallo, 2008; El Enshasy et al., 2013). The scent profiles of truffles vary in intensity and encompass a range of aromatic notes from earthy, cheesy, garlicky, and leathery to vanilla, creamy, dusty, and gasoline-like (Xiao et al., 2015; Patel et al., 2017; Mustafa et al., 2020). This aromatic range is due to the variations in volatile composition produced by different truffle varieties. For example, the aroma of white truffle (*Tuber magnatum*) is characterized by the presence of bismethylthiomethane, which exudes a sulfury, garlic-like aroma, whereas dimethylsulfide contributes a sulfurous note to the black truffle (*Tuber melanosporum*) (Fiecchi et al., 1967; Talou et al., 1989; Schmidberger and Schieberle, 2017). It is believed that truffles evolved to produce these volatiles as an ecological mechanism to attract animals and disseminate spores (Allen and Bennett, 2021). For example, Talou et al. (1990) demonstrated that dogs and pigs were attracted to dimethyl sulfide.

A volatile profiling study utilizing several *Tuber* truffles collected in Turkey identified 60 volatiles across numerous classes of aromatic compounds. Terpenes were the most abundant chemical class, including limonene and *p*-cymene, and aldehydes were the second most prevalent volatile class found in the samples (Korkmaz et al., 2023). Volatile compounds can also be used as chemical markers to monitor the storage quality and shelf life of truffles. Savini et al. (2017) identified three volatiles associated with freshly packaged *Tuber melanosporum*: 2,2-dimethyl-4-octenal, 2-methyl-3-methylidenecyclopentane-1-carbaldehyde, and octenal. Two other compounds were identified as markers for the decay of raw, refrigerated black truffles: 2-propyl-butanoate and 2-nonen-1-ol. The complex bacterial communities that colonize truffle fruiting bodies also contribute to their characteristic aroma. Splivallo et al. (2015) reported that the bacteria isolated from *Tuber borchii* fruiting bodies are able to generate sulfur -containing volatiles such as thiophene derivatives. Future investigations should explore how the microbiome surrounding truffles leads to aroma formation in truffles and other symbiotic fungi.

9. Nanotechnology

Nanotechnology can be used as a novel technology to optimize the truffle farming industry. Nanomaterials such as silver nanoparticles (AgNPs) may encapsulate beneficial microorganisms that gradually release into the mycorrhizal environment to stimulate the growth and development of the rhizosphere (Fratoddi et al., 2021). Owaid et al. (2022) mycofabricated silver nanoparticles using a desert truffle, *Picoa* sp. and evaluated their antifungal activity against pathogenic fungi. The AgNPs demonstrated significant inhibition of fungal pathogens, including *Pythium* sp. and *Aspergillus niger*. A common problem in truffle cultivation is iron deficiency in the alkaline, calcareous soils. Leonardi et al. (2020) biosynthesized ferric hydroxide nanoparticles to test their effects on the mycorrhiza formation and plant growth of *Quercus robur* seedlings inoculated with *Tuber borchii* (the bianchetto truffle). Low doses of the iron nanofertilizers increased the plant height and reduced the leaf chlorosis of *Q. robur* seedlings while also stimulating the *T. borchii* mycorrhizal colonization and increasing the ectomycorrhizal mantle thickness.

10. Antimicrobial Activity

Extracts from *Tirmania nivea* showed remarkable inhibitory activity against several bacterial species. The minimum inhibitory concentration (MIC) values ranged from 0.36 to 1.32 mg/ml (Hamza et al., 2016). Another desert truffle variety, *Tirmania pinoyi*, demonstrated significant antimicrobial activity against Gram-negative bacteria, Gram-positive bacteria, and the fungal pathogen *Candida albicans* (Neggaz and Fortas, 2013). Tejedor-Calvo et al. (2021) tested several *Tuber* and *Terfezia* species against different bacteria, which exhibited slight inhibition of microbial growth. The aqueous extract of *Terfezia magnusii* displayed an inhibition zone of 19.7 mm for *Staphylococcus aureus* and 8.1 mm for *Micrococcus luteus*. Similarly, the methanol extract of *Tuber gennadii* demonstrated an inhibition zone of 8.1 mm against *Shigella flexneri*. The data also revealed that the aqueous extract of *Tuber melanosporum* inhibited *Staphylococcus aureus* with an inhibition zone of 8.7 mm, whereas the ethyl acetate extract of *Tuber aestivum* inhibited *Listeria monocytogenes* with an inhibition zone of 8.5 mm.

11. Conclusion

Truffle species from a variety of regions and cultures represent a unique source of bioactive compounds for different industrial applications. Well known and appreciated for their value in the culinary and gastronomic markets, truffle fungi are also nutritious edible mushrooms that produce specialized metabolites such as exopolysaccharides, lectins, melanins, and mycosterols, which can support and advance human health. More studies should continue to classify the biochemical diversity within truffle fungi to better understand the metabolic pathways responsible for their bioactivity. Given the antibacterial, antifungal, and antioxidant properties of truffle species, future research may explore the promising biomedical and biotechnological applications of truffle extracts and isolates.

References

Águeda, B., Zambonelli, A. and Molina, R. (2014). Tuber 2013: scientific advances in sustainable truffle culture. Mycorrhiza, 24(Suppl 1): 1–4.

Akyüz, M. (2013). Nutritive value, flavonoid content and radical scavenging activity of the truffle (*Terfezia boudieri* Chatin). J. Soil Sci. Pl. Nutr., 13(1): 143–151.

Allen, K. and Bennett, J.W. (2021). Tour of truffles: aromas, aphrodisiacs, adaptogens, and more. Mycobiology, 49(3): 201–212.

Amicucci, A., Barbieri, E., Sparvoli, V., Gioacchini, A.M., Calcabrini, C. et al. (2018). Microbial and pigment profile of the reddish patch occurring within Tuber magnatum ascomata. Fungal biology, 122(12): 1134–1141.

Benaceur, F., Chaibi, R., Berrabah, F., Neifar, A., Leboukh, M. et al. (2020). Purification and characterization of latent polyphenol oxidase from truffles (*Terfezia arenaria*). Int. J. Biol. Macromol., 145: 885–893.

Bouatia, M., Touré, H.A., Cheikh, A., Eljaoudi, R., Rahali, Y. et al. (2018). Analysis of nutrient and antinutrient content of the truffle (*Tirmania pinoyi*) from Morocco. Int. Food Res. J., 25(1): 174–178.

Carrizo, M.E., Capaldi, S., Perduca, M., Irazoqui, F.J., Nores et al. (2005). The antineoplastic lectin of the common edible mushroom (*Agaricus bisporus*) has two binding sites, each specific for a different configuration at a single epimeric hydroxyl. J. Biol. Chem., 280(11): 10614–10623.

Çayan, F., Tel-Çayan, G., Deveci, E., Duru, M.E., Türk, M. et al. (2022). A detailed study on multifaceted bioactivities of the extracts and isolated compounds from truffle *Reddellomyces parvulosporus*. Int. J. Food Sci. Technol., 57(3): 1411–1419.

Cerigini, E., Palma, F., Barbieri, E., Buffalini, M., Stocchi, V. et al. (2008). The *Tuber borchii* fruiting body-specific protein TBF-1, a novel lectin which interacts with associated Rhizobium species. FEMS Microbiol. Lett., 284(2): 197–203.

Chen, C.C., Nargotra, P., Kuo, C.H. and Liu, Y.C. (2023). High-Molecular-Weight exopolysaccharides production from *Tuber borchii* cultivated by submerged fermentation. Int. J. Mol. Sci., 24(5): 4875.

Corrêa, R.C., Peralta, R.M., Bracht, A. and Ferreira, I.C. (2017). The emerging use of mycosterols in food industry along with the current trend of extended use of bioactive phytosterols. Tr. Food Sci. Technol., 67: 19–35.

De Angelis, F., Arcadi, A., Marinelli, F., Paci, M., Botti, D. et al. (1996). Partial structures of truffle melanins. Phytochemistry, 43(5): 1103–1106.

Donot, F., Fontana, A., Baccou, J.C. and Schorr-Galindo, S. (2012). Microbial exopolysaccharides: Main examples of synthesis, excretion, genetics and extraction. Carbohyd. Polym., 87(2): 951–962.

Eisenman, H.C. and Casadevall, A. (2012). Synthesis and assembly of fungal melanin. Appl. Microbiol. Biotechnol., 93: 931–940.

El Enshasy, H., Elsayed, E.A., Aziz, R. and Wadaan, M.A. (2013). Mushrooms and Truffles: Historical Biofactories for Complementary Medicine in Africa and in the Middle East. Evid. Based Compl. Alt. Med., 2013, 620451–10.

Fiecchi, A., Kienle, M.G., Scala, A. and Cabella, P. (1967). Bis-methylthiomethane, an odorous substance from white truffle, *Tuber magnatum* Pico. Tetrahed. Lett., 8(18): 1681–1682.

Ford, W.W. (1911). The distribution of haemolysins, agglutinins and poisons in fungi, especially the Amanitas, the Entolomas, the Lactarius and the Inocybes. J. Pharmacol. Exp. Ther., 2: 285–318.

Fratoddi, I., Bodo, E., Cerra, S., Salamone, T.A., Hajareh Haghighi, F. et al. (2021). Truffle farming and nanomaterials: a new technology for the optimization of the mycorrhization process and release of "helper" microorganisms. In: IRIS/UNIROMA Research Catalogue1/04, Publication in conference proceedings/04f Poster

Gajos, M., Ryszka, F. and Geistlinger, J. (2014). The therapeutic potential of truffle fungi: a patent survey. Acta Mycol., 49(2): 305–318.

Gurung, N., Ray, S., Bose, S. and Rai, V. (2013). A Broader View: Microbial Enzymes and Their Relevance in Industries, Medicine, and Beyond. BioMed Res. Int., 2013, 329121-18.

Hamid, L.L. and Al-Meani, S.L. (2021). Extraction and purification of a lectins from Iraqi Truffle (*Terfezia* sp.). Egypt. J. Chem., 64(6): 2983–2987.

Hamidi, M., Okoro, O.V., Milan, P.B., Khalili, M.R., Samadian, H. et al. (2022). Fungal exopolysaccharides: Properties, sources, modifications, and biomedical applications. Carbohyd. Polym., 284: 119152.

Hamza, A., Jdir, H. and Zouari, N. (2016). Nutritional, antioxidant and antibacterial properties of *Tirmania nivea*, a wild edible desert truffle from Tunisia arid zone. Med. Aromat. Pl., 5(4): DOI: 10.4172/2167-0412.1000258

Hilszczańska, D., Siebyła, M., Horak, J., Król, M., Podsadni, P. et al. (2016). Comparison of chemical composition in *Tuber aestivum* Vittad. of different geographical origin. Chem. Biodivers., 13(12): 1617–1629.

Jacobs, R. (2019). The Truffle Underground: A Tale of Mystery, Mayhem, and Manipulation in the Shadowy Market of the World›s Most Expensive Fungus. Clarkson Potter Publishers, USA, p. 288.

Kaliyaperumal, M., Kezo, K. and Gunaseelan, S. (2018). A Global Overview of Edible Mushrooms. pp. 15–56. In: Singh, B.P., Lallawmsanga, Passari, A. (eds.). *Biology of Macrofungi*. Springer International Publishing AG.

Kanmani, P. and Lim, S.T. (2013). Synthesis and structural characterization of silver nanoparticles using bacterial exopolysaccharide and its antimicrobial activity against food and multidrug resistant pathogens. Proc. Biochem., 48(7): 1099–1106.

Kivrak, Ş. and Kivrak, İ. (2018). Investigation of Chemical Composition and Nutritional Value of Truffle Mushroom (*Tuber nitidum* Vittad.). Süleyman Demirel Üniversitesi Fen Bilimleri Enstitüsü Dergisi, 22: 339–344.

Korkmaz, C., Hellal, K., Taş Küçükaydın, M., Çayan, F., Küçükaydın, S. et al. (2023). Volatile Compound Profiling of Seven *Tuber* Species Using HS-SPME-GC-MS and Classification by a Chemometric Approach. ACS Omega, 8(37): 34111–34119.

Lee, H., Nam, K., Zahra, Z. and Farooqi, M.Q.U. (2020). Potentials of truffles in nutritional and medicinal applications: A review. Fungal Biol. Biotechnol., 7: 1–17.

Leonardi, P., Lugli, F., Iotti, M., Puliga, F., Piana, F. et al. (2020). Effects of biogenerated ferric hydroxides nanoparticles on truffle mycorrhized plants. Mycorrhiza, 30(2): 211–219.

Liu, X., Zhang, Z., Xu, L., Shi, W., Sun, Q. et al. (2020). Dioscorea saponin transforms the structure of truffle exo-polysaccharide and enhances its antioxidant activity. LWT, 127: 109417. DOI: 10.1016/j.lwt.2020.109417

Mahapatra, S. and Banerjee, D. (2013). Fungal exopolysaccharide: production, composition and applications. Microbiology Insights, 6: MBI-S10957.

Mahdi, L.H., Alsaadi, L.G., Mater, H.N., Kadhem, B.M., Zwain, L.A. et al. (2020). Antibacterial, antiviral and anticarcinogenic effect of a novel lectin characterized and purified from Terfezia claveryi. Biochem. Cell. Arch., 20(0972-5075):735-741

Miranda, M., Bonfigli, A., Zarivi, O., Ragnelli, A.M., Pacioni, G. et al (1992). Truffle tyrosinase: Properties and activity. Pl. Sci., 81(2): 175–182.

Miranda, M., Zarivi, O., Bonfigli, A., Amicarelli, F., Aimola, P. et al. (1997). Melanogenesis, tyrosinase expression, and reproductive differentiation in black and white truffles (Ascomycotina). Pigment Cell Research, 10(1–2): 46–53.

Miranda, M., Zarivi, O., Bonfigli, A., Porretta, R., Aimola, P. et al. (1996). White truffles, like black ones, are tyrosinase positive. Plant Science, 120(1): 29–36.

Mustafa, A.M., Angeloni, S., Nzekoue, F.K., Abouelenein, D., Sagratini, G. et al. (2020). An overview on truffle aroma and main volatile compounds. Molecules, 25(24): 5948. DOI: 10.3390/molecules25245948

Nadim, M., Deshaware, S., Saidi, N., Abd-Elhakeem, M.A., Ojamo, H. et al. (2015). Extracellular enzymatic activity of *Tuber maculatum* and *Tuber aestivum* mycelia. Adv. Microbiol., 5(5): 523–530.

Neggaz, S. and Fortas, Z. (2013). Tests of antibiotic properties of Algerian desert truffle against bacteria and fungi. J. Life Sci., 7(3): 259–266.

Osemwegie, O.O., Adetunji, C.O., Ayeni, E.A., Adejobi, O.I., Arise, R.O. et al. (2020). Exopolysaccharides from bacteria and fungi: current status and perspectives in Africa. Heliyon, 6: e04205. DOI: 10.1016/j.heliyon.2020.e04205

Owaid, M.N., Rabeea, M.A., Aziz, A.A., Jameel, M.S., Dheyab, M.A. et al. (2022). Mycogenic fabrication of silver nanoparticles using *Picoa*, Pezizales, characterization and their antifungal activity. Environ. Nanotechnol. Monitor. Manage., 17: 100612.

Pacioni, G., Ragnelli, A.M. and Miranda, M. (1995). Truffle development and interactions with the biotic environment: Molecular aspects. Biotechnology of Ectomycorrhizae: Molecular Approaches, 213–227.

Pacioni, G., Rapino, C., Zarivi, O., Falconi, A., Leonardi, M. et al. (2015). Truffles contain endocannabinoid metabolic enzymes and anandamide. Phytochemistry, 110: 104–110.

Patel, S. (2012). Food, health and agricultural importance of truffles: A review of current scientific literature. Curr. Tr. Biotechnol. Pharm., 6(1): 15–27.

Patel, S., Rauf, A., Khan, H., Khalid, S. and Mubarak, M.S. et al. (2017). Potential health benefits of natural products derived from truffles: A review. Tr. Food Sci. Technol., 70: 1–8.

Reyna, S. and Garcia-Barreda, S. (2014). Black truffle cultivation: a global reality. Forest Systems, 23(2): 317–328.

Savini, S., Loizzo, M.R., Tundis, R., Mozzon, M., Foligni, R. et al. (2017). Fresh refrigerated *Tuber melanosporum* truffle: Effect of the storage conditions on the antioxidant profile, antioxidant activity and volatile profile. Eur. Food Res. Technol., 243: 2255–2263.

Sawaya, W.N., Al-Shalhat, A., Al-Sogair, A. and Al-Mohammad, M. (1985). Chemical composition and nutritive value of truffles of Saudi Arabia. J. Food Sci., 50(2): 450–453.

Schmidberger, P.C. and Schieberle, P. (2017). Characterization of the key aroma compounds in white Alba truffle (*Tuber magnatum* pico) and Burgundy truffle (*Tuber uncinatum*) by means of the sensomics approach. J. Agric. Food Chem., 65(42): 9287–9296.

Shavit, E. (2013). The history of desert truffle use. pp. 217–241. In: Kagan-Zur, V., Roth-Berjerano, N.k, Sirit, Y. and Morte, A. (eds.). *Desert Truffles: Phylogeny, Physiology, Distribution and Domestication*. Springer Berlin Heidelberg, Germany.

Siachoono, S., Shakachite, O., Muyenga, A. and Bwalya, J. (2015). Under ground treasure: a preliminary inquiry into the ecology and distribution of Zambian truffles. Int. J. Biol., 8(2): 1–8.

Sommer, K. and Vetter, W. (2020). Gas chromatography with mass spectrometry detection and characterization of 27 sterols in two truffle (*Tuber*) species. J. Food Comp. Anal., 94: 103650.

Splivallo, R. (2008). Biological significance of truffle secondary metabolites. pp. 141–165. In: Karlovsky, P. (ed.). *Secondary Metabolites in Soil Ecology*, Volumer 14, *Soil Biology*, Springer-Verlag Berlin Heidelberg, Germany.

Splivallo, R., Deveau, A., Valdez, N., Kirchhoff, N., Frey-Klett, P. et al. (2015). Bacteria associated with truffle-fruiting bodies contribute to truffle aroma. Environ. Microbiol., 17(8): 2647–2660.

Swan, G.A. (1974). Structure, Chemistry, and Biosynthesis of the Melanins. pp. 521–582. In: Herz, W., Grisebach, H. and Kirby, G.W. (eds.). *Progress in the Chemistry of Organic Natural Products*, Volume 31, Springer, Vienna.

Talou, T., Delmas, M. and Gaset, A. (1989). Analysis of headspace volatiles from entire black truffle (*Tuber melanosporum*). J. Sci. Food Agric., 48(1): 57–62.

Talou, T., Gaset, A., Delmas, M., Kulifaj, M., Montant, C. et al. (1990). Dimethyl sulphide: the secret for black truffle hunting by animals? Mycol. Res., 94(2): 277–278.

Tang, Y., Li, H.M. and Tang, Y.J. (2012). Comparison of sterol composition between *Tuber* fermentation mycelia and natural fruiting bodies. Food Chem., 132(3): 1207–1213.

Tejedor-Calvo, E., Amara, K., Reis, F.S., Ba-rros, L., Martins, A. et al. (2021). Chemical composition and evaluation of antioxidant, antimicrobial and antiproliferative activities of *Tuber* and *Terfezia* truffles. Food Res. Int., 140, 110071. DOI: 10.1016/j.foodres.2020.110071

Tejedor-Calvo, E., Morales, D., Marco, P., Sánchez, S., Garcia-Barreda, S. et al. (2020). Screening of bioactive compounds in truffles and evaluation of pressurized liquid extractions (PLE) to obtain fractions with biological activities. Food Res. Int., 132: 109054. DOI: 10.1016/j.foodres.2020.109054

Thomas, P., Elkhateeb, W. and Daba, G. (2021). Industrial applications of truffles and truffle-like fungi. pp. 82–88. In: Sridhar, K.R. and Deshmukh, S.K. (eds.). *Advances in Macrofungi - Diversity, Ecology and Biotechnology*. CRC Press, Boca Raton, USA.

Varrot, A., Basheer, S.M. and Imberty, A. (2013). Fungal lectins: structure, function and potential applications. Curr. Opin. Str. Biol., 23(5): 678–685.

Veeraraghavan, V.P., Hussain, S., Balakrishna, J.P., Dhawale, L., Kullappan, M. et al. (2022). A Comprehensive and Critical Review on Ethnopharmacological Importance of Desert Truffles: *Terfezia claveryi*, *Terfezia boudieri*, and *Tirmania nivea*. Food Rev. Int., 38(S1): 846–865.

Velíšek, J. and Cejpek, K. (2011). Pigments of higher fungi—a review. Cz. J. Food Sci., 29(2): 87–102.

Wahiba, B., Wafaà, T., Asmaà, K., Bouziane, A. and Mohammed, B. et al. (2016). Nutritional and antioxidant profile of red truffles (*Terfezia claveryi*) and white truffle (*Tirmania nivea*) from southwestern of Algeria. Der Pharmacia Letter, 8: 134–141.

Wang, S. and Marcone, M.F. (2011). The biochemistry and biological properties of the world›s most expensive underground edible mushroom: Truffles. Food Res. Int., 44(9): 2567–2581.

Weijn, A., Bastiaan-Net, S., Wichers, H.J. and Mes, J.J. (2013). Melanin biosynthesis pathway in *Agaricus bisporus* mushrooms. Fungal Gen. Biol., 55: 42–53.

Wu, Z., Meenu, M. and Xu, B. (2021). Nutritional value and antioxidant activity of Chinese black truffle (*Tuber indicum*) grown in different geographical regions in China. LWT, 135: 110226. DOI. org/10.1016/j.lwt.2020.110226

Xiao, D.R., Liu, R.S., He, L., Li, H.M., Tang, Y.L. et al. (2015). Aroma improvement by repeated freeze-thaw treatment during *Tuber melanosporum* fermentation. Sci. Rep., 5(1): 17120. DOI. org/10.1038/srep17120

Yan, X., Wang, Y., Sang, X. and Fan, L. (2017). Nutritional value, chemical composition and antioxidant activity of three *Tuber* species from China. AMB Express, 7(1): 1–8.

Zhang, Z., Wang, J., Liu, L., Shun, Q., Shi, W. et al. (2019). The optimum conditions and mechanism for increasing exo-polysaccharide production of *Truffles melanosporum* by Dioscorea saponins. LWT, 107: 331–339.

Chapter 17

Industrial Perspectives on Truffles and Their Products

Nirali Dedhia

1. Introduction

From the time of their discovery and beyond, the use of truffles has always been aimed at elevating the culinary appeal of the food. A savory flavor profile rich in earthy and musty notes is contributed by truffles of various species that are found and cultivated across different continents. This has, by and large, been the industrial relevance of truffles. The premium character of these fungi is reflected in the various controversies and tiffs with respect to which variety is the most exquisite (Rittersma, 2012). The *Tuber magnatum* Pico or white Alba truffle from Italy fetches the highest price for their gastronomic contributions (Čejka et al., 2023). The *Tuber* species is dominant in Northern Europe and the Himalayan regions of Asia, while the *Terefezia* and *Tirmania* species dominate the arid as well as semi-arid Mediterranean region (Kaliyaperumal et al., 2018).

Truffle cells, like any other microbial, plant, or animal cell, are well differentiated and comprise various cell organelles. These are made up of micro- and macromolecules that have some form of significance to humans. This significance in terms of therapeutic potential, culinary potential, or processing aid can be fathomed and explored. Truffle constituents, including their celebrated aromas and volatiles, polysaccharides, enzymes, lectins, pigments, and bioactive metabolites, are widely explored in selected truffle species. The therapeutic potential of truffles, which are low in aroma and volatiles, has been widely explored relatively and has demonstrated promising results. An understanding of the different molecules synthesized in the life cycle of truffles helps us understand the value of truffles beyond their aroma. The

Department of Food Engineering and Technology, Institute of Chemical Technology, Mumbai, India.
Email: nj.dedhia@ictmumbai.edu.in

nutritional, nutraceutical, and biocatalytic potential of truffles and their constituents is a growing area of research. Substantial research in these aspects has been conducted by researchers across the globe. An understanding of these is essential to celebrate truffles from a holistic point of view. This will help us explore the untapped potential of the ectomycorrhizal species that belong to this group. Figure 1 gives a holistic view of the different uses, research, and commercials related to the cultivation of truffles all across the globe. If we consider a circular economy approach, then each component extracted judiciously from the truffles can have applications and improve the economic gains with minimal wastage.

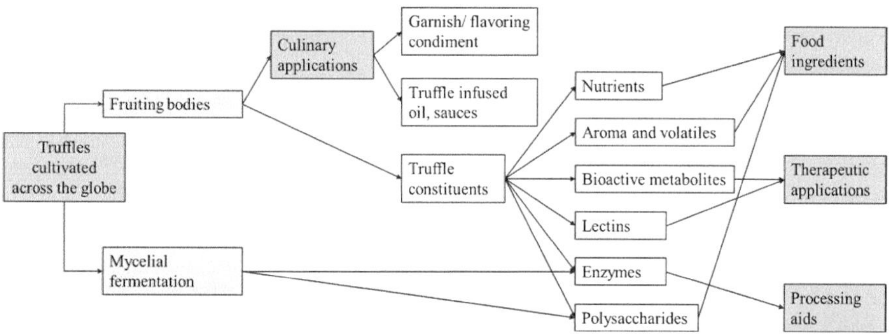

Fig. 1. A gamut of applications of truffles and their constituents towards the enhancement of human life.

2. Truffle Industry

The truffle industry is dominated by the Tuber species, which are produced and sold for their aroma profile. At the start of this decade, *Tuber magnatum* fetched a price of EUR 1000–1500/kg due to its peculiar and attractive aroma, while *Tuber aestivum* fetched a price ranging from EUR 25–70/kg in Spain to EUR 200–600/kg in Switzerland (Oliach et al., 2021). In European countries such as France, Italy, Spain, and Australia, truffles are an emerging 'global industry' due to their increased production (Čejka et al., 2022). The economic activity of the truffle sector generates tens of millions of euros per year (Masiero et al., 2016). So much so, that truffle cultivation is now being experimented with in the arid regions of Saudi Arabia as an agri-business opportunity (Bajaj et al., 2021). A study reports the different marketing strategies that can be adopted to improve the sale of a luxury food item such as truffles (Milanesi et al., 2020). These symbolize the premium place that truffles hold in the food industry, despite their lower scale of production per year.

Apart from the direct benefits of the sale of intentionally or accidentally produced truffles of different varieties, there are other revenue-generating avenues. These include the sale of truffle-associated products such as truffle oils and butters (Tejedor-Calvo, García-Barreda, et al., 2023), truffle tourism, and an increase in land value in rural areas (Samils et al., 2008). There is an auxillary branch that works towards research on different aspects of truffles, such as their improved cultivation (Čejka et al., 2020) as well as their nutraceutical potential (Marathe et al., 2020; Shah et al., 2020).

Truffles are fresh produce and highly perishable, with a short postharvest shelf life. Additionally, truffles are harvested at the point of full maturity, and senescence occurs rapidly after harvest (Murcia et al., 2014). Enhancing the shelf-life of these high-priced agricultural commodities using low temperature storage, drying, or modified-atmospheric storage, along with modern preservation techniques, will be helpful. Hence, this aspect has been widely researched and adopted by farmers and the food processing industry. This also creates a processing industry catering to the shelf life enhancement of truffles. Different strategies have been developed and discussed here. Surface decontamination can reduce the initial microbial load and enhance the shelf life when combined with a suitable packaging and storage technique (Susana Rivera et al., 2011). Vacuum-, low-oxygen-, and nitrogen-packaged truffles under refrigerated conditions have been reported to have a prolonged shelf-life of 2 weeks, mainly due to a delay in the biochemical activities (Savini et al., 2017). Edible and biodegradable packaging has been explored and reported to enhance shelf life by almost 21 days with few changes in texture and volatile profile. This extension may be attributed to changes in the atmosphere and respiration rates (Tejedor-Calvo et al., 2024). Similar results have been reported for modified atmospheric storage (Rivera et al., 2010). Of the different drying techniques, freeze-drying has been reported to be the best among those studied with respect to the retention of the overall original aroma of fresh truffle (Campo et al., 2017). The higher cost of the freeze-drying process may be justified by the cost of the raw material, depending on the economies of scale. Freezing freshly harvested and cleaned truffles may seem like an easy alternative to extend shelf life, but it has been reported to bring about changes in the aroma profile of black truffles (Culleré et al., 2013). Truffle processing, sauce, and preservation by autoclaving are practices in the industry. The addition of nutraceuticals such as linolenic acid has been reported to enhance the shelf-life of truffle sauce (Tirillini et al., 2019).

3. Truffle Constituents

To obtain an understanding of the industrial significance of truffles, it is important to appreciate their constituents, such as: Truffle nutrients; Bioactive metabolites; Polysaccharides (considered separately from nutrients due to the literature focus); Pigments; Lectins (considered separately from nutrient proteins due to their therapeutic effect); Enzymes; Aromas and volatiles

3.1 Truffle Nutrients

The nutritional potential of truffles is related to the constituent proteins, lipids, carbohydrates, and minerals in different truffle species. The *Terfezia, Tirmania*, and *Tuber* species have been explored for their nutritional content. An understanding of the elaborate nutritional profile is important as it increases the industrial applications of truffles as a whole or their constituents as food ingredients. Table 1 summarizes the major nutrient content reported for truffles.

The presence of minerals such as calcium, magnesium, sodium, potassium, phosphorus, iron, manganese, copper, zinc, and selenium, among others, has also

Table 1. Nutritional value of selected truffles (dry mass basis; -, data not reported).

Proximal value (g/100 g)	*Picoa juniperi*	*Terfezia olbiensis*	*Terfezia claveryi*	*Tirmania nivea*	*Tuber aestivum*	*Tuber brumale*	*Tuber indicum*	*Tuber magnatum*	*Tuber melanosporum*
Moisture	76.93	79.98±1.41	83.14±1.09	68±02	-	-	-	-	-
Protein	4.69	35.87±1.02	32.27±0.99	11.97±0.03	19.9±0.4	22.12±0.04	16.7±0.1	24.15±0.06	22.43±0.06
Lipid	3.40	3.17±0.09	2.78±0.24	2.15±0.093	0.99±0.04	2.28±0.02	2.6±0.2	1.52±0.06	2.3±0.3
Carbohydrates	-	48.42±0.99	45.91±0.67	-	79.1±0.2	75.60±0.05	80.7±0.2	74.32±0.09	75.3±0.2
Sugars	2.25	8.23±1.01	6.57±0.87	5.01±0.09	2.81±0.08	2.8±0.1	2.1 ± 0.1	0.23±0.01	1.44±0.05
Ash	1.50	15.29±0.46	14.22±0.3	5.5 ±0.65	0.0074±0.0001	0.010±0.001	0.0075±0.0001	0.0112±0.0005	0.00793±0.00003
Origin	Turkey	Turkey	Turkey	Algeria	Spain	Spain	China	Italy	Spain
Reference	Akyüz and Kirbag (2018)	Kıvrak (2015)	Kıvrak (2015)	Hamza et al. (2016); Wahiba et al. (2016)	Tejedor-Calvo et al. (2021)	Tejedor-Calvo et al. (2021)	Tejedor-Calvo et al. (2021)	Tejedor-Calvo et al. (2021)	Tejedor-Calvo et al. (2021)

been identified and reported for selected truffles (Akyüz and Kirbag, 2018; Hamza et al., 2016). These may be attributed to the mineral bioaccumulation property of truffles (Shah et al., 2021). Reports on other micronutrients such as vitamin C, tocopherols, and vitamin A also add to the strength of the nutrient potential of truffles. *Tuber, Terfezia*, and *Picoa* species have been reported to be rich in vitamins A, E, and C, just like fruits and vegetables. This will help add to their antioxidant properties. Truffles, on a dry-weight basis, show good nutritional potential (Akyüz and Kirbag, 2018; Tejedor-Calvo et al., 2021). Hence, truffles, after their flavor extraction, can serve as a nutrition source. Studies have shown the occurrence of polyunsaturated fatty acids, low sugar content, and reasonable protein content in truffles, making them nutritious food sources.

3.2 Bioactive Metabolites

The inclination of consumers towards natural products in preventive and therapeutic approaches to disease management has led to a dynamic shift in the never-ending search for newer sources of bioactive metabolites. These can be extracted from natural sources or cultivated by fermentative techniques for better commercial gains. Consequently, truffles have also been explored for the presence and therapeutic benefits of their bioactive metabolites. The major bioactive metabolites associated with truffles are their phenolic and flavonoid compounds. Many researchers have reported these bioactive metabolites for different truffle species. There is much data correlating the truffle phenolics and flavonoids, along with ascorbic acid and carotenoids, to the demonstrated biopotential from *in vitro*, cell lines, and *in vivo* assays. This biopotential has demonstrated antioxidant, antiinflammatory (Marathe et al., 2020), antidiabetic (Zhang et al., 2020), anticancer (Dahham et al., 2018), and antimicrobial effects (Khaled et al., 2014; Sallam et al., 2022) effects. Truffle extracts using hot water or organic solvents such as ethanol, methanol, petroleum ether, hexane, and chloroform have been studied. With the change in solvent of extraction, variation in biopotential properties has been starkly observed. This can be attributed to the different bioactives extracted by the solvents based on their structure and, hence, their preferential solubility. Many compounds have been identified qualitatively as well as quantitatively and reported by researchers. Truffle mycelium has been cultivated via fermentative techniques and explored for its bioactive content and potential. *T. borchii*, *T. maculatum* (Shah et al., 2021), and *T. magnatum* (Hsu et al., 2022) mycelial extracts have been quantified for their phenolic and flavonoid content and have demonstrated *in vitro* antioxidant capacities.

Water extracts of *T. aestivum* have been found to contain cinnamic acid, while methanol extracts have been found to contain hydroxycinnamic acid, 3,4-dimethoxycinnamic acid, and kaempferol, which are responsible for the antioxidant activity (Shah et al., 2020). Gallic acid, amentoflavone, quercetin, quercetin-3-*o*-glucoside, hyperoside, rutin, epicatechin, and catechin have been found in water and/or methanol extracts of T. magnatum (Beara et al., 2014). Chlorogenic acid, quinic acid, trans-ferulic acid, quercetin-3-*o*-rhamonoside, naringenin, and apigenin are found in the water extracts of *Terfezia bouideri* (Nouiri et al., 2021). Ethanolic extracts of *Terfezia claveryi* have been reported for bioactive metabolites such as hydroxybenzoic acid, vanillin, vanillic acid, trans-cinnamic acid, syringic acid, and rutin, among others (Al-Atassi et al., 2022). The biopotential is evident through quenching of free radicals, increasing activity or inhibition of enzymes in cell lines, and notable changes in inflammation in animal models (Nouiri et al., 2021). The chemical structure of the phenols and flavonoids rich in hydroxyl groups, ring structures, and/or organic acid groups is responsible for the demonstrated activities. Figure 2 gives a schematic chemical reaction between the DPPH (2,2-Diphenyl-1-picrylhydrazyl) free radicals and chlorogenic acid from truffles for the radical scavenging effect of the latter. This is an *in vitro* assay technique to estimate the antioxidant properties of a single or group of compounds.

While performing the various assays to affirm the biopotential, researchers have generally compared it with standard and popular pharmaceutical drugs or antioxidants in most cases. This indicates that, apart from their aroma appeal, even if the appeal is lower, truffles present a good opportunity to be used for nutraceutical properties (Al-Laith, 2010). They also have the potential to be natural antioxidants that can be used by the food and cosmetic industries. These avenues can be commercially worked up to increase the economic gains from truffle cultivation in even the remotest of geographical locations, where other industries are difficult to function. The antioxidant capacities of extracts comparable to oil-soluble antioxidants (Shah et al., 2020) can be useful in improving the shelf-life or preservation of foods to replace synthetic antioxidants currently used. Truffle-infused oils may have an enhanced shelf life. Thus, apart from the taste profile of truffle-infused oil, it may also have a preservation angle.

The biopotential expands the arena of use of truffles as nutraceuticals for preventive and therapeutic purposes. It also expands applications as a natural food and/or cosmetic additive because of its antioxidant effect. The bioactive metabolites

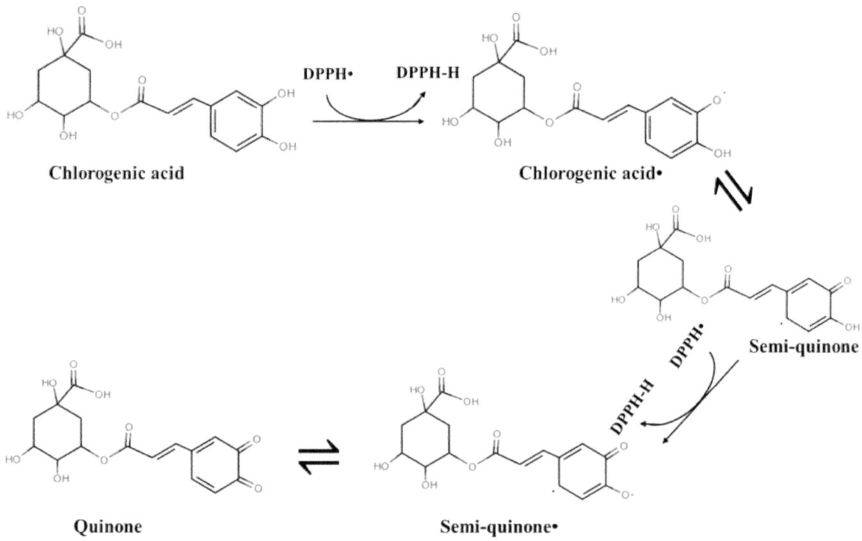

Fig. 2. Demonstration of *in vitro* antioxidant property by chemical reaction between the DPPH (2,2-Diphenyl-1-picrylhydrazyl) free radicals and chlorogenic acid from truffles.

from truffle mycelium increase the scope for commercial mycelium fermentation and bioactive extraction for nutraceutical, food, and/or cosmetic applications.

3.3 Polysaccharides

Polysaccharides are ubiquitously present in nature and have a wide variety of applications. These applications are predominantly in the food, pharmaceutical, cosmetics, paper, textile, and petroleum industries, to name a few. Polysaccharides are generally responsible for making up the cell-wall structure. Upon extraction, they have a similar role in providing structural integrity (such as water holding, viscosity, gelling, film formation, and bio-adhesiveness) to the matrix in which they are added. Apart from these benefits, polysaccharides have also been widely reported for therapeutic potential due to their antidiabetic, anticancer, antioxidant, antitumor, and antimicrobial activities (Dedhia et al., 2022). These are mainly the motivating reasons for the exploration of polysaccharides from various natural sources, including truffles. Carbohydrates make up the major portion of the dry matter of truffles, as evident from Table 1. An understanding of polysaccharides becomes important from a nutritional and functional ingredient point of view. Truffle polysaccharide extraction and structure elucidation have been explored by researchers. Polysaccharide extraction has been reported in an aqueous medium, and hence only water-soluble truffle polysaccharide fractions are evaluated (Miao et al., 2011).

Truffle polysaccharides have been widely studied for their antioxidant, antiinflammatory properties. They have also been evaluated for thickening properties in food systems. The polysaccharides from *Tuber aestivum* and *T. melanosporum* have demonstrated antioxidant properties by free radical scavenging as well as increasing the secretion level of intracellular antioxidant enzymes such as superoxide dismutase and catalase in IPEC-J2 cells. These polysaccharides were mostly made up of mannose and glucose in different ratios (Guan et al., 2022). The monosaccharides formed from the polysaccharides of *T. aestivum* have glucose (>90%), rhamnose, and mannose sugars (Bhotmange et al., 2017). *Tuber rufum* polysaccharides were found to be made up of D-glucose, D-galactose, and L-fucose. The structure has a backbone of two (1 → 6)-α-d-galactopyranosyl, one (1 → 4)-α-d-glucopyranosyl, two (1 → 6)-β-d-glucopyranosyl, and one (1 → 4)-β-d-glucopyranosyl residues, out of which (1 → 4)-α-d-glucopyranosyl residue was branched at the O-2 position with terminal α-l-fucopyranosyl residue and at the O-6 position with terminal α-d-galactopyranosyl residue. It has been reported to protect cells from damage due to oxidation (Pattanayak et al., 2017). The water-soluble polysaccharides from the Chinese truffle *Tuber huidongense* have two fractions. One of the fractions is composed of glucose, mannose, and galactose units, while the other is composed of glucose units. Both these fractions have demonstrated in vitro antioxidant properties (Chen et al., 2016). Water-soluble extracts of Chinese truffles have demonstrated properties of increased viscosity and reduction of syneresis in frozen yogurt (Miao et al., 2011). The desert truffle, *Terfezia claveryi* polysaccharides, have been found to demonstrate in vitro anti-carcinogenic activity (Bawadekji, 2018).

Truffle mycelial production by fermentation has been performed to evaluate the polysaccharide content as well as optimize conditions for polysaccharide production. A study reports improved polysaccharide production by fermentation compared to those in the fruiting bodies (Zhao et al., 2014). It has also been noted as an efficient technique for tuber polysaccharide production (Sun et al., 2020; Tang et al., 2008). The submerged fermentation technique can be scaled-up, and higher production rates of specific truffle polysaccharides can be achieved. Truffle polysaccharides with confirmed therapeutic or functional properties can hence be synthesized on an industrial scale.

3.4 Pigments

Pigments are the natural colouring materials that are abundantly present in the cells of microorganisms as well as higher plants and animals. These are widely explored as a natural substitute for synthetic colours used in food, pharmaceuticals, and cosmetic products (Jurić et al., 2022). Pigments present in the truffles are either their own or from the microorganisms thriving on them. Red-coloured pigments of carotenoid compounds have been reported in *Tuber magnatum* ascomata (Amicucci et al., 2018). Pigment changes leading to changes in colour of ascocarps due to maturation were corroborated with changes in the total phenolics and flavonoid content in *Tuber aestivum* (Shah et al., 2020). Pigments in truffles due to the polymerization reaction of *o*-quinones oxidation products of *o*-diphenols (Gouzi et al., 2013) have been less explored to date. It remains an area of indepth investigation. Fermentative pigment production from truffle mycelia is also not reported to the best of our literature search. This can be an area of research to evaluate the industrial significance of truffles from yet another facet.

3.5 Enzymes

Enzymes are considered to be key biocatalysts for various synthesis processes, even on a commercial scale. Enzymes from a few varieties of truffles have been extracted or produced via fermentative techniques and evaluated for their reactivity. Tyrosinase enzyme occurrence and production have been reported in most Tuber species. It has been attributed to the activity of pigment and structured wall formation (Zarivi et al., 2011). *Tuber melanosporum* has been evaluated for the endocannabinoid enzyme system due to the presence of dark pigments (Pacioni et al., 2015). Lipoxygenase, phosphatase, and esterase from *Terfezia claveryi* (Navarro-Ródenas et al., 2009; Pérez-Gilabert, Morte et al., 2005; Pérez-Gilabert, Sánchez-Felipe, et al., 2005) and polyphenol oxidase from *Terefezia arenaria* (Benaceur et al., 2020) have been found and characterized. The detection of different enzymes paves the way for mycelial production to increase production and demonstrate further applications. Mycelial production of amylase, xylanase, laccase, lipase, peroxidase, cellulase, and catalase (Bedade et al., 2017a, b, 2018) has been reported for *Tuber maculatum*, whereas amylase, peroxidase, and catalase have been reported for *Tuber aestivum* (Nadim et al., 2015). Further studies on the purification of these enzymes will explore their potential for commercial applications. These enzymes find applications as processing aids in the food, textile, leather, pharmaceutical, and cosmetic industries. They function during processes to reduce viscosity, increase extraction yields, and provide specific textures (Zhang et al., 2018).

3.6 Lectins

Lectins are versatile proteins of non-immune origin that bind reversibly and specifically to carbohydrates. Fungi have been worked upon for the past decade for their lectins (Varrot et al., 2013). Novel sources of lectins are continually explored for their unique specificity and potential for biomedical and biotechnological applications. These can find applications for human, animal, and plant health. Lectins extracted and purified from *Terfezia claveryi* have been reported for potential applications in antiviral, antibacterial, and anticarcinogenic effects (Mahdi et al., 2020). *Tuber borchii* lectins have also been reported for the same effects (Cerigini et al., 2008).

3.7 Aroma and Volatiles

Truffle aroma and volatile compounds are of prime importance for their detection (by pigs and dogs), identification, and classification. The volatile organic compounds in truffles are sensed by trained pigs and dogs. Subsequently they are excavated from the soil (Mustafa et al., 2020). The truffle aroma and volatiles are made up of alcohols, aldehydes, esters, acids, and sulfur-containing compounds, among others. These compounds are associated with greasy, creamy, earthy, woody, and mushroomy notes that are characteristics of truffles of *Terfezia*, *Tirmania*, and *Tuber* species. The volatile profile varies with species as well as with the maturity and geographical location of cultivation of the same species (Niimi et al., 2021; Shah et al., 2020). Along with its indigenous biochemistry, the bacterial community associated with truffles has been reported to be involved in the synthesis of sulfur-containing volatiles from precursors synthesized by truffles (Splivallo et al., 2015). Table 2 lists the common aroma and volatile compounds reported in different truffle species.

Changes in the composition of these volatile compounds have been noted during storage (Choo et al., 2021). Hence, appropriate preservation techniques are significant for reducing losses and improving the life of truffles during storage and transportation. Truffle aromas and volatiles can be preserved by extracting or isolating them from the perishable truffle matrix. Their extraction with the help of supercritical fluid technology (SCF), which uses carbon dioxide in its supercritical state for the extraction of non-polar compounds, has been attempted and has shown optimistic results with respect to the quality of the product obtained (Tejedor-Calvo et al., 2021). Alternately, the truffle oils can be SCF extracted and entrapped in other oils or appropriate food matrices for various functional benefits (Tejedor-Calvo et al., 2023). The remainder of the material after extraction can be used for its nutrient purpose or biopotential. Such techniques to extract aroma from low-value truffles can have potential applications in the flavouring of foods by natural means. This may have a nutrient purpose or biopotential. Such techniques to extract aroma from low-

Table 2. Aromatic and volatile compounds reported in different truffle species (-, not reported).

Aroma and volatile compounds	*Terfezia bouderi*	*Tirmania nivea*	*Tuber aestivum*	*Tuber indicum*	*Tuber magnatum*	*Tuber melanosporum*
Sulfur compounds	**Dimethyl sulfide**	**Dimethyl sulfide**	**Dimethyl sulfide**	**Dimethyl sulfide**	**Dimethyl sulfide**	**Dimethyl sulfide**
	-	-	-	-	Bis(methylthio) methane (=2,4-Dithiapentane)	-
	-	-	-	-	Methyl (methylthio) methyl disulfide	-
	-	-	Dimethyl disulfide	-	Dimethyl disulfide	Dimethyl disulfide
	-	-	Dimethyl trisulfide	-	Dimethyl trisulfide	Dimethyl trisulfide
	-	-	Sulfinylbismethane	-	-	Sulfinylbismethane
	-	-	-	-	Methanethiol	Methanethiol
	-	-	Methylsulfanylmethane	Methylsulfanylmethane	Methylsulfanylmethane	Methylsulfanylmethane
Alcohols	**1-Octen-3-ol**	**1-Octen-3-ol**	**1-Octen-3-ol**	**1-Octen-3-ol**	**1-Octen-3-ol**	**1-Octen-3-ol**
	3-Octanol	-	3-Octanol	3-Octanol	-	3-Octanol
	1-Hexanol	1-Hexanol	1-Hexanol	-	-	1-Hexanol
Esters	**Ethyl acetate**	**Ethyl acetate**	**Ethyl acetate**	**Ethyl acetate**	**Ethyl acetate**	**Ethyl acetate**
	-	-	-	-	-	Butyl-2-methylbutanoate
	-	-	Butan-2-yl formate	Butan-2-yl formate	-	Butan-2-yl formate
Aldehydes	2-Octenal	2-Octenal	2-Octenal	2-Octenal	2-Octenal	2-Octenal
		-	Acetaldehyde	-	Acetaldehyde	Acetaldehyde
	Nonanal	-	Nonanal	-	Nonanal	Nonanal
	Hexanal	Hexanal	Hexanal	-	Hexanal	Hexanal
	-	-	3-Methyl-butanal	3-Methyl-butanal	3-Methyl-butanal	3-Methyl-butanal
Ketones	**2-Butanone**	**2-Butanone**	**2-Butanone**	**2-Butanone**	**2-Butanone**	**2-Butanone**
	3-Octanone	-	3-Octanone	3-Octanone	3-Octanone	3-Octanone
	-	-	2-Propanone	-	-	2-Propanone
	3-Octen-2-one	3-Octen-2-one	-	-	3-Octen-2-one	3-Octen-2-one
Aromatic compounds	**Benzeneacetaldehyde**	**Benzeneacetaldehyde**	**Benzeneacetaldehyde**	**Benzeneacetaldehyde**	**Benzeneacetaldehyde**	**Benzeneacetaldehyde**
	-	-	1-Methoxy-3-methylbenzene	1-Methoxy-3-methylbenzene	1-Methoxy-3-methylbenzene	1-Methoxy-3-methylbenzene
	-	Benzaldehyde	Benzaldehyde		Benzaldehyde	Benzaldehyde
	-	-	1,4-Dimethoxybenzene	1,4-Dimethoxybenzene	1,4-Dimethoxybenzene	1,4-Dimethoxybenzene
Reference	Farag et al. (2021); Kamle et al. (2017)	Farag et al. (2021); Kamle et al. (2017)	Mustafa et al. (2020); Shah et al. (2020)	Mustafa et al. (2020); Shah et al. (2020)	Mustafa et al. (2020); Shah et al. (2020)	Mustafa et al. (2020); Shah et al. (2020)

value truffles can have potential applications in the flavouring of foods by natural means. This may be a competitive substitute for the synthetic truffle flavours and, at the same time, valorize the low-value truffles. Studies have been conducted to encapsulate the truffle volatiles to be used as flavouring agents in foods (Phong et al., 2022). The microencapsulation technology helps to envelope the volatile oils in a protective edible shell that will release the aroma upon ingestion of the food containing the microcapsules (Dedhia and Shah, 2023). The same can be explored in depth to preserve, stabilize and specifically release the truffle aroma and volatile compounds in foods. Apart from the techniques of extraction and encapsulation of the volatiles, an understanding of the biosynthesis of truffle aromas (Splivallo and Culleré, 2016) adds to another dimension of commercial-scale production and subsequent sale.

4. Conclusion

Truffles, depending on their aroma profile, are classified as high-value or low-value. Studies conducted over decades across different regions of the globe indicate the industrial significance of truffles beyond their culinary appeal. Thus, truffles as such or in dried forms can be consumed for nutritional purposes due to their favourable amino acid, polysaccharide, and lipid profiles. Their extracts containing bioactive metabolites have been widely reported for nutraceutical applications. Truffle polysaccharides have also been widely reported for their nutraceutical applications and as thickening agents in food products. Polysaccharides, enzymes, lectins, and pigments from the fruiting bodies or through mycelial production show promising results for various industrial applications. Their applications mainly lie in the domain of manufacturing aids, food ingredients, or nutraceuticals. Truffle aroma and volatiles can be extracted to improve their shelf-life and then incorporate the concentrated flavours into foods. Alternatively, the biosynthesis of truffle flavours can help production at larger scales. Most of the low-value truffles and desert truffles have been explored for opportunities apart from their culinary appeal. Studies on the extraction and formulation of truffle flavours are an area of research that needs further development. This will help valorize the low-value truffles. Mycelial fermentation is a promising pathway for the production of significant quantities of desired truffle constituents. This technique needs to be well explored for the production of lectins, bioactive metabolites, enzymes, and polysaccharides. Maximizing the utilization of every constituent of truffles will pave the way for a circular economy, reduce losses and wastage, and also help valorize certain lesser-explored truffles.

References

Akyüz, M. and Kirbag, S. (2018). Nutritive value of desert truffles species of genera *Terfezia* and *Picoa* (Ascomycetes) from arid and semiarid regions of Eastern Turkey. Int. J. Med. Mushrooms, 20(11): 1097–1106. https://doi.org/10.1615/IntJMedMushrooms.2018028796

Al-Atassi, A.Z., Al-Khatib, R.A. and Othman, M.A. (2022). Comparison of phenolic contents and antioxidant activities for black and white desert truffles spread in Syria. Iraqi J. Pharmaceut. Sci., 31(2): 184–192. https://doi.org/10.31351/vol31iss2pp184-192

Al-Laith, A.A.A. (2010). Antioxidant components and antioxidant/antiradical activities of desert truffle (*Tirmania nivea*) from various Middle Eastern origins. J. Food Comp. Anal., 23(1): 15–22. https://doi.org/10.1016/j.jfca.2009.07.005

Amicucci, A., Barbieri, E., Sparvoli, V., Gioacchini, A.M., Calcabrini, C. et al. (2018). Microbial and pigment profile of the reddish patch occurring within Tuber magnatum ascomata. Fungal Biol., 122(12): 1134–1141. https://doi.org/10.1016/j.funbio.2018.07.007

Bajaj, S.R., Marathe, S.J., Grebenc, T., Zambonelli, A., Shamekh, S. et al. (2021). First report of European truffle ectomycorrhiza in the semi-arid climate of Saudi Arabia. 3 Biotech, 11(1): 24. https://doi.org/10.1007/s13205-020-02559-w

Bawadekji, A. (2018). Evaluation of some *in vitro* anti-carcinogenic activities of polysaccharides extracted from ascomata of the desert Truffle *Terfezia claveryi* Chatin. J. Appl. Environ. Biol. Sci., 8(2): 152–159.

Beara, I. N., Lesjak, M.M., Cetojević-Simin, D.D., Marjanović, Z.S., Ristić, J.D. et al. (2014). Phenolic profile, antioxidant, anti-inflammatory and cytotoxic activities of black (*Tuber aestivum* Vittad.) and white (*Tuber magnatum* Pico) truffles. Food Chem., 165: 460–466. https://doi.org/10.1016/j.foodchem.2014.05.116

Bedade, D., Berezina, O., Singhal, R., Deska, J., Shamekh, S. et al. (2017a). Extracellular xylanase production from a new xylanase producer *Tuber maculatum* mycelium under submerged fermentation and its characterization. Biocat. Agric. Biotechnol., 11: 288–293. https://doi.org/10.1016/j.bcab.2017.07.008

Bedade, D., Deska, J., Bankar, S., Bejar, S., Singhal, R. et al. (2018). Fermentative production of extracellular amylase from novel amylase producer, *Tuber maculatum* mycelium, and its characterization. Prep. Biochem. Biotechnol., 48(6): 549–555. https://doi.org/10.1080/10826068.2018.1476876

Bedade, D.K., Singhal, R.S., Turunen, O., Deska, J., Shamekh, S. et al. (2017b). Biochemical characterization of extracellular cellulase from *Tuber maculatum* mycelium produced under submerged fermentation. Appl. Biochem. Biotechnol., 181(2): 772–783. https://doi.org/10.1007/s12010-016-2248-8

Benaceur, F., Chaibi, R., Berrabah, F., Neifar, A., Leboukh, M. et al. (2020). Purification and characterization of latent polyphenol oxidase from truffles (*Terfezia arenaria*). Int. J. Biol. Macromol., 145: 885–893. https://doi.org/10.1016/j.ijbiomac.2019.09.126

Bhotmange, D.U., Wallenius, J.H., Singhal, R.S. and Shamekh, S.S. (2017). Enzymatic extraction and characterization of polysaccharide from *Tuber aestivum*. Bioactive Carbohydrates and Dietary Fibre, 10: 1–9. https://doi.org/10.1016/j.bcdf.2017.02.001

Campo, E., Marco, P., Oria, R., Blanco, D., Venturini, M.E. et al. (2017). What is the best method for preserving the genuine black truffle *(Tuber melanosporum)* aroma? An olfactometric and sensory approach. LWT, 80: 84–91. https://doi.org/10.1016/j.lwt.2017.02.009

Čejka, T., Isaac, E.L., Oliach, D., Martínez-Peña, F., Egli, S. et al. (2022). Risk and reward of the global truffle sector under predicted climate change. Environ. Res. Lett., 17(2): 024001. https://doi.org/10.1088/1748-9326/ac47c4

Čejka, T., Trnka, M. and Büntgen, U. (2023). Sustainable cultivation of the white truffle (*Tuber magnatum*) requires ecological understanding. Mycorrhiza, 33(5): 291–302. https://doi.org/10.1007/s00572-023-01120-w

Čejka, T., Trnka, M., Krusic, P. J., Stobbe, U., Oliach, D. et al. (2020). Predicted climate change will increase the truffle cultivation potential in central Europe. Sci. Rep., 10(1): Article 1. https://doi.org/10.1038/s41598-020-76177-0

Cerigini, E., Palma, F., Barbieri, E., Buffalini, M., Stocchi, V. et al. (2008). The Tuber borchii fruiting body-specific protein TBF-1, a novel lectin which interacts with associated *Rhizobium* species. FEMS Microbiol. Lett., 284(2): 197–203. https://doi.org/10.1111/j.1574-6968.2008.01197.x

Chen, G., Zhang, S., Ran, C., Wang, L. and Kan, J. (2016). Extraction, characterization and antioxidant activity of water-soluble polysaccharides from *Tuber huidongense*. Int. J. Biol. Macromol., 91: 431–442. https://doi.org/10.1016/j.ijbiomac.2016.05.108

Choo, K., Bollen, M., Dykes, G. and Coorey, R. (2021). Aroma-volatile profile and its changes in Australian grown black Périgord truffle (*Tuber melanosporum*) during storage. Int. J. Food Sci. Technol., 56(11): 5762–5776. https://doi.org/10.1111/ijfs.15171

Culleré, L., Ferreira, V., Venturini, M.E., Marco, P., Blanco, D. et al. (2013). Chemical and sensory effects of the freezing process on the aroma profile of black truffles (*Tuber melanosporum*). Food Chem., 136(2): 518–525. https://doi.org/10.1016/j.foodchem.2012.08.030

Dahham, S.S., Al-Rawi, S.S., Ibrahim, A.H., Abdul Majid, A.S., Abdul Majid, A.M.S. et al. (2018). Antioxidant, anticancer, apoptosis properties and chemical composition of black truffle *Terfezia claveryi*. Saudi J. Biol. Sci., 25(8): 1524–1534. https://doi.org/10.1016/j.sjbs.2016.01.031

Dedhia, N., Marathe, S.J. and Singhal, R.S. (2022). Food polysaccharides: A review on emerging microbial sources, bioactivities, nanoformulations and safety considerations. Carbohyd. Polym., 287: 119355. https://doi.org/10.1016/j.carbpol.2022.119355

Dedhia, N. and Shah, N. (2023). Polyphenols, phytosterols, aromatics, and essential oils. pp. 255–275. In: Prakash, S., Bhandari, B.R. and Gaiani, C. (eds.). Engineering Plant-Based Food Systems. Academic Press. https://doi.org/10.1016/B978-0-323-89842-3.00016-6

Farag, M.A., Fathi, D., Shamma, S., Shawkat, M.S.A., Shalabi, S.M. et al. (2021). Comparative metabolome classification of desert truffles *Terfezia claveryi* and *Terfezia boudieri via* its aroma and nutrients profile. LWT, 142: 111046. https://doi.org/10.1016/j.lwt.2021.111046

Gouzi, H., Depagne, C., Benmansour, A. and Coradin, T. (2013). First extraction of polyphenol oxidase from edible desert truffle (*Terfezia leonis* Tul.) and its thermal behavior. Eur. Food Res. Technol., 237(5): 721–729. https://doi.org/10.1007/s00217-013-2040-8

Guan, T., Wei, X., Xu, P., Chen, K., Zou, Y. et al. (2022). Comparison of structural and antioxidant activity of polysaccharide extracted from truffles. J. Food Sci., 87(7): 2999–3012. https://doi.org/10.1111/1750-3841.16207

Hamza, A., Zouari, N., Zouari, S., Jdir, H., Zaidi, S. et al. (2016). Nutraceutical potential, antioxidant and antibacterial activities of *Terfezia boudieri* Chatin, a wild edible desert truffle from Tunisia arid zone. Arab. J. Chem., 9(3): 383–389. https://doi.org/10.1016/j.arabjc.2013.06.015

Hsu, J.-Y., Chen, M.-H., Lai, Y.-S. and Chen, S.-D. (2022). Antioxidant profile and biosafety of white truffle mycelial products obtained by solid-state fermentation. Molecules, 27(1): 109. Article 1. https://doi.org/10.3390/molecules27010109

Jurić, S., Jurić, M., Król-Kilińska, Ż., Vlahoviček-Kahlina, K., Vinceković, M. et al. (2022). Sources, stability, encapsulation and application of natural pigments in foods. Food Rev. Int., 38(8): 1735–1790. https://doi.org/10.1080/87559129.2020.1837862

Kaliyaperumal, M., Kezo, K. and Gunaseelan, S. (2018). A global overview of edible mushrooms. pp. 15–56. In: Singh, B., Lallawmsanga and Passari, A.K. (eds.). Biology of Macrofungi. Springer International Publishing. https://doi.org/10.1007/978-3-030-02622-6_2

Kamle, M., Bar, E., Lewinsohn, D., Shavit, E., Roth-Bejerano, N. et al. (2017). Characterization of morphology, volatile profiles, and molecular markers in edible desert truffles from the Negev Desert. J. Agric. Food Chem., 65(14): 2977–2983. https://doi.org/10.1021/acs.jafc.6b04063

Khaled, J.M., Alharbi, N.S. and Shine, K. (2014). Efficacy of crude alcoholic extracts from culture broth of desert truffles against some clinical microbial isolates. J. Pure Appl. Microbiol., 8(5): 3871–3877.

Kıvrak, İ. (2015). Analytical methods applied to assess chemical composition, nutritional value and in vitro bioactivities of *Terfezia olbiensis* and *Terfezia claveryi* from Turkey. Food Anal. Meth., 8(5): 1279–1293. https://doi.org/10.1007/s12161-014-0009-2

Mahdi, L.H., Alsaadi, L.G., Mater, H.N., Kadhem, B.M., Zwain, L.A.H. et al. (2020). Antibacterial, antiviral and anticarcinogenic effect of a novel lectin characterized and purified from *Terfezia claveryi*. Biochemical & Cellular Archives | EBSCOhost. Bioch. Cell. Arch., 20 (0972–5075): 735–741. https://openurl.ebsco.com/EPDB%3Agcd%3A7%3A17540247/detailv2?sid=ebsco%3Aplink%3Ascholar&id=ebsco%3Agcd%3A144593581&crl=c

Marathe, S.J., Hamzi, W., Bashein, A.M., Deska, J., Seppänen-Laakso, T. et al. (2020). Anti-angiogenic and anti-inflammatory activity of the summer truffle (*Tuber aestivum* Vittad.) extracts and a correlation with the chemical constituents identified therein. Food Res. Int., 137: 109699. https://doi.org/10.1016/j.foodres.2020.109699

Masiero, M., Pettenella, D. and Secco, L. (2016). From failure to value: Economic valuation for a selected set of products and services from Mediterranean forests. For. Syst., 25(1): e051. https://doi.org/10.5424/fs/2016251-08160

Miao, Y.-Z., Lin, Q., Cao, Y., He, G.-H., Qiao, D.-R. et al. (2011). Extraction of water-soluble polysaccharides (WSPS) from Chinese truffle and its application in frozen yogurt. Carb. Polym., 86(2): 566–573. https://doi.org/10.1016/j.carbpol.2011.04.071

Milanesi, M., Gigliotti, M. and Runfola, A. (2020). The international marketing strategy of luxury food smEs: The case of truffle. J. Food Prod. Market., 26(9): 600–618. https://doi.org/10.1080/10454446.2020.1854916

Murcia, M., Maggi, L., Hussain, G., Jiménez-Monreal, A., Martínez-Tomé, M. et al (2014). Preservation of truffles. pp. 299–321. In: Kagan-Zur, V., Roth-Bejerano, N., Sitrit, Y., Morte, A. (eds) Desert Truffles. Soil Biology, vol 38. Springer, Berlin, Heidelberg. https://doi.org/10.1007/978-3-642-40096-4_19

Mustafa, A.M., Angeloni, S., Nzekoue, F.K., Abouelenein, D., Sagratini, G. et al. (2020). An overview on truffle aroma and main volatile compounds. Molecules, 25(24): 5948. https://doi.org/10.3390/molecules25245948

Nadim, M., Deshaware, S., Saidi, N., Abd-Elhakeem, M.A., Ojamo, H. et al (2015). Extracellular enzymatic activity of *Tuber maculatum* and *Tuber aestivum* mycelia. Adv. Microbiol., 5(7): 523–530. https://doi.org/10.4236/aim.2015.57054

Navarro-Ródenas, A., Morte, A. and Pérez-Gilabert, M. (2009). Partial purification, characterisation and histochemical localisation of alkaline phosphatase from ascocarps of the edible desert truffle *Terfezia claveryi* Chatin. Pl. Biol., 11(5): 678–685. https://doi.org/10.1111/j.1438-8677.2008.00172.x

Niimi, J., Deveau, A. and Splivallo, R. (2021). Geographical-based variations in white truffle *Tuber magnatum* aroma is explained by quantitative differences in key volatile compounds. New Phytol., 230(4): 1623–1638. https://doi.org/10.1111/nph.17259

Nouiri, E., Ben Ali, R., Ghali, R., Araoud, M., El May, M.V. et al (2021). Protective and curative effects of aqueous extract of *Terfezia Boudieri* (edible desert truffle species) against paracetamol acute toxicity in the rat. Nutr. Canc., 73(1): 113–123. https://doi.org/10.1080/01635581.2020.1742359

Oliach, D., Vidale, E., Brenko, A., Marois, O., Andrighetto, N. et al. (2021). Truffle market evolution: an application of the Delphi method. Forests, 12(9): 1174. https://doi.org/10.3390/f12091174

Pacioni, G., Rapino, C., Zarivi, O., Falconi, A., Leonardi, M. et al. (2015). Truffles contain endocannabinoid metabolic enzymes and anandamide. Phytochem., 110: 104–110. https://doi.org/10.1016/j.phytochem.2014.11.012

Pattanayak, M., Samanta, S., Maity, P., Manna, D.K., Sen, I.K. et al. (2017). Polysaccharide of an edible truffle *Tuber rufum*: Structural studies and effects on human lymphocytes. Int. J. Biol. Macromol., 95: 1037–1048. https://doi.org/10.1016/j.ijbiomac.2016.10.092

Pérez-Gilabert, M., Morte, A., Ávila-González, R. and García-Carmona, F. (2005). Characterization and histochemical localization of nonspecific esterase from ascocarps of desert truffle (*Terfezia claveryi* Chatin). J. Agric. Food Chem., 53(14): 5754–5759. https://doi.org/10.1021/jf050334d

Pérez-Gilabert, M., Sánchez-Felipe, I. and García-Carmona, F. (2005). Purification and partial characterization of lipoxygenase from desert truffle (*Terfezia claveryi* Chatin) ascocarps. J. Agric. Food Chem., 53(9): 3666–3671. https://doi.org/10.1021/jf048087l

Phong, W.N., Al-Salami, H., Gibberd, M.R., Dykes, G.A., Payne, A.D. et al (2022). Comparative evaluation of encapsulation using β-cyclodextrin versus freeze-drying for better retention and stabilizing of black Périgord truffle (*Tuber melanosporum*) aroma. J. Food Sci., 87(8): 3482–3495. https://doi.org/10.1111/1750-3841.16241

Rittersma, R.C. (2012). Industrialized delicacies: The Rise of the umbrian truffle business. Gastronomica, 12(3): 87–93. https://doi.org/10.1525/GFC.2012.12.3.87

Rivera, C.S., Blanco, D., Salvador, M.L. and Venturini, M.E. (2010). Shelf-life extension of fresh Tuber aestivum and *Tuber melanosporum* truffles by modified atmosphere packaging with microperforated films. J. Food Sci., 75(4): E225–E233. https://doi.org/10.1111/j.1750-3841.2010.01602.x

Sallam, F.E.A., El-mokadem, M.T., Mekawey, A.A.I. and Saker, E.A.E. (2022). Egyptian truffles as a source of antimicrobial and antioxidant agents. J. Sci. Res. Sci., 39(2): 49–61. https://doi.org/10.21608/jsrs.2022.275787

Samils, N., Olivera, A., Danell, E., Alexander, S.J., Fischer, C., Colinas, C. et al. (2008). The socioeconomic impact of truffle cultivation in Rural Spain. Econ. Bot., 62(3): 331–340. https://doi.org/10.1007/s12231-008-9030-y

Savini, S., Loizzo, M.R., Tundis, R., Mozzon, M., Foligni, R. et al. (2017). Fresh refrigerated Tuber melanosporum truffle: Effect of the storage conditions on the antioxidant profile, antioxidant activity and volatile profile. Eur. Food Res. Technol., 243(12): 2255–2263. https://doi.org/10.1007/s00217-017-2927-x

Shah, N., Marathe, S.J., Croce, D., Ciardi, M., Longo, V. et al. (2021). An investigation of the antioxidant potential and bioaccumulated minerals in *Tuber borchii* and *Tuber maculatum* mycelia obtained by submerged fermentation. Arch. Microbiol., 204(1): 64. https://doi.org/10.1007/s00203-021-02717-5

Shah, N., Usvalampi, A., Chaudhary, S., Seppänen-Laakso, T., Marathe, S. et al. (2020). An investigation on changes in composition and antioxidant potential of mature and immature summer truffle (*Tuber aestivum*). Eur. Food Res. Technol., 246(4): 723–731. https://doi.org/10.1007/s00217-020-03438-7

Splivallo, R. and Culleré, L. (2016). The smell of truffles: from aroma biosynthesis to product quality. pp. 393–407. In: Zambonelli, A., Iotti, M., Murat, C. (eds.). True Truffle (*Tuber* spp.) in the World. Soil Biology, 47: Springer, Cham. https://doi.org/10.1007/978-3-319-31436-5_23

Splivallo, R., Deveau, A., Valdez, N., Kirchhoff, N., Frey-Klett, P. et al. (2015). Bacteria associated with truffle-fruiting bodies contribute to truffle aroma. Environ. Microbiol., 17(8): 2647–2660. https://doi.org/10.1111/1462-2920.12521

Sun, Q., Zhang, Z., Xu, L., Shi, W., Liu, X. and Wang, F. (2020). Increasing production of truffle polysaccharides in the solid-state fermentation of *Tuber melanosporum* by diosgenin based on orthogonal matrix and nonlinear regression analysis. Food Sci. Technol. Res., 26(4): 487–494. https://doi.org/10.3136/fstr.26.487

Susana Rivera, C., Venturini, M.E., Oria, R. and Blanco, D. (2011). Selection of a decontamination treatment for fresh *Tuber aestivum* and *Tuber melanosporum* truffles packaged in modified atmospheres. Food Control, 22(3–4): 626–632. https://doi.org/10.1016/j.foodcont.2010.10.015

Tang, Y.-J., Zhu, L.-L., Li, D.-S., Mi, Z.-Y. and Li, H.-M. et al. (2008). Significance of inoculation density and carbon source on the mycelial growth and *Tuber* polysaccharides production by submerged fermentation of Chinese truffle *Tuber sinense*. Proc. Biochem., 43(5): 576–586. https://doi.org/10.1016/j.procbio.2008.01.021

Tejedor-Calvo, E., Amara, K., Reis, F.S., Barros, L., Martins, A. et al. (2021). Chemical composition and evaluation of antioxidant, antimicrobial and antiproliferative activities of *Tuber* and *Terfezia* truffles. Food Res. Int., 140: 110071. https://doi.org/10.1016/j.foodres.2020.110071

Tejedor-Calvo, E., Baquero-Aznar, V., Vega-Diez, S., Salvador, M.L., Sanz, M.Á. et al. (2024). Potential of a smart gelatine hydrogel-based package to extend fresh black truffle (*Tuber melanosporum*) shelf-life preserving its aroma profile. Food Hydrocoll., 151: 109874. https://doi.org/10.1016/j.foodhyd.2024.109874

Tejedor-Calvo, E., García-Barreda, S., Felices-Mayordomo, M., Blanco, D., Sánchez, S. et al. (2023). Truffle flavored commercial products veracity and sensory analysis from truffle and non-truffle consumers. Food Control, 145: 109424. https://doi.org/10.1016/j.foodcont.2022.109424

Tejedor-Calvo, E., García-Barreda, S., Sánchez, S., Morales, D., Soler-Rivas, C. et al. (2021). Supercritical CO2 extraction method of aromatic compounds from truffles. LWT, 150: 111954. https://doi.org/10.1016/j.lwt.2021.111954

Tejedor-Calvo, E., Marco, P., Spègel, P. and Soler-Rivas, C. (2023). Extraction and trapping of truffle flavoring compounds into food matrices using supercritical CO2. Food Res. Int., 164, 112422. https://doi.org/10.1016/j.foodres.2022.112422

Tirillini, B., Maggi, F., Venanzoni, R. and Angelini, P. (2019). Enhanced duration of truffle sauce preservation due to addition of linoleic acid. J. Food Quality, 2019(1): 8421898. https://doi.org/10.1155/2019/8421898

Varrot, A., Basheer, S.M. and Imberty, A. (2013). Fungal lectins: Structure, function and potential applications. Curr. Opin. Str. Biol., 23(5): 678–685. https://doi.org/10.1016/j.sbi.2013.07.007

Wahiba, B., Tehami, W., Kattaia, A., Bouziane, A. and Mohammed, B. et al. (2016). Nutritional and antioxidant profile of red truffles (*Terfezia claveryi*) and white truffle (*Tirmania nivea*) from southwestern of Algeria. Der Pharmacia Lettre, 8(17): 134–141.

Zarivi, O., Bonfigli, A., Colafarina, S., Aimola, P., Ragnelli, A.M. et al. (2011). Tyrosinase expression during black truffle development: From free living mycelium to ripe fruit body. Phytochem., 72(18): 2317–2324. https://doi.org/10.1016/j.phytochem.2011.08.025

Zhang, T., Jayachandran, M., Ganesan, K. and Xu, B. (2020). The black truffle, *Tuber melanosporum* (Ascomycetes), ameliorates hyperglycemia and regulates insulin signaling pathway in STZ-induced diabetic rats. Int. J. Med. Mushrooms, 22(11): 1057–1066. https://doi.org/10.1615/IntJMedMushrooms.2020036355

Zhang, Y., He, S. and Simpson, B.K. (2018). Enzymes in food bioprocessing - Novel food enzymes, applications, and related techniques. Curr. Opin. Food Sci., 19: 30–35. https://doi.org/10.1016/j.cofs.2017.12.007

Zhao, W., Wang, X.-H., Li, H.-M., Wang, S.-H., Chen, T. et al. (2014). Isolation and characterization of polysaccharides with the antitumor activity from Tuber fruiting bodies and fermentation system. Appl. Microbiol. Biotechnol., 98(5): 1991–2002. https://doi.org/10.1007/s00253-013-5379-7

Chapter 18

Truffle Products

From Food to Pharmaceutical Industries

Hesham Ali El Enshasy,[1,2,3] *Rozaliana Abdul Karim,*[1]
Intan Baizura Mohamed,[1] *Muktiningsih Nurjayadi,*[4]
Dalia Sukmawati[5] and *Tony Peng*[6,*]

1. Introduction

Macrofungi such as mushrooms and truffles have been widely used worldwide for centuries. They have been recognized as high-value natural products and used as high-value food. In addition, they have been recognized as part of traditional or folk medicine in different cultures (El Enshasy et al., 2013; Veerarahavan et al., 2021). Truffles are very unique in nature and rare compared to mushrooms. They are characterized by a globose and potato-like structure. The truffle textures are varied, with a smooth, reticulate or wrinkled surface in the absence of stalks and gills. They can be classified into two categories: tuber truffles (true truffles) and desert truffles.

[1] Innovation Centre of Agritechnology in Advanced Bioprocessing (ICA), Universiti Teknologi Malayisa (UTM), Pagoh Campus, Johor, Malaysia.
[2] Faculty of Chemical Engineering, Universiti Teknologi Malaysia, 81310, UTM Skudai, Johor, Malaysia.
[3] City of Scientific Research and Technology Applications, New Burg Al Arab, Alexandria, Egypt.
[4] Chemistry Department, Faculty of Mathematics and Natural Sciences, Universitas Negeri Jakarta, Indonesia.
[5] Biology Department, Faculty of Mathematics and Natural Sciences, Universitas Negeri Jakarta, Indonesia.
[6] GK Bio International Sdn. Bhd., Petaling Jaya, Selangor, Malaysia.
* Corresponding author: hesham@utm.my

In Europe, most truffles are *Tuber* sp., such as *T. melanosporum, T. aesticum,* and *T. magnatum*. In the Middle East and North Africa, desert truffle species such as *Terfezia* and *Tirmania* are more common, with two main groups and a traditional name (Kam'ah).

The two common types of desert truffles can be differentiated by their color and have either a white color (*Terfezia claveryi*) or a red color (*Terfezia boudieri*), which are collected mainly in the summer. Truffles have been considered to be wild, natural, and non-cultivable macrofungi for many years. In general, truffles are characterized by their specific aroma, which makes them an attractive high value ingredient in the food industry. The collection of this type of subterrestrial organism needs experts who have trained animals such as dogs and pigs to recognize their specific aroma based on the presence of specific volatile compounds. With the increased awareness of the high value of truffles, the market size for truffles has grown in the last few years. The truffles market is considered the fastest-growing market among natural products . The truffle products market reached about USD 0.91 billion in 2022 and is expected to grow at a CAGR of about 8.5%, which is considered high compared to other natural product markets. The growth in market demand has been driven by the advancement of truffle-based product formulations and the development of novel product market niches . Currently, the main truffle-based products are mainly in the food, nutraceutical and cosmeceutical sectors. However, much research related to bioactive molecule safety and efficacy assessment for the treatment of different diseases has been published in recent years. These include *in vitro* and *in vivo* studies, as well as a few clinical trials. This chapter envisages truffle bioactive compounds, clinical trials, importance (cosmeceuticals, food industry, health benefits), truffle products, the truffle market and truffle quality control.

2. Truffle Bioactive Compounds and Health Benefits

Various studies have been published for the nutritional content and chemical profile of truffles (Patel et al., 2017; Elsyed et al., 2021; Farag et al., 2021). Truffles have been considered a high-nutrition food based on their high protein content, which can be between 20–27% of dry matter. In addition, they are rich in fat, dietary fibres, and essential amino acids (valine, serine, isoleucine, threonine, methionine and phenylalanine) as well as minerals from macro- and micro-elements (K, P, Zn, Cu, Fe, Mg, and Mn). It also contains many organic compounds, such as ascorbic acid, ketones, aldehydes, and alcohols. It has also been reported that some truffles can have ergosteroids and brassicasterol (Patel et al., 2017). Truffle aroma is one of the major criteria used for truffle quality and value. More than 200 volatile organic compounds (VOCs) have been identified in different types of truffles using different techniques such as solid phase microextraction gas chromatography (SPME) and GC-MS-MS.

However, most of the volatile compounds involved in the sensory odor of humans are those containing sulphur. In general, the truffle's aroma can vary from earthy, garlicky, cheese-like, leathery, creamy, dusty, vanilla-like, to gasoline-like. Therefore, VOCs can also be considered one of the fingerprints of some truffle groups. For example, whereas 2-octanal can be found in both white and black truffles, 3-methyl-4 ,5-dihyrothiophene is the major component of VOCs in white truffles (Splivallo and Ebeler, 2015; Kanchiswamy et al., 2015). However, recent studies have reported that the characteristic aroma is not mainly produced by the fungus . Truffle microbiomes also play a significant role in the production of the signature aroma of truffles (Vahdatzadeh et al., 2015; Allen and Bennett, 2021). Due to the presence of various functional chemicals, many authors have reported that truffles possess general antioxidant properties (Chen et al., 2016; Yan et al., 2017). In addition, they have been reported to have many health benefits, such as antidepressant , antimicrobial , immunomodulation , and anticancer effects (Elsayed et al., 2014; Lee et al., 2020).

3. Commercialized Truffle-Based Products in the Food Industry

Truffle-based products have gained significant popularity in the food and nutraceutical markets due to their unique flavor, aroma and potential health benefits. The special characteristics of truffles have led to their incorporation into various food products, offering consumers a gourmet experience. Truffles, such as *Tuber magnatum Pico* (white truffle) and *Tuber melanosporum Vittad.* (black truffle), are highly valued in many countries for their distinct flavors and aromas (Mello et al., 2006). These valued fungi are known for their special characteristics, making them sought after ingredients in the culinary world. Besides, they have been acknowledged for their nutritional and medicinal properties, leading to their integration into various foods and nutraceutical products (Lee et al., 2020). Research on truffles has explored their biochemical composition, aromatic qualities and nutritional benefits, shedding light on their potential applications in medicine and functional foods (Lee et al., 2020). Truffles are rich in volatile organic compounds that contribute to their unique aroma, making them not only a culinary delicacy but also a source of bioactive compounds with potential health-promoting properties (Torregiani et al., 2017; Mustafa et al., 2020). The high nutritional value of truffles, along with their antioxidant, anti-inflammatory and antimicrobial properties, has positioned them as valuable ingredients in the nutraceutical market.

Genomic analysis of truffles has provided insights into their evolutionary origins and symbiotic mechanisms, offering a deeper understanding of the processes governing fruit body formation and symbiotic relationships (Martin et al., 2010). By unravelling the genetic traits of truffles, researchers aim to enhance crop production efficiency and develop sustainable cultivation practices to meet the growing demand for these high quality fungi. The authenticity and quality of truffle based products are crucial to the market. Studies have focused on the authentication of truffle flavors in oils and processed products to ensure accurate labelling and quality standards. The extraction of aroma volatiles from truffles is a crucial aspect of creating truffle based products with authentic flavors. Metabolic profiling and stable isotope ratio analysis have been employed to differentiate between commercial and homemade truffle flavored oils, enhancing consumer trust and product integrity (Wernig et al., 2018). Techniques such as supercritical carbon dioxide extraction have shown promise in preserving the aroma of truffles, allowing for the development of natural truffle based products for year-round consumption (Phong et al., 2022). By utilizing advanced extraction methods, manufacturers can capture the essence of truffles in various food applications.

In the market, truffle based products have gained popularity among consumers for their distinct flavor profile and potential health benefits. Truffle infused oils, sauces, spreads and seasonings have become sought after ingredients in gourmet cuisine, adding a touch of luxury and sophistication to dishes (Das et al., 2011). Table 1 shows several lists of companies that provide a diverse range of truffle products to cater to various culinary needs, enhance flavors and offer gourmet experiences globally. For instance, Urbani Truffles offers a wide range of truffle products, including truffle oils, sauces, dairy products, salts, vinegar, honey, preserved truffles, frozen truffles, dried truffles, creams and sauces, condiments, cold cuts, cheese, pasta and snacks. Truffle oil products range from White Truffle Oil, Black Truffle Oil, White Truffle Extra-Virgin Olive Oil, Black Truffle Extra-Virgin Olive Oil, Organic White Truffle Extra Virgin Olive Oil and Organic Black Truffle Extra Virgin Olive Oil. The truffle sauces from the same company also have a wide range of registered products, such as White Truffles and Porcini, Black Truffles and Mushrooms, Truffles and Mushroom Sauce, White Truffles and Mushrooms, Cream and Truffles, Pesto and Truffles, Artichokes and Truffles, Red Pesto and Truffles, Organic White Truffle Sauce and Organic Black Truffle Sauce. These products provide a convenient way for individuals to enjoy the distinctive flavor of truffles in their everyday meals.

Truffle-based products are often associated with gourmet cuisine and are used to elevate the taste of dishes in high-end restaurants and home kitchens alike. Additionally, the development of truffle-based products has also led to advancements in food preservation techniques. Modified atmosphere packaging has been employed to extend the shelf life of fresh truffles, ensuring that their quality and flavor are maintained over an extended period of time (Rivera et al., 2010). By implementing innovative packaging solutions, truffle producers can deliver high quality products to consumers while prolonging the freshness of the truffles.

Furthermore, the market evolution of truffles has prompted the development of strategies to strengthen the link between truffles, tourism, and gastronomy, enhancing the overall truffle sector. Initiatives to increase consumer awareness and consumption of truffles, along with the development of tourism workshops for truffle farmers, aim to promote sustainable truffle culture and market growth (Oliach et al., 2021). The cultivation of truffles has been a subject of intense research to meet the growing demand in the global market and to address concerns related to the overexploitation of natural truffle resources. In recent years, there has been a growing interest in the production of truffle based products through innovative methods such as submerged fermentation. Research is being conducted to explore semi-artificial cultivation techniques to meet the increasing global demand for truffles while addressing concerns about the overexploitation of natural truffle resources (Tang et al., 2015). These efforts aim to ensure the sustainability of truffle production and maintain the ecological balance of truffle ecosystems (Büntgen et al., 2017).

Overall, truffle-based products continue to captivate food enthusiasts and chefs worldwide, offering a luxurious and indulgent culinary experience with their unique flavors and aromas. With ongoing research, technological advancements and sustainable practices in truffle cultivation and product development, the market for

Table 1. Examples of registered truffle-formulated products on the international market.

Registered Product Name	Company	Website
1. Truffle Oil 2. Truffle Sauce 3. Truffle Dairy 4. Truffle Salt 5. Truffle Vinegar 6. Truffle Honey	Urbani Truffles, USA	https://www.urbani.com
1. Preserved Truffles 2. Frozen Truffles 3. Dried Truffles 4. Cream and Sauces 5. Truffle Oil 6. Truffle Creams 7. Truffle Condiments 8. Truffle Cold Cuts 9. Truffle Cheese 10. Truffle Pasta 11. Truffle Sauces 12. Truffle Honey	Bernardini Tartufi, Italy	https://bernardinitartufi.com
1. Truffle Sauces 2. Truffle Oil 3. Truffle Sea Salt 4. Truffle Butter 5. Truffle Honey 6. Truffle Slices	Craft Truffles, Singapore	https://crafttruffles.com
1. Truffle Sauce 2. Truffle Oil 3. Truffle Mayo 4. Truffle Pasta 5. Truffle Salt	Truff, USA	https://www.truff.com
1. Truffle Sauce 2. Truffle Butter 3. Creams with Truffle 4. Truffle Grill 5. Truffle Honey 6. Truffle Pasta 7. Truffle Pate 8. Truffle Pesto 9. Truffle Preserved Truffles 10. Truffle Salt 11. Truffle Snacks 12. Truffle Condiments 13. Truffle Oil	Truffles & Co., Malta	https://trufflesandco.com
1. White Truffle Spray 2. Truffle BBQ Sauce 3. Grill Master BB Sauce 4. Truffle Risotto	Turci, Italy	https://turcibrand.com
1. Fresh Truffles 2. Preserved Truffle 3. Truffle Butter 4. Truffle Condiments 5. Truffle Snacks & Pasta 6. Truffle Oil	TruffleHunter, UK	https://www.trufflehunter.co.uk

Table 1. contd. ...

Table 1. contd. ...

Registered Product Name	Company	Website
1. Truffle Creams 2. Truffle Salts 3. Truffle Sauce 4. Truffle Oil 5. Truffle Vinegar 6. Truffle Honey 7. Truffle Cheese	INAUDI, Italy	https://www.inaudi.com
1. Truffle Infused Oil 2. Truffle Honey 3. Truffle Zest 4. Truffle Sea Salts 5. Truffle Sauce	Sabatino Tartufi, Canada	https://www.sabatinotartufi.com
1. Truffle Oil and Vinegar 2. Truffle Pates & Creams 3. Truffle Carpaccio 4. Truffle Honey 5. Truffle Salt 6. Truffle Powder	La Rustichella, Italy	https://www.larustichellatruffles.com
1. Truffle Oil 2. Truffle Butter 3. Truffle Peelings 4. Truffle Juice	D'Artagnan, USA	https://www.dartagnan.com/
1. Truffle Snack & Appetizer 2. Truffle Seasonings & Sauces 3. Truffle Oils 4. Preserved Truffles 5. Truffle Condiments	TartufLanghe, Italy	https://www.tartuflanghe.com/
1. Truffle Oil 2. Truffle Sauce 3. Truffle Sea Salt 4. Truffle Honey	Trufflin NYC, USA	https://www.trufflin-nyc.com/
1. Truffle Salt 2. Truffle Honey 3. Truffle Oil 4. Truffle Sauce 5. Truffle Wines	Truffle Hill, Australia	https://www.trufflehill.com.au/
1. Truffle Oil	Oregon Truffle Oil, USA	https://www.oregontruffleoil.com/

high-quality truffle-based products is expected to expand further, catering to the growing demand for unique and premium food offerings.

4. Versatility of Truffle Extracts in Cosmeceuticals

Truffles contain many bioactive compounds that benefit the skin (Hyde et al., 2010). Besides, their luxurious scent puts truffle extracts in demand in the high end perfume market (Thomas et al., 2021). Due to their unique properties, it has become significant for industry players to include truffles in their cosmetic products (Wu et al., 2016). Studies have shown that several significant species of truffles are commonly used in cosmetics, perfumes, and personal care products. Cosmetic companies widely incorporate truffle extracts into various products, such as moisturizers, serums, hair care, and fragrances (Table 2). Thus, many researchers have been interested in studying the bioactive compounds of truffles, which make them valuable for cosmetic applications. Recently, they have patented the use of truffle extract in cosmetic formulations. For example, Portolan et al. (2016) invented an aqueous extract of *Tuber melanosporumin*, a skincare formulation that is used explicitly for skin treatment. Meanwhile, Coquet and Chabert (2018) introduced a new method for extracting *Tuber Magnatum* for cosmetic preparation. Therefore, it proved that truffles have demand in the cosmetic industry (Hyde et al., 2010; Wu et al., 2016; Thomas et al., 2021).

Truffles have numerous nutritional compounds and bioactivities that make them effective antioxidant , anticancer, and antimicrobial agents (Chauhan et al., 2021; Gouzi et al., 2013; Dahham et al., 2018). Truffle-based cosmetic products can offer antioxidant benefits because of their functional compounds, such as ascorbic acid and terpenoids. Elkhateeb et al. (2024) explored the nutritional composition and biological activities of different species of truffles. They reported that truffles possess antioxidant properties. Antioxidants are essential in cosmetic and personal care products as they can help neutralize the free radicals in the body, reducing the risk of chronic diseases and aging (Brudzyńska et al., 2022). Truffle based skin care products are claimed to have rich antioxidant benefits. International brands like Scientific, D'Alba, and NuMe choose truffles for their aromatic flavor and nutritional value. For example, SKINTIFIC Truffle Biome Skin Reborn Cream Gel Moisturizer can deeply repair skin, relieve breakouts and signs of irritation, and maintain the skin barrier. Antioxidants can help protect the skin from free radicals so that it looks healthy.

Cayan et al. (2021) found several bioactive compounds that show antioxidant properties in the truffle, *Reddellomyces parvulosporus*. The functional compounds are brassicasterol, mannitol, ergosterol peroxide, and fumaric acid. These compounds show promising results in anti-tyrosinase, anti-urease, and anti-butyrylcholinesterase activity. It can be applied to cosmetic products such as skin depigmentation and anti-aging agents. Many cosmetic companies claim their products have anti-aging effects and enhance skin health. Truffoire, Skin & Co. Roma, and Nume Lab released a wide range of skincare products that affected consumers. Truffle is a natural ingredient with numerous bioactive constituents beneficial in cosmeceuticals (Cayan et al., 2021).

Besides, truffles have been reported to have an abundance of polysaccharides. Polysaccharides from truffles have also gained attention in the cosmetic industry due to their unique properties that are favourable for their application in cosmetic products. A group of polysaccharides like glucose, pectin and its derivatives can enhance viscosity and boost the stability of cosmetic emulsions. It can thicken and gelate solutions, which are important in cosmetic formulations (Li et al., 2020). Truffle extract is not only used in skincare and personal care products but is also found in fragrances. The aromatic properties of the white truffle (*Tuber magnatum*) and the black Périgord truffle (*Tuber melanosporum*) have attracted attention for potential perfume development (Splivallo et al. 2011; Allen and Bennett et al.,

2021). Truffle extract has become a popular natural ingredient in high end perfumes worldwide.

Truffle releases volatile organic compounds, such as sulfur compounds and alcohols, which add to their unique scent (Allen & Bennett et al., 2021). Splivallo et al. 2011 explained further about the production of sulfur-containing volatile organic compounds (VOCs) such as dimethyl sulfide, methional, and 2-Methyl-4,5-dihydrothiophene, which are responsible for the release of the aroma. In addition to the release of VOCs, Islam et al. (2013) discovered several novel proteins found in *Tuber melanosporum* involved in producing the aroma profile. The distinctive and attractive aroma, musky, earthy, and sometimes aphrodisiac-like scent of truffles, has sparked interest in their potential application in perfumes. International companies like Tomford, Scent Trunk, Scent Perfique Ltd. and Jo have released truffle-based perfumes with exclusive and iconic scents. The use of rare and precious ingredients makes the perfumes sophisticated and luxurious.

5. Truffles in the Pharmaceutical Market

Based on its natural source and wide use as a high-value food, it has also been accepted by the community as medicine. Truffles were one of the components of traditional medicines in different cultures. It has been used in the treatment of different types of chronic and non-chronic diseases and in folk and complementary medicine (Mandeel and Al-Laith, 2007; El Enshasy et al., 2013). However, as with any natural product in traditional use, truffles have been used in the form of crude water extract. Thus, their biotherapeutic activities are attributed to a combination of chemical cocktails giving them a natural balanced composition. Registration of pharmaceutical products with therapeutic claims needs to follow a long process (RD, biomolecule determination, *in vitro* test, *in vivo* test, clinical trials phases I, II, III, and IV). This long process of product registration is to ensure the safety of the formulated product (including cytotoxicity, drug-drug interaction, target group of patients) and efficacy (in terms of dose requirement and treatment to cure a specific disease). This long drug registration journey can take more than 10 years.

However, in the part related to bioactive molecules discovery, much research and data elucidate the chemical structures of bioactive compounds and support the functionality of bioactive molecules at the *in vitro* level. Based on this data, many clinical trials have been carried out in recent years. The medicinal value of truffles has been reported in studies and proven to be anti-inflammatory potent (Dawood et al., 2023; Souna et al., 2023; immunomodulatory (Al Obaydi et al., 2020), and antimicrobial (Dib-Bellahouel and Fortas, 2011; Casarica et al., 2016), The anticancer properties of truffle extracts obtained either from a wild collection or fermentation broth have also been reported by many authors (Zhao et al., 2014; Attia et al., 2018). The black desert truffle (*Terfezia claveryi*) extract exhibits anticancer activity and initiates apoptosis of cancer cells (Dahham et al., 2018). Another study reported the potential use of the same truffle for the preparation of silver nanoparticles, which exhibit anticancer properties against breast cancer cells (Khadri et al., 2017). Research on desert truffles also showed that different solvent extracts of the desert truffles such as *Tirmania nivea* and *Terfazia claveryi*, exhibited antimicrobial and

Table 2. List of truffle-based cosmetic products in the international market.

Registered product name	Cosmetic claim	Company	Website
SKINTIFIC Truffle Biome Skin Reborn Cream Gel Moisturizer	Deeply repair skin; Relieve breakouts and signs of irritation Balance skin microflora; Maintain skin barrier and smooth skin	Skintific, Indonesia	https://skintific.com
Truffle Therapy Face Toner	Hydrate and alleviate blemished, dull or uneven skin; Antioxidant properties - help neutralize free radicals in the skin; Enhance skin health and radiance	Skin & Co Roma, Italy	https://skinandco.com
Black Truffle Deep Renewal Vitamin C	Illuminates the skin and reduces the visibility of sun damage, including dark spots; Stimulates collagen production; Reduces the appearance of fine lines and wrinkles; Protects the skin from damage from free radicals; Helps promote circulation to brighten dull skin, revealing a healthy and radiant complexion	Truffoire, USA	https://truffoire.com
Crème Ancienne White Truffle Face Serum	Anti-aging effect; Skin rejuvenation	Fresh Inc, UK	https://www.fresh.com
Black Truffle Luxury Golden Beauty Foundation	Hydrates skin; Removes blemishes and dark spots	Guerniss, Bangladesh	https://www.guerniss.com
Trufflesque	Helps to retain moisture in the skin; Promotes collagen production; Brightens skin tone, firms, and smoothens skin	TempleSpa, UK	https://www.templespa.com
Absolute Radiance Eye Serum	Anti-aging ; Rejuvenates skin; Hydrates and moisturizes ; Reduces dark spots and hyperpigmentation	Nume Lab, Switzerland	https://www.nume-lab.com
Beaurunch Truffle Consommé Ampoule	Soothing effect; Anti-aging effect; Prevents formation of free radicals; Smoothens fine wrinkles; Promotes skin regeneration	Na, on-Gyeol, Hungary	https://www.naongyeol.com
d'Alba White Truffle Oil Serum Body Cleanser	Skin Moisturizer ; Antioxidant agent	D'Alba Piedmont, Italy	https://dalba.com
Black Truffle. Whitening Anti Age Skin Treatment with Rose Stem Cells	Anti-aging effect	Christopher Hanlon®, Australia	https://christopherhanlon.com

Truffle Face Cream	Reduces fine lines; Enhances elasticity; Moisturizes skin	Savor//beauty, New York	https://www.savorbeauty.com
Deluxe Codic Fensine Serum	Recharges skin; Moisturizes and nourishes skin	Tegoder Cosmetics, UK	https://tegoder.co.uk
Trufflus Cream	Improves skin tone and prevents dark marks; Helps to reduce wrinkles and enhances elasticity associated with aging; Boosts skin elasticity through cell growth and hydration.	Loar group LLC, US	https://loarus.com
NuMe White Truffle Conditioner	Antioxidant effect	NuMe ABV Group, Inc., US	https://numehair.com
Truffle Olive Oil Shampoo	Prevents hair fall; Strengthen hair; Restores hair shine; Indulgent herbal aroma	Mi Bliss, India	https://mi-bliss.com
Truffecta Indulge Bundle	Replenishes and refreshes all hair types; Promotes an optimal scalp environment to purify strands; Improves hair elasticity and strength; Restores hair shine	TruffLuv, North America	https://truffluv.com
Previa White Truffle Regenerating Reconstruct Treatment	Renews the hair fibre, making the hair healthy; Anti-aging effect	Previa, Italy	https://www.previa.it
Replenishing Truffle Leave in Conditioner Infused with Biotin & Ginseng	Promotes an optimal scalp environment to purify strands; Renews the hair fibre, making the hair healthy	Karma Beauty, USA	https://karmahairbeauty.com
Premium Line Black Truffle	Stimulating agent for hair growth; Renew the hair fibre, making the hair healthy; Restores hair shine	Messinian Spa, Greece	https://www.messinianspa.gr
White Truffle Nutrient - Rich Moisture Shampoo	Prevents hair fall; Strengthens hair; Restores hair shine	Calabasas beauty Group LLC, USA	https://fortheloveofhair.com
Black Orchid Eau De Parfum	Iconic scents' Long lasting fragrance longevity; Luxurious sensory experience	Tomford, USA	https://www.tomfordbeauty.com
Miss Layla	Exclusive scent	Scent Trunk, USA	https://scenttrunk.com
Black Truffle & Tonka Bean Dreamz Fragrance Collection	Irresistible gourmand floral scent	Scent Perfique Ltd, UK	https://www.scentperfique.com
Kavvabis Truffle Perfume	Iconic scents; Luxurious sensory experience	The Dam, Australia	https://thedam.raintree.com.au
Red Truffle 21	Blends its earthy richness; Long lasting fragrance longevity; Luxurious sensory experience	Jo, UK	https://www.joloves.com

anticancer activity against breast cancer, HepG2, and L929 cell lines (El Sayed et al., 2019a, 2019b). However, compared to mushrooms, very few clinical trials have been carried out for the bioactive compounds of truffles. This is based on scarce available data about the natural products of truffles compared to other bioresources.

6. Truffles Quality Control and Authentication

Truffles, as a highly sought after delicacy, necessitate stringent quality control measures to ensure their authenticity and safety in the market, maintaining consumer trust and upholding industry standards. Quality control and authentication processes are essential for verifying the origin, purity and overall quality of truffles and truffle-based products. Various analytical methods have been developed to authenticate truffles and assess their quality, ranging from molecular techniques to volatile compound analysis. One such method involves the use of Gaschromatography-Mass Spectrometry (GC-MS) coupled with solid-phase microextraction, which allows for the rapid identification of truffle species based on their volatile compounds (Gioacchini et al., 2005). This approach has proven effective in distinguishing different truffle species, aiding quality control and authentication processes.

In addition to GC-MS, molecular tools such as Polymerase Chain Reaction (PCR) have been optimized for the identification of specific truffle species, like Tuber melanosporum, throughout the truffle production chain, thereby enhancing quality control and ensuring accurate species identification (Séjalon-Delmas et al., 2000). PCR primers specific for the genus Tuber have been developed to reveal the presence of various truffle species in truffle grounds, emphasizing the importance of molecular tools in identifying and characterizing truffle species. By utilizing these specific primers and DNA analysis, this method enhances quality control in the truffle industry, ensuring the accurate identification of truffle species from production to the final commercial products (Zampieri et al., 2009). Such molecular techniques play a crucial role in maintaining the integrity and authenticity of truffles on the market.

The unique aroma of truffles, highly valued in culinary settings, is attributed to a complex mixture of volatile organic compounds (VOCs) present in these fungi. These VOCs, such as alcohols, ketones, aldehydes, and aromatic and sulfur compounds, contribute to the distinct scent that distinguishes various truffle species (Balivo, 2024). Analyzing the VOC profile of truffles is crucial for the precise identification and characterization of truffle aromas. Other analytical methods, such as Proton-Transfer-Reaction Mass Spectrometry (PTR-MS) as well as electronic nose sensing (EN), are commonly employed to analyze the VOC profile of truffles, enabling accurate identification and characterization of truffle aromas (Mustafa et al., 2020). By analyzing the VOCs emitted by truffles, PTR-MS can be utilized for quality control purposes, allowing for the evaluation of truffle freshness and aroma intensity, which are key indicators of truffle quality. Meanwhile, by utilizing EN technology, which mimics the human olfactory system, truffle aromas can be effectively characterized and differentiated based on their unique VOC compositions, contributing to quality control and authentication processes in the truffle industry.

Authentication of truffle products involves verifying their origin, species, and quality through advanced analytical techniques. Methods such as Fourier

Transform Near-Infrared (FT-NIR) spectroscopy combined with chemometrics have been developed to differentiate between commercially relevant truffle species, offering a reliable approach for authentication in the truffle market (Segelke et al., 2020). Additionally, advancements in mass spectrometry techniques, such as Matrix-Assisted Laser Desorption/Ionization Time-Of-Flight Mass Spectrometry (MALDI-TOF MS) biotyping has been utilized for rapid and reliable identification of commercial truffles, offering a cost-effective alternative to traditional molecular methods (Karkouri et al., 2019). By analyzing fungal proteins and utilizing MALDI-TOF MS, truffle authentication can be efficiently carried out, contributing to fraud control and quality assurance in the truffle market (Klein et al., 2020). This method has also been widely used in clinical laboratories for the accurate identification of microorganisms, showcasing high accuracy in microorganism identification (Liu et al., 2019).

Furthermore, stable isotope ratio analysis has been recognized as a powerful tool for differentiating truffle species and determining their geographical origin, contributing significantly to the authentication and quality control of truffle products in the market (Segelke et al., 2020). By analyzing stable carbon, nitrogen, and hydrogen isotope ratios, researchers can differentiate between truffles from different countries of origin and species, providing valuable insights into their authenticity and quality (Krauß and Vetter, 2020). This analytical approach has shown high potential for addressing authentication issues related to truffles, providing a reliable means of quality control and ensuring the accurate labeling of truffle products based on their origin and species. By leveraging stable isotope analysis, researchers can provide consumers with reliable information about the origin and characteristics of truffles, fostering trust and transparency in the industry.

Preservation techniques are also significant in maintaining truffle quality and authenticity. While refrigeration is a widely accepted method for preserving truffles, irradiation has been explored as an efficient method to preserve the biochemical properties of fresh truffles, ensuring their quality and authenticity over an extended period of time. Studies have shown that irradiation is believed to be one of the most efficient methods for preserving the freshness and quality of truffles, making it a promising technique for ensuring the authenticity of truffle products (Wang & Marcone, 2011). In addition to irradiation, other preservation techniques such as freeze-drying, freezing, hot air drying and canning have been studied to maintain the physicochemical properties of truffles, aiming to extend their shelf life while preserving their sensory characteristics. Refrigeration combined with packaging is commonly used for short-term storage of truffles, but exploring alternative preservation techniques like irradiation and freeze-drying can offer extended shelf life and maintain the sensory attributes of truffles, meeting consumer expectations for fresh and high-quality products (Phong et al., 2022).

Appropriate preservation methods such as irradiation, freeze-drying, and refrigeration help truffle producers ensure that their products meet market demands for fresh, high quality truffles while preserving their unique characteristics. Moreover, techniques like modified atmosphere packaging (MAP) have been employed to extend the shelf life of fresh truffles, preserving their quality and reducing microbial spoilage (Rivera et al., 2010). MAP relies on modifying the atmosphere inside the package to reduce the respiration of the product, thereby slowing down its degradation rate and extending its shelf life (Guillard et al., 2012). By creating controlled atmospheric conditions using microperforated films, the growth of spoilage microorganisms can be inhibited, leading to prolonged shelf life and enhanced sensory attributes of truffles. This preservation method plays a crucial role in improving overall quality control measures in truffle storage and distribution, ensuring that truffle producers can safeguard the integrity of their products and maintain their sensory attributes (Francis et al., 2007). This preservation method also, when combined with other techniques like refrigeration, offers an effective solution for truffle producers to maintain the integrity and authenticity of their products on the market.

In the context of truffle cultivation, the quality assessment of truffle-inoculated seedlings is also essential for ensuring successful production and maintaining industry standards. Parameters such as plant health, homogeneity and the percentage of inoculated ectomycorrhizae are evaluated to determine the quality of truffle-inoculated seedlings, aiding in certification processes and promoting the growth of truffle cultivation practices. By examining the mycorrhization level in seedlings from commercial nurseries and adjusting quality parameters based on certification experiences, the authenticity and quality of truffle seedlings can be improved, leading to more successful cultivation practices (Donnini et al., 2014). This assessment ensures that truffle seedlings are healthy, well-inoculated, and capable of forming symbiotic relationships with truffle fungi, which are crucial for successful truffle production. The presence of well-colonized truffle-inoculated seedlings is a key factor in agroforestry systems planted for truffle production (Ferrara et al., 2023). The integration of advanced analytical methods, molecular tools and quality assessment methods plays an essential role in truffle quality control and authentication, which are essential for ensuring the integrity and authenticity of truffles in the food and nutraceutical markets. By leveraging these innovative approaches, the truffle industry can uphold standards of authenticity, traceability and safety, ensuring that consumers receive genuine and high-quality truffle products. Besides, through continuous research and technological advancements, the field of truffle quality control is evolving to meet the demands of a competitive market while safeguarding the reputation and integrity of this valued culinary delicacy.

7. Conclusion and Future Perspectives

Truffles can be considered one of the most highly promising biofactories for natural products of high market value. However, the main limitation for the market growth of truffles is their availability. So far, the main source of truffles is the natural collection. However, during the last few years, some research has demonstrated the possible cultivation of truffles and successful attempts have been made to develop new technology for truffle farming. This reduces the threat of the limitations of raw materials and makes truffles a renewable bioresource to secure business growth and sustainability. In addition, with the increased research about the health benefits of truffle bioactive molecules, the new market, especially in the nutraceuticals and cosmeceuticals industries, will grow with new health and therapeutic claims. However, research, especially that related to *in vivo* studies and clinical trials, needs to be conducted to push truffles as alternative prescribed medicines. This, for sure, will increase the market value of truffles and their share in natural medicinal products.

References

Al Obaydi, M.F., Hamed, W.M., Al Kury, L.T. and Talib, W.H. (2020). *Terfezia boudieri*: A desert truffle with anticancer and immunomodulatory activities. Front. Nut., 7: 38.

Allen, K. and Bennett, J.W. (2021). Tour of truffles: aromas, aphrodisiacs, adaptogens, and more. Mycobiology, 49(3): 201–212.

Attia W.Y., El-Naggar, R.E., Bawadekji, A. and Al Ali, M. (2018). Evaluation of some in vitro anti-carcinogenic activities of polysaccharides extracted from ascomata of the desert truffle *Terfezia claveryi* chatin. J. Appl. Environ. Biol. Sci., 8(4): 152–159.

Balivo, A., De Falco, E., Branca, L., Caputo, M., Sacchi, R. et al. (2024). Odour fingerprints of black (*Tuber mesentericum*) and bianchetto (*Tuber borchii*) truffles from different areas of the Campania region. Horticulturae, 10: 557. https://doi.org/10.3390/ horticulturae10060557

Brudzynska, P., Kurzawa, M., Sionkowska, A. and Grisel, M. (2022). Antioxidant activity of plant-derived colorants for potential cosmetic application. Cosmetics, 9(4): 81. https://doi.org/10.3390/cosmetics9040081

Büntgen, U., Bagi, I., Fekete, O., Molinier, V., Peter, M. et al. (2017). New insights into the complex relationship between weight and maturity of Burgundy truffles (*Tuber aestivum*). PLoS One, 12(1): E0170375.

Casarica A., Moscovici, M., Daas, M., Nicu, I., Panteli, M. et al. (2016). Purified extract from brown truffles of the species *Terfezia claveryl* chatin and its antimicrobial activity. Farmacia, 64: 298–301.

Çayan, F., Tel-Çayan, G., Deveci, E., Duru, M.E., Turk, M. et al. (2021). A detailed study on multifaceted bioactivities of the extracts and isolated compounds from truffle *Reddellomyces parvulosporus*. Int. J. Food Sci. Technol., 57(3): 1411–1419.

Chauhan, O.P., Vijay, V., Pandey, A.K. and Semwal, A.D. (2021). Biochemical and health properties of truffles. Def. Life Sci. J., 6(3): 251–258.

Chen, G., Zhang, S., Ran, C., Wang, L., Kan, J. et al. (2016). Extraction, characterization and antioxidant activity of water-soluble polysaccharides from *Tuber nuidongense.* Int. J. Biol. Macromol., 91: 431–442.

Coquet, C. and Chabert, R. (2017). April 14. FR3065172A1 - Cosmetic preparation containing white truffle extract and associated cosmetic process - google patents. https://patents.google.com/patent/FR3065172A1/en

Dahham, S.S., Al-Rawi, S.S., Ibrahim, A.H., Abdul Majid, A.S., Abdul Majid, A.M.S. et al. (2018). Antioxidant, anticancer, apoptosis properties and chemical composition of black truffle *Terfezia claveryi*. Saudi J. Biol. Sci., 25(8): 1524–1534.

Das, L., Bhaumik, E., Raychaudhuri, U. and Chakraborty, R. (2012). Role of nutraceuticals in human health. J. Food Sci. Technol., 49: 173–183.

Dawood A.S., Sedeek, M.S., Farag, M.A. and Abdelnaser, A. (2023). *Terfezia boudieri* and *Terfezia claveryi* inhibit the LPS/IFN-γ-mediated inflammation in RAW 264.7 macrophages through an Nrf2-independent mechanism. Sci. Rep., 13: 10106.

Dibe-Bellahouel, S. and Fortas, Z. (2011). Antibacterial activity of various fractions of ethyl acetate extract from the desert truffle, *Tirmania pinoyi*, preliminary analyzed by gas chromatography-mass spectrometry (GC-MS). Afr. J. Biotechnol., 10: 9694–9699.

Donnini, D., Benucci, G.M.N., Bencivenga, M. and Falini, L.B. (2014). Quality assessment of truffle-inoculated seedlings in Italy: proposing revised parameters for certification. For. Syst., 23(2): 385–393.

El Enshasy, H., Elsayed, E.A., Aziz, R. and Wadaan, M.A. (2013). Mushrooms and truffles historical biofactories for complementary medicine in Africa and in the middle east. Evid. Complement. Altern. Med., 2013: 1–10.

El Karkouri, K., Couderc, C., Decloquement, P., Abeille, A. and Raoult, D. et al. (2019). Rapid MALDI-TOF MS identification of commercial truffles. Sci. Rep., 9(1): 17686.

Elkhateeb, W., Daba, G., Somasekhar, T. and Narayana, G. (2024). The precious truffles: bioactive compounds as a source of various biological activities. Environ. Sci., 3(2): 40–50.

Elsayed E.A., Al-Sahli, F., Barakat, I.A., El Enshasy, H.A., Wadaan, M.A. et al. (2019a). Assessment of *in vitro* antimicrobial and anti-breast cancer activities of extracts isolated from desert truffles in Saudi Arabia. J. Sci. Appl. Res., 78: 419–425.

Elsayed E.A., Alsahli, F.D., Barakat, I.A., El Enshasy, H.A., Wadaan, M.A. et al. (2019b). Cytotoxic activities of different solvent extracts of *Tirmania nivea* and *Terfazia claveryi* against HepG2 and L929 cells. J. Sci. Appl. Res., 78: 454–457.

Elsayed E.A., Abdel Alsaheb, R., Ho, T., Hanapi, S.Z., Sukmawati, D. et al. (2021). Truffles: A potential source of bioactive compounds. pp. 163–174. In: Sridhar, K.R. and Deshmukh, S.K. (eds.). Advances in Macrofungi: Industrial Avenues and Prospects, CRC Press, Boca Raton, USA.

Farag, M.A., Fathi, D., Shamma, S., Shawkat, M.S.A., Shalabi, S.M. et al. (2021). Comparative metabolome classification of desert trufles *Terfezia claveryi* and *Terfezia boudieri* via its aroma and nutrients profile. LWT, 142: 111046.

Ferrara, G., Lombardini, L., Mazzeo, A. and Bruno, G.L. (2023). Evaluation of Pecan [*Carya illinoinensis* (Wangenh.) K. Koch] cultivars for possible cultivation for both fruit and truffle production in the Puglia region, Southeastern Italy. Horticulturae, 9(2): 261.

Francis, G.A., Scollard, J., Meally, A., Bolton, D.J., Gahan, C.G.M. et al. 2007. The glutamate decarboxylase acid resistance mechanism affects survival of *Listeria monocytogenes* LO28 in modified atmosphere-packaged foods. J. Appl. Microbiol., 103(6): 2316–2324.

Gajos, M., Ryszka F. and Geistlinger, J. 2014. The therapeutic potential of truffle fungi: a patent survey. Acta Mycol., 1(1): 305–318.

Gioacchini, A.M., Menotta, M., Bertini, L., Rossi, I., Zeppa, S. et al. (2005). Solid phase microextraction gas chromatography/mass spectrometry: a new method for species identification of truffles. Rapid Commun. Mass Spectrom., 19(17): 2365–2370.

Gouzi, H., Leboukh, M. and Bouchouka, E. (2013). Antioxidant and antiradical properties of methanolic extracts from Algerian wild edible desert truffles (*Terfezia* and *Tirmania, Ascomycetes*). Int. J. Med. Mushr., 15(5): 471–486.

Guillard, V., Guillaume, C. and Destercke, S. (2012). Parameter uncertainties and error propagation in modified atmosphere packaging modelling. Postharvest Biol. Technol., 67:154–166.

Hyde, K.D., Bahkali, A.H. and Moslem, M.A. (2010). Fungi—an unusual source for cosmetics. Fungal Divers., 43(1): 1–9.

Islam, M.T., Mohamedali, A., Garg, G., Khan, J.M., Gorse, A. et al. (2013). Unlocking the puzzling biology of the black perigord truffle *Tuber melanosporum*. J. Proteome Res., 12(12): 5349–5356.

Kanchiswany, C.N, Malnoy, M. and Maffel, M.E. (2015). Chemical diversity of microbial volatiles and their potential for plant growth and productivity. Front. Plant Sci., 6: 151.

Khadri, H, Aldebasi, Y.H. and Riazunnisa, K. (2017). Truffle mediated (*Terfezia claveryi*) synthesis of silver nanoparticles and its potential cytotoxicity in human breast cancer cells (MCF-7). Afr. J. Biotechnol., 16: 1278–1284.

Khalifa, S.A., Farag, M.A., Yosri, N., Sabir, J.S., Saeed, A. et al. (2019). Truffles: from islamic culture to chemistry, pharmacology, and food trends in recent times. Trends Food Sci. Technol., 91:193–218.

Klein, M., Kielhauser, R. and Dilger, T. (2020). Optimized rapid and reliable identification of *Tuber* spp. by matrix-assisted laser desorption ionization time-of-flight mass spectrometry. J. Mass Spectrom., 55(12): 4655.

Kraub, S. and Vetter, W. (2020). Geographical and species differentiation of truffles (*Tuber* spp.) by means of stable isotope ratio analysis of light elements (H, C, and N). J. Agric. Food Chem., 68(49): 14386–14392.

Lee, H., Nam, K., Zahra, Z. and Farooqi, M.Q.U. (2020). Potentials of truffles in nutritional and medicinal applications: a review. Fungal Biol Biotechnol., 7(1): 9

Li, X., Zhu, Z., Ye, L., Kang, Z., Zhang, X. et al. 2020. Comparison of the partial structure and antioxidant activity of polysaccharides from two species of chinese truffles. Molecules, 25(18): 4345.

Li, Y., Shan, M., Zhu, Z., Mao, X., Yan, M. et al. (2019). Application of MALDI-TOF MS to rapid identification of anaerobic bacteria. BMC Infect. Dis., 19: 1–11.

Lindqvist, A. (2012). Preference and gender associations of perfumes applied on human skin. J. Sens. Stud., 27(6): 490–497.

Mandeel, Q.A. and A.A.A. Al-Laith. 2007. Ethnomycological aspects of the desert truffle amongn native Bahraini and non-Bahraini peoples of the Kingdom of Bahrain. J. Ehtnopharmcol. 110: 118-129.
Martin, F., Kohler, A., Murat, C., Balestrini, R., Coutinho, P.M. et al. (2010). Perigord black truffle genome uncovers evolutionary origins and mechanisms of symbiosis. Nature, 464(7291): 1033–1038.
Mello, A., Murat, C. and Bonfante, P. (2006). Truffles: much more than a prized and local fungal delicacy. FEMS Microbiol. Lett., 260(1): 1–8.
Mustafa, A.M., Angeloni, S., Nzekoue, F.K., Abouelenein, D., Sagratini, G. et al. (2020). An overview on truffle aroma and main volatile compounds. Molecules, 25(24): 5948.
Oliach, D., Vidale, E., Brenko, A., Marois, O., Andrighetto, N. et al. (2021). Truffle market evolution: an application of the Delphi method. Forests, 12(9): 1174.
Patel, S., Rauf, A., Khan, H., Khalid, S., Mubarak, M.S. et al. (2017). Potential health benefits of natural products derived from truffles: a review. Trends Food Sci. Technol., 70: 1–8.
Phong, W.N., Gibberd, M.R., Payne, A.D., Dykes, G.A., Coorey, R. et al. (2022). Methods used for extraction of plant volatiles have potential to preserve truffle aroma: a review. Compr. Rev. Food Sci. Food Saf., 21(2): 1677–1701.
Portolan, F., Lequoy, V., Botto, J., Mantelin, J., Imbert, I. et al. (2014), July 8. US20170202773A1 - aqueous extract of truffle and cosmetic composition thereof - google patents. https://patents.google.com/patent/US20170202773A1/en
Rivera, C.S., Blanco, D., Salvador, M.L. and Venturini, M.E. (2010). Shelf-life extension of fresh *Tuber aestivum* and *Tuber melanosporum* truffles by modified atmosphere packaging with microperforated films. J. Food Sci., 75(4): 225–233.
Rovero, P., Malgapo, D.M.H., Sparavigna, A., Beilin, G., Wong, V. et al. (2022). The clinical evidence-based paradigm of topical anti-aging skincare formulations enriched with bio-active peptide SAL-III (KP1) as collagen modulator: from bench to bedside. Clin. Cosmet. Investig. Dermatol., 15: 2693–2703.
Segelke, T., Schelm, S., Ahlers, C. and Fischer, M. (2020). Food authentication: Truffle (*Tuber* spp.) species differentiation by FT-NIR and chemometrics. Foods, 9(7): 922.
Séjalon-Delmas, N., Roux, C., Martins, M., Kulifaj, M., Bécard, G. et al. (2000). Molecular tools for the identification of *Tuber melanosporum* in agroindustry. J. Agric. Food Chem., 48(6): 2608–2613.
Souna, M., Tefiani, C., Azzi, R., Chaouche, T.M., Habi, S et al. (2023). Mychochemical constituents and anti-inflammatory activity of *Terfezia claveryi* chatin from Algeria. Biol. Life Sci. Forum., 27: 19.
Splivallo, R. and Ebeler, S.E. (2015). Sulfure volatiles of microbial origin are key contributors to human sensed truffle aroma. Appl. Microbiol. Biotechnol., 99: 2584–2592.
Splivallo, R., Ottonello, S., Mello, A. and Karlovsky, P. (2011). Truffle volatiles: from chemical ecology to aroma biosynthesis. New Phytol., 189:688–699.
Tang, Y.J., Liu, R.S. and Li, H.M. (2015). Current progress on truffle submerged fermentation: a promising alternative to its fruiting bodies. Appl. Microbiol. Biotechnol., 99: 2041–2053.
Thomas, P., Elkhateeb, W. and Daba, G. (2021). Industrial applications of truffles and truffle-like fungi. pp. 82–88. In: Sridhar, K.R. and Deshmukh, S.K. (eds.). Advances Macrofungi: Industrial Avenues and Prospects. CRC Press, Boca Raton, FL, USA.
Torregiani, E., Lorier, S., Sagratini, G., Maggi, F., Vittori, S. et al. (2017). Comparative analysis of the volatile profile of 20 commercial samples of truffles, truffle sauces, and truffle flavored oils by using HS-SPME-GC-MS. Food Anal. Methods, 10: 1857–1869.
Vahdatzadeh, M., Deveau, A. and Splivallo, R. (2015). The role of the microbiome of truffles in aroma formation: a Meta-Analysis approach. Appl. Environ. Microbiol., 81(20): 6946–6952
Veeraraghavan, V.P., Hussain, S., Papayya Balakrishna, J., Shawale, L., Kullappan, M. et al. (2021). A comprehensive and critical review on Ethnopharmacological importance of desert truffles. *Terfezia claveryi, Terfezia boudieri,* and *Tirmania nivea.* Food Rev. Int. 38 (supl. 1): 846–865.
Wang, S. and Marcone, M.F. (2011). The biochemistry and biological properties of the world›s most expensive underground edible mushroom: Truffles. Food Res. Int., 44(9): 2567–2581.
Wernig, F., Buegger, F., Pritsch, K. and Splivallo, R. (2018). Composition and authentication of commercial and home-made white truffle-flavored oils. Food Control. 87:9–16.
Wu, Y., Choi, M., Li, J., Yang, H., Shin, H. et al. (2016). Mushroom cosmetics: the present and future. Cosmet. Toiletries., 3(3): 22

Yan X., Wang, Y., Sang, X. and Fan, L. (2017). Nutritional value, chemical composition and antioxidant activity of three Tuber species from China. AMB Express, 7: 136.

Zampieri, E., Mello, A., Bonfante, P. and Murat, C. (2009). PCR primers specific for the genus *Tuber* reveal the presence of several truffle species in a truffle-ground. FEMS Microbiol. Lett., 297(1): 67–72.

Zhao, W., Wang, X., Li, H., Wang, S., Chen, T., Yuan, Z. et al. (2014). Isolation and characterization of polysaccharides with the antitumor activity from Tuber fruiting bodies and fermentation system. Appl. Microbiol. Biotechnol., 98(5): 1991–2002.

Index